BNi Building News™

GENERAL CONSTRUCTION 1993 COSTBOOK

THIRD EDITION

Edited by

A.M. FOGARTY & ASSOCIATES

Estimating Consultants
To The Construction Industry
For Over 35 Years

BNi Building News™

Los Angeles • Boston

BNI Publications, Inc.
3055 OVERLAND AVENUE
LOS ANGELES, CA 90034
(310) 202-7775

77 WEXFORD STREET
NEEDHAM HEIGHTS, MA 02194
(617) 455-1466

A. M. Fogarty & Assoc., Inc.
175 DERBY STREET, SUITE 5
HINGHAM, MA 02043
(617) 749-7272

ISBN 1-55701-070-6

Distributor to the trade, Macmillan Publishing Company, 866 Third Avenue, New York, New York 10022.

PREFACE

For the past 47 years, Building News has been dedicated to providing construction professionals with timely and reliable information. Based on this experience, our staff has researched and compiled thousands of up-to-the-minute costs for the **Building News 1993 Costbooks.** This book is an essential reference for contractors, engineers, architects, facilities managers — any construction professional who must provide an estimate on any type of building project.

Whether working up a preliminary estimate or submitting a formal bid, the costs listed here can quickly and easily be tailored to your needs. All costs are based on national averages, while a table of modifiers is provided for regional adjustments. Overhead and profit are included in all costs.

Complete man-hour tables follow the unit costs to provide data on typical durations of specific tasks. This information can be used to schedule projects as well as to determine specific labor costs based on local labor rates.

All data is categorized according to the MASTERFORMAT of the Construction Specifications Institute (CSI). This industry standard provides an all-inclusive checklist to ensure that no element of a project is overlooked. In addition, to make specific items even easier to locate, there is a complete alphabetical index.

This costbook contains an appendix with reference charts and tables taken from an array of sources. Text explains the costs in certain categories and provides helpful pointers that should be taken into account with every estimate.

A section on square foot costs provides an overview of project costs for different building types — commercial, residential, etc. — with summaries of the actual projects. Square foot costs are invaluable for making budget estimates and checking prices when time is a factor. For a complete source of square foot costs refer to the **Building News Square Foot 1993 Costbook.**

The "Features in this Book" section presents a clear overview of the many features of this book. Included is an explanation of the data, sample page layout and discussion of how to best use the information in the book.

Of course, all buildings and construction projects are unique. The costs provided in this book are based on averages from well-managed projects with good labor productivity under normal working conditions (eight hours a day). Other circumstances affecting costs such as overtime, unusual working conditions, savings from buying bulk quantities for large projects, and unusual or hidden costs must be factored in as they arise.

The data provided in this book is for estimating purposes only. Check all applicable federal, state and local codes and regulations for specific requirements.

About the Editors

As a leader in construction cost estimating since 1972, A.M. Fogarty & Associates has estimated literally billions of dollars worth of construction projects. Over 250 years of combined experience provides constantly and accuracy, assuming you the most reliable information that is so valuable to the success of your business. In addition to construction cost estimating, A.M. Fogarty & Associates provides consulting services in the areas of construction management, scheduling and planning, specifications and building evaluation.

BNi Building News™

TABLE OF CONTENTS

CSI MASTERFORMAT

All data in the Costbook pages and Man-Hour tables is organized according to the CSI MASTERFORMAT — the industry standard numbering/classification system. The data is divided into the 16 divisions as shown below. The five digit numbers within each division correspond to the MASTERFORMAT Broadscope designations. Each section is further broken down into MASTERFORMAT Mediumscope designations. These numbers, in most cases, are the same as those used in architectural and engineering specifications.

The construction estimating information in this book is divided into two main sections: Costbook Pages and Man-Hour Tables. Each is organized to the 16 divisions of the CSI MASTERFORMAT as shown below. In addition, there are estensive Construction Reference Tables, Geographic Cost Modifiers, Square Foot Tables and a detailed Index. Sample pages with graphic explanations are included before the Costbook pages, Man-Hour Tables and Square Foot Tables. These explanations along with the discussions below will provide a good understanding of what is included in this book and how it can best be used for construction estimating.

GENERAL REQUIREMENTS	Division 1
Summary of Work	01010
Allowances	01020
Measurement and Payment	01025
Alternates/Alternatives	01030
Modification Procedures	01035
Coordination	01040
Field Engineering	01050
Regulatory Requirements	01060
Identification Systems	01070
References	01090
Special Project Procedures	01100
Project Meetings	01200
Submittals	01300
Quality Control	01400
Construction Facilities & Temporary Controls	01500
Material and Equipment	01600
Facility Startup/Commissioning	01650
Contract Closeout	01700
Maintenance	01800

SITEWORK	Division 2
Subsurface Investigation	02010
Demolition	02050
Site Preparation	02100
Dewatering	02140
Shoring and Underpinning	02150
Excavation Support Systems	02160
Cofferdams	02170
Earthwork	02200
Tunneling	02300
Piles and Caissons	02350
Railroad Work	02450
Marine Work	02480
Paving and Surfacing	02500
Utility Piping Materials	02600
Water Distribution	02660
Fuel and Steam Distribution	02680
Sewerage and Drainage	02700
Restoration of Underground Pipe	02760
Ponds and Reservoirs	02770
Power and Communications	02780
Site Improvements	02800
Landscaping	02900

CONCRETE	Division 3
Concrete Formwork	03100
Concrete Reinforcement	03200
Concrete Accessories	03250
Cast-in-Place Concrete	03300
Concrete Curing	03370
Precast Concrete	03400
Cementitious Decks and Toppings	03500
Grout	03600
Concrete Restoration and Cleaning	03700
Mass Concrete	03800

MASONRY	Division 4
Mortar and Masonry Grout	04100
Masonry Accessories	04150
Unit Masonry	04200
Stone	04400
Masonry Restoration and Cleaning	04500
Refractories	04550
Corrosion Resistant Masonry	04600
Simulated Masonry	04700

FEATURES IN THIS BOOK

The construction estimating information in this book is divided into two main sections: Costbook Pages and Man-Hour Tables. Each section is organized according to the 16 divisions of the CSI MASTERFORMAT as shown on the previous pages. In addition, there are extensive Supporting Construction Reference tables, Geographic Costs Modifiers, Square Foot tables and a detailed Index.

Sample pages with graphic explanations are included before the Costbook pages and Man-Hour tables. These explanations along with the discussions below, will provide a good understanding of what is included in this book and how it can best be used in construction estimating.

Material Costs

The material costs used in this book represent national averages for prices that a contractor would expect to pay plus an allowance for freight (if applicable), handling and storage. These costs reflect neither the lowest or highest prices, but rather a typical average cost over time. Periodic fluctuations in availability and in certain commodities (eg. copper, lumber) can significantly affect local material pricing. In the final estimating and bidding stages of a project when the highest degree of accuracy is required, it is best to check local, current prices.

Labor Costs

Labor costs include the basic wage, plus commonly applicable taxes, insurance and markups for overhead and profit. The labor rates used here to develop the costs are typical average prevailing wage rates. Rates for different trades are used where appropriate for each type of work.

Taxes and insurance which are most often applied to labor rates include employer-paid Social Security/Medicare taxes (FICA), Worker's Compensation insurance, state and federal unemployment taxes, and business insurance. Fixed government rates as well as average allowances are included in the labor costs. However, most of these items vary significantly from state to state and within states. For more specific data, local agencies and sources should be consulted.

Equipment Costs

Costs for various types and pieces of equipment are included in Division 1 - General Requirements and can be included in an estimate when required either as a total "Equipment" category or with specific appropriate trades. Costs for equipment are included when appropriate in the installation of costs in the Costbook pages.

Overhead And Profit

Included in the labor costs are allowances for overhead and profit for the contractor/employer whose workers are performing the specific tasks. No cost allowances or fees are included for management of subcontractors by the general contractor or construction manager. These costs, where appropriate, must be added to the costs as listed in the book.

The allowance for overhead is included to account for office overhead, the contractors' typical costs of doing business. These costs normally include in-house office staff salaries and benefits, office rent and operating expenses, professional fees, vehicle costs and other operating costs which are not directly applicable to specific jobs. It should be noted for this book that office overhead as included should be distinguished from project overhead, the General Requirements (CSI Division 1) which are specific to particular projects. Project overhead should be included on an item by item basis for each job.

Depending on the trade, an allowance of 10-15 percent is incorporated into the labor/installation costs to account for typical profit of the installing contractor. See Division 1, General Requirements, for a more detailed review of typical profit allowances.

Adjustments to Costs

The costs as presented in this book attempt to represent national averages. Costs, however, vary among regions, states and even between adjacent localities.

In order to more closely approximate the probable costs for specific locations throughout the U.S., a table of Geographic Cost Modifiers is provided. These adjustment factors are used to modify costs obtained from this book to help account for regional variations of construction costs. Whenever local current costs are known, whether material or equipment prices or labor rates, they should be used if more accuracy is required.

Man-Hour Tables

The man-hour data used to develop the labor costs are listed in the second main section of this book, the Man-Hour Tables. These productivities represent typical installation labor for thousands of construction items. The data takes into account all activities involved in normal construction under commonly experienced working conditions such as site movement, material handling, start-up, etc. As with the Costbook pages, these items are listed according to the CSI MASTERFORMAT.

Square Foot Tables

Included as an additional reference are Square Foot Tables which list hundreds of actual projects for dozens of building types each with associated building size and total square foot building cost. This data provides an overview of construction costs by building type. These costs are for actual projects. The variations within similar building types may be due, among other factors, to size, location, quality and specified components, materials and processes. Depending upon all such factors, specific building costs can vary significantly and may not necessarily fall within the range of costs as presented.

The data has been updated to reflect current construction costs and is a partial summary of the projects and costs presented in the **Building News Square Foot 1993 Costbook.** The source of this data is *Design Cost & Data* magazine which has over 25 years of history in tracking, collecting and publishing construction cost data.

Editors' Note: The **Building News 1993 Costbooks** are intended to provide accurate, reliable, average costs and typical productivities for thousands of common construction components. The data is developed and compiled from various industry sources, including government, manufacturers, suppliers and working professionals. The intent of the information is to provide assistance and guidelines to construction professionals in estimating. The user should be aware that local conditions, material and labor availability and cost variations, economic considerations, weather, local codes and regulations, etc., all affect the actual cost of construction. These and other such factors must be considered and incorporated into any and all construction estimates.

Sample Costbook Page

In order to best use the information in this book, please review this sample page and read the "Features In This Book" section.

CSI MASTERFORMAT Division

CSI Broadscope Category

CSI Mediumscope Category (First 5 Digits)

03 CONCRETE				
PLACING CONCRETE	UNIT	MAT.	INST.	TOTAL
03380.05 BEAM CONCRETE				
By crane	C.Y.	65.00	38.50	103.50
By pump	"	65.00	35.00	100.00
By hand buggy	"	65.00	20.40	85.40
Bond beam, 3000# concrete				
By pump				
8" high				
4" wide	L.F.	0.15	0.77	0.92
6" wide	"	0.36	0.87	1.23
8" wide	"	0.47	0.96	1.43
10" wide	"	0.63	1.05	1.68
12" wide	"	0.84	1.20	2.04
16" high				
8" wide	L.F.	1.15	1.20	2.35
10" wide	"	1.55	1.35	2.90
12" wide	"	2.05	1.60	3.65
By crane				
8" high				
4" wide	L.F.	0.18	0.84	1.02
6" wide	"	0.36	0.92	1.28
8" wide	"	0.47	0.96	1.43
10" wide	"	0.63	1.05	1.68
12" wide	"	0.84	1.20	2.04
16" high				
8" wide	L.F.	1.15	1.20	2.35
10" wide	"	1.55	1.30	2.85
12" wide	"	2.05	1.50	3.55
03380.15 COLUMN CONCRETE				
Columns				
2500# or 3000# concrete				
By crane	C.Y.	56.50	35.00	91.50
By pump	"	56.50	32.10	88.60
3500# or 4000# concrete				
By crane	C.Y.	60.50	35.00	95.50
By pump	"	60.50	32.10	92.60
5000# concrete				
By crane	C.Y.	65.00	35.00	100.00
By pump	"	65.00	32.10	97.10
03380.20 ELEVATED SLAB CONCRETE				
Elevated slab				
2500# or 3000# concrete				
By crane	C.Y.	56.50	19.25	75.75
By pump	"	56.50	14.80	71.30
By hand buggy	"	56.50	20.40	76.90
3500# or 4000# concrete				
By crane	C.Y.	60.50	19.25	79.75
By pump	"	60.50	14.80	75.30
By hand buggy	"	60.50	20.40	80.90
5000# concrete				
By crane	C.Y.	65.00	19.25	84.25
By pump	"	65.00	14.80	79.80

Material Cost
Material costs represent average contractor prices plus an allowance for freight, handling and storage.

Installation Cost
Installation costs include basic wage rates, markups for taxes, insurance overhead and profit and also include equipment costs where appropriate.

Detailed Descriptions
Complete descriptions of items may include information listed above a particular line. Review of the whole category is recommended for a complete description.

Total Cost
The total cost is the sum of material and installation costs. This total represents typical contractors' costs including overhead and profit, but does not include markups for the general contractor or construction management fees.

Unit of Measurement
Each item (and cost) is defined in terms of the common estimating unit. All costs are listed in dollars per unit.

BNi Building News™

01 GENERAL

REQUIREMENTS	UNIT	MAT.	INST.	TOTAL
01020.10 ALLOWANCES				
Overhead				
$20,000 project				
Minimum	PCT.			15.00
Average	"			20.00
Maximum	"			40.00
$100,000 project				
Minimum	PCT.			12.00
Average	"			15.00
Maximum	"			25.00
$500,000 project				
Minimum	PCT.			10.00
Average	"			12.00
Maximum	"			20.00
$1,000,000 project				
Minimum	PCT.			6.00
Average	"			10.00
Maximum	"			12.00
$10,000,000 project				
Minimum	PCT.			1.65
Profit				
$20,000 project				
Minimum	PCT.			10.00
Average	"			15.00
Maximum	"			25.00
$100,000 project				
Minimum	PCT.			10.00
Average	"			12.00
Maximum	"			20.00
$500,000 project				
Minimum	PCT.			5.00
Average	"			10.00
Maximum	"			15.00
$1,000,000 project				
Minimum	PCT.			3.00
Average	"			8.00
Maximum	"			15.00
Professional fees				
Architectural				
$100,000 project				
Minimum	PCT.			5.00
Average	"			10.00
Maximum	"			20.00
$500,000 project				
Minimum	PCT.			5.00
Average	"			8.00
Maximum	"			12.00
$1,000,000 project				
Minimum	PCT.			3.50
Average	"			7.00
Maximum	"			10.00
Taxes				
Sales tax				

01 GENERAL

REQUIREMENTS	UNIT	MAT.	INST.	TOTAL
01020.10 ALLOWANCES				
Minimum	PCT.			4.00
Average	"			5.00
Maximum	"			8.50
Unemployment				
Minimum	PCT.			2.00
Average	"			5.00
Maximum	"			7.00
Social security (FICA)	"			
Average	"			
01050.10 FIELD STAFF				
Superintendent				
Minimum	YEAR			44,000
Average	"			68,200
Maximum	"			99,000
Field engineer				
Minimum	YEAR			38,500
Average	"			57,200
Maximum	"			82,500
Foreman				
Minimum	YEAR			27,500
Average	"			40,700
Maximum	"			55,000
Bookkeeper/timekeeper				
Minimum	YEAR			13,200
Average	"			20,900
Maximum	"			27,500
Watchman				
Minimum	YEAR			11,000
Average	"			13,750
Maximum	"			16,500
01310.10 SCHEDULING				
Scheduling for				
$100,000 project				
Minimum	PCT.			1.00
Average	"			2.00
Maximum	"			4.00
$500,000 project				
Minimum	PCT.			0.40
Average	"			1.00
Maximum	"			2.00
$1,000,000 project				
Minimum	PCT.			0.30
Average	"			0.80
Maximum	"			1.50
Scheduling software, not including computer				
Minimum	EA.			500.00
Average	"			2,500
Maximum	"			40,000

01 GENERAL

REQUIREMENTS	UNIT	MAT.	INST.	TOTAL
01330.10 SURVEYING				
Surveying				
Small crew	DAY	0.00	424.00	424.00
Average crew	"	0.00	642.00	642.00
Large crew	"	0.00	848.00	848.00
Lot lines and boundaries				
Minimum	ACRE	0.00	291.00	291.00
Average	"	0.00	618.00	618.00
Maximum	"	0.00	1,020	1,020
01380.10 JOB REQUIREMENTS				
Job photographs, small jobs				
Minimum	EA.			55.00
Average	"			110.00
Maximum	"			220.00
Large projects				
Minimum	EA.			385.00
Average	"			550.00
Maximum	"			1,100
01410.10 TESTING				
Testing concrete, per test				
Minimum	EA.			12.10
Average	"			22.00
Maximum	"			33.00
Soil, per test				
Minimum	EA.			33.00
Average	"			44.00
Maximum	"			66.00
Welding, per test				
Minimum	EA.			11.00
Average	"			22.00
Maximum	"			66.00
01500.10 TEMPORARY FACILITIES				
Barricades, temporary				
Highway				
Concrete	L.F.	6.60	2.10	8.70
Wood	"	1.75	0.85	2.60
Steel	"	2.10	0.71	2.81
Pedestrian barricades				
Plywood	L.F.	1.20	0.71	1.91
Chain link fence	"	1.75	0.71	2.46
Trailers, general office type, per month				
Minimum	EA.			165.00
Average	"			242.00
Maximum	"			550.00
Crew change trailers, per month				
Minimum	EA.			66.00
Average	"			77.00
Maximum	"			110.00

01 GENERAL

REQUIREMENTS

01505.10 MOBILIZATION

REQUIREMENTS	UNIT	MAT.	INST.	TOTAL
Equipment mobilization				
Bulldozer				
Minimum	EA.			110.00
Average	"			220.00
Maximum	"			330.00
Backhoe/front-end loader				
Minimum	EA.			55.00
Average	"			110.00
Maximum	"			220.00
Crane, crawler type				
Minimum	EA.			1,100
Average	"			2,750
Maximum	"			5,500
Truck crane				
Minimum	EA.			275.00
Average	"			412.00
Maximum	"			660.00
Pile driving rig				
Minimum	EA.			5,500
Average	"			11,000
Maximum	"			16,500

01525.10 CONSTRUCTION AIDS

REQUIREMENTS	UNIT	MAT.	INST.	TOTAL
Scaffolding/staging, rent per month				
Measured by lineal feet of base				
10' high	L.F.			6.60
20' high	"			11.00
30' high	"			16.50
40' high	"			19.80
50' high	"			23.10
Measured by square foot of surface				
Minimum	S.F.			0.44
Average	"			0.55
Maximum	"			0.88
Safety nets, heavy duty, per job				
Minimum	S.F.			0.22
Average	"			0.28
Maximum	"			0.55
Tarpaulins, fabric, per job				
Minimum	S.F.			0.11
Average	"			0.22
Maximum	"			0.55

01570.10 SIGNS

REQUIREMENTS	UNIT	MAT.	INST.	TOTAL
Construction signs, temporary				
Signs, 2' x 4'				16.50
Minimum	EA.			19.80
Average	"			44.00
Maximum	"			165.00
Signs, 4' x 8'				27.50

01 GENERAL

REQUIREMENTS	UNIT	MAT.	INST.	TOTAL
01570.10 SIGNS				
Minimum	EA.			38.50
Average	"			110.00
Maximum	"			440.00
Signs, 8' x 8'				
Minimum	EA.			44.00
Average	"			165.00
Maximum	"			1,650
01600.10 EQUIPMENT				
Air compressor				
60 cfm				
By day	EA.			55.00
By week	"			165.00
By month	"			495.00
300 cfm				
By day	EA.			99.00
By week	"			297.00
By month	"			891.00
600 cfm				
By day	EA.			330.00
By week	"			990.00
By month	"			2,970
Air tools, per compressor, per day				
Minimum	EA.			22.00
Average	"			27.50
Maximum	"			38.50
Generators, 5 kw				
By day	EA.			44.00
By week	"			132.00
By month	"			396.00
Heaters, salamander type, per week				
Minimum	EA.			55.00
Average	"			82.50
Maximum	"			110.00
Pumps, submersible				
50 gpm				
By day	EA.			44.00
By week	"			132.00
By month	"			396.00
100 gpm				
By day	EA.			55.00
By week	"			165.00
By month	"			495.00
500 gpm				
By day	EA.			88.00
By week	"			264.00
By month	"			792.00
Diaphragm pump, by week				
Minimum	EA.			88.00
Average	"			110.00
Maximum	"			132.00

01 GENERAL

REQUIREMENTS	UNIT	MAT.	INST.	TOTAL
01600.10 EQUIPMENT				
Pickup truck				
By day	EA.			77.00
By week	"			231.00
By month	"			693.00
Dump truck				
6 cy truck				
By day	EA.			220.00
By week	"			660.00
By month	"			1,980
10 cy truck				
By day	EA.			253.00
By week	"			759.00
By month	"			2,200
16 cy truck				
By day	EA.			275.00
By week	"			825.00
By month	"			2,420
Backhoe, track mounted				
1/2 cy capacity				
By day	EA.			440.00
By week	"			1,320
By month	"			3,960
1 cy capacity				
By day	EA.			715.00
By week	"			2,150
By month	"			6,430
2 cy capacity				
By day	EA.			990.00
By week	"			2,970
By month	"			8,800
3 cy capacity				
By day	EA.			1,650
By week	"			4,950
By month	"			14,850
Backhoe/loader, rubber tired				
1/2 cy capacity				
By day	EA.			275.00
By week	"			825.00
By month	"			2,480
3/4 cy capacity				
By day	EA.			330.00
By week	"			990.00
By month	"			2,970
Bulldozer				
75 hp				
By day	EA.			440.00
By week	"			1,320
By month	"			3,960
200 hp				
By day	EA.			880.00
By week	"			2,640
By month	"			7,920

01 GENERAL

REQUIREMENTS	UNIT	MAT.	INST.	TOTAL
01600.10 EQUIPMENT				
400 hp				
By day	EA.			1,320
By week	"			3,960
By month	"			11,880
Cranes, crawler type				
15 ton capacity				
By day	EA.			440.00
By week	"			1,320
By month	"			3,960
25 ton capacity				
By day	EA.			550.00
By week	"			1,650
By month	"			4,950
50 ton capacity				
By day	EA.			990.00
By week	"			2,970
By month	"			8,910
100 ton capacity				
By day	EA.			1,650
By week	"			4,950
By month	"			14,850
Truck mounted, hydraulic				
15 ton capacity				
By day	EA.			440.00
By week	"			1,320
By month	"			3,960
Loader, rubber tired				
1 cy capacity				
By day	EA.			385.00
By week	"			1,160
By month	"			3,960
2 cy capacity				
By day	EA.			550.00
By week	"			1,650
By month	"			4,950
3 cy capacity				
By day	EA.			880.00
By week	"			2,640
By month	"			7,920
01740.10 BONDS				
Performance bonds				
Minimum	PCT			1.00
Average	"			2.00
Maximum	"			3.00

02 SITEWORK

SOIL TESTS

02010.10 SOIL BORING

SOIL TESTS	UNIT	MAT.	INST.	TOTAL
Borings, uncased, stable earth				
2-1/2" dia.	L.F.	0.00	13.75	13.75
4" dia.	"	0.00	15.70	15.70
Cased, including samples				
2-1/2" dia.	L.F.	0.00	18.35	18.35
4" dia.	"	0.00	31.40	31.40
Drilling in rock				
No sampling	L.F.	0.00	28.90	28.90
With casing and sampling	"	0.00	36.70	36.70
Test pits				
Light soil	EA.	0.00	183.00	183.00
Heavy soil	"	0.00	275.00	275.00

DEMOLITION

02060.10 BUILDING DEMOLITION

DEMOLITION	UNIT	MAT.	INST.	TOTAL
Building, complete with disposal				
Wood frame	C.F.	0.00	0.15	0.15
Concrete	"	0.00	0.23	0.23
Steel frame	"	0.00	0.30	0.30
Rubbish handling				
Load in dumpster or truck				
Minimum	C.F.	0.00	0.47	0.47
Maximum	"	0.00	0.71	0.71
For use of elevators, add				
Minimum	C.F.	0.00	0.11	0.11
Maximum	"	0.00	0.21	0.21
Rubbish hauling				
Hand loaded on trucks, 2 mile trip	C.Y.	0.00	16.80	16.80
Machine loaded on trucks, 2 mile trip	"	0.00	11.50	11.50

HIGHWAY DEMOLITION

02065.10 PAVEMENT DEMOLITION

HIGHWAY DEMOLITION	UNIT	MAT.	INST.	TOTAL
Bituminous pavement, up to 3" thick				
On streets	S.Y.	0.00	4.40	4.40
On pipe trench	"	0.00	2.75	2.75

02 SITEWORK

HIGHWAY DEMOLITION	UNIT	MAT.	INST.	TOTAL
02065.10 PAVEMENT DEMOLITION				
Concrete pavement, 6" thick				
No reinforcement	S.Y.	0.00	7.35	7.35
With wire mesh	"	0.00	11.00	11.00
With rebars	"	0.00	13.75	13.75
9" thick				
No reinforcement	S.Y.	0.00	9.15	9.15
With wire mesh	"	0.00	13.75	13.75
With rebars	"	0.00	18.35	18.35
12" thick				
No reinforcement	S.Y.	0.00	11.00	11.00
With wire mesh	"	0.00	15.70	15.70
With rebars	"	0.00	22.00	22.00
Sidewalk, 4" thick, with disposal	"	0.00	3.65	3.65
Removal of pavement markings by waterblasting	S.F.	0.00	0.10	0.10
02065.15 SAW CUTTING PAVEMENT				
Pavement, bituminous				
2" thick	L.F.	0.00	0.80	0.80
3" thick	"	0.00	1.00	1.00
4" thick	"	0.00	1.25	1.25
5" thick	"	0.00	1.35	1.35
6" thick	"	0.00	1.45	1.45
Concrete pavement, with wire mesh				
4" thick	L.F.	0.00	1.55	1.55
5" thick	"	0.00	1.65	1.65
6" thick	"	0.00	1.80	1.80
8" thick	"	0.00	2.00	2.00
10" thick	"	0.00	2.20	2.20
Plain concrete, unreinforced				
4" thick	L.F.	0.00	1.35	1.35
5" thick	"	0.00	1.55	1.55
6" thick	"	0.00	1.65	1.65
8" thick	"	0.00	1.80	1.80
10" thick	"	0.00	2.00	2.00
02065.80 CURB & GUTTER				
Removal, plain concrete curb	L.F.	0.00	3.65	3.65
Plain concrete curb and 2' gutter	"	0.00	4.90	4.90
02065.85 GUARDRAILS				
Remove standard guardrail				
Steel	L.F.	0.00	5.50	5.50
Wood	"	0.00	4.40	4.40
02075.80 CORE DRILLING				
Concrete				
6" thick				
3" dia.	EA.	0.00	19.05	19.05
4" dia.	"	0.00	22.25	22.25
6" dia.	"	0.00	26.70	26.70

02 SITEWORK

HIGHWAY DEMOLITION

HIGHWAY DEMOLITION	UNIT	MAT.	INST.	TOTAL
02075.80 CORE DRILLING				
8" dia.	EA.	0.00	44.50	44.50
8" thick				
3" dia.	EA.	0.00	26.70	26.70
4" dia.	"	0.00	33.40	33.40
6" dia.	"	0.00	38.10	38.10
8" dia.	"	0.00	53.50	53.50
10" thick				
3" dia.	EA.	0.00	33.40	33.40
4" dia.	"	0.00	38.10	38.10
6" dia.	"	0.00	44.50	44.50
8" dia.	"	0.00	67.00	67.00
12" thick				
3" dia.	EA.	0.00	44.50	44.50
4" dia.	"	0.00	53.50	53.50
6" dia.	"	0.00	67.00	67.00
8" dia.	"	0.00	89.00	89.00

SITE DEMOLITION

SITE DEMOLITION	UNIT	MAT.	INST.	TOTAL
02105.10 CATCH BASINS/MANHOLES				
Abandon catch basin or manhole (fill with sand)	EA.	0.00	220.00	220.00
Remove and reset frame and cover	"	0.00	110.00	110.00
Remove catch basin, to 10' deep				
Masonry	EA.	0.00	275.00	275.00
Concrete	"	0.00	367.00	367.00
02105.20 FENCES				
Remove fencing				
Chain link, 8' high				
For disposal	L.F.	0.00	1.00	1.00
For reuse	"	0.00	2.55	2.55
Wood				
4' high	EA.	0.00	0.68	0.68
6' high	"	0.00	0.82	0.82
8' high	"	0.00	1.00	1.00
Masonry				
8" thick				
4' high	EA.	0.00	2.05	2.05
6' high	"	0.00	2.55	2.55
8' high	"	0.00	2.90	2.90
12" thick				
4' high	EA.	0.00	3.40	3.40
6' high	"	0.00	4.10	4.10

02 SITEWORK

SITE DEMOLITION	UNIT	MAT.	INST.	TOTAL
02105.20 FENCES				
8' high	EA.	0.00	5.10	5.10
12' high	"	0.00	6.80	6.80
02105.30 HYDRANTS				
Remove fire hydrant	EA.	0.00	183.00	183.00
Remove and reset fire hydrant	"	0.00	550.00	550.00
02105.42 DRAINAGE PIPING				
Remove drainage pipe, not including excavation				
12" dia.	L.F.	0.00	4.60	4.60
18" dia.	"	0.00	5.80	5.80
24" dia.	"	0.00	7.35	7.35
36" dia.	"	0.00	9.15	9.15
02105.43 GAS PIPING				
Remove welded steel pipe, not including excavation				
4" dia.	L.F.	0.00	7.20	7.20
5" dia.	"	0.00	11.50	11.50
6" dia.	"	0.00	14.35	14.35
8" dia.	"	0.00	23.00	23.00
10" dia.	EA.	0.00	28.70	28.70
02105.45 SANITARY PIPING				
Remove sewer pipe, not including excavation				
4" dia.	L.F.	0.00	4.60	4.60
6" dia.	"	0.00	5.20	5.20
8" dia.	"	0.00	5.75	5.75
10" dia.	"	0.00	6.05	6.05
12" dia.	"	0.00	6.40	6.40
15" dia.	"	0.00	6.75	6.75
18" dia.	"	0.00	7.65	7.65
24" dia.	"	0.00	9.55	9.55
30" dia.	"	0.00	11.50	11.50
36" dia.	"	0.00	14.35	14.35
02105.48 WATER PIPING				
Remove water pipe, not including excavation				
4" dia.	L.F.	0.00	5.20	5.20
6" dia.	"	0.00	5.45	5.45
8" dia.	"	0.00	6.05	6.05
10" dia.	"	0.00	6.40	6.40
12" dia.	"	0.00	6.75	6.75
14" dia.	"	0.00	7.20	7.20
16" dia.	"	0.00	7.65	7.65
18" dia.	"	0.00	8.20	8.20
20" dia.	"	0.00	8.85	8.85
Remove valves				
6"	EA.	0.00	57.50	57.50
10"	"	0.00	64.00	64.00

02 SITEWORK

SITE DEMOLITION	UNIT	MAT.	INST.	TOTAL
02105.48 WATER PIPING				
14"	EA.	0.00	72.00	72.00
18"	"	0.00	95.50	95.50
02105.60 UNDERGROUND TANKS				
Remove underground storage tank, and backfill				
50 to 250 gals	EA.	0.00	383.00	383.00
600 gals	"	0.00	383.00	383.00
1000 gals	"	0.00	574.00	574.00
4000 gals	"	0.00	919.00	919.00
5000 gals	"	0.00	919.00	919.00
10,000 gals	"	0.00	1,530	1,530
12,000 gals	"	0.00	1,910	1,910
15,000 gals	"	0.00	2,300	2,300
20,000 gals	"	0.00	2,870	2,870
02105.66 SEPTIC TANKS				
Remove septic tank				
1000 gals	EA.	0.00	95.50	95.50
2000 gals	"	0.00	115.00	115.00
5000 gals	"	0.00	144.00	144.00
15,000 gals	"	0.00	1,150	1,150
25,000 gals	"	0.00	1,530	1,530
40,000 gals	"	0.00	2,300	2,300
02105.80 WALLS, EXTERIOR				
Concrete wall				
Light reinforcing				
6" thick	S.F.	0.00	7.35	7.35
8" thick	"	0.00	7.85	7.85
10" thick	"	0.00	9.15	9.15
12" thick	"	0.00	11.00	11.00
Masonry				
No reinforcing				
8" thick	S.F.	0.00	3.15	3.15
12" thick	"	0.00	3.70	3.70
16" thick	"	0.00	4.40	4.40
Horizontal reinforcing				
8" thick	S.F.	0.00	3.55	3.55
12" thick	"	0.00	3.95	3.95
16" thick	"	0.00	5.00	5.00
Remove concrete headwall				
15" pipe	EA.	0.00	55.00	55.00
18" pipe	"	0.00	61.00	61.00
24" pipe	"	0.00	73.50	73.50
30" pipe	"	0.00	78.50	78.50
36" pipe	"	0.00	84.50	84.50
48" pipe	"	0.00	91.50	91.50
60" pipe	"	0.00	110.00	110.00

02 SITEWORK

SITE DEMOLITION

SITE DEMOLITION	UNIT	MAT.	INST.	TOTAL
02110.10 CLEARING AND GRUBBING				
Clear wooded area				
Light density	ACRE	0.00	2,750	2,750
Medium density	"	0.00	3,670	3,670
Heavy density	"	0.00	4,400	4,400
02110.50 TREE CUTTING & CLEARING				
Cut trees and clear out stumps				
9" to 12" dia.	EA.	0.00	221.00	221.00
To 24" dia.	"	0.00	276.00	276.00
24" dia. and up	"	0.00	367.00	367.00
Loading and trucking				
For machine load, per load, round trip				
1 mile	EA.	0.00	44.00	44.00
3 mile	"	0.00	50.00	50.00
5 mile	"	0.00	55.00	55.00
10 mile	"	0.00	73.50	73.50
20 mile	"	0.00	110.00	110.00
Hand loaded, round trip				
1 mile	EA.	0.00	100.00	100.00
3 mile	"	0.00	114.00	114.00
5 mile	"	0.00	133.00	133.00
10 mile	"	0.00	160.00	160.00
20 mile	"	0.00	200.00	200.00
Tree trimming for pole line construction				
Light cutting	L.F.	0.00	0.55	0.55
Medium cutting	"	0.00	0.73	0.73
Heavy cutting	"	0.00	1.10	1.10

DEWATERING

DEWATERING	UNIT	MAT.	INST.	TOTAL
02144.10 WELLPOINT SYSTEMS				
Pumping, gas driven, 50' hose				
3" header pipe	DAY	0.00	400.00	400.00
6" header pipe	"	0.00	400.00	400.00
Wellpoint system per job				
6" header pipe	L.F.	0.88	1.60	2.48
8" header pipe	"	1.10	2.00	3.10
10" header pipe	"	1.30	2.65	3.95
Jetting wellpoint system				
14' long	EA.	27.50	26.70	54.20
18' long	"	33.00	33.30	66.30
Sand filter for wellpoints	L.F.	1.30	0.67	1.97
Replacement of wellpoint components	EA.	0.00	8.00	8.00

02 SITEWORK

SHORING AND UNDERPINNING

	UNIT	MAT.	INST.	TOTAL
02162.10 TRENCH SHEETING				
Closed timber, including pull and salvage, excavation				
8' deep	S.F.	1.60	3.80	5.40
10' deep	"	1.60	3.80	5.40
12' deep	"	1.60	4.05	5.65
14' deep	"	1.60	4.25	5.85
16' deep	"	1.60	4.55	6.15
18' deep	"	1.60	5.60	7.20
20' deep	"	1.60	5.40	7.00
02170.10 COFFERDAMS				
Cofferdam, steel, pull and salvage, driven from shore				
15' deep	S.F.	11.00	8.70	19.70
20' deep	"	11.00	8.15	19.15
25' deep	"	11.00	7.65	18.65
30' deep	"	11.00	7.20	18.20
40' deep	"	11.00	6.80	17.80
Driven from barge				
20' deep	S.F.	11.00	9.40	20.40
30' deep	"	11.00	8.70	19.70
40' deep	"	11.00	8.15	19.15
50' deep	"	11.00	7.65	18.65

EARTHWORK

	UNIT	MAT.	INST.	TOTAL
02210.10 HAULING MATERIAL				
Haul material by 10 cy dump truck, round trip distance				
1 mile	C.Y.	0.00	2.35	2.35
2 mile	"	0.00	2.80	2.80
5 mile	"	0.00	3.80	3.80
10 mile	"	0.00	4.20	4.20
20 mile	"	0.00	4.65	4.65
30 mile	"	0.00	5.60	5.60
Site grading, cut & fill, sandy clay, 200' haul, 75 hp dozer	"	0.00	1.70	1.70
Spread topsoil by equipment on site	"	0.00	1.85	1.85
Site grading (cut and fill to 6") less than 1 acre				
75 hp dozer	C.Y.	0.00	2.80	2.80
1.5 cy backhoe/loader	"	0.00	4.20	4.20
02210.30 BULK EXCAVATION				
Excavation, by small dozer				
Large areas	C.Y.	0.00	0.80	0.80
Small areas	"	0.00	1.35	1.35
Trim banks	"	0.00	2.00	2.00

EARTHWORK

02210.30 BULK EXCAVATION

EARTHWORK	UNIT	MAT.	INST.	TOTAL
Drag line				
1-1/2 cy bucket				
Sand or gravel	C.Y.	0.00	1.85	1.85
Light clay	"	0.00	2.45	2.45
Heavy clay	"	0.00	2.75	2.75
Unclassified	"	0.00	2.95	2.95
2 cy bucket				
Sand or gravel	C.Y.	0.00	1.70	1.70
Light clay	"	0.00	2.20	2.20
Heavy clay	"	0.00	2.45	2.45
Unclassified	"	0.00	2.60	2.60
2-1/2 cy bucket				
Sand or gravel	C.Y.	0.00	1.55	1.55
Light clay	"	0.00	2.00	2.00
Heavy clay	"	0.00	2.20	2.20
Unclassified	"	0.00	2.30	2.30
3 cy bucket				
Sand or gravel	C.Y.	0.00	1.35	1.35
Light clay	"	0.00	1.85	1.85
Heavy clay	"	0.00	2.00	2.00
Unclassified	"	0.00	2.10	2.10
Hydraulic excavator				
1 cy capacity				
Light material	C.Y.	0.00	1.90	1.90
Medium material	"	0.00	2.30	2.30
Wet material	"	0.00	2.85	2.85
Blasted rock	"	0.00	3.30	3.30
1-1/2 cy capacity				
Light material	C.Y.	0.00	0.69	0.69
Medium material	"	0.00	0.93	0.93
Wet material	"	0.00	1.10	1.10
Blasted rock	"	0.00	1.40	1.40
2 cy capacity				
Light material	C.Y.	0.00	0.62	0.62
Medium material	"	0.00	0.79	0.79
Wet material	"	0.00	0.93	0.93
Blasted rock	"	0.00	1.10	1.10
Wheel mounted front-end loader				
7/8 cy capacity				
Light material	C.Y.	0.00	1.45	1.45
Medium material	"	0.00	1.70	1.70
Wet material	"	0.00	1.95	1.95
Blasted rock	"	0.00	2.35	2.35
1-1/2 cy capacity				
Light material	C.Y.	0.00	0.79	0.79
Medium material	"	0.00	0.85	0.85
Wet material	"	0.00	0.93	0.93
Blasted rock	"	0.00	1.00	1.00
2-1/2 cy capacity				
Light material	C.Y.	0.00	0.65	0.65
Medium material	"	0.00	0.69	0.69
Wet material	"	0.00	0.74	0.74

02 SITEWORK

EARTHWORK

EARTHWORK	UNIT	MAT.	INST.	TOTAL
02210.30 BULK EXCAVATION				
Blasted rock	C.Y.	0.00	0.79	0.79
3-1/2 cy capacity				
Light material	C.Y.	0.00	0.62	0.62
Medium material	"	0.00	0.65	0.65
Wet material	"	0.00	0.69	0.69
Blasted rock	"	0.00	0.74	0.74
6 cy capacity				
Light material	C.Y.	0.00	0.37	0.37
Medium material	"	0.00	0.40	0.40
Wet material	"	0.00	0.43	0.43
Blasted rock	"	0.00	0.46	0.46
Track mounted front-end loader				
1-1/2 cy capacity				
Light material	C.Y.	0.00	0.98	0.98
Medium material	"	0.00	1.05	1.05
Wet material	"	0.00	1.15	1.15
Blasted rock	"	0.00	1.30	1.30
2-3/4 cy capacity				
Light material	C.Y.	0.00	0.56	0.56
Medium material	"	0.00	0.62	0.62
Wet material	"	0.00	0.69	0.69
Blasted rock	"	0.00	0.79	0.79
02220.10 BORROW				
Borrow fill, F.O.B. at pit				11.00
Sand, haul to site, round trip				
10 mile	C.Y.	3.85	5.55	9.40
20 mile	"	3.85	9.25	13.10
30 mile	"	3.85	13.90	17.75
Place borrow fill and compact				
Less than 1 in 4 slope	C.Y.	3.85	2.80	6.65
Greater than 1 in 4 slope	"	3.85	3.70	7.55
02220.20 GRAVEL AND STONE				
F.O.B. PLANT				
No. 21 crusher run stone	C.Y.			10.15
No. 26 crusher run stone	"			9.35
No. 57 stone	"			10.45
No. 67 gravel	"			3.85
No. 68 stone	"			9.90
No. 78 stone	"			9.90
No. 78 gravel, (pea gravel)	"			7.70
No. 357 or B-3 stone	"			10.45
Structural & foundation backfill				
No. 21 crusher run stone	TON			8.80
No. 26 crusher run stone	"			9.05
No. 57 stone	"			10.10
No. 67 gravel	"			4.40
No. 68 stone	"			9.60
No. 78 stone	"			9.60
No. 78 gravel, (pea gravel)	"			7.40
No. 357 or B-3 stone	"			10.15

02 SITEWORK

EARTHWORK

EARTHWORK	UNIT	MAT.	INST.	TOTAL
02220.40 BUILDING EXCAVATION				
Structural excavation, unclassified earth				
3/8 cy backhoe	C.Y.	0.00	7.85	7.85
3/4 cy backhoe	"	0.00	5.85	5.85
1 cy backhoe	"	0.00	4.90	4.90
Foundation backfill and compaction by machine	"	0.00	11.75	11.75
02220.50 UTILITY EXCAVATION				
Trencher, sandy clay, 8" wide trench				
18" deep	L.F.	0.00	0.93	0.93
24" deep	"	0.00	1.05	1.05
36" deep	"	0.00	1.20	1.20
Trench backfill, 95% compaction				
Tamp by hand	C.Y.	0.00	13.25	13.25
Vibratory compaction	"	0.00	10.60	10.60
Trench backfilling, with borrow sand, place & compact	"	4.10	10.60	14.70
02220.60 TRENCHING				
Trenching and continuous footing excavation				
By gradall				
1 cy capacity				
Light soil	C.Y.	0.00	1.70	1.70
Medium soil	"	0.00	1.80	1.80
Heavy/wet soil	"	0.00	1.95	1.95
Loose rock	"	0.00	2.15	2.15
Blasted rock	"	0.00	2.25	2.25
By hydraulic excavator				
1/2 cy capacity				
Light soil	C.Y.	0.00	1.95	1.95
Medium soil	"	0.00	2.15	2.15
Heavy/wet soil	"	0.00	2.35	2.35
Loose rock	"	0.00	2.60	2.60
Blasted rock	"	0.00	2.95	2.95
1 cy capacity				
Light soil	C.Y.	0.00	1.30	1.30
Medium soil	"	0.00	1.40	1.40
Heavy/wet soil	"	0.00	1.50	1.50
Loose rock	"	0.00	1.60	1.60
Blasted rock	"	0.00	1.70	1.70
1-1/2 cy capacity				
Light soil	C.Y.	0.00	1.15	1.15
Medium soil	"	0.00	1.25	1.25
Heavy/wet soil	"	0.00	1.30	1.30
Loose rock	"	0.00	1.40	1.40
Blasted rock	"	0.00	1.50	1.50
2 cy capacity				
Light soil	C.Y.	0.00	1.10	1.10
Medium soil	"	0.00	1.15	1.15
Heavy/wet soil	"	0.00	1.25	1.25
Loose rock	"	0.00	1.30	1.30
Blasted rock	"	0.00	1.40	1.40
2-1/2 cy capacity				

02 SITEWORK

EARTHWORK	UNIT	MAT.	INST.	TOTAL
02220.60 TRENCHING				
Light soil	C.Y.	0.00	1.00	1.00
Medium soil	"	0.00	1.05	1.05
Heavy/wet soil	"	0.00	1.10	1.10
Loose rock	"	0.00	1.15	1.15
Blasted rock	"	0.00	1.25	1.25
Trencher, chain, 1' wide to 4' deep				
Light soil	C.Y.	0.00	1.00	1.00
Medium soil	"	0.00	1.15	1.15
Heavy soil	"	0.00	1.35	1.35
Hand excavation				
Bulk, wheeled 100'				
Normal soil	C.Y.	0.00	23.55	23.55
Sand or gravel	"	0.00	21.20	21.20
Medium clay	"	0.00	30.30	30.30
Heavy clay	"	0.00	42.40	42.40
Loose rock	"	0.00	53.00	53.00
Trenches, up to 2' deep				
Normal soil	C.Y.	0.00	26.50	26.50
Sand or gravel	"	0.00	23.55	23.55
Medium clay	"	0.00	35.30	35.30
Heavy clay	"	0.00	53.00	53.00
Loose rock	"	0.00	70.50	70.50
Trenches, to 6' deep				
Normal soil	C.Y.	0.00	30.30	30.30
Sand or gravel	"	0.00	26.50	26.50
Medium clay	"	0.00	42.40	42.40
Heavy clay	"	0.00	70.50	70.50
Loose rock	"	0.00	106.00	106.00
Backfill trenches				
With compaction				
By hand	C.Y.	0.00	17.65	17.65
By 60 hp tracked dozer	"	0.00	1.05	1.05
By 200 hp tracked dozer	"	0.00	0.65	0.65
By small front-end loader	"	0.00	1.20	1.20
Spread dumped fill or gravel, no compaction				
6" layers	S.Y.	0.00	0.67	0.67
12" layers	"	0.00	0.80	0.80
Compaction in 6" layers				
By hand with air tamper	S.Y.	0.00	0.51	0.51
Backfill trenches, sand bedding, no compaction				
By hand	C.Y.	4.95	17.65	22.60
By small front-end loader	"	4.95	1.70	6.65
02220.70 ROADWAY EXCAVATION				
Roadway excavation				
1/4 mile haul	C.Y.	0.00	1.10	1.10
2 mile haul	"	0.00	1.85	1.85
5 mile haul	"	0.00	2.80	2.80
Excavation of open ditches	"	0.00	0.79	0.79
Trim banks, swales or ditches	S.Y.	0.00	0.93	0.93
Bulk swale excavation by dragline				
Small jobs	C.Y.	0.00	2.75	2.75

02 SITEWORK

EARTHWORK	UNIT	MAT.	INST.	TOTAL
02220.70 ROADWAY EXCAVATION				
Large jobs	C.Y.	0.00	1.55	1.55
Spread base course	"	0.00	1.40	1.40
Roll and compact	"	0.00	1.85	1.85
02220.71 BASE COURSE				
Base course, crushed stone				
3" thick	S.Y.	1.95	0.28	2.23
4" thick	"	2.65	0.30	2.95
6" thick	"	3.85	0.33	4.18
8" thick	"	4.95	0.37	5.32
10" thick	"	6.15	0.40	6.55
12" thick	"	6.95	0.46	7.41
Base course, bank run gravel				
4" deep	S.Y.	1.20	0.29	1.49
6" deep	"	1.75	0.32	2.07
8" deep	"	2.50	0.35	2.85
10" deep	"	3.20	0.37	3.57
12" deep	"	3.85	0.43	4.28
Prepare and roll sub base				
Minimum	S.Y.	0.00	0.28	0.28
Maximum	"	0.00	0.46	0.46
02220.90 HAND EXCAVATION				
Excavation				
To 2' deep				
Normal soil	C.Y.	0.00	23.55	23.55
Sand and gravel	"	0.00	21.20	21.20
Medium clay	"	0.00	26.50	26.50
Heavy clay	"	0.00	30.30	30.30
Loose rock	"	0.00	35.30	35.30
To 6' deep				
Normal soil	C.Y.	0.00	30.30	30.30
Sand and gravel	"	0.00	26.50	26.50
Medium clay	"	0.00	35.30	35.30
Heavy clay	"	0.00	42.40	42.40
Loose rock	"	0.00	53.00	53.00
Backfilling foundation without compaction, 6" lifts	"	0.00	12.75	12.75
Compaction of backfill around structures or in trench				
By hand with air tamper	C.Y.	0.00	15.15	15.15
By hand with vibrating plate tamper	"	0.00	14.15	14.15
1 ton roller	"	0.00	21.00	21.00
Miscellaneous hand labor				
Trim slopes, sides of excavation	S.F.	0.00	0.04	0.04
Trim bottom of excavation	"	0.00	0.04	0.04
Excavation around obstructions and services	C.Y.	0.00	70.50	70.50
02240.05 SOIL STABILIZATION				
Straw bale secured with rebar	L.F.	1.10	0.71	1.81
Filter barrier, 18" high filter fabric	"	1.10	2.10	3.20
Sediment fence, 36" fabric with 6" mesh	"	2.50	2.65	5.15
Soil stabilization with tar paper, burlap, straw and stakes	S.F.	0.22	0.03	0.25

EARTHWORK

02240.30 GEOTEXTILE

EARTHWORK	UNIT	MAT.	INST.	TOTAL
Filter cloth, light reinforcement				
Woven				
12'-6" wide x 50' long	S.F.	0.02	0.03	0.05
Various lengths	"	0.03	0.03	0.06
Non-woven				
14'-8" wide x 430' long	S.F.	0.11	0.03	0.14
Various lengths	"	0.17	0.03	0.20

02270.10 SLOPE PROTECTION

EARTHWORK	UNIT	MAT.	INST.	TOTAL
Gabions, stone filled				
6" deep	S.Y.	13.75	10.00	23.75
9" deep	"	17.60	11.45	29.05
12" deep	"	23.10	13.35	36.45
18" deep	"	29.70	16.00	45.70
36" deep	"	53.00	26.70	79.70

02270.40 RIPRAP

EARTHWORK	UNIT	MAT.	INST.	TOTAL
Riprap				
Crushed stone blanket, max size 2-1/2"	TON	17.60	30.30	47.90
Stone, quarry run, 300 lb. stones	"	24.00	27.90	51.90
400 lb. stones	"	24.00	25.90	49.90
500 lb. stones	"	22.00	24.20	46.20
750 lb. stones	"	24.00	22.70	46.70
Dry concrete riprap in bags 3" thick, 80 lb. per bag	BAG	3.30	1.50	4.80

02280.20 SOIL TREATMENT

EARTHWORK	UNIT	MAT.	INST.	TOTAL
Soil treatment, termite control pretreatment, under slabs		0.07	0.11	0.18
Under slabs	S.F.	0.07	0.12	0.19
By walls	"	0.07	0.14	0.21

02290.30 WEED CONTROL

EARTHWORK	UNIT	MAT.	INST.	TOTAL
Weed control, bromicil, 15 lb./acre, wettable powder	ACRE	165.00	102.00	267.00
Vegetation control, by application of plant killer	S.Y.	0.02	0.08	0.10
Weed killer, lawns and fields	"	0.11	0.04	0.15

TUNNELING

02300.10 PIPE JACKING

TUNNELING	UNIT	MAT.	INST.	TOTAL
Pipe casing, horizontal jacking				
18" dia.	L.F.	44.00	40.40	84.40

TUNNELING

02300.10 PIPE JACKING	UNIT	MAT.	INST.	TOTAL
21" dia.	L.F.	53.00	43.20	96.20
24" dia.	"	55.00	45.40	100.40
27" dia.	"	61.50	45.40	106.90
30" dia.	"	68.00	47.80	115.80
36" dia.	"	77.00	52.00	129.00
42" dia.	"	110.00	57.00	167.00
48" dia.	"	132.00	60.50	192.50

PILES AND CAISSONS

02360.50 PRESTRESSED PILING	UNIT	MAT.	INST.	TOTAL
Prestressed concrete piling, less than 60' long				
10" sq.	L.F.	6.95	2.55	9.50
12" sq.	"	10.10	2.65	12.75
14" sq.	"	10.75	2.70	13.45
16" sq.	"	12.40	2.75	15.15
18" sq.	"	17.35	3.00	20.35
20" sq.	"	23.10	3.05	26.15
24" sq.	"	27.70	3.15	30.85
More than 60' long				
12" sq.	L.F.	10.10	2.20	12.30
14" sq.	"	11.00	2.20	13.20
16" sq.	"	11.30	2.25	13.55
18" sq.	"	17.35	2.30	19.65
20" sq.	"	23.10	2.35	25.45
24" sq.	"	27.70	2.40	30.10
Straight cylinder, less than 60' long				
12" dia.	L.F.	9.80	2.75	12.55
14" dia.	"	13.30	2.85	16.15
16" dia.	"	16.15	2.90	19.05
18" dia.	"	18.50	3.00	21.50
20" dia.	"	24.25	3.05	27.30
24" dia.	"	30.00	3.15	33.15
More than 60' long				
12" dia.	L.F.	9.80	2.20	12.00
14" dia.	"	13.30	2.25	15.55
16" dia.	"	16.15	2.30	18.45
18" dia.	"	18.50	2.35	20.85
20" dia.	"	24.25	2.40	26.65
24" dia.	"	30.00	2.45	32.45
Concrete sheet piling				
12" thick x 20' long	S.F.	10.40	6.10	16.50
25' long	"	10.40	5.55	15.95

PILES AND CAISSONS

PILES AND CAISSONS	UNIT	MAT.	INST.	TOTAL
02360.50 PRESTRESSED PILING				
30' long	S.F.	10.40	5.10	15.50
35' long	"	10.40	4.70	15.10
40' long	"	10.40	4.35	14.75
16" thick x 40' long	"	13.85	3.40	17.25
45' long	"	13.85	3.20	17.05
50' long	"	13.85	3.05	16.90
55' long	"	13.85	2.90	16.75
60' long	"	13.85	2.75	16.60
02360.60 STEEL PILES				
H-section piles				
8x8				
36 lb/ft				
30' long	L.F.	12.60	5.10	17.70
40' long	"	12.60	4.05	16.65
50' long	"	12.60	3.40	16.00
10x10				
42 lb/ft				
30' long	L.F.	14.85	5.10	19.95
40' long	"	14.85	4.05	18.90
50' long	"	14.85	3.40	18.25
57 lb/ft				
30' long	L.F.	19.45	5.10	24.55
40' long	"	19.45	4.05	23.50
50' long	"	19.45	3.40	22.85
12x12				
53 lb/ft				
30' long	L.F.	17.15	5.55	22.70
40' long	"	17.15	4.35	21.50
50' long	"	17.15	3.40	20.55
74 lb/ft				
30' long	L.F.	24.00	5.55	29.55
40' long	"	24.00	4.35	28.35
50' long	"	24.00	3.40	27.40
14x14				
73 lb/ft				
40' long	L.F.	24.00	5.55	29.55
50' long	"	24.00	4.35	28.35
60' long	"	24.00	3.40	27.40
89 lb/ft				
40' long	L.F.	30.90	5.55	36.45
50' long	"	30.90	4.35	35.25
60' long	"	30.90	3.40	34.30
102 lb/ft				
40' long	L.F.	34.30	5.55	39.85
50' long	"	34.30	4.35	38.65
60' long	"	34.30	3.40	37.70
117 lb/ft				
40' long	L.F.	40.00	5.80	45.80
50' long	"	40.00	4.50	44.50
60' long	"	40.00	3.50	43.50
Splice				

PILES AND CAISSONS

PILES AND CAISSONS	UNIT	MAT.	INST.	TOTAL
02360.60 STEEL PILES				
8"	EA.	57.00	34.00	91.00
10"	"	63.00	40.80	103.80
12"	"	86.00	40.80	126.80
14"	"	109.00	51.00	160.00
Tapered friction piles, with fluted steel casing, up to 50'				
With 4000 psi concrete no reinforcing				
12" dia.	L.F.	14.85	3.05	17.90
14" dia.	"	17.15	3.15	20.30
16" dia.	"	20.60	3.20	23.80
18" dia.	"	22.90	3.60	26.50
02360.65 STEEL PIPE PILES				
Concrete filled, 3000# concrete, up to 40'				
8" dia.	L.F.	12.85	4.35	17.20
10" dia.	"	15.85	4.50	20.35
12" dia.	"	19.80	4.70	24.50
14" dia.	"	21.80	4.90	26.70
16" dia.	"	24.75	5.10	29.85
18" dia.	"	27.70	5.30	33.00
Pipe piles, non-filled				
8" dia.	L.F.	10.90	3.40	14.30
10" dia.	"	12.85	3.50	16.35
12" dia.	"	15.85	3.60	19.45
14" dia.	"	16.85	3.80	20.65
16" dia.	"	19.80	3.95	23.75
18" dia.	"	21.80	4.05	25.85
Splice				
8" dia.	EA.	41.60	40.80	82.40
10" dia.	"	42.60	40.80	83.40
12" dia.	"	44.50	51.00	95.50
14" dia.	"	62.50	51.00	113.50
16" dia.	"	95.00	68.00	163.00
18" dia.	"	119.00	68.00	187.00
Standard point				
8" dia.	EA.	71.50	40.80	112.30
10" dia.	"	99.00	40.80	139.80
12" dia.	"	134.00	51.00	185.00
14" dia.	"	193.00	51.00	244.00
16" dia.	"	287.00	68.00	355.00
18" dia.	"	292.00	68.00	360.00
Heavy duty point				
8" dia.	EA.	104.00	51.00	155.00
10" dia.	"	144.00	51.00	195.00
12" dia.	"	168.00	68.00	236.00
14" dia.	"	218.00	68.00	286.00
16" dia.	"	277.00	81.50	358.50
18" dia.	"	346.00	81.50	427.50
02360.70 STEEL SHEET PILING				
Steel sheet piling, driven and pulled, 12" wide				
20' long	S.F.	11.00	6.10	17.10

PILES AND CAISSONS

PILES AND CAISSONS	UNIT	MAT.	INST.	TOTAL
02360.70 STEEL SHEET PILING				
35' long	S.F.	11.00	4.35	15.35
50' long	"	11.00	3.05	14.05
Over 50' long	"	11.00	2.75	13.75
02360.80 WOOD AND TIMBER PILES				
Treated wood piles, 12" butt, 8" tip				
25' long	L.F.	6.05	6.10	12.15
30' long	"	6.40	5.10	11.50
35' long	"	6.40	4.35	10.75
40' long	"	6.40	3.80	10.20
12" butt, 7" tip				
40' long	L.F.	7.05	3.80	10.85
45' long	"	7.05	3.40	10.45
50' long	"	8.25	3.05	11.30
55' long	"	8.25	2.75	11.00
60' long	"	8.25	2.55	10.80
02360.90 PILE TESTING				
Pile test				
50 ton to 100 ton	EA.			13,200
To 200 ton	"			19,800
To 300 ton	"			22,000
To 400 ton	"			33,000
To 600 ton	"			38,500
02380.10 CAISSONS				
Caisson, including 3000# concrete, in stable ground				
18" dia.	L.F.	10.30	12.20	22.50
24" dia.	"	17.15	12.70	29.85
30" dia.	"	27.50	15.25	42.75
36" dia.	"	37.80	17.45	55.25
48" dia.	"	65.00	20.35	85.35
60" dia.	"	108.00	27.70	135.70
72" dia.	"	172.00	33.90	205.90
84" dia.	"	212.00	43.60	255.60
Wet ground, casing required but pulled				
18" dia.	L.F.	10.30	15.25	25.55
24" dia.	"	17.15	16.95	34.10
30" dia.	"	27.50	19.05	46.55
36" dia.	"	37.80	20.35	58.15
48" dia.	"	65.00	25.40	90.40
60" dia.	"	108.00	33.90	141.90
72" dia.	"	158.00	51.00	209.00
84" dia.	"	212.00	76.50	288.50
Soft rock				
18" dia.	L.F.	10.30	43.60	53.90
24" dia.	"	17.15	76.50	93.65
30" dia.	"	27.50	102.00	129.50
36" dia.	"	37.80	153.00	190.80
48" dia.	"	65.00	203.00	268.00

02 SITEWORK

PILES AND CAISSONS

02380.10 CAISSONS	UNIT	MAT.	INST.	TOTAL
60" dia.	L.F.	108.00	305.00	413.00
72" dia.	"	172.00	339.00	511.00
84" dia.	"	212.00	381.00	593.00

RAILROAD WORK

02450.10 RAILROAD WORK	UNIT	MAT.	INST.	TOTAL
Rail				
90 lb	L.F.	11.55	0.44	11.99
100 lb	"	14.15	0.44	14.59
115 lb	"	15.85	0.44	16.29
132 lb	"	19.25	0.44	19.69
Rail relay				
90 lb	L.F.	3.95	0.44	4.39
100 lb	"	5.65	0.44	6.09
115 lb	"	6.80	0.44	7.24
132 lb	"	8.50	0.44	8.94
New angle bars, per pair				
90 lb	EA.	51.00	0.55	51.55
100 lb	"	56.50	0.55	57.05
115 lb	"	73.50	0.55	74.05
132 lb	"	85.00	0.55	85.55
Angle bar relay				
90 lb	EA.	22.65	0.55	23.20
100 lb	"	23.80	0.55	24.35
115 lb	"	24.95	0.55	25.50
132 lb	"	26.10	0.55	26.65
New tie plates				
90 lb	EA.	5.65	0.39	6.04
100 lb	"	6.25	0.39	6.64
115 lb	"	6.50	0.39	6.89
132 lb	"	7.35	0.39	7.74
Tie plate relay				
100 lb	EA.	3.15	0.39	3.54
115 lb	"	3.10	0.39	3.49
132 lb	"	3.95	0.39	4.34
Track accessories				
Wooden cross ties, 8'	EA.	26.10	2.75	28.85
Concrete cross ties, 8'	"	181.00	5.50	186.50
Tie plugs, 5"	"	11.35	0.28	11.63
Track bolts and nuts, 1"	"	2.85	0.28	3.13
Lockwashers, 1"	"	0.63	0.18	0.81
Track spikes, 6"	"	0.68	1.10	1.78

RAILROAD WORK	UNIT	MAT.	INST.	TOTAL
02450.10 RAILROAD WORK				
Wooden switch ties	B.F.	0.90	0.28	1.18
Rail anchors	EA.	2.85	1.00	3.85
Ballast	TON	8.50	5.50	14.00
Gauge rods	EA.	22.65	4.40	27.05
Compromise splice bars	"	283.00	7.35	290.35
Turnout				
90 lb	EA.	7,930	1,100	9,030
100 lb	"	8,500	1,100	9,600
110 lb	"	9,060	1,100	10,160
115 lb	"	9,290	1,100	10,390
132 lb	"	10,200	1,100	11,300
Turnout relay				
90 lb	EA.	5,100	1,100	6,200
100 lb	"	5,670	1,100	6,770
110 lb	"	59.00	1,100	1,159
115 lb	"	6,230	1,100	7,330
132 lb	"	6,800	1,100	7,900
Railroad track in place, complete				
New rail				
90 lb	L.F.	102.00	11.00	113.00
100 lb	"	102.00	11.00	113.00
110 lb	"	96.50	11.00	107.50
115 lb	"	106.00	11.00	117.00
132 lb	"	108.00	11.00	119.00
Rail relay				
90 lb	L.F.	56.50	11.00	67.50
100 lb	"	62.50	11.00	73.50
110 lb	"	62.50	11.00	73.50
115 lb	"	68.00	11.00	79.00
132 lb	"	70.00	11.00	81.00
No. 8 turnout				
90 lb	EA.	20,390	1,470	21,860
100 lb	"	22,660	1,470	24,130
110 lb	"	24,930	1,470	26,400
115 lb	"	25,490	1,470	26,960
132 lb	"	26,060	1,470	27,530
No. 8 turnout relay				
90 lb	EA.	16,990	1,470	18,460
100 lb	"	16,990	1,470	18,460
110 lb	"	16,990	1,470	18,460
115 lb	"	19,260	1,470	20,730
132 lb	"	19,260	1,470	20,730
Railroad crossings, asphalt, based on 8" thick x 20'				
Including track and approach				
12' roadway	EA.	453.00	275.00	728.00
15' roadway	"	510.00	314.00	824.00
18' roadway	"	623.00	367.00	990.00
21' roadway	"	702.00	440.00	1,142
24' roadway	"	793.00	550.00	1,343
Precast concrete inserts				
12' roadway	EA.	680.00	110.00	790.00
15' roadway	"	793.00	138.00	931.00

RAILROAD WORK

RAILROAD WORK	UNIT	MAT.	INST.	TOTAL
02450.10 RAILROAD WORK				
18' roadway	EA.	963.00	183.00	1,146
21' roadway	"	1,190	220.00	1,410
24' roadway	"	1,360	244.00	1,604
Molded rubber, with headers				
12' roadway	EA.	3,970	110.00	4,080
15' roadway	"	5,100	138.00	5,238
18' roadway	"	6,000	183.00	6,183
21' roadway	"	6,800	220.00	7,020
24' roadway	"	7,930	244.00	8,174

PAVING AND SURFACING

PAVING AND SURFACING	UNIT	MAT.	INST.	TOTAL
02510.20 ASPHALT SURFACES				
Asphalt wearing surface, for flexible pavement				
1" thick	S.Y.	1.90	1.00	2.90
1-1/2" thick	"	2.90	1.25	4.15
2" thick	"	3.30	1.55	4.85
3" thick	"	4.95	2.05	7.00
Bituminous sidewalk, no base				
2" thick	S.Y.	4.10	1.30	5.40
3" thick	"	4.70	1.35	6.05
02520.10 CONCRETE PAVING				
Concrete paving, reinforced, 5000 psi concrete				
6" thick	S.Y.	15.75	9.60	25.35
7" thick	"	16.15	10.15	26.30
8" thick	"	19.90	10.95	30.85
9" thick	"	19.90	11.75	31.65
10" thick	"	23.95	12.80	36.75
11" thick	"	24.25	13.85	38.10
12" thick	"	26.60	15.25	41.85
15" thick	"	32.30	19.05	51.35
Concrete paving, for pipe trench, reinforced				
7" thick	S.Y.	16.15	11.00	27.15
8" thick	"	18.25	12.20	30.45
9" thick	"	19.90	13.75	33.65
10" thick	"	21.95	15.70	37.65
Fibrous concrete				
5" thick	S.Y.	14.75	11.75	26.50
8" thick	"	18.75	12.70	31.45
Roller compacted concrete, (RCC), place and compact				
8" thick	S.Y.	18.25	15.25	33.50
12" thick	"	26.60	19.05	45.65

02 SITEWORK

PAVING AND SURFACING	UNIT	MAT.	INST.	TOTAL
02520.10 CONCRETE PAVING				
Steel edge forms up to				
12" deep	L.F.	0.46	0.68	1.14
15" deep	"	0.58	0.82	1.40
Paving finishes				
Belt dragged	S.Y.	0.00	1.00	1.00
Curing	"	0.23	0.20	0.43
02545.10 ASPHALT REPAIR				
Coal tar emulsion seal coat, rubber additive, fuel resistant	S.Y.	0.71	0.29	1.00
Bituminous surface treatment, single	"	0.66	0.20	0.86
Double	"	0.88	0.02	0.90
Bituminous prime coat	"	0.44	0.03	0.47
Tack coat	"	0.22	0.02	0.24
Crack sealing, concrete paving	L.F.	0.33	0.14	0.47
Bituminous paving for pipe trench, 4" thick	S.Y.	4.40	7.35	11.75
Polypropylene, nonwoven paving fabric	"	0.55	0.10	0.65
Rubberized asphalt	"	0.88	1.85	2.73
Asphalt slurry seal	"	2.20	1.20	3.40
02580.10 PAVEMENT MARKINGS				
Pavement line marking, paint				
4" wide	L.F.	0.02	0.05	0.07
6" wide	"	0.04	0.11	0.15
8" wide	"	0.07	0.17	0.24
Reflective paint, 4" wide	"	0.18	0.17	0.35
Airfield markings, retro-reflective				
White	L.F.	0.29	0.17	0.46
Yellow	"	0.31	0.17	0.48
Preformed tape, 4" wide				
Inlaid reflective	L.F.	0.87	0.03	0.90
Reflective paint	"	0.58	0.05	0.63
Thermoplastic				
White	L.F.	0.23	0.10	0.33
Yellow	"	0.24	0.10	0.34
12" wide, thermoplastic, white	"	0.69	0.29	0.98
Directional arrows, reflective preformed tape	EA.	58.00	20.40	78.40
Messages, reflective preform tape (per letter)	"	28.90	10.20	39.10
Handicap symbol, preformed tape	"	52.00	20.40	72.40
Parking stall painting	"	0.78	4.10	4.88

UTILITIES

	UNIT	MAT.	INST.	TOTAL
02605.30 MANHOLES				
Precast sections, 48" dia.				
Base section	EA.	88.00	91.50	179.50
1'0" riser	"	98.00	95.50	193.50
1'4" riser	"	24.20	76.50	100.70
2'8" riser	"	30.80	82.00	112.80
4'0" riser	"	48.40	88.50	136.90
2'8" cone top	"	98.00	95.50	193.50
Precast manholes, 48" dia.	"	61.50	115.00	176.50
4' deep	EA.	187.00	230.00	417.00
6' deep	"	264.00	287.00	551.00
7' deep	"	314.00	328.00	642.00
8' deep	"	363.00	383.00	746.00
10' deep	"	390.00	460.00	850.00
Cast-in-place, 48" dia., with frame and cover				
5' deep	EA.	363.00	574.00	937.00
6' deep	"	478.00	656.00	1,134
8' deep	"	698.00	766.00	1,464
10' deep	"	814.00	919.00	1,733
Brick manholes, 48" dia. with cover, 8" thick				
4' deep	EA.	478.00	263.00	741.00
6' deep	"	600.00	292.00	892.00
8' deep	"	770.00	329.00	1,099
10' deep	"	957.00	376.00	1,333
12' deep	"	1,200	439.00	1,639
14' deep	"	1,460	526.00	1,986
Inverts for manholes				
Single channel	EA.	71.50	105.00	176.50
Triple channel	"	82.50	132.00	214.50
Frames and covers, 24" diameter				
300 lb	EA.	236.00	21.20	257.20
400 lb	"	248.00	23.55	271.55
500 lb	"	286.00	30.30	316.30
Watertight, 350 lb	"	297.00	70.50	367.50
For heavy equipment, 1200 lb	"	649.00	106.00	755.00
Steps for manholes				
7" x 9"	EA.	8.35	4.25	12.60
8" x 9"	"	10.75	4.70	15.45
Curb inlet, 4' throat, cast in place				
12"-30" pipe	EA.	231.00	574.00	805.00
36"-48" pipe	"	253.00	656.00	909.00
Raise exist frame and cover, repaving	"	0.00	230.00	230.00
02610.10 CAST IRON FLANGED PIPE				
Cast iron flanged sections				
4" pipe, with one bolt set				
3' section	EA.	104.00	10.45	114.45
4' section	"	110.00	11.50	121.50
5' section	"	121.00	12.75	133.75
6' section	"	132.00	14.35	146.35
8' section	"	154.00	16.40	170.40
10' section	"	170.00	23.00	193.00

02 SITEWORK

UTILITIES	UNIT	MAT.	INST.	TOTAL
02610.10 CAST IRON FLANGED PIPE				
12' section	EA.	187.00	38.30	225.30
15' section	"	214.00	57.50	271.50
18' section	"	242.00	76.50	318.50
6" pipe, with one bolt set				
3' section	EA.	132.00	11.50	143.50
4' section	"	143.00	13.50	156.50
5' section	"	154.00	15.30	169.30
6' section	"	165.00	17.65	182.65
8' section	"	187.00	25.50	212.50
10' section	"	209.00	28.70	237.70
12' section	"	231.00	38.30	269.30
15' section	"	264.00	57.50	321.50
18' section	"	319.00	82.00	401.00
8" pipe, with one bolt set				
3' section	EA.	176.00	14.35	190.35
4' section	"	198.00	16.40	214.40
5' section	"	209.00	19.15	228.15
6' section	"	220.00	23.00	243.00
8' section	"	253.00	32.80	285.80
10' section	"	286.00	38.30	324.30
12' section	"	308.00	57.50	365.50
15' section	"	363.00	76.50	439.50
18' section	"	429.00	95.50	524.50
10" pipe, with one bolt set				
3' section	EA.	231.00	14.75	245.75
4' section	"	264.00	16.90	280.90
5' section	"	286.00	19.80	305.80
6' section	"	308.00	23.95	331.95
8' section	"	341.00	34.80	375.80
10' section	"	374.00	41.00	415.00
12' section	"	429.00	64.00	493.00
15' section	"	473.00	82.00	555.00
18' section	"	550.00	115.00	665.00
12" pipe, with one bolt set				
3' section	EA.	308.00	15.95	323.95
4' section	"	330.00	18.55	348.55
5' section	"	363.00	22.10	385.10
6' section	"	385.00	26.10	411.10
8' section	"	434.00	38.30	472.30
10' section	"	484.00	44.20	528.20
12' section	"	539.00	72.00	611.00
15' section	"	616.00	95.50	711.50
18' section	"	693.00	128.00	821.00
02610.11 CAST IRON FITTINGS				
Mechanical joint, with 2 bolt kits				
90 deg bend				
4"	EA.	60.50	14.15	74.65
6"	"	77.00	16.30	93.30
8"	"	104.00	21.20	125.20
10"	"	160.00	30.30	190.30

UTILITIES

02610.11 CAST IRON FITTINGS

	UNIT	MAT.	INST.	TOTAL
12"	EA.	231.00	42.40	273.40
14"	"	319.00	53.00	372.00
16"	"	352.00	70.50	422.50
45 deg bend				
4"	EA.	49.50	14.15	63.65
6"	"	71.50	16.30	87.80
8"	"	104.00	21.20	125.20
10"	"	132.00	30.30	162.30
12"	"	187.00	42.40	229.40
14"	"	231.00	53.00	284.00
16"	"	297.00	70.50	367.50
Tee, with 3 bolt kits				
4" x 4"	EA.	93.50	21.20	114.70
6" x 6"	"	121.00	26.50	147.50
8" x 8"	"	176.00	35.30	211.30
10" x 10"	"	242.00	53.00	295.00
12" x 12"	"	319.00	70.50	389.50
Wye, with 3 bolt kits				
6" x 6"	EA.	160.00	26.50	186.50
8" x 8"	"	236.00	35.30	271.30
10" x 10"	"	346.00	53.00	399.00
12" x 12"	"	484.00	70.50	554.50
Reducer, with 2 bolt kits				
6" x 4"	EA.	60.50	26.50	87.00
8" x 6"	"	82.50	35.30	117.80
10" x 8"	"	138.00	53.00	191.00
12" x 10"	"	148.00	70.50	218.50
Flanged, 90 deg bend, 125 lb.				
4"	EA.	38.50	17.65	56.15
6"	"	60.50	21.20	81.70
8"	"	88.00	26.50	114.50
10"	"	154.00	35.30	189.30
12"	"	220.00	53.00	273.00
14"	"	440.00	70.50	510.50
16"	"	660.00	70.50	730.50
Tee				
4"	EA.	77.00	26.50	103.50
6"	"	110.00	30.30	140.30
8"	"	160.00	35.30	195.30
10"	"	314.00	42.40	356.40
12"	"	418.00	53.00	471.00
14"	"	957.00	70.50	1,028
16"	"	1,380	106.00	1,486

02610.13 GATE VALVES

	UNIT	MAT.	INST.	TOTAL
Gate valve, (AWWA) mechanical joint, with adjustable box				
4" valve	EA.	336.00	38.30	374.30
6" valve	"	390.00	46.00	436.00
8" valve	"	468.00	57.50	525.50
10" valve	"	682.00	67.50	749.50
12" valve	"	946.00	82.00	1,028

UTILITIES

UTILITIES	UNIT	MAT.	INST.	TOTAL
02610.13 GATE VALVES				
14" valve	EA.	2,310	95.50	2,406
16" valve	"	3,080	104.00	3,184
18" valve	"	4,180	115.00	4,295
Flanged, with box, post indicator (AWWA)				
4" valve	EA.	286.00	46.00	332.00
6" valve	"	352.00	52.00	404.00
8" valve	"	440.00	64.00	504.00
10" valve	"	682.00	76.50	758.50
12" valve	"	946.00	95.50	1,042
14" valve	"	2,310	115.00	2,425
16" valve	"	3,080	144.00	3,224
02610.15 WATER METERS				
Water meter, displacement type				
1"	EA.	121.00	30.00	151.00
1-1/2"	"	440.00	33.30	473.30
2"	"	660.00	37.50	697.50
02610.17 CORPORATION STOPS				
Stop for flared copper service pipe				
3/4"	EA.	14.30	15.00	29.30
1"	"	20.90	16.65	37.55
1-1/4"	"	60.50	20.00	80.50
1-1/2"	"	71.50	25.00	96.50
2"	"	104.00	30.00	134.00
02610.40 DUCTILE IRON PIPE				
Ductile iron pipe, cement lined, slip-on joints				
4"	L.F.	6.35	3.20	9.55
6"	"	7.85	3.40	11.25
8"	"	10.30	3.60	13.90
10"	"	14.15	3.85	18.00
12"	"	17.35	4.60	21.95
14"	"	21.95	5.75	27.70
16"	"	25.40	6.40	31.80
18"	"	28.90	7.20	36.10
20"	"	32.30	8.20	40.50
Mechanical joint pipe				
4"	L.F.	6.95	4.40	11.35
6"	"	8.65	4.80	13.45
8"	"	11.55	5.20	16.75
10"	"	15.00	5.75	20.75
12"	"	19.25	7.65	26.90
14"	"	24.20	8.85	33.05
16"	"	26.60	10.45	37.05
18"	"	30.00	11.50	41.50
20"	"	34.70	12.75	47.45
Fittings, mechanical joint				
90 degree elbow				
4"	EA.	86.00	14.15	100.15
6"	"	116.00	16.30	132.30

02 SITEWORK

UTILITIES	UNIT	MAT.	INST.	TOTAL
02610.40 DUCTILE IRON PIPE				
8"	EA.	165.00	21.20	186.20
10"	"	242.00	30.30	272.30
12"	"	330.00	42.40	372.40
14"	"	506.00	53.00	559.00
16"	"	572.00	70.50	642.50
18"	"	880.00	85.00	965.00
20"	"	1,010	106.00	1,116
45 degree elbow				
4"	EA.	87.00	14.15	101.15
6"	"	104.00	16.30	120.30
8"	"	143.00	21.20	164.20
10"	"	198.00	30.30	228.30
12"	"	275.00	42.40	317.40
14"	"	418.00	53.00	471.00
16"	"	495.00	70.50	565.50
18"	"	715.00	106.00	821.00
20"	"	803.00	106.00	909.00
Tee				
4"x3"	EA.	121.00	26.50	147.50
4"x4"	"	132.00	26.50	158.50
6"x3"	"	154.00	30.30	184.30
6"x4"	"	160.00	30.30	190.30
6"x6"	"	170.00	30.30	200.30
8"x4"	"	214.00	35.30	249.30
8"x6"	"	248.00	35.30	283.30
8"x8"	"	231.00	35.30	266.30
10"x4"	"	297.00	42.40	339.40
10"x6"	"	319.00	42.40	361.40
10"x8"	"	328.00	42.40	370.40
10"x10"	"	363.00	42.40	405.40
12"x4"	"	440.00	53.00	493.00
12"x6"	"	341.00	53.00	394.00
12"x8"	"	380.00	53.00	433.00
12"x10"	"	434.00	53.00	487.00
12"x12"	"	448.00	56.50	504.50
14"x4"	"	550.00	60.50	610.50
14"x6"	"	588.00	60.50	648.50
14"x8"	"	600.00	60.50	660.50
14"x10"	"	622.00	60.50	682.50
14"x12"	"	638.00	65.00	703.00
14"x14"	"	654.00	65.00	719.00
16"x4"	"	715.00	70.50	785.50
16"x6"	"	732.00	70.50	802.50
16"x8"	"	644.00	70.50	714.50
16"x10"	"	658.00	70.50	728.50
16"x12"	"	632.00	70.50	702.50
16"x14"	"	676.00	70.50	746.50
16"x16"	"	693.00	70.50	763.50
18"x6"	"	825.00	77.00	902.00
18"x8"	"	842.00	77.00	919.00
18"x10"	"	858.00	77.00	935.00
18"x12"	"	874.00	77.00	951.00

UTILITIES	UNIT	MAT.	INST.	TOTAL
02610.40 DUCTILE IRON PIPE				
18"x14"	EA.	984.00	77.00	1,061
18"x16"	"	979.00	77.00	1,056
18"x18"	"	1,090	77.00	1,167
20"x6"	"	1,040	85.00	1,125
20"x8"	"	1,050	85.00	1,135
20"x10"	"	1,080	85.00	1,165
20"x12"	"	1,100	85.00	1,185
20"x14"	"	1,120	85.00	1,205
20"x16"	"	1,320	85.00	1,405
20"x18"	"	1,380	85.00	1,465
20"x20"	"	1,410	85.00	1,495
Cross				
4"x3"	EA.	147.00	35.30	182.30
4"x4"	"	165.00	35.30	200.30
6"x3"	"	170.00	42.40	212.40
6"x4"	"	183.00	42.40	225.40
6"x6"	"	206.00	42.40	248.40
8"x4"	"	233.00	47.10	280.10
8"x6"	"	253.00	47.10	300.10
8"x8"	"	284.00	47.10	331.10
10"x4"	"	319.00	53.00	372.00
10"x6"	"	341.00	53.00	394.00
10"x8"	"	375.00	53.00	428.00
10"x10"	"	442.00	53.00	495.00
12"x4"	"	407.00	60.50	467.50
12"x6"	"	462.00	60.50	522.50
12"x8"	"	451.00	60.50	511.50
12"x10"	"	522.00	65.00	587.00
12"x12"	"	561.00	65.00	626.00
14"x4"	"	534.00	70.50	604.50
14"x6"	"	606.00	70.50	676.50
14"x8"	"	632.00	70.50	702.50
14"x10"	"	679.00	70.50	749.50
14"x12"	"	726.00	77.00	803.00
14"x14"	"	788.00	77.00	865.00
16"x4"	"	690.00	85.00	775.00
16"x6"	"	723.00	85.00	808.00
16"x8"	"	759.00	85.00	844.00
16"x10"	"	803.00	85.00	888.00
16"x12"	"	842.00	85.00	927.00
16"x14"	"	908.00	85.00	993.00
16"x16"	"	968.00	85.00	1,053
18"x6"	"	891.00	94.00	985.00
18"x8"	"	924.00	94.00	1,018
18"x10"	"	966.00	94.00	1,060
18"x12"	"	1,020	94.00	1,114
18"x14"	"	1,200	94.00	1,294
18"x16"	"	1,280	94.00	1,374
18"x18"	"	1,350	94.00	1,444
20"x6"	"	1,070	101.00	1,171
20"x8"	"	1,110	101.00	1,211
20"x10"	"	1,140	101.00	1,241

UTILITIES	UNIT	MAT.	INST.	TOTAL
02610.40 DUCTILE IRON PIPE				
20"x12"	EA.	1,200	101.00	1,301
20"x14"	"	1,260	101.00	1,361
20"x16"	"	1,480	101.00	1,581
20"x18"	"	1,560	106.00	1,666
20"x20"	"	1,650	106.00	1,756
02610.60 PLASTIC PIPE				
PVC, class 150 pipe				
4" dia.	L.F.	2.75	2.85	5.60
6" dia.	"	5.40	3.10	8.50
8" dia.	"	8.70	3.30	12.00
10" dia.	"	12.65	3.60	16.25
12" dia.	"	18.70	3.85	22.55
Schedule 40 pipe				
1-1/2" dia.	L.F.	1.65	1.25	2.90
2" dia.	"	1.95	1.35	3.30
2-1/2" dia.	"	2.15	1.40	3.55
3" dia.	"	3.10	1.50	4.60
4" dia.	"	4.40	1.75	6.15
6" dia.	"	7.45	2.10	9.55
90 degree elbows				
1"	EA.	0.55	3.55	4.10
1-1/2"	"	0.99	3.55	4.54
2"	"	1.20	3.85	5.05
2-1/2"	"	3.40	4.25	7.65
3"	"	4.55	4.70	9.25
4"	"	6.85	5.30	12.15
6"	"	35.20	7.05	42.25
45 degree elbows				
1"	EA.	0.77	3.55	4.32
1-1/2"	"	1.20	3.55	4.75
2"	"	1.65	3.85	5.50
2-1/2"	"	3.20	4.25	7.45
3"	"	4.70	4.70	9.40
4"	"	7.60	5.30	12.90
6"	"	36.30	7.05	43.35
Tees				
1"	EA.	0.85	4.25	5.10
1-1/2"	"	1.05	4.25	5.30
2"	"	2.10	4.70	6.80
2-1/2"	"	4.30	5.30	9.60
3"	"	6.35	6.05	12.40
4"	"	9.35	7.05	16.40
6"	"	35.20	8.50	43.70
Couplings				
1"	EA.	0.50	3.55	4.05
1-1/2"	"	0.61	3.55	4.16
2"	"	1.05	3.85	4.90
2-1/2"	"	2.10	4.25	6.35
3"	"	3.60	4.70	8.30
4"	"	5.15	5.30	10.45

UTILITIES	UNIT	MAT.	INST.	TOTAL
02610.60 PLASTIC PIPE				
6"	EA.	12.65	7.05	19.70
Drainage pipe				
PVC schedule 80				
1" dia.	L.F.	0.88	1.25	2.13
1-1/2" dia.	"	1.10	1.25	2.35
ABS, 2" dia.	"	1.45	1.35	2.80
2-1/2" dia.	"	2.10	1.40	3.50
3" dia.	"	2.50	1.50	4.00
4" dia.	"	3.35	1.75	5.10
6" dia.	"	5.60	2.10	7.70
8" dia.	"	7.60	3.00	10.60
10" dia.	"	9.35	3.60	12.95
12" dia.	"	16.50	3.85	20.35
90 degree elbows				
1"	EA.	1.65	3.55	5.20
1-1/2"	"	2.10	3.55	5.65
2"	"	2.55	3.85	6.40
2-1/2"	"	6.15	4.25	10.40
3"	"	6.40	4.70	11.10
4"	"	10.55	5.30	15.85
6"	"	24.20	7.05	31.25
45 degree elbows				
1"	EA.	2.65	3.55	6.20
1-1/2"	"	3.40	3.55	6.95
2"	"	4.20	3.85	8.05
2-1/2"	"	8.25	4.25	12.50
3"	"	8.80	4.70	13.50
4"	"	15.95	5.30	21.25
6"	"	38.50	7.05	45.55
Tees				
1"	EA.	1.75	4.25	6.00
1-1/2"	"	5.70	4.25	9.95
2"	"	7.05	4.70	11.75
2-1/2"	"	8.25	5.30	13.55
3"	"	9.00	6.05	15.05
4"	"	15.95	7.05	23.00
6"	"	33.00	8.50	41.50
Couplings				33.00
1"	EA.	1.30	3.55	4.85
1-1/2"	"	2.40	3.55	5.95
2"	"	3.40	3.85	7.25
2-1/2"	"	7.70	4.25	11.95
3"	"	7.70	4.70	12.40
4"	"	8.25	5.30	13.55
6"	"	13.75	7.05	20.80
Pressure pipe				
PVC, class 200 pipe				
3/4"	L.F.	0.88	1.05	1.93
1"	"	0.99	1.10	2.09
1-1/4"	"	1.10	1.20	2.30
1-1/2"	"	1.30	1.25	2.55
2"	"	1.45	1.35	2.80

UTILITIES	UNIT	MAT.	INST.	TOTAL
02610.60 PLASTIC PIPE				
2-1/2"	L.F.	2.10	1.40	3.50
3"	"	2.70	1.50	4.20
4"	"	4.60	1.75	6.35
6"	"	8.80	2.10	10.90
8"	"	13.20	3.30	16.50
90 degree elbows				
3/4"	EA.	0.44	3.55	3.99
1"	"	0.55	3.55	4.10
1-1/4"	"	0.88	3.55	4.43
1-1/2"	"	1.05	3.55	4.60
2"	"	2.10	3.85	5.95
2-1/2"	"	3.05	4.25	7.30
3"	"	6.85	4.70	11.55
4"	"	17.60	5.30	22.90
6"	"	35.20	7.05	42.25
8"	"	66.00	10.60	76.60
45 degree elbows				
3/4"	EA.	0.66	3.55	4.21
1"	"	0.83	3.55	4.38
1-1/4"	"	1.20	3.55	4.75
1-1/2"	"	1.45	3.55	5.00
2"	"	2.05	3.85	5.90
2-1/2"	"	3.30	4.25	7.55
3"	"	7.70	4.70	12.40
4"	"	14.30	5.30	19.60
6"	"	35.20	7.05	42.25
8"	"	71.50	10.60	82.10
Tees				
3/4"	EA.	0.50	4.25	4.75
1"	"	0.71	4.25	4.96
1-1/4"	"	1.05	4.25	5.30
1-1/2"	"	1.45	4.25	5.70
2"	"	2.10	4.70	6.80
2-1/2"	"	3.30	5.30	8.60
3"	"	10.20	6.05	16.25
4"	"	14.30	7.05	21.35
6"	"	50.50	8.50	59.00
8"	"	104.00	11.80	115.80
Couplings				
3/4"	EA.	0.33	3.55	3.88
1"	"	0.50	3.55	4.05
1-1/4"	"	0.66	3.55	4.21
1-1/2"	"	0.71	3.55	4.26
2"	"	0.99	3.85	4.84
2-1/2"	"	2.10	4.25	6.35
3"	"	3.30	4.70	8.00
4"	"	4.70	4.70	9.40
6"	"	14.30	5.30	19.60
8"	"	26.40	7.05	33.45

UTILITIES	UNIT	MAT.	INST.	TOTAL
02610.90 VITRIFIED CLAY PIPE				
Vitrified clay pipe, extra strength				
6" dia.	L.F.	2.75	5.20	7.95
8" dia.	"	3.30	5.45	8.75
10" dia.	"	5.05	5.75	10.80
12" dia.	"	7.25	7.65	14.90
15" dia.	"	13.20	11.50	24.70
18" dia.	"	19.80	12.75	32.55
24" dia.	"	36.30	16.40	52.70
30" dia.	"	61.50	23.00	84.50
36" dia.	"	88.00	32.80	120.80
02630.10 TAPPING SADDLES & SLEEVES				
Tapping saddle, tap size to 2"				
4" saddle	EA.	35.20	10.60	45.80
6" saddle	"	41.80	13.25	55.05
8" saddle	"	48.40	17.65	66.05
10" saddle	"	57.00	21.20	78.20
12" saddle	"	68.00	30.30	98.30
14" saddle	"	77.00	42.40	119.40
Tapping sleeve				
4x4	EA.	385.00	14.15	399.15
6x4	"	495.00	16.30	511.30
6x6	"	506.00	16.30	522.30
8x4	"	517.00	21.20	538.20
8x6	"	539.00	21.20	560.20
10x4	"	825.00	46.00	871.00
10x6	"	1,100	46.00	1,146
10x8	"	1,120	46.00	1,166
10x10	"	1,160	47.90	1,208
12x4	"	1,190	47.90	1,238
12x6	"	1,230	52.00	1,282
12x8	"	1,260	57.50	1,318
12x10	"	1,320	64.00	1,384
12x12	"	1,380	72.00	1,452
Tapping valve, mechanical joint				
4" valve	EA.	352.00	144.00	496.00
6" valve	"	418.00	191.00	609.00
8" valve	"	638.00	287.00	925.00
10" valve	"	1,010	383.00	1,393
12" valve	"	1,650	574.00	2,224
Tap hole in pipe				
4" hole	EA.	0.00	26.50	26.50
6" hole	"	0.00	42.40	42.40
8" hole	"	0.00	70.50	70.50
10" hole	"	0.00	85.00	85.00
12" hole	"	0.00	106.00	106.00
02640.15 VALVE BOXES				
Valve box, adjustable, for valves up to 20"				
3' deep	EA.	88.00	7.05	95.05
4' deep	"	88.00	8.50	96.50
5' deep	"	88.00	10.60	98.60

UTILITIES	UNIT	MAT.	INST.	TOTAL
02640.19 THRUST BLOCKS				
Thrust block, 3000# concrete				
1/4 c.y.	EA.	58.50	45.20	103.70
1/2 c.y.	"	82.50	54.00	136.50
3/4 c.y.	"	93.50	90.50	184.00
1 c.y.	"	116.00	181.00	297.00
02645.10 FIRE HYDRANTS				
Standard, 3 way post, 6" mechanical joint				
2' deep	EA.	858.00	383.00	1,241
4' deep	"	924.00	460.00	1,384
6' deep	"	979.00	574.00	1,553
8' deep	"	1,060	656.00	1,716
02665.10 CHILLED WATER SYSTEMS				
Chilled water pipe, 2" thick insulation, w/casing				
Align and tack weld on sleepers				
1-1/2" dia.	L.F.	9.05	1.05	10.10
3" dia.	"	14.70	1.65	16.35
4" dia.	"	17.00	2.30	19.30
6" dia.	"	19.25	2.85	22.10
8" dia.	"	26.10	3.30	29.40
10" dia.	"	34.00	3.85	37.85
12" dia.	"	39.70	4.60	44.30
14" dia.	"	56.50	5.00	61.50
16" dia.	"	71.50	5.75	77.25
Align and tack weld on trench bottom				
18" dia.	L.F.	73.50	6.40	79.90
20" dia.	"	99.50	7.20	106.70
Preinsulated fittings				
Align and tack weld on sleepers				
Elbows				
1-1/2"	EA.	181.00	18.75	199.75
3"	"	238.00	30.00	268.00
4"	"	306.00	37.50	343.50
6"	"	442.00	50.00	492.00
8"	"	612.00	60.00	672.00
Tees				
1-1/2"	EA.	283.00	20.00	303.00
3"	"	385.00	33.30	418.30
4"	"	487.00	42.90	529.90
6"	"	680.00	60.00	740.00
8"	"	906.00	75.00	981.00
Reducers				
3"	EA.	295.00	25.00	320.00
4"	"	499.00	30.00	529.00
6"	"	668.00	37.50	705.50
8"	"	702.00	50.00	752.00
Anchors, not including concrete				
4"	EA.	125.00	37.50	162.50
6"	"	181.00	37.50	218.50

02 SITEWORK

UTILITIES

	UNIT	MAT.	INST.	TOTAL
02665.10 CHILLED WATER SYSTEMS				
Align and tack weld on trench bottom				
Elbows				
10"	EA.	793.00	72.00	865.00
12"	"	963.00	82.00	1,045
14"	"	1,130	88.50	1,219
16"	"	1,250	95.50	1,346
18"	"	1,420	104.00	1,524
20"	"	1,700	115.00	1,815
Tees				
10"	EA.	1,250	72.00	1,322
12"	"	1,640	82.00	1,722
14"	"	1,700	88.50	1,789
16"	"	1,810	95.50	1,906
18"	"	2,040	104.00	2,144
20"	"	2,380	115.00	2,495
Reducers				
10"	EA.	963.00	47.90	1,011
12"	"	1,080	52.00	1,132
14"	"	1,250	57.50	1,308
16"	"	1,470	64.00	1,534
18"	"	1,700	72.00	1,772
20"	"	1,810	82.00	1,892
Anchors, not including concrete				
10"	EA.	283.00	47.90	330.90
12"	"	312.00	52.00	364.00
14"	"	351.00	57.50	408.50
16"	"	510.00	64.00	574.00
18"	"	651.00	72.00	723.00
20"	"	963.00	82.00	1,045
02670.10 WELLS				
Domestic water, drilled and cased				
4" dia.	L.F.	7.70	108.00	115.70
6" dia.	"	8.25	124.00	132.25
8" dia.	"	10.45	162.00	172.45
02685.10 GAS DISTRIBUTION				
Gas distribution lines				
Polyethylene, 60 psi coils				
1-1/4" dia.	L.F.	0.71	2.00	2.71
1-1/2" dia.	"	0.99	2.15	3.14
2" dia.	"	1.20	2.50	3.70
3" dia.	"	2.30	3.00	5.30
30' pipe lengths				
3" dia.	L.F.	2.50	3.35	5.85
4" dia.	"	4.10	3.75	7.85
6" dia.	"	5.50	5.00	10.50
8" dia.	"	10.45	6.00	16.45
Steel, schedule 40, plain end				
1" dia.	L.F.	2.30	2.50	4.80
2" dia.	"	3.85	2.75	6.60

UTILITIES

02685.10 GAS DISTRIBUTION	UNIT	MAT.	INST.	TOTAL
3" dia.	L.F.	7.20	3.00	10.20
4" dia.	"	9.90	7.65	17.55
5" dia.	"	19.25	8.20	27.45
6" dia.	"	21.45	9.55	31.00
8" dia.	"	31.90	10.45	42.35
Natural gas meters, direct digital reading, threaded				
250 cfh @ 5 lbs	EA.	93.50	60.00	153.50
425 cfh @ 10 lbs	"	220.00	60.00	280.00
800 cfh @ 20 lbs	"	319.00	75.00	394.00
1000 cfh @ 25 lbs	"	946.00	75.00	1,021
1,400 cfh @ 100 lbs	"	2,150	100.00	2,250
2,300 cfh @ 100 lbs	"	3,080	150.00	3,230
5,000 cfh @ 100 lbs	"	4,620	300.00	4,920
Gas pressure regulators				
Threaded				
3/4"	EA.	40.00	37.50	77.50
1"	"	41.80	50.00	91.80
1-1/4"	"	44.00	50.00	94.00
1-1/2"	"	275.00	50.00	325.00
2"	"	286.00	60.00	346.00
Flanged				
3"	EA.	825.00	75.00	900.00
4"	"	1,160	100.00	1,260

02690.10 STORAGE TANKS	UNIT	MAT.	INST.	TOTAL
Oil storage tank, underground				
Steel				
500 gals	EA.	440.00	144.00	584.00
1,000 gals	"	803.00	191.00	994.00
4,000 gals	"	2,190	383.00	2,573
5,000 gals	"	2,910	574.00	3,484
10,000 gals	"	4,120	1,150	5,270
Fiberglass, double wall				
550 gals	EA.	1,650	191.00	1,841
1,000 gals	"	2,090	191.00	2,281
2,000 gals	"	2,480	287.00	2,767
4,000 gals	"	3,300	574.00	3,874
6,000 gals	"	3,850	766.00	4,616
8,000 gals	"	5,280	1,150	6,430
10,000 gals	"	6,880	1,440	8,320
12,000 gals	"	8,250	1,910	10,160
15,000 gals	"	9,620	2,550	12,170
20,000 gals	"	11,550	2,870	14,420
Above ground				
Steel				
275 gals	EA.	170.00	115.00	285.00
500 gals	"	605.00	191.00	796.00
1,000 gals	"	902.00	230.00	1,132
1,500 gals	"	1,350	287.00	1,637
2,000 gals	"	1,760	383.00	2,143
5,000 gals	"	4,840	574.00	5,414

02 SITEWORK

UTILITIES	UNIT	MAT.	INST.	TOTAL
02690.10 STORAGE TANKS				
Fill cap	EA.	22.00	30.00	52.00
Vent cap	"	22.00	30.00	52.00
Level indicator	"	55.00	30.00	85.00
02695.80 STEAM METERS				
In-line turbine, direct reading, 300 lb, flanged				
2"	EA.	2,640	37.50	2,678
3"	"	2,860	50.00	2,910
4"	"	3,190	60.00	3,250
Threaded, 2"				
5" line	EA.	5,170	300.00	5,470
6" line	"	5,280	300.00	5,580
8" line	"	5,390	300.00	5,690
10" line	"	5,610	300.00	5,910
12" line	"	5,780	300.00	6,080
14" line	"	6,050	300.00	6,350
16" line	"	6,320	300.00	6,620

SEWERAGE AND DRAINAGE	UNIT	MAT.	INST.	TOTAL
02720.10 CATCH BASINS				
Standard concrete catch basin				
Cast in place, 3'8" x 3'8", 6" thick wall				
2' deep	EA.	187.00	287.00	474.00
3' deep	"	253.00	287.00	540.00
4' deep	"	330.00	383.00	713.00
5' deep	"	385.00	383.00	768.00
6' deep	"	440.00	460.00	900.00
4'x4', 8" thick wall, cast in place				
2' deep	EA.	198.00	287.00	485.00
3' deep	"	275.00	287.00	562.00
4' deep	"	368.00	383.00	751.00
5' deep	"	429.00	383.00	812.00
6' deep	"	473.00	460.00	933.00
Frames and covers, cast iron				
Round				
24" dia.	EA.	165.00	53.00	218.00
26" dia.	"	182.00	53.00	235.00
28" dia.	"	220.00	53.00	273.00
Rectangular				
23"x23"	EA.	143.00	53.00	196.00
27"x20"	"	176.00	53.00	229.00
24"x24"	"	170.00	53.00	223.00

SEWERAGE AND DRAINAGE

02720.10 CATCH BASINS

SEWERAGE AND DRAINAGE	UNIT	MAT.	INST.	TOTAL
26"x26"	EA.	187.00	53.00	240.00
Curb inlet frames and covers				
27"x27"	EA.	352.00	53.00	405.00
24"x36"	"	209.00	53.00	262.00
24"x25"	"	192.00	53.00	245.00
24"x22"	"	165.00	53.00	218.00
20"x22"	"	220.00	53.00	273.00
Airfield catch basin frame and grating, galvanized				
2'x4'	EA.	418.00	53.00	471.00
2'x2'	"	286.00	53.00	339.00

02720.40 STORM DRAINAGE

SEWERAGE AND DRAINAGE	UNIT	MAT.	INST.	TOTAL
Concrete pipe				
Plain, bell and spigot joint, class II				
6" pipe	L.F.	2.95	5.20	8.15
8" pipe	"	3.40	5.75	9.15
10" pipe	"	3.50	6.05	9.55
12" pipe	"	4.30	6.40	10.70
15" pipe	"	5.35	6.75	12.10
18" pipe	"	7.25	7.20	14.45
21" pipe	"	9.05	7.65	16.70
24" pipe	"	11.35	8.20	19.55
Reinforced, class III, tongue and groove joint				
12" pipe	L.F.	6.60	6.40	13.00
15" pipe	"	7.90	6.75	14.65
18" pipe	"	10.00	7.20	17.20
21" pipe	"	13.20	7.65	20.85
24" pipe	"	16.05	8.20	24.25
27" pipe	"	18.70	8.85	27.55
30" pipe	"	21.00	9.55	30.55
36" pipe	"	31.90	10.45	42.35
42" pipe	"	40.70	11.50	52.20
48" pipe	"	57.00	12.75	69.75
54" pipe	"	75.00	14.35	89.35
60" pipe	"	88.00	16.40	104.40
66" pipe	"	99.00	19.15	118.15
72" pipe	"	132.00	23.00	155.00
Flared end-section, concrete				
12" pipe	L.F.	39.30	6.40	45.70
15" pipe	"	44.90	6.75	51.65
18" pipe	"	50.50	7.20	57.70
24" pipe	"	61.50	8.20	69.70
30" pipe	"	75.00	9.55	84.55
36" pipe	"	103.00	10.45	113.45
42" pipe	"	112.00	11.50	123.50
48" pipe	"	123.00	12.75	135.75
54" pipe	"	135.00	14.35	149.35
Corrugated metal pipe, coated, paved invert				
16 ga.				
8" pipe	L.F.	5.45	3.85	9.30
10" pipe	"	7.60	3.95	11.55
12" pipe	"	8.70	4.10	12.80

SEWERAGE AND DRAINAGE

02720.40 STORM DRAINAGE

SEWERAGE AND DRAINAGE	UNIT	MAT.	INST.	TOTAL
15" pipe	L.F.	10.75	4.40	15.15
18" pipe	"	13.05	4.80	17.85
21" pipe	"	15.40	5.20	20.60
24" pipe	"	18.15	5.75	23.90
30" pipe	"	25.90	6.40	32.30
36" pipe	"	33.00	7.20	40.20
12 ga., 48" pipe	"	53.50	8.20	61.70
10 ga.				
60" pipe	L.F.	77.00	9.55	86.55
72" pipe	"	102.00	11.50	113.50
Galvanized or aluminum, plain				
16 ga.				
8" pipe	L.F.	4.95	3.85	8.80
10" pipe	"	6.60	3.95	10.55
12" pipe	"	7.70	4.10	11.80
15" pipe	"	10.10	4.40	14.50
18" pipe	"	12.10	4.80	16.90
24" pipe	"	17.05	5.75	22.80
30" pipe	"	24.20	6.40	30.60
36" pipe	"	30.80	7.20	38.00
12 ga., 48" pipe	"	50.50	8.20	58.70
10 ga., 60" pipe	"	71.50	9.55	81.05
Galvanized or aluminum, coated oval arch				
16 ga.				
17" x 13"	L.F.	14.95	5.20	20.15
21" x 15"	"	18.15	5.75	23.90
14 ga.				
28" x 20"	L.F.	28.30	6.40	34.70
35" x 24"	"	36.30	8.20	44.50
12 ga.				
42" x 29"	L.F.	49.50	9.55	59.05
57" x 38"	"	66.00	11.50	77.50
64" x 43"	"	77.00	12.10	89.10
Oval arch culverts, plain				
16 ga.				60.50
17" x 13"	L.F.	88.00	5.20	93.20
21" x 15"	"	14.45	5.75	20.20
14 ga.				
28" x 20"	L.F.	23.55	6.40	29.95
35" x 24"	"	29.50	8.20	37.70
12 ga.				
57" x 38"	L.F.	55.00	9.55	64.55
64" x 43"	"	60.50	11.50	72.00
71" x 47"	"	88.00	12.10	100.10
Nestable corrugated metal pipe				
16 ga.				
10" pipe	L.F.	7.50	3.95	11.45
12" pipe	"	9.95	4.10	14.05
15" pipe	"	12.45	4.40	16.85
18" pipe	"	14.75	4.80	19.55
24" pipe	"	20.40	5.75	26.15
30" pipe	"	23.80	6.40	30.20

SEWERAGE AND DRAINAGE

SEWERAGE AND DRAINAGE	UNIT	MAT.	INST.	TOTAL
02720.40 STORM DRAINAGE				
14 ga., 36" pipe	L.F.	28.30	7.20	35.50
Headwalls, cast in place, 30 deg wingwall				
12" pipe	EA.	231.00	68.00	299.00
15" pipe	"	275.00	68.00	343.00
18" pipe	"	346.00	77.50	423.50
24" pipe	"	517.00	77.50	594.50
30" pipe	"	627.00	90.50	717.50
36" pipe	"	688.00	136.00	824.00
42" pipe	"	858.00	136.00	994.00
48" pipe	"	913.00	181.00	1,094
54" pipe	"	1,030	226.00	1,256
60" pipe	"	1,170	271.00	1,441
4" cleanout for storm drain				
4" pipe	EA.	302.00	26.50	328.50
6" pipe	"	368.00	26.50	394.50
8" pipe	"	517.00	26.50	543.50
Connect new drain line				
To existing manhole	EA.	58.50	70.50	129.00
To new manhole	"	47.30	42.40	89.70
02720.70 UNDERDRAIN				
Drain tile, clay				
6" pipe	L.F.	2.05	2.45	4.50
8" pipe	"	4.00	2.55	6.55
12" pipe	"	7.45	2.75	10.20
Porous concrete, standard strength				
6" pipe	L.F.	1.90	2.45	4.35
8" pipe	"	2.85	2.55	5.40
12" pipe	"	6.30	2.75	9.05
15" pipe	"	7.15	3.05	10.20
18" pipe	"	9.35	3.65	13.00
Corrugated metal pipe, perforated type				
6" pipe	L.F.	5.40	2.75	8.15
8" pipe	"	6.30	2.90	9.20
10" pipe	"	7.45	3.05	10.50
12" pipe	"	1.15	3.25	4.40
18" pipe	"	13.45	3.45	16.90
Perforated clay pipe				
6" pipe	L.F.	1.95	3.15	5.10
8" pipe	"	2.85	3.25	6.10
12" pipe	"	6.30	3.35	9.65
Drain tile, concrete				
6" pipe	L.F.	1.85	2.45	4.30
8" pipe	"	2.85	2.55	5.40
12" pipe	"	5.95	2.75	8.70
Perforated rigid PVC underdrain pipe				
4" pipe	L.F.	0.86	1.85	2.71
6" pipe	"	1.60	2.20	3.80
8" pipe	"	2.85	2.45	5.30
10" pipe	"	3.55	2.75	6.30
12" pipe	"	6.00	3.15	9.15

02 SITEWORK

SEWERAGE AND DRAINAGE

SEWERAGE AND DRAINAGE	UNIT	MAT.	INST.	TOTAL
02720.70 UNDERDRAIN				
Underslab drainage, crushed stone				
3" thick	S.F.	0.13	0.37	0.50
4" thick	"	0.18	0.42	0.60
6" thick	"	0.20	0.46	0.66
8" thick	"	0.23	0.48	0.71
Plastic filter fabric for drain lines	"	0.03	0.20	0.23
Gravel fill in trench, crushed or bank run, 1/2" to 3/4"	C.Y.	13.75	27.50	41.25
02730.10 SANITARY SEWERS				
Clay				
6" pipe	L.F.	3.65	3.85	7.50
8" pipe	"	4.55	4.10	8.65
10" pipe	"	6.50	4.40	10.90
12" pipe	"	8.60	4.80	13.40
PVC				
4" pipe	L.F.	0.83	2.85	3.68
6" pipe	"	1.70	3.00	4.70
8" pipe	"	2.30	3.20	5.50
10" pipe	"	3.75	3.40	7.15
12" pipe	"	5.60	3.60	9.20
Cleanout				
4" pipe	EA.	4.20	26.50	30.70
6" pipe	"	8.90	26.50	35.40
8" pipe	"	27.20	26.50	53.70
Connect new sewer line				
To existing manhole	EA.	49.50	70.50	120.00
To new manhole	"	33.00	42.40	75.40
02740.10 DRAINAGE FIELDS				
Perforated PVC pipe, for drain field				
4" pipe	L.F.	0.83	2.55	3.38
6" pipe	"	1.55	2.75	4.30
02740.50 SEPTIC TANKS				
Septic tank, precast concrete				
1000 gals	EA.	456.00	191.00	647.00
2000 gals	"	936.00	287.00	1,223
5000 gals	"	3,850	574.00	4,424
25,000 gals	"	15,400	2,300	17,700
40,000 gals	"	22,000	3,830	25,830
Leaching pit, precast concrete, 72" diameter				
3' deep	EA.	275.00	144.00	419.00
6' deep	"	330.00	164.00	494.00
8' deep	"	440.00	191.00	631.00
02760.10 PIPELINE RESTORATION				
Relining existing water main				
6" dia.	L.F.	3.75	16.15	19.90
8" dia.	"	4.30	17.00	21.30

SEWERAGE AND DRAINAGE

02760.10 PIPELINE RESTORATION

	UNIT	MAT.	INST.	TOTAL
10" dia.	L.F.	4.85	17.95	22.80
12" dia.	"	5.35	19.00	24.35
14" dia.	"	5.65	20.20	25.85
16" dia.	"	5.95	21.55	27.50
18" dia.	"	6.25	23.10	29.35
20" dia.	"	7.15	24.85	32.00
24" dia.	"	7.70	26.90	34.60
36" dia.	"	8.25	32.30	40.55
48" dia.	"	9.35	35.90	45.25
72" dia.	"	11.90	40.40	52.30
Replacing in line gate valves				
6" valve	EA.	402.00	215.00	617.00
8" valve	"	660.00	269.00	929.00
10" valve	"	991.00	323.00	1,314
12" valve	"	1,750	404.00	2,154
16" valve	"	3,960	462.00	4,422
18" valve	"	6,050	539.00	6,589
20" valve	"	8,250	646.00	8,896
24" valve	"	11,550	808.00	12,358
36" valve	"	31,900	1,080	32,980

SITE IMPROVEMENTS

02810.40 LAWN IRRIGATION

	UNIT	MAT.	INST.	TOTAL
Residential system, complete				
Minimum	ACRE			11,000
Maximum	"			16,500
Commercial system, complete				
Minimum	ACRE			13,200
Maximum	"			19,800
Components				
Pop-up head	EA.	8.80	13.25	22.05
Impact head	"	8.80	13.25	22.05
Rotary impact head	"	24.20	26.50	50.70
Shrub head	"	2.30	13.25	15.55
Hose bibb	"	4.30	26.50	30.80

02830.10 CHAIN LINK FENCE

	UNIT	MAT.	INST.	TOTAL
Chain link fence, 9 ga., galvanized, with posts 10' o.c.				
4' high	L.F.	4.20	1.45	5.65
5' high	"	5.15	1.85	7.00
6' high	"	6.30	2.55	8.85
7' high	"	6.60	3.15	9.75
8' high	"	7.65	4.10	11.75

SITE IMPROVEMENTS

02830.10 CHAIN LINK FENCE

SITE IMPROVEMENTS	UNIT	MAT.	INST.	TOTAL
For barbed wire with hangers, add				
3 strand	L.F.	1.45	1.00	2.45
6 strand	"	2.40	1.70	4.10
Corner or gate post, 3" post				
4' high	EA.	45.10	6.80	51.90
5' high	"	49.50	7.55	57.05
6' high	"	55.00	8.85	63.85
7' high	"	66.00	10.20	76.20
8' high	"	69.50	11.35	80.85
4" post				
4' high	EA.	77.00	7.55	84.55
5' high	"	91.50	8.85	100.35
6' high	"	101.00	10.20	111.20
7' high	"	110.00	11.35	121.35
8' high	"	121.00	12.75	133.75
Gate with gate posts, galvanized, 3' wide				
4' high	EA.	44.00	51.00	95.00
5' high	"	57.00	68.00	125.00
6' high	"	69.50	68.00	137.50
7' high	"	81.50	102.00	183.50
8' high	"	93.00	102.00	195.00
Fabric, galvanized chain link, 2" mesh, 9 ga.				
4' high	L.F.	1.75	0.68	2.43
5' high	"	2.10	0.82	2.92
6' high	"	2.40	1.00	3.40
8' high	"	3.40	1.35	4.75
Line post, no rail fitting, galvanized, 2-1/2" dia.				
4' high	EA.	13.20	5.85	19.05
5' high	"	14.30	6.35	20.65
6' high	"	15.95	6.80	22.75
7' high	"	18.15	8.15	26.30
8' high	"	19.80	10.20	30.00
1-7/8" H beam				
4' high	EA.	18.15	5.85	24.00
5' high	"	20.35	6.35	26.70
6' high	"	24.20	6.80	31.00
7' high	"	27.50	8.15	35.65
8' high	"	29.70	10.20	39.90
2-1/4" H beam				
4' high	EA.	13.20	5.85	19.05
5' high	"	16.50	6.35	22.85
6' high	"	18.70	6.80	25.50
7' high	"	22.00	8.15	30.15
8' high	"	25.30	10.20	35.50
Vinyl coated, 9 ga., with posts 10' o.c.				
4' high	L.F.	3.95	1.45	5.40
5' high	"	4.40	1.85	6.25
6' high	"	5.05	2.55	7.60
7' high	"	5.85	3.15	9.00
8' high	"	6.70	4.10	10.80
For barbed wire w/hangers, add				
3 strand	L.F.	1.45	1.00	2.45

SITE IMPROVEMENTS	UNIT	MAT.	INST.	TOTAL
02830.10 CHAIN LINK FENCE				
6 Strand	L.F.	2.40	1.70	4.10
Corner, or gate post, 4' high				
3" dia.	EA.	55.00	6.80	61.80
4" dia.	"	86.00	6.80	92.80
6" dia.	"	99.00	8.15	107.15
Gate, with posts, 3' wide				
4' high	EA.	55.00	51.00	106.00
5' high	"	67.00	68.00	135.00
6' high	"	78.00	68.00	146.00
7' high	"	90.00	102.00	192.00
8' high	"	104.00	102.00	206.00
Line post, no rail fitting, 2-1/2" dia.				
4' high	EA.	23.10	5.85	28.95
5' high	"	30.80	6.35	37.15
6' high	"	37.40	6.80	44.20
7' high	"	44.00	8.15	52.15
8' high	"	48.40	10.20	58.60
Corner post, no top rail fitting, 4" dia.				
4' high	EA.	93.50	6.80	100.30
5' high	"	102.00	7.55	109.55
6' high	"	121.00	8.85	129.85
7' high	"	132.00	10.20	142.20
8' high	"	143.00	11.35	154.35
Fabric, vinyl, chain link, 2" mesh, 9 ga.				
4' high	L.F.	3.20	0.68	3.88
5' high	"	3.85	0.82	4.67
6' high	"	4.50	1.00	5.50
8' high	"	6.05	1.35	7.40
Swing gates, galvanized, 4' high				
Single gate				
3' wide	EA.	126.00	51.00	177.00
4' wide	"	138.00	51.00	189.00
Double gate				
10' wide	EA.	324.00	81.50	405.50
12' wide	"	319.00	81.50	400.50
14' wide	"	336.00	81.50	417.50
16' wide	"	368.00	81.50	449.50
18' wide	"	396.00	117.00	513.00
20' wide	"	418.00	117.00	535.00
22' wide	"	462.00	117.00	579.00
24' wide	"	473.00	136.00	609.00
26' wide	"	495.00	136.00	631.00
28' wide	"	528.00	163.00	691.00
30' wide	"	561.00	163.00	724.00
5' high				
Single gate				
3' wide	EA.	121.00	68.00	189.00
4' wide	"	138.00	68.00	206.00
Double gate				
10' wide	EA.	319.00	102.00	421.00
12' wide	"	341.00	102.00	443.00
14' wide	"	594.00	102.00	696.00

SITE IMPROVEMENTS	UNIT	MAT.	INST.	TOTAL
02830.10 CHAIN LINK FENCE				
16' wide	EA.	385.00	102.00	487.00
18' wide	"	407.00	117.00	524.00
20' wide	"	451.00	117.00	568.00
22' wide	"	473.00	117.00	590.00
24' wide	"	495.00	136.00	631.00
26' wide	"	517.00	136.00	653.00
28' wide	"	572.00	163.00	735.00
30' wide	"	594.00	163.00	757.00
6' high				
Single gate				
3' wide	EA.	138.00	68.00	206.00
4' wide	"	148.00	68.00	216.00
Double gate				
10' wide	EA.	330.00	102.00	432.00
12' wide	"	374.00	102.00	476.00
14' wide	"	396.00	102.00	498.00
16' wide	"	429.00	102.00	531.00
18' wide	"	456.00	117.00	573.00
20' wide	"	473.00	117.00	590.00
22' wide	"	506.00	117.00	623.00
24' wide	"	539.00	136.00	675.00
26' wide	"	561.00	136.00	697.00
28' wide	"	616.00	163.00	779.00
30' wide	"	638.00	163.00	801.00
7' high				
Single gate				
3' wide	EA.	154.00	102.00	256.00
4' wide	"	170.00	102.00	272.00
Double gate				
10' wide	EA.	396.00	136.00	532.00
12' wide	"	429.00	136.00	565.00
14' wide	"	462.00	136.00	598.00
16' wide	"	495.00	136.00	631.00
18' wide	"	528.00	163.00	691.00
20' wide	"	561.00	163.00	724.00
22' wide	"	594.00	163.00	757.00
24' wide	"	638.00	204.00	842.00
26' wide	"	682.00	204.00	886.00
28' wide	"	720.00	255.00	975.00
30' wide	"	814.00	255.00	1,069
8' high				
Single gate				
3' wide	EA.	165.00	102.00	267.00
4' wide	"	176.00	102.00	278.00
Double gate				
10' wide	EA.	385.00	136.00	521.00
12' wide	"	418.00	136.00	554.00
14' wide	"	451.00	136.00	587.00
16' wide	"	484.00	136.00	620.00
18' wide	"	517.00	163.00	680.00
20' wide	"	539.00	163.00	702.00
22' wide	"	572.00	163.00	735.00

SITE IMPROVEMENTS

02830.10 CHAIN LINK FENCE

	UNIT	MAT.	INST.	TOTAL
24' wide	EA.	594.00	204.00	798.00
26' wide	"	638.00	204.00	842.00
28' wide	"	704.00	255.00	959.00
30' wide	"	726.00	255.00	981.00
Vinyl coated swing gates, 4' high				
Single gate				
3' wide	EA.	176.00	51.00	227.00
4' wide	"	187.00	51.00	238.00
Double gate				
10' wide	EA.	418.00	81.50	499.50
12' wide	"	528.00	81.50	609.50
14' wide	"	561.00	81.50	642.50
16' wide	"	627.00	81.50	708.50
18' wide	"	704.00	117.00	821.00
20' wide	"	836.00	117.00	953.00
22' wide	"	924.00	117.00	1,041
24' wide	"	990.00	136.00	1,126
26' wide	"	1,040	136.00	1,176
28' wide	"	1,210	163.00	1,373
30' wide	"	1,260	163.00	1,423
5' high				
Single gate				
3' wide	EA.	192.00	68.00	260.00
4' wide	"	214.00	68.00	282.00
Double gate				
10' wide	EA.	566.00	102.00	668.00
12' wide	"	605.00	102.00	707.00
14' wide	"	676.00	102.00	778.00
16' wide	"	704.00	102.00	806.00
18' wide	"	847.00	117.00	964.00
20' wide	"	924.00	117.00	1,041
22' wide	"	1,010	117.00	1,127
24' wide	"	1,160	136.00	1,296
26' wide	"	1,210	136.00	1,346
28' wide	"	1,320	163.00	1,483
30' wide	"	1,430	163.00	1,593
6' high				
Single gate				
3' wide	EA.	204.00	68.00	272.00
4' wide	"	214.00	68.00	282.00
Double gate				
10' wide	EA.	506.00	102.00	608.00
12' wide	"	578.00	102.00	680.00
14' wide	"	660.00	102.00	762.00
16' wide	"	770.00	102.00	872.00
18' wide	"	858.00	117.00	975.00
20' wide	"	924.00	117.00	1,041
22' wide	"	1,000	117.00	1,117
24' wide	"	1,120	136.00	1,256
26' wide	"	1,190	136.00	1,326
28' wide	"	1,320	163.00	1,483
30' wide	"	1,430	163.00	1,593

SITE IMPROVEMENTS	UNIT	MAT.	INST.	TOTAL
02830.10 CHAIN LINK FENCE				
7' high				
Single gate				
3' wide	EA.	226.00	102.00	328.00
4' wide	"	280.00	102.00	382.00
Double gate				
10' wide	EA.	566.00	136.00	702.00
12' wide	"	671.00	136.00	807.00
14' wide	"	759.00	136.00	895.00
16' wide	"	858.00	136.00	994.00
18' wide	"	968.00	163.00	1,131
20' wide	"	1,100	163.00	1,263
22' wide	"	1,210	163.00	1,373
24' wide	"	1,320	204.00	1,524
26' wide	"	1,430	204.00	1,634
28' wide	"	1,540	255.00	1,795
30' wide	"	1,650	255.00	1,905
8' high				
Single gate				
3' wide	EA.	176.00	102.00	278.00
4' wide	"	220.00	102.00	322.00
Double gate				
10' wide	EA.	544.00	136.00	680.00
12' wide	"	660.00	136.00	796.00
14' wide	"	770.00	136.00	906.00
16' wide	"	858.00	136.00	994.00
18' wide	"	968.00	163.00	1,131
20' wide	"	1,130	163.00	1,293
22' wide	"	1,210	163.00	1,373
24' wide	"	1,320	204.00	1,524
28' wide	"	1,590	204.00	1,794
30' wide	"	1,740	255.00	1,995
Motor operator for gates, no wiring	"			3,300
Drilling fence post holes				
In soil				
By hand	EA.	0.00	10.20	10.20
By machine auger	"	0.00	6.35	6.35
In rock				
By jackhammer	EA.	0.00	84.50	84.50
By rock drill	"	0.00	25.40	25.40
Aluminum privacy slats, installed vertically	S.F.	0.55	0.51	1.06
Post hole, dig by hand	EA.	0.00	13.60	13.60
Set fence post in concrete	"	5.50	10.20	15.70
02830.70 RECREATIONAL COURTS				
Walls, galvanized steel				
8' high	L.F.	7.65	4.25	11.90
10' high	"	9.15	4.70	13.85
12' high	"	10.45	5.60	16.05
Vinyl coated				
8' high	L.F.	7.35	4.25	11.60
10' high	"	9.05	4.70	13.75
12' high	"	10.25	5.60	15.85

02 SITEWORK

SITE IMPROVEMENTS	UNIT	MAT.	INST.	TOTAL
02830.70 RECREATIONAL COURTS				
Gates, galvanized steel				
Single, 3' transom				
3'x7'	EA.	180.00	102.00	282.00
4'x7'	"	191.00	117.00	308.00
5'x7'	"	258.00	136.00	394.00
6'x7'	"	280.00	163.00	443.00
Double, 3' transom				
10'x7'	EA.	460.00	408.00	868.00
12'x7'	"	561.00	453.00	1,014
14'x7'	"	673.00	510.00	1,183
Double, no transom				
10'x10'	EA.	471.00	340.00	811.00
12'x10'	"	561.00	408.00	969.00
14'x10'	"	645.00	453.00	1,098
Vinyl coated				
Single, 3' transom				
3'x7'	EA.	477.00	102.00	579.00
4'x7'	"	516.00	117.00	633.00
5'x7'	"	482.00	136.00	618.00
6'x7'	"	516.00	163.00	679.00
Double, 3'				
10'x7'	EA.	1,290	408.00	1,698
12'x7'	"	1,400	453.00	1,853
14'x7'	"	1,460	510.00	1,970
Double, no transom				
10'x10'	EA.	1,290	340.00	1,630
12'x10'	"	1,350	408.00	1,758
14'x10'	"	1,460	453.00	1,913
Baseball backstop, regulation				
Galvanized	EA.			4,400
Vinyl coated	"			6,050
Softball backstop, regulation				
14' high				
Galvanized	EA.			4,070
Vinyl coated	"			6,050
18' high				
Galvanized	EA.			4,840
Vinyl coated	"			7,040
20' high				
Galvanized	EA.			5,720
Vinyl coated	"			8,250
22' high				
Galvanized	EA.			6,600
Vinyl coated	"			9,680
24' high				
Galvanized	EA.			8,030
Vinyl coated	"			13,200
Wire and miscellaneous metal fences				
Chicken wire, post 4' o.c.				
2" mesh				
4' high	L.F.	0.78	1.00	1.78
6' high	"	0.90	1.35	2.25

SITE IMPROVEMENTS	UNIT	MAT.	INST.	TOTAL
02830.70 RECREATIONAL COURTS				
Galvanized steel				
12 gauge, 2" by 4" mesh, posts 5' o.c.				
3' high	L.F.	1.35	1.00	2.35
5' high	"	1.90	1.25	3.15
14 gauge, 1" by 2" mesh, posts 5' o.c.				
3' high	L.F.	1.10	1.00	2.10
5' high	"	1.80	1.25	3.05
Vinyl coated				
12'high	L.F.	10.25	5.60	15.85
02840.30 GUARDRAILS				
Pipe bollard, steel pipe, concrete filled, painted				
6" dia.	EA.	88.00	17.00	105.00
8" dia.	"	126.00	25.50	151.50
12" dia.	"	209.00	68.00	277.00
Corrugated steel, guardrail, galvanized	L.F.	15.40	1.85	17.25
End section, wrap around or flared	EA.	44.00	20.40	64.40
Timber guardrail, 4" x 8"	L.F.	17.60	1.35	18.95
Guard rail, 3 cables, 3/4" dia.				
Steel posts	L.F.	7.15	13.75	20.90
Wood posts	"	6.80	12.20	19.00
Steel box beam				
6" x 6"	L.F.	34.10	6.10	40.20
6" x 8"	"	36.30	1.25	37.55
Concrete posts	EA.	16.50	10.20	26.70
Barrel type impact barrier	"	297.00	20.40	317.40
Light shield, 6' high	L.F.	17.60	4.10	21.70
02840.40 PARKING BARRIERS				
Timber, treated, 8' long				
4" x 4"	EA.	6.60	17.00	23.60
6" x 6"	"	13.20	20.40	33.60
Precast concrete, 4' long, with dowels				
12" x 6"	EA.	16.50	10.20	26.70
12" x 8"	"	18.70	11.35	30.05
02840.60 SIGNAGE				
Traffic signs				
Reflectorized signs per OSHA standards, including post				
Stop, 24"x24"	EA.	38.10	13.60	51.70
Yield, 30" triangle	"	23.55	13.60	37.15
Speed limit, 12"x18"	"	23.55	13.60	37.15
Directional, 12"x18"	"	33.70	13.60	47.30
Exit, 12"x18"	"	33.70	13.60	47.30
Entry, 12"x18"	"	33.70	13.60	47.30
Warning, 24"x24"	"	47.10	13.60	60.70
Informational, 12"x18"	"	11.20	13.60	24.80
Handicap parking, 12"x18"	"	15.70	13.60	29.30

02 SITEWORK

SITE IMPROVEMENTS

02860.40 RECREATIONAL FACILITIES

	UNIT	MAT.	INST.	TOTAL
Bleachers, outdoor, portable, per seat				
10 tiers				
Minimum	EA.	18.30	6.85	25.15
Maximum	"	36.60	9.15	45.75
20 tiers				
Minimum	EA.	25.20	6.45	31.65
Maximum	"	43.50	8.45	51.95
Grandstands, fixed, wood seat, steel frame, per seat				
15 tiers				
Minimum	EA.	36.60	11.00	47.60
Maximum	"	59.50	18.35	77.85
30 tiers				
Minimum	EA.	45.80	10.00	55.80
Maximum	"	80.00	15.70	95.70
Seats				
Seat backs only				
Fiberglass	EA.	20.60	2.05	22.65
Steel and wood seat	"	32.00	2.05	34.05
Seat restoration, fiberglass on wood				
Seats	EA.	13.75	4.10	17.85
Plain bench, no backs	"	8.00	1.70	9.70
Benches				
Park, precast concrete with backs				
4' long	EA.	606.00	68.00	674.00
8' long	"	1,260	102.00	1,362
Fiberglass, with backs				
4' long	EA.	480.00	51.00	531.00
8' long	"	961.00	68.00	1,029
Wood, with backs and fiberglass supports				
4' long	EA.	217.00	51.00	268.00
8' long	"	229.00	68.00	297.00
Steel frame, 6' long				
All steel	EA.	200.00	51.00	251.00
Hardwood boards	"	137.00	51.00	188.00
Players bench (no back), steel frame, fir seat, 10' long	"	149.00	68.00	217.00
Backstops				
Handball or squash court, outdoor				
Wood	EA.			17,600
Masonry	"			17,600
Soccer goal posts	PAIR			1,320
Running track				
Gravel and cinders over stone base	S.Y.	5.25	2.75	8.00
Rubber-cork base resilient pavement	"	7.15	22.00	29.15
For colored surfaces, add	"	4.60	2.20	6.80
Colored rubberized asphalt	"	9.70	27.50	37.20
Artificial resilient mat over asphalt	"	22.90	55.00	77.90
Tennis courts				
Bituminous pavement, 2-1/2" thick	S.Y.	7.20	6.85	14.05
Colored sealer, acrylic emulsion				
3 coats	S.Y.	3.65	1.35	5.00
For 2 color seal coating, add	"	0.68	0.20	0.88
For preparing old courts, add	"	1.60	0.14	1.74

SITE IMPROVEMENTS

02860.40 RECREATIONAL FACILITIES

SITE IMPROVEMENTS	UNIT	MAT.	INST.	TOTAL
Net, nylon, 42' long	EA.	229.00	25.50	254.50
Paint markings on asphalt, 2 coats	"	57.00	204.00	261.00
Complete court with fence, etc., bituminous				
Minimum	EA.			8,800
Average	"			14,850
Maximum	"			19,250
Clay court				
Minimum	EA.			9,350
Average	"			12,100
Maximum	"			17,270
Playground equipment				
Basketball backboard				
Minimum	EA.	400.00	51.00	451.00
Maximum	"	715.00	58.50	773.50
Bike rack, 10' long	"	286.00	40.80	326.80
Golf shelter, fiberglass	"	1,430	51.00	1,481
Ground socket for movable posts				
Minimum	EA.	66.50	12.75	79.25
Maximum	"	128.00	12.75	140.75
Horizontal monkey ladder, 14' long	"	458.00	34.00	492.00
Posts, tether ball	"	217.00	10.20	227.20
Multiple purpose, 10' long	"	217.00	20.40	237.40
See-saw, steel				
Minimum	EA.	372.00	81.50	453.50
Average	"	686.00	102.00	788.00
Maximum	"	972.00	136.00	1,108
Slide				
Minimum	EA.	669.00	163.00	832.00
Maximum	"	1,120	185.00	1,305
Swings, plain seats				
8' high				
Minimum	EA.	486.00	136.00	622.00
Maximum	"	858.00	157.00	1,015
12' high				
Minimum	EA.	744.00	157.00	901.00
Maximum	"	1,260	227.00	1,487

02870.10 PREFABRICATED PLANTERS

SITE IMPROVEMENTS	UNIT	MAT.	INST.	TOTAL
Concrete precast, circular				
24" dia., 18" high	EA.	162.00	20.40	182.40
42" dia., 30" high	"	208.00	25.50	233.50
Fiberglass, circular				
36" dia., 27" high	EA.	289.00	10.20	299.20
60" dia., 39" high	"	751.00	11.35	762.35
Tapered, circular				
24" dia., 36" high	EA.	231.00	9.25	240.25
40" dia., 36" high	"	404.00	10.20	414.20
Square				
2' by 2', 17" high	EA.	156.00	9.25	165.25
4' by 4', 39" high	"	647.00	11.35	658.35
Rectangular				
4' by 1', 18" high	EA.	219.00	10.20	229.20

LANDSCAPING

02910.10 SHRUB & TREE MAINTENANCE

LANDSCAPING	UNIT	MAT.	INST.	TOTAL
Moving shrubs on site				
12" ball	EA.	0.00	25.50	25.50
24" ball	"	0.00	34.00	34.00
3' high	"	0.00	20.40	20.40
4' high	"	0.00	22.65	22.65
5' high	"	0.00	25.50	25.50
18" spread	"	0.00	29.10	29.10
30" spread	"	0.00	34.00	34.00
Moving trees on site				
24" ball	EA.	0.00	55.00	55.00
48" ball	"	0.00	73.50	73.50
Trees				
3' high	EA.	0.00	22.00	22.00
6' high	"	0.00	24.45	24.45
8' high	"	0.00	27.50	27.50
10' high	"	0.00	36.70	36.70
Palm trees				
7' high	EA.	0.00	27.50	27.50
10' high	"	0.00	36.70	36.70
20' high	"	0.00	110.00	110.00
40' high	"	0.00	220.00	220.00
Guying trees				
4" dia.	EA.	5.50	10.20	15.70
8" dia.	"	5.50	12.75	18.25

02920.10 TOPSOIL

TOPSOIL	UNIT	MAT.	INST.	TOTAL
Spread topsoil, with equipment				
Minimum	C.Y.	0.00	5.55	5.55
Maximum	"	0.00	6.95	6.95
By hand				
Minimum	C.Y.	0.00	20.40	20.40
Maximum	"	0.00	25.50	25.50
Area preparation for seeding (grade, rake and clean)				
Square yard	S.Y.	0.00	0.16	0.16
By acre	ACRE	0.00	816.00	816.00
Remove topsoil and stockpile on site				
4" deep	C.Y.	0.00	4.65	4.65
6" deep	"	0.00	4.25	4.25
Spreading topsoil from stock pile				
By loader	C.Y.	0.00	5.05	5.05
By hand	"	0.00	55.50	55.50
Top dress by hand	S.Y.	0.00	0.56	0.56
Place imported top soil				
By loader				
4" deep	S.Y.	0.00	0.56	0.56
6" deep	"	0.00	0.62	0.62
By hand				
4" deep	S.Y.	0.00	2.25	2.25
6" deep	"	0.00	2.55	2.55
Plant bed preparation, 18" deep				
With backhoe/loader	S.Y.	0.00	1.40	1.40
By hand	"	0.00	2.05	2.05

02 SITEWORK

LANDSCAPING	UNIT	MAT.	INST.	TOTAL
02930.30 SEEDING				
Mechanical seeding, 175 lb/acre				
By square yard	S.Y.	0.45	0.05	0.50
By acre	ACRE	1,020	254.00	1,274
450 lb/acre				
By square yard	S.Y.	0.85	0.06	0.91
By acre	ACRE	1,810	318.00	2,128
Seeding by hand, 10 lb per 100 s.y.				
By square yard	S.Y.	0.51	0.07	0.58
By acre	ACRE	1,360	340.00	1,700
Reseed disturbed areas	S.F.	0.29	0.10	0.39
02950.10 PLANTS				
Euonymus coloratus, 18" (purple wintercreeper)	EA.	1.35	3.40	4.75
Hedera Helix, 2-1/4" pot (English ivy)	"	0.55	3.40	3.95
Liriope muscari, 2" clumps	"	2.40	2.05	4.45
Santolina, 12"	"	2.75	2.05	4.80
Vinca major or minor, 3" pot	"	0.44	2.05	2.49
Cortaderia argentia, 2 gallon (pampas grass)	"	8.80	2.05	10.85
Ophiopogan japonicus, 1 quart (4" pot)	"	2.40	2.05	4.45
Ajuga reptans, 2-3/4" pot (carpet bugle)	"	0.44	2.05	2.49
Pachysandra terminalis, 2-3/4" pot (Japanese spurge)	"	0.61	2.05	2.66
02950.30 SHRUBS				
Juniperus conferia litoralis, 18"-24" (Shore Juniper)	EA.	19.80	8.15	27.95
Horizontalis plumosa, 18"-24" (Andorra Juniper)	"	20.90	8.15	29.05
Sabina tamar-iscfolia-tamarix juniper, 18"-24"	"	20.90	8.15	29.05
Chin San Hose, 18"-24" (San Hose Juniper)	"	20.90	8.15	29.05
Sargenti, 18"-24" (Sargent's Juniper)	"	19.80	8.15	27.95
Nandina domestica, 18"-24" (Heavenly Bamboo)	"	13.20	8.15	21.35
Raphiolepis Indica Springtime, 18"-24" Indian Hawthorn	"	14.30	8.15	22.45
Osmanthus Heterophyllus Gulftide, 18"-24" (Osmanthus)	"	15.40	8.15	23.55
Ilex Cornuta Burfordi Nana, 18"-24" (Dwarf Burford Holly)	"	17.60	8.15	25.75
Glabra, 18"-24" (Inkberry Holly)	"	16.50	8.15	24.65
Azalea, Indica types, 18"-24"	"	18.70	8.15	26.85
Kurume types, 18"-24"	"	20.90	8.15	29.05
Berberis Julianae, 18"-24" (Wintergreen Barberry)	"	12.10	8.15	20.25
Pieris Japonica Japanese, 18"-24" (Japanese Pieris)	"	12.10	8.15	20.25
Ilex Cornuta Rotunda, 18"-24" (Dwarf Chinese Holly)	"	14.30	8.15	22.45
Juniperus Horizontalis Plumosa, 24"-30" (Andorra Juniper)	"	13.20	10.20	23.40
Rhodopendrow Hybrids, 24"-30"	"	35.20	10.20	45.40
Aucuba Japonica Varigata, 24"-30" (Gold Dust Aucuba)	"	12.10	10.20	22.30
Ilex Crenata Willow Leaf, 24"-30" (Japanese Holly)	"	13.20	10.20	23.40
Cleyera Japonica, 30"-36" (Japanese Cleyera)	"	15.40	12.75	28.15
Pittosporum Tobira, 30"-36"	"	17.60	12.75	30.35
Prumus Laurocerasus, 30"-36"	"	33.00	12.75	45.75
Ilex Cornuta Burfordi, 30"-36" (Burford Holly)	"	17.60	12.75	30.35
Abelia Grandiflora, 24"-36" (Yew Podocarpus)	"	12.10	10.20	22.30
Podocarpos Macrophylla, 24"-36" (Yew Podocarpus)	"	19.80	10.20	30.00
Pyracantha Coccinea Lalandi, 3'-4' (Firethorn)	"	11.00	12.75	23.75
Photinia Frazieri, 3'-4' (Red Photinia)	"	17.60	12.75	30.35
Forsythia Suspensa, 3'-4' (Weeping Forsythia)	"	11.00	12.75	23.75

02 SITEWORK

LANDSCAPING

LANDSCAPING	UNIT	MAT.	INST.	TOTAL
02950.30 SHRUBS				
Camellia Japonica, 3'-4' (Common Camellia)	EA.	19.80	12.75	32.55
Juniperus Chin Torulosa, 3'-4' (Hollywood Juniper)	"	20.90	12.75	33.65
Cupressocyparis Leylandi, 3'-4'	"	17.60	12.75	30.35
Ilex Opaca Fosteri, 5'-6' (Foster's Holly)	"	71.50	17.00	88.50
Opaca, 5'-6' (American Holly)	"	104.00	17.00	121.00
Nyrica Cerifera, 4'-5' (Southern Wax Myrtles)	"	22.00	14.55	36.55
Ligustrum Japonicum, 4'-5' (Japanese Privet)	"	17.60	14.55	32.15
02950.60 TREES				
Cornus Florida, 5'-6' (White flowering Dogwood)	EA.	49.50	17.00	66.50
Prunus Serrulata Kwanzan, 6'-8' (Kwanzan Cherry)	"	55.00	20.40	75.40
Caroliniana, 6'-8' (Carolina Cherry Laurel)	"	64.00	20.40	84.40
Cercis Canadensis, 6'-8' (Eastern Redbud)	"	44.00	20.40	64.40
Koelreuteria Paniculata, 8'-10' (Goldenrain tree)	"	77.00	25.50	102.50
Acer Platanoides, 1-3/4"-2" (11'-13') (Norway Maple)	"	104.00	34.00	138.00
Rubrum, 1-3/4"-2" (11'-13') (Red Maple)	"	77.00	34.00	111.00
Saccharum, 1-3/4"-2" (Sugar Maple)	"	138.00	34.00	172.00
Fraxinus Pennsylvanica, 1-3/4"-2" Laneolata-Green Ash	"	66.00	34.00	100.00
Celtis Occidentalis, 1-3/4"-2" (American Hackberry)	"	99.00	34.00	133.00
Glenditsia Triacantos Inermis, 2"	"	88.00	34.00	122.00
Prunus Cerasifera 'Thundercloud', 6'-8'	"	53.00	20.40	73.40
Yeodensis, 6'-8' (Yoshino Cherry)	"	55.00	20.40	75.40
Lagerstroemia Indica, 8'-10' (Crapemyrtle)	"	88.00	25.50	113.50
Crataegus Phaenopyrum, 8'-10' Washington Hawthorn	"	138.00	25.50	163.50
Quercus Borealis, 1-3/4"-2" (Northern Red Oak)	"	82.50	34.00	116.50
Quercus Acutissima, 1-3/4"-2" (8'-10') (Sawtooth Oak)	"	77.00	34.00	111.00
Saliz Babylonica, 1-3/4"-2" (Weeping Willow)	"	38.50	34.00	72.50
Tilia Cordata Greenspire, 1-3/4"-2" (10'-12')	"	165.00	34.00	199.00
Malus, 2"-2-1/2" (8'-10') (Flowering Crabapple)	"	82.50	34.00	116.50
Platanus Occidentalis, (12'-14')	"	126.00	40.80	166.80
Pyrus Calleryana Bradford, 2"-2-1/2" (Bradford Pear)	"	99.00	34.00	133.00
Quercus Palustris, 2"-2-1/2" (12'-14') (Pin Oak)	"	110.00	34.00	144.00
Phellos, 2-1/2"-3" (Willow Oak)	"	121.00	40.80	161.80
Nigra, 2"-2-1/2" (Water Oak)	"	104.00	34.00	138.00
Magnolia Soulangeana, 4'-5' (Saucer Magnolia)	"	60.50	17.00	77.50
Grandiflora, 6'-8' (Southern Magnolia)	"	82.50	20.40	102.90
Cedrus Deodara, 10'-12' (Deodare Cedar)	"	138.00	34.00	172.00
Ginkgo Biloba, 10'-12' (2"-2-1/2") (Maidenhair Tree)	"	126.00	34.00	160.00
Pinus Thunbergi, 5'-6' (Japanese Black Pine)	"	49.50	17.00	66.50
Strobus, 6'-8' (White Pine)	"	55.00	20.40	75.40
Taeda, 6'-8' (Loblolly Pine)	"	46.20	20.40	66.60
Quercus Virginiana, 2"-2-1/2" (live oak)	"	121.00	40.80	161.80
02970.10 FERTILIZING				
Fertilizing (23#/1000 sf)				
By square yard	S.Y.	0.02	0.06	0.08
By acre	ACRE	77.00	318.00	395.00
Liming (70#/1000 sf)				
By square yard	S.Y.	0.02	0.08	0.10
By acre	ACRE	71.50	423.00	494.50

LANDSCAPING

02980.10 LANDSCAPE ACCESSORIES

LANDSCAPING	UNIT	MAT.	INST.	TOTAL
Steel edging, 3/16" x 4"	L.F.	0.33	0.25	0.58
Landscaping stepping stones, 15"x15", white	EA.	3.30	1.00	4.30
Wood chip mulch	C.Y.	27.50	13.60	41.10
2" thick	S.Y.	1.65	0.41	2.06
4" thick	"	3.20	0.58	3.78
6" thick	"	4.75	0.74	5.49
Gravel mulch, 3/4" stone	C.Y.	19.80	20.40	40.20
White marble chips, 1" deep	S.F.	0.66	0.20	0.86
Peat moss				
2" thick	S.Y.	2.20	0.45	2.65
4" thick	"	4.30	0.68	4.98
6" thick	"	6.25	0.85	7.10
Landscaping timbers, treated lumber				
4" x 4"	L.F.	0.66	0.68	1.34
6" x 6"	"	1.30	0.73	2.03
8" x 8"	"	2.30	0.85	3.15

FORMWORK	UNIT	MAT.	INST.	TOTAL
03110.05 BEAM FORMWORK				
Beam forms, job built				
Beam bottoms				
1 use	S.F.	3.10	4.30	7.40
2 uses	"	1.50	4.10	5.60
3 uses	"	1.20	3.95	5.15
4 uses	"	0.96	3.80	4.76
5 uses	"	102.00	3.70	105.70
Beam sides				
1 use	S.F.	2.20	2.90	5.10
2 uses	"	1.05	2.70	3.75
3 uses	"	0.96	2.55	3.51
4 uses	"	0.86	2.45	3.31
5 uses	"	0.90	2.35	3.25
03110.10 BOX CULVERT FORMWORK				
Box culverts, job built				
6' x 6'				
1 use	S.F.	1.75	2.55	4.30
2 uses	"	0.99	2.45	3.44
3 uses	"	0.83	2.35	3.18
4 uses	"	0.71	2.25	2.96
5 uses	"	0.61	2.15	2.76
8' x 12'				
1 use	S.F.	1.75	2.15	3.90
2 uses	"	0.99	2.05	3.04
3 uses	"	0.83	2.00	2.83
4 uses	"	0.71	1.90	2.61
5 uses	"	0.61	1.85	2.46
03110.15 COLUMN FORMWORK				
Column, square forms, job built				
8" x 8" columns				
1 use	S.F.	2.40	5.20	7.60
2 uses	"	1.20	4.95	6.15
3 uses	"	0.99	4.75	5.74
4 uses	"	0.88	4.60	5.48
5 uses	"	0.84	4.45	5.29
12" x 12" columns				
1 use	S.F.	2.30	4.70	7.00
2 uses	"	1.10	4.50	5.60
3 uses	"	0.88	4.35	5.23
4 uses	"	0.77	4.20	4.97
5 uses	"	0.72	4.10	4.82
16" x 16" columns				
1 use	S.F.	2.00	4.30	6.30
2 uses	"	1.05	4.15	5.20
3 uses	"	0.83	4.00	4.83
4 uses	"	0.71	3.90	4.61
5 uses	"	0.61	3.80	4.41
24" x 24" columns				

FORMWORK	UNIT	MAT.	INST.	TOTAL
03110.15 COLUMN FORMWORK				
1 use	S.F.	1.85	3.95	5.80
2 uses	"	0.88	3.85	4.73
3 uses	"	0.71	3.75	4.46
4 uses	"	0.61	3.60	4.21
5 uses	"	0.55	3.50	4.05
36" x 36" columns				
1 use	S.F.	1.85	3.65	5.50
2 uses	"	0.88	3.55	4.43
3 uses	"	0.71	3.50	4.21
4 uses	"	0.61	3.40	4.01
5 uses	"	0.55	3.30	3.85
Round fiber forms, 1 use				
10" dia.	L.F.	2.75	5.15	7.90
12" dia.	"	3.30	5.25	8.55
14" dia.	"	4.40	5.45	9.85
16" dia.	"	5.50	5.70	11.20
18" dia.	"	8.80	6.10	14.90
24" dia.	"	11.00	6.60	17.60
30" dia.	"	16.50	7.15	23.65
36" dia.	"	20.90	7.80	28.70
42" dia.	"	39.60	8.55	48.15
03110.18 CURB FORMWORK				
Curb forms				
Straight, 6" high				
1 use	L.F.	1.30	2.55	3.85
2 uses	"	0.73	2.45	3.18
3 uses	"	0.51	2.35	2.86
4 uses	"	0.45	2.25	2.70
5 uses	"	0.40	2.15	2.55
Curved, 6" high				
1 use	L.F.	1.35	3.20	4.55
2 uses	"	0.78	3.05	3.83
3 uses	"	0.56	2.85	3.41
4 uses	"	0.51	2.75	3.26
5 uses	"	0.45	2.60	3.05
03110.20 ELEVATED SLAB FORMWORK				
Elevated slab formwork				
Slab, with drop panels				
1 use	S.F.	2.05	2.15	4.20
2 uses	"	1.05	2.00	3.05
3 uses	"	0.95	1.90	2.85
4 uses	"	0.84	1.85	2.69
5 uses	"	0.74	1.85	2.59
Floor slab, hung from steel beams				
1 use	S.F.	1.90	2.10	4.00
2 uses	"	0.99	1.90	2.89
3 uses	"	0.89	1.85	2.74
4 uses	"	0.78	1.75	2.53

03 CONCRETE

FORMWORK

FORMWORK	UNIT	MAT.	INST.	TOTAL
03110.20 ELEVATED SLAB FORMWORK				
5 uses	S.F.	0.68	1.80	2.48
Floor slab, with pans or domes				
1 use	S.F.	2.75	2.45	5.20
2 uses	"	1.25	2.25	3.50
3 uses	"	1.05	2.15	3.20
4 uses	"	0.95	2.05	3.00
5 uses	"	1.20	2.10	3.30
Equipment curbs, 12" high				
1 use	L.F.	1.70	2.70	4.40
2 uses	"	0.76	2.45	3.21
3 uses	"	0.56	2.35	2.91
4 uses	"	0.47	2.25	2.72
5 uses	"	0.71	2.25	2.96
03110.25 EQUIPMENT PAD FORMWORK				
Equipment pad, job built				
1 use	S.F.	2.10	3.40	5.50
2 uses	"	1.25	3.20	4.45
3 uses	"	0.99	3.00	3.99
4 uses	"	0.77	2.85	3.62
5 uses	"	0.66	2.70	3.36
03110.35 FOOTING FORMWORK				
Wall footings, job built, continuous				
1 use	S.F.	0.88	2.70	3.58
2 uses	"	0.54	2.45	2.99
3 uses	"	0.41	2.35	2.76
4 uses	"	0.34	2.25	2.59
5 uses	"	0.38	2.25	2.63
Column footings, spread				
1 use	S.F.	1.05	3.40	4.45
2 uses	"	0.64	3.05	3.69
3 uses	"	0.43	2.85	3.28
4 uses	"	0.34	2.70	3.04
5 uses	"	0.42	2.70	3.12
03110.50 GRADE BEAM FORMWORK				
Grade beams, job built				
1 use	S.F.	1.80	2.60	4.40
2 uses	"	0.94	2.45	3.39
3 uses	"	0.71	2.35	3.06
4 uses	"	0.61	2.25	2.86
5 uses	"	0.54	2.15	2.69
03110.53 PILE CAP FORMWORK				
Pile cap forms, job built				
Square				
1 use	S.F.	1.75	3.20	4.95
2 uses	"	1.05	3.05	4.10
3 uses	"	0.83	2.85	3.68

FORMWORK	UNIT	MAT.	INST.	TOTAL
03110.53 PILE CAP FORMWORK				
4 uses	S.F.	0.71	2.70	3.41
5 uses	"	0.61	2.55	3.16
Triangular				
1 use	S.F.	1.95	3.65	5.60
2 uses	"	1.10	3.45	4.55
3 uses	"	0.99	3.20	4.19
4 uses	"	0.83	3.05	3.88
5 uses	"	0.66	2.85	3.51
03110.55 SLAB/MAT FORMWORK				
Mat foundations, job built				
1 use	S.F.	1.70	3.25	4.95
2 uses	"	0.88	3.05	3.93
3 uses	"	0.66	2.85	3.51
4 uses	"	0.55	2.70	3.25
5 uses	"	0.48	2.60	3.08
Edge forms				
6" high				
1 use	L.F.	1.55	2.45	4.00
2 uses	"	0.88	2.25	3.13
3 uses	"	0.66	2.15	2.81
4 uses	"	0.55	2.05	2.60
5 uses	"	0.44	2.10	2.54
12" high				
1 use	L.F.	1.45	2.70	4.15
2 uses	"	0.88	2.45	3.33
3 uses	"	0.66	2.35	3.01
4 uses	"	0.55	2.25	2.80
5 uses	"	0.41	2.25	2.66
Formwork for openings				
1 use	S.F.	2.10	5.40	7.50
2 uses	"	1.20	4.70	5.90
3 uses	"	0.99	4.30	5.29
4 uses	"	0.77	3.95	4.72
5 uses	"	0.66	3.85	4.51
03110.60 STAIR FORMWORK				
Stairway forms, job built				
1 use	S.F.	2.75	5.15	7.90
2 uses	"	1.55	4.70	6.25
3 uses	"	1.20	4.30	5.50
4 uses	"	1.10	3.95	5.05
5 uses	"	0.94	3.65	4.59
Stairs, elevated				
1 use	S.F.	3.30	5.15	8.45
2 uses	"	1.75	4.30	6.05
3 uses	"	1.55	3.65	5.20
4 uses	"	1.30	3.45	4.75
5 uses	"	1.10	3.20	4.30

FORMWORK

03110.65 WALL FORMWORK

FORMWORK	UNIT	MAT.	INST.	TOTAL
Wall forms, exterior, job built				
Up to 8' high wall				
1 use	S.F.	1.75	2.70	4.45
2 uses	"	0.95	2.45	3.40
3 uses	"	0.84	2.35	3.19
4 uses	"	0.74	2.25	2.99
5 uses	"	0.63	2.25	2.88
Over 8' high wall				
1 use	S.F.	1.90	3.40	5.30
2 uses	"	1.10	3.05	4.15
3 uses	"	0.99	2.85	3.84
4 uses	"	0.89	2.70	3.59
5 uses	"	0.78	2.70	3.48
Over 16' high wall				
1 use	S.F.	2.00	3.65	5.65
2 uses	"	1.20	3.45	4.65
3 uses	"	1.10	3.20	4.30
4 uses	"	0.99	3.05	4.04
5 uses	"	0.89	2.85	3.74
Radial wall forms				
1 use	S.F.	2.00	3.95	5.95
2 uses	"	1.20	3.65	4.85
3 uses	"	1.10	3.45	4.55
4 uses	"	0.99	3.20	4.19
5 uses	"	0.89	3.05	3.94
Retaining wall forms				
1 use	S.F.	1.65	2.85	4.50
2 uses	"	0.89	2.70	3.59
3 uses	"	0.78	2.55	3.33
4 uses	"	0.68	2.45	3.13
5 uses	"	0.57	2.35	2.92
Radial retaining wall forms				
1 use	S.F.	1.95	4.30	6.25
2 uses	"	1.15	3.95	5.10
3 uses	"	1.05	3.65	4.70
4 uses	"	0.95	3.45	4.40
5 uses	"	0.84	3.20	4.04
Column pier and pilaster				
1 use	S.F.	2.00	5.40	7.40
2 uses	"	1.20	4.70	5.90
3 uses	"	1.10	4.30	5.40
4 uses	"	0.99	3.95	4.94
5 uses	"	0.89	3.85	4.74
Interior wall forms				
Up to 8' high				
1 use	S.F.	1.75	2.45	4.20
2 uses	"	0.95	2.25	3.20
3 uses	"	0.84	2.15	2.99
4 uses	"	0.74	2.05	2.79
5 uses	"	0.63	2.10	2.73
Over 8' high				
1 use	S.F.	1.90	3.00	4.90

FORMWORK	UNIT	MAT.	INST.	TOTAL
03110.65 WALL FORMWORK				
2 uses	S.F.	1.10	2.70	3.80
3 uses	"	0.99	2.55	3.54
4 uses	"	0.89	2.45	3.34
5 uses	"	0.78	2.45	3.23
Over 16' high				
1 use	S.F.	2.00	3.20	5.20
2 uses	"	1.20	3.05	4.25
3 uses	"	1.10	2.85	3.95
4 uses	"	0.99	2.70	3.69
5 uses	"	0.89	2.55	3.44
Radial wall forms				
1 use	S.F.	2.00	3.45	5.45
2 uses	"	1.20	3.20	4.40
3 uses	"	1.10	3.05	4.15
4 uses	"	0.99	2.85	3.84
5 uses	"	0.89	2.70	3.59
Curved wall forms, 24" sections				
1 use	S.F.	2.00	5.15	7.15
2 uses	"	1.20	4.70	5.90
3 uses	"	1.10	4.30	5.40
4 uses	"	0.99	3.95	4.94
5 uses	"	0.89	3.65	4.54
PVC form liner, per side, smooth finish				
1 use	S.F.	3.65	2.15	5.80
2 uses	"	2.00	2.05	4.05
3 uses	"	1.65	2.00	3.65
4 uses	"	1.30	1.85	3.15
5 uses	"	1.05	1.70	2.75
03110.90 MISCELLANEOUS FORMWORK				
Keyway forms (5 uses)				
2 x 4	L.F.	0.11	1.30	1.41
2 x 6	"	0.17	1.45	1.62
Bulkheads				
Walls, with keyways				
2 piece	L.F.	1.85	2.45	4.30
3 piece	"	1.95	2.55	4.50
Elevated slab, with keyway				
2 piece	L.F.	2.25	2.15	4.40
3 piece	"	3.10	2.35	5.45
Ground slab, with keyway				
2 piece	L.F.	2.25	1.85	4.10
3 piece	"	3.10	2.00	5.10
Chamfer strips				
Wood				
1/2" wide	L.F.	0.11	0.57	0.68
3/4" wide	"	0.17	0.57	0.74
1" wide	"	0.20	0.57	0.77
PVC				
1/2" wide	L.F.	0.51	0.57	1.08
3/4" wide	"	0.56	0.57	1.13
1" wide	"	0.73	0.57	1.30

FORMWORK

03110.90 MISCELLANEOUS FORMWORK

FORMWORK	UNIT	MAT.	INST.	TOTAL
Radius				
1"	L.F.	0.51	0.61	1.12
1-1/2"	"	1.20	0.61	1.81
Reglets				
Galvanized steel, 24 ga.	L.F.	0.85	1.05	1.90
Metal formwork				
Straight edge forms				
4" high	L.F.	0.11	1.60	1.71
6" high	"	0.11	1.70	1.81
8" high	"	0.13	1.85	1.98
12" high	"	0.18	2.00	2.18
16" high	"	0.22	2.15	2.37
Curb form, S-shape				
12" x				
1'-6"	L.F.	0.22	3.65	3.87
2'	"	0.25	3.45	3.70
2'-6"	"	0.29	3.20	3.49
3'	"	0.33	2.85	3.18

REINFORCEMENT

03210.05 BEAM REINFORCING

REINFORCEMENT	UNIT	MAT.	INST.	TOTAL
Beam-girders				
#3 - #4	TON	696.00	738.00	1,434
#5 - #6	"	630.00	590.00	1,220
#7 - #8	"	550.00	487.00	1,037
#9 - #10	"	550.00	418.00	968.00
#11 - #12	"	550.00	390.00	940.00
#13 - #14	"	550.00	366.00	916.00
Galvanized				
#3 - #4	TON	1,130	731.00	1,861
#5 - #6	"	1,070	585.00	1,655
#7 - #8	"	1,030	487.00	1,517
#9 - #10	"	1,030	418.00	1,448
#11 - #12	"	1,030	390.00	1,420
#13 - #14	"	1,030	366.00	1,396
Bond beams				
#3 - #4	TON	638.00	975.00	1,613
#5 - #6	"	578.00	731.00	1,309
#7 - #8	"	550.00	650.00	1,200
Galvanized				
#3 - #4	TON	1,130	975.00	2,105
#5 - #6	"	1,070	731.00	1,801
#7 - #8	"	1,030	650.00	1,680

REINFORCEMENT	UNIT	MAT.	INST.	TOTAL
03210.10 BOX CULVERT REINFORCING				
Box culverts				
#3 - #4	TON	638.00	366.00	1,004
#5 - #6	"	578.00	325.00	903.00
#7 - #8	"	550.00	292.00	842.00
#9 - #10	"	550.00	266.00	816.00
#11 - #12	"	550.00	244.00	794.00
Galvanized				
#3 - #4	TON	1,130	366.00	1,496
#5 - #6	"	1,070	325.00	1,395
#7 - #8	"	1,030	292.00	1,322
#9 - #10	"	1,030	266.00	1,296
#11 - #12	"	1,030	244.00	1,274
03210.15 COLUMN REINFORCING				
Columns				
#3 - #4	TON	696.00	843.00	1,539
#5 - #6	"	630.00	656.00	1,286
#7 - #8	"	550.00	585.00	1,135
#9 - #10	"	550.00	532.00	1,082
#11 - #12	"	550.00	487.00	1,037
#13 - #14	"	550.00	450.00	1,000
#15 - #16	"	550.00	418.00	968.00
Galvanized				
#3 - #4	TON	1,130	835.00	1,965
#5 - #6	"	1,070	650.00	1,720
#7 - #8	"	1,030	585.00	1,615
#9 - #10	"	1,030	532.00	1,562
#11 - #12	"	1,030	487.00	1,517
#13 - #14	"	1,030	450.00	1,480
#15 - #16	"	1,030	418.00	1,448
Spirals				
8" to 24" dia.	TON	770.00	731.00	1,501
24" to 48" dia.	"	770.00	650.00	1,420
48" to 84" dia.	"	770.00	585.00	1,355
03210.20 ELEVATED SLAB REINFORCING				
Elevated slab				
#3 - #4	TON	638.00	384.00	1,022
#5 - #6	"	578.00	341.00	919.00
#7 - #8	"	550.00	292.00	842.00
#9 - #10	"	550.00	266.00	816.00
#11 - #12	"	550.00	244.00	794.00
Galvanized				
#3 - #4	TON	1,130	384.00	1,514
#5 - #6	"	1,070	341.00	1,411
#7 - #8	"	1,030	292.00	1,322
#9 - #10	"	1,030	266.00	1,296
#11 - #12	"	1,030	244.00	1,274

REINFORCEMENT

	UNIT	MAT.	INST.	TOTAL
03210.25 EQUIP. PAD REINFORCING				
Equipment pad				
#3 - #4	TON	638.00	614.00	1,252
#5 - #6	"	578.00	559.00	1,137
#7 - #8	"	550.00	512.00	1,062
#9 - #10	"	550.00	473.00	1,023
#11 - #12	"	550.00	439.00	989.00
03210.35 FOOTING REINFORCING				
Footings				
Grade 50				
#3 - #4	TON	638.00	512.00	1,150
#5 - #6	"	578.00	439.00	1,017
#7 - #8	"	550.00	366.00	916.00
#9 - #10	"	550.00	325.00	875.00
Grade 60				
#3 - #4	TON	660.00	512.00	1,172
#5 - #6	"	600.00	439.00	1,039
#7 - #8	"	578.00	366.00	944.00
#9 - #10	"	578.00	325.00	903.00
Grade 70				
#3 - #4	TON	682.00	487.00	1,169
#5 - #6	"	622.00	418.00	1,040
#7 - #8	"	600.00	366.00	966.00
#9 - #10	"	600.00	325.00	925.00
#11- #12	"	600.00	292.00	892.00
Straight dowels, 24" long				
1" dia.	EA.	2.00	2.90	4.90
3/4" dia.	"	1.75	2.90	4.65
5/8" dia.	"	1.55	2.45	4.00
1/2" dia.	"	1.10	2.10	3.20
03210.45 FOUNDATION REINFORCING				
Foundations				
#3 - #4	TON	696.00	492.00	1,188
#5 - #6	"	630.00	422.00	1,052
#7 - #8	"	550.00	366.00	916.00
#9 - #10	"	550.00	325.00	875.00
#11 - #12	"	550.00	292.00	842.00
Galvanized				
#3 - #4	TON	1,130	487.00	1,617
#5 - #6	"	1,070	418.00	1,488
#7 - #8	"	1,030	366.00	1,396
#9 - #10	"	1,030	325.00	1,355
#11 - #12	"	1,030	292.00	1,322
03210.50 GRADE BEAM REINFORCING				
Grade beams				
#3 - #4	TON	696.00	454.00	1,150
#5 - #6	"	630.00	394.00	1,024
#7 - #8	"	550.00	344.00	894.00
#9 - #10	"	550.00	308.00	858.00
#11 - #12	"	550.00	278.00	828.00

03 CONCRETE

REINFORCEMENT	UNIT	MAT.	INST.	TOTAL
03210.50 GRADE BEAM REINFORCING				
Galvanized				
#3 - #4	TON	1,130	450.00	1,580
#5 - #6	"	1,070	390.00	1,460
#7 - #8	"	1,030	344.00	1,374
#9 - #10	"	1,030	308.00	1,338
#11 - #12	"	1,030	278.00	1,308
03210.53 PILE CAP REINFORCING				
Pile caps				
#3 - #4	TON	638.00	731.00	1,369
#5 - #6	"	578.00	650.00	1,228
#7 - #8	"	550.00	585.00	1,135
#9 - #10	"	550.00	532.00	1,082
#11 - #12	"	550.00	487.00	1,037
Galvanized				
#3 - #4	TON	1,130	731.00	1,861
#5 - #6	"	1,070	650.00	1,720
#7 - #8	"	1,030	585.00	1,615
#9 - #10	"	1,030	532.00	1,562
#11 - #12	"	1,030	487.00	1,517
03210.55 SLAB/MAT REINFORCING				
Bars, slabs				
#3 - #4	TON	638.00	512.00	1,150
#5 - #6	"	578.00	439.00	1,017
#7 - #8	"	550.00	384.00	934.00
#9 - #10	"	550.00	325.00	875.00
#11 - #12	"	550.00	292.00	842.00
Galvanized				
#3 - #4	TON	1,130	512.00	1,642
#5 - #6	"	1,070	439.00	1,509
#7 - #8	"	1,030	366.00	1,396
#9 - #10	"	1,030	325.00	1,355
#11 - #12	"	1,030	292.00	1,322
Wire mesh, slabs				
Galvanized				
4x4				
W1.4xW1.4	S.F.	0.24	0.20	0.44
W2.0xW2.0	"	0.30	0.21	0.51
W2.9xW2.9	"	0.42	0.22	0.64
W4.0xW4.0	"	0.62	0.25	0.87
6x6				
W1.4xW1.4	S.F.	0.17	0.15	0.32
W2.0xW2.0	"	0.22	0.16	0.38
W2.9xW2.9	"	0.30	0.17	0.47
W4.0xW4.0	"	0.46	0.20	0.66
Standard				
2x2				
W.9xW.9	S.F.	0.18	0.19	0.37
4x4				
W1.4xW1.4	S.F.	0.12	0.19	0.31
W2.0xW2.0	"	0.15	0.21	0.36

REINFORCEMENT	UNIT	MAT.	INST.	TOTAL
03210.55 **SLAB/MAT REINFORCING**				
W2.9xW2.9	S.F.	0.21	0.22	0.43
W4.0xW4.0	"	0.32	0.24	0.56
6x6				
W1.4xW1.4	S.F.	0.08	0.15	0.23
W2.0xW2.0	"	0.11	0.16	0.27
W2.9xW2.9	"	0.15	0.17	0.32
W4.0xW4.0	"	0.22	0.19	0.41
03210.60 **STAIR REINFORCING**				
Stairs				
#3 - #4	TON	638.00	585.00	1,223
#5 - #6	"	578.00	487.00	1,065
#7 - #8	"	550.00	418.00	968.00
#9 - #10	"	550.00	366.00	916.00
Galvanized				
#3 - #4	TON	1,130	585.00	1,715
#5 - #6	"	1,070	487.00	1,557
#7 - #8	"	1,030	418.00	1,448
#9 - #10	"	1,030	366.00	1,396
03210.65 **WALL REINFORCING**				
Walls				
#3 - #4	TON	638.00	439.00	1,077
#5 - #6	"	578.00	384.00	962.00
#7 - #8	"	550.00	325.00	875.00
#9 - #10	"	550.00	292.00	842.00
Galvanized				
#3 - #4	TON	1,130	439.00	1,569
#5 - #6	"	1,070	384.00	1,454
#7 - #8	"	1,030	325.00	1,355
#9 - #10	"	1,030	292.00	1,322
Masonry wall (horizontal)				
#3 - #4	TON	638.00	1,230	1,868
#5 - #6	"	578.00	1,020	1,598
Galvanized				
#3 - #4	TON	1,130	1,230	2,360
#5 - #6	"	1,070	1,020	2,090
Masonry wall (vertical)				
#3 - #4	TON	638.00	1,540	2,178
#5 - #6	"	578.00	1,230	1,808
Galvanized				
#3 - #4	TON	1,130	1,540	2,670
#5 - #6	"	1,070	1,230	2,300

ACCESSORIES	UNIT	MAT.	INST.	TOTAL
03250.40 CONCRETE ACCESSORIES				
Expansion joint, poured				
Asphalt				
1/2" x 1"	L.F.	0.34	0.41	0.75
1" x 2"	"	1.05	0.44	1.49
Liquid neoprene, cold applied				
1/2" x 1"	L.F.	0.92	0.42	1.34
1" x 2"	"	3.65	0.45	4.10
Polyurethane, 2 parts				
1/2" x 1"	L.F.	0.95	0.68	1.63
1" x 2"	"	3.85	0.74	4.59
Rubberized asphalt, cold				
1/2" x 1"	L.F.	0.20	0.41	0.61
1" x 2"	"	0.75	0.44	1.19
Hot, fuel resistant				
1/2" x 1"	L.F.	0.95	0.41	1.36
1" x 2"	"	3.85	0.44	4.29
Expansion joint, premolded, in slabs				
Asphalt				
1/2" x 6"	L.F.	0.26	0.51	0.77
1" x 12"	"	0.52	0.68	1.20
Cork				
1/2" x 6"	L.F.	0.73	0.51	1.24
1" x 12"	"	2.90	0.68	3.58
Neoprene sponge				
1/2" x 6"	L.F.	1.05	0.51	1.56
1" x 12"	"	3.95	0.68	4.63
Polyethylene foam				
1/2" x 6"	L.F.	0.28	0.51	0.79
1" x 12"	"	1.70	0.68	2.38
Polyurethane foam				
1/2" x 6"	L.F.	0.48	0.51	0.99
1" x 12"	"	1.20	0.68	1.88
Polyvinyl chloride foam				
1/2" x 6"	L.F.	1.25	0.51	1.76
1" x 12"	"	4.60	0.68	5.28
Rubber, gray sponge				
1/2" x 6"	L.F.	1.75	0.51	2.26
1" x 12"	"	6.85	0.68	7.53
Asphalt felt control joints or bond breaker, screed joints				
4" slab	L.F.	0.47	0.41	0.88
6" slab	"	0.64	0.45	1.09
8" slab	"	0.86	0.51	1.37
10" slab	"	1.20	0.58	1.78
Keyed cold expansion and control joints, 24 ga.				
4" slab	L.F.	0.42	1.25	1.67
5" slab	"	0.51	1.25	1.76
6" slab	"	0.59	1.35	1.94
8" slab	"	0.74	1.45	2.19
10" slab	"	0.86	1.55	2.41
Waterstops				
Polyvinyl chloride				
Ribbed				

ACCESSORIES

03250.40 CONCRETE ACCESSORIES

ACCESSORIES	UNIT	MAT.	INST.	TOTAL
3/16" thick x				
4" wide	L.F.	0.62	1.00	1.62
6" wide	"	0.75	1.15	1.90
1/2" thick x				
9" wide	L.F.	2.60	1.25	3.85
Ribbed with center bulb				
3/16" thick x 9" wide	L.F.	2.40	1.25	3.65
3/8" thick x 9" wide	"	1.85	1.25	3.10
Dumbbell type, 3/8" thick x 6" wide	"	1.85	1.15	3.00
Plain, 3/8" thick x 9" wide	"	3.60	1.25	4.85
Center bulb, 3/8" thick x 9" wide	"	3.70	1.25	4.95
Rubber				
Flat dumbbell				
3/8" thick x				
6" wide	L.F.	4.75	1.15	5.90
9" wide	"	7.60	1.25	8.85
Center bulb				
3/8" thick x				
6" wide	L.F.	4.75	1.15	5.90
9" wide	"	7.40	1.25	8.65
Vapor barrier				
4 mil polyethylene	S.F.	0.02	0.07	0.09
6 mil polyethylene	"	0.03	0.07	0.10
Gravel porous fill, under floor slabs, 3/4" stone	C.Y.	11.20	34.00	45.20
Reinforcing accessories				
Beam bolsters				
1-1/2" high, plain	L.F.	0.31	0.29	0.60
Galvanized	"	0.38	0.29	0.67
3" high				
Plain	L.F.	0.76	0.37	1.13
Galvanized	"	0.80	0.37	1.17
Slab bolsters				
1" high				
Plain	L.F.	0.22	0.15	0.37
Galvanized	"	0.29	0.15	0.44
2" high				
Plain	L.F.	0.29	0.16	0.45
Galvanized	"	0.33	0.16	0.49
Chairs, high chairs				
3" high				
Plain	EA.	0.34	0.73	1.07
Galvanized	"	0.45	0.73	1.18
5" high				
Plain	EA.	0.45	0.77	1.22
Galvanized	"	0.56	0.77	1.33
8" high				
Plain	EA.	0.92	0.84	1.76
Galvanized	"	1.25	0.84	2.09
12" high				
Plain	EA.	1.95	1.00	2.95
Galvanized	"	2.35	0.97	3.32
Continuous, high chair				

ACCESSORIES

03250.40 CONCRETE ACCESSORIES	UNIT	MAT.	INST.	TOTAL
3" high				
Plain	L.F.	0.40	0.19	0.59
Galvanized	"	0.50	0.19	0.69
5" high				
Plain	L.F.	0.62	0.21	0.83
Galvanized	"	0.80	0.21	1.01
8" high				
Plain	L.F.	0.83	0.22	1.05
Galvanized	"	1.10	0.22	1.32
12" high				
Plain	L.F.	1.80	0.24	2.04
Galvanized	"	2.35	0.24	2.59

CAST-IN-PLACE CONCRETE

03300.10 CONCRETE ADMIXTURES	UNIT	MAT.	INST.	TOTAL
Concrete admixtures				
Water reducing admixture	GAL			9.35
Set retarder	"			16.50
Air entraining agent	"			5.50

03350.10 CONCRETE FINISHES	UNIT	MAT.	INST.	TOTAL
Floor finishes				
Broom	S.F.	0.00	0.29	0.29
Screed	"	0.00	0.25	0.25
Darby	"	0.00	0.25	0.25
Steel float	"	0.00	0.34	0.34
Granolithic topping				
1/2" thick	S.F.	0.17	0.93	1.10
1" thick	"	0.28	1.00	1.28
2" thick	"	0.51	1.15	1.66
Wall finishes				
Burlap rub, with cement paste	S.F.	0.06	0.34	0.40
Float finish	"	0.08	0.51	0.59
Etch with acid	"	0.24	0.34	0.58
Sandblast				
Minimum	S.F.	0.06	0.51	0.57
Maximum	"	0.22	0.51	0.73
Bush hammer				
Green concrete	S.F.	0.00	1.00	1.00
Cured concrete	"	0.00	1.55	1.55
Break ties and patch holes	"	0.00	0.41	0.41
Carborundum				
Dry rub	S.F.	0.00	0.68	0.68

03 CONCRETE

CAST-IN-PLACE CONCRETE

CAST-IN-PLACE CONCRETE	UNIT	MAT.	INST.	TOTAL
03350.10 CONCRETE FINISHES				
Wet rub	S.F.	0.00	1.00	1.00
Floor hardeners				
Metallic				
Light service	S.F.	0.14	0.25	0.39
Heavy service	"	0.61	0.34	0.95
Non-metallic				
Light service	S.F.	0.11	0.25	0.36
Heavy service	"	0.50	0.34	0.84
Rusticated concrete finish				
Beveled edge	L.F.	0.22	1.15	1.37
Square edge	"	0.24	1.45	1.69
Solid board concrete finish				
Standard	S.F.	0.55	1.70	2.25
Rustic	"	0.50	2.05	2.55
03360.10 PNEUMATIC CONCRETE				
Pneumatic applied concrete (gunite)				
2" thick	S.F.	2.25	1.35	3.60
3" thick	"	3.10	1.85	4.95
4" thick	"	3.65	2.20	5.85
Finish surface				
Minimum	S.F.	0.00	1.30	1.30
Maximum	"	0.00	2.55	2.55
03370.10 CURING CONCRETE				
Sprayed membrane				
Slabs	S.F.	0.03	0.04	0.07
Walls	"	0.07	0.05	0.12
Curing paper				
Slabs	S.F.	0.03	0.05	0.08
Walls	"	0.03	0.06	0.09
Burlap				
7.5 oz.	S.F.	0.04	0.07	0.11
12 oz.	"	0.07	0.07	0.14

PLACING CONCRETE

PLACING CONCRETE	UNIT	MAT.	INST.	TOTAL
03380.05 BEAM CONCRETE				
Beams and girders				
2500# or 3000# concrete				
By crane	C.Y.	61.50	38.50	100.00
By pump	"	61.50	35.00	96.50

PLACING CONCRETE

PLACING CONCRETE	UNIT	MAT.	INST.	TOTAL
03380.05 BEAM CONCRETE				
By hand buggy	C.Y.	61.50	20.40	81.90
3500# or 4000# concrete				
By crane	C.Y.	60.50	38.50	99.00
By pump	"	60.50	35.00	95.50
By hand buggy	"	60.50	20.40	80.90
5000# concrete				
By crane	C.Y.	65.00	38.50	103.50
By pump	"	65.00	35.00	100.00
By hand buggy	"	65.00	20.40	85.40
Bond beam, 3000# concrete				
By pump				
8" high				
4" wide	L.F.	0.15	0.77	0.92
6" wide	"	0.36	0.87	1.23
8" wide	"	0.47	0.96	1.43
10" wide	"	0.63	1.05	1.68
12" wide	"	0.84	1.20	2.04
16" high				
8" wide	L.F.	1.15	1.20	2.35
10" wide	"	1.55	1.35	2.90
12" wide	"	2.05	1.60	3.65
By crane				
8" high				
4" wide	L.F.	0.18	0.84	1.02
6" wide	"	0.36	0.92	1.28
8" wide	"	0.47	0.96	1.43
10" wide	"	0.63	1.05	1.68
12" wide	"	0.84	1.20	2.04
16" high				
8" wide	L.F.	1.15	1.20	2.35
10" wide	"	1.55	1.30	2.85
12" wide	"	2.05	1.50	3.55
03380.15 COLUMN CONCRETE				
Columns				
2500# or 3000# concrete				
By crane	C.Y.	61.50	35.00	96.50
By pump	"	61.50	32.10	93.60
3500# or 4000# concrete				
By crane	C.Y.	60.50	35.00	95.50
By pump	"	60.50	32.10	92.60
5000# concrete				
By crane	C.Y.	65.00	35.00	100.00
By pump	"	65.00	32.10	97.10
03380.20 ELEVATED SLAB CONCRETE				
Elevated slab				
2500# or 3000# concrete				
By crane	C.Y.	56.50	20.00	76.50
By pump	"	56.50	15.40	71.90
By hand buggy	"	56.50	21.20	77.70
3500# or 4000# concrete				

03 CONCRETE

PLACING CONCRETE	UNIT	MAT.	INST.	TOTAL
03380.20 ELEVATED SLAB CONCRETE				
By crane	C.Y.	60.50	19.25	79.75
By pump	"	60.50	14.80	75.30
By hand buggy	"	60.50	20.40	80.90
5000# concrete				
By crane	C.Y.	65.00	20.00	85.00
By pump	"	65.00	15.40	80.40
By hand buggy	"	65.00	21.20	86.20
Topping				
2500# or 3000# concrete				
By crane	C.Y.	56.50	19.25	75.75
By pump	"	56.50	14.80	71.30
By hand buggy	"	56.50	20.40	76.90
3500# or 4000# concrete				
By crane	C.Y.	60.50	19.25	79.75
By pump	"	60.50	14.80	75.30
By hand buggy	"	60.50	20.40	80.90
5000# concrete				
By crane	C.Y.	65.00	19.25	84.25
By pump	"	65.00	14.80	79.80
By hand buggy	"	65.00	20.40	85.40
03380.25 EQUIPMENT PAD CONCRETE				
Equipment pad				
2500# or 3000# concrete				
By chute	C.Y.	56.50	7.05	63.55
By pump	"	56.50	28.60	85.10
By crane	"	56.50	33.40	89.90
3500# or 4000# concrete				
By chute	C.Y.	60.50	7.05	67.55
By pump	"	60.50	28.60	89.10
By crane	"	60.50	33.40	93.90
5000# concrete				
By chute	C.Y.	65.00	7.05	72.05
By pump	"	65.00	28.60	93.60
By crane	"	65.00	33.40	98.40
03380.35 FOOTING CONCRETE				
Continuous footing				
2500# or 3000# concrete				
By chute	C.Y.	56.50	7.05	63.55
By pump	"	56.50	25.00	81.50
By crane	"	56.50	28.60	85.10
3500# or 4000# concrete				
By chute	C.Y.	60.50	6.80	67.30
By pump	"	60.50	24.05	84.55
By crane	"	60.50	27.50	88.00
5000# concrete				
By chute	C.Y.	65.00	7.05	72.05
By pump	"	65.00	25.00	90.00
By crane	"	65.00	28.60	93.60
Spread footing				

PLACING CONCRETE	UNIT	MAT.	INST.	TOTAL
03380.35 FOOTING CONCRETE				
2500# or 3000# concrete				
Under 5 cy				
By chute	C.Y.	56.50	7.05	63.55
By pump	"	56.50	26.70	83.20
By crane	"	56.50	30.80	87.30
Over 5 cy				
By chute	C.Y.	56.50	5.10	61.60
By pump	"	56.50	22.65	79.15
By crane	"	56.50	25.70	82.20
3500# or 4000# concrete				
Under 5 c.y.				
By chute	C.Y.	60.50	6.80	67.30
By pump	"	60.50	25.70	86.20
By crane	"	60.50	29.60	90.10
Over 5 c.y.				
By chute	C.Y.	60.50	5.10	65.60
By pump	"	60.50	22.65	83.15
By crane	"	60.50	25.70	86.20
5000# concrete				
Under 5 c.y.				
By chute	C.Y.	65.00	7.05	72.05
By pump	"	65.00	26.70	91.70
By crane	"	65.00	30.80	95.80
Over 5 c.y.				
By chute	C.Y.	65.00	5.10	70.10
By pump	"	65.00	22.65	87.65
By crane	"	65.00	25.70	90.70
03380.50 GRADE BEAM CONCRETE				
Grade beam				
2500# or 3000# concrete				
By chute	C.Y.	61.50	6.80	68.30
By crane	"	61.50	27.50	89.00
By pump	"	61.50	24.05	85.55
By hand buggy	"	61.50	20.40	81.90
3500# or 4000# concrete				
By chute	C.Y.	60.50	6.80	67.30
By crane	"	60.50	27.50	88.00
By pump	"	60.50	24.05	84.55
By hand buggy	"	60.50	20.40	80.90
5000# concrete				
By chute	C.Y.	65.00	6.80	71.80
By crane	"	65.00	27.50	92.50
By pump	"	65.00	24.05	89.05
By hand buggy	"	65.00	20.40	85.40
03380.53 PILE CAP CONCRETE				
Pile cap				
2500# or 3000 concrete				
By chute	C.Y.	56.50	6.80	63.30
By crane	"	56.50	32.10	88.60

PLACING CONCRETE	UNIT	MAT.	INST.	TOTAL
03380.53 PILE CAP CONCRETE				
By pump	C.Y.	56.50	27.50	84.00
By hand buggy	"	56.50	20.40	76.90
3500# or 4000# concrete				
By chute	C.Y.	60.50	6.80	67.30
By crane	"	60.50	32.10	92.60
By pump	"	60.50	27.50	88.00
By hand buggy	"	60.50	20.40	80.90
5000# concrete				
By chute	C.Y.	65.00	6.80	71.80
By crane	"	65.00	32.10	97.10
By pump	"	65.00	27.50	92.50
By hand buggy	"	65.00	20.40	85.40
03380.55 SLAB/MAT CONCRETE				
Slab on grade				
2500# or 3000# concrete				
By chute	C.Y.	56.50	5.30	61.80
By crane	"	56.50	16.70	73.20
By pump	"	56.50	14.30	70.80
By hand buggy	"	56.50	14.15	70.65
3500# or 4000# concrete				
By chute	C.Y.	60.50	5.10	65.60
By crane	"	60.50	16.05	76.55
By pump	"	60.50	13.75	74.25
By hand buggy	"	60.50	13.60	74.10
5000# concrete				
By chute	C.Y.	65.00	5.30	70.30
By crane	"	65.00	16.70	81.70
By pump	"	65.00	14.30	79.30
By hand buggy	"	65.00	14.15	79.15
Foundation mat				
2500# or 3000# concrete, over 20 cy				
By chute	C.Y.	56.50	4.10	60.60
By crane	"	56.50	13.75	70.25
By pump	"	56.50	12.00	68.50
By hand buggy	"	56.50	10.20	66.70
03380.58 SIDEWALKS				
Walks, cast in place with wire mesh, base not incl.				
4" thick	S.F.	0.85	0.68	1.53
5" thick	"	1.15	0.82	1.97
6" thick	"	1.35	1.00	2.35
03380.60 STAIR CONCRETE				
Stairs				
2500# or 3000# concrete				
By chute	C.Y.	56.50	6.80	63.30
By crane	"	56.50	32.10	88.60
By pump	"	56.50	27.50	84.00
By hand buggy	"	56.50	20.40	76.90
3500# or 4000# concrete				
By chute	C.Y.	60.50	6.80	67.30

PLACING CONCRETE

PLACING CONCRETE	UNIT	MAT.	INST.	TOTAL
03380.60 STAIR CONCRETE				
By crane	C.Y.	60.50	32.10	92.60
By pump	"	60.50	27.50	88.00
By hand buggy	"	60.50	20.40	80.90
5000# concrete				
By chute	C.Y.	65.00	6.80	71.80
By crane	"	65.00	32.10	97.10
By pump	"	65.00	27.50	92.50
By hand buggy	"	65.00	20.40	85.40
03380.65 WALL CONCRETE				
Walls				
2500# or 3000# concrete				
To 4'				
By chute	C.Y.	56.50	6.05	62.55
By crane	"	56.50	33.40	89.90
By pump	"	56.50	30.80	87.30
To 8'				
By crane	C.Y.	56.50	36.40	92.90
By pump	"	56.50	33.40	89.90
To 16'				
By crane	C.Y.	56.50	38.50	95.00
By pump	"	56.50	35.00	91.50
Over 16'				
By crane	C.Y.	56.50	42.80	99.30
By pump	"	56.50	38.50	95.00
3500# or 4000# concrete				
To 4'				
By chute	C.Y.	60.50	5.85	66.35
By crane	"	60.50	32.10	92.60
By pump	"	60.50	29.60	90.10
To 8'				
By crane	C.Y.	60.50	35.00	95.50
By pump	"	60.50	32.10	92.60
To 16'				
By crane	C.Y.	60.50	38.50	99.00
By pump	"	60.50	35.00	95.50
Over 16'				
By crane	C.Y.	60.50	42.80	103.30
By pump	"	60.50	38.50	99.00
5000# concrete				
To 4'				
By chute	C.Y.	65.00	5.85	70.85
By crane	"	65.00	32.10	97.10
By pump	"	65.00	29.60	94.60
To 8'				
By crane	C.Y.	65.00	35.00	100.00
By pump	"	65.00	32.10	97.10
To 16'				
By crane	C.Y.	65.00	38.50	103.50
By pump	"	65.00	35.00	100.00
Filled block (CMU)				
3000# concrete, by pump				

PLACING CONCRETE	UNIT	MAT.	INST.	TOTAL
03380.65 WALL CONCRETE				
4" wide	S.F.	0.21	1.45	1.66
6" wide	"	0.47	1.65	2.12
8" wide	"	0.74	2.00	2.74
10" wide	"	0.99	2.35	3.34
12" wide	"	1.25	2.85	4.10
Pilasters, 3000# concrete	C.F.	2.85	40.00	42.85
Wall cavity, 2" thick, 3000# concrete	S.F.	0.53	1.35	1.88

PRECAST CONCRETE	UNIT	MAT.	INST.	TOTAL
03400.10 PRECAST BEAMS				
Prestressed, double tee, 24" deep, 8' wide				
35' span				
115 psf	S.F.	3.95	0.51	4.46
140 psf	"	4.35	0.51	4.86
40' span				
80 psf	S.F.	3.95	0.54	4.49
143 psf	"	4.35	0.54	4.89
45' span				
50 psf	S.F.	3.95	0.47	4.42
70 psf	"	4.35	0.47	4.82
100 psf	"	4.35	0.47	4.82
130 psf	"	4.70	0.47	5.17
50' span				
75 psf	S.F.	4.35	0.42	4.77
100 psf	"	4.70	0.42	5.12
Precast beams, girders and joists				
1000 lb/lf live load				
10' span	L.F.	41.20	10.15	51.35
20' span	"	46.80	6.10	52.90
30' span	"	53.00	5.10	58.10
3000 lb/lf live load				
10' span	L.F.	42.90	10.15	53.05
20' span	"	52.50	6.10	58.60
30' span	"	64.50	5.10	69.60
5000 lb/lf live load				
10' span	L.F.	42.90	10.15	53.05
20' span	"	59.00	6.10	65.10
30' span	"	70.00	5.10	75.10
03400.20 PRECAST COLUMNS				
Prestressed concrete columns				
10" x 10"				

PRECAST CONCRETE

03400.20 PRECAST COLUMNS

PRECAST CONCRETE	UNIT	MAT.	INST.	TOTAL
10' long	EA.	120.00	61.00	181.00
15' long	"	180.00	63.50	243.50
20' long	"	240.00	68.00	308.00
25' long	"	300.00	72.50	372.50
30' long	"	360.00	76.50	436.50
12" x 12"				
20' long	EA.	346.00	76.50	422.50
25' long	"	433.00	82.50	515.50
30' long	"	520.00	87.00	607.00
16" x 16"				
20' long	EA.	617.00	76.50	693.50
25' long	"	772.00	82.50	854.50
30' long	"	924.00	87.00	1,011
20" x 20"				
20' long	EA.	963.00	80.50	1,044
25' long	"	1,210	84.50	1,295
30' long	"	1,440	89.50	1,530
24" x 24"				
20' long	EA.	1,390	84.50	1,475
25' long	"	1,730	89.50	1,820
30' long	"	2,080	95.50	2,176
28" x 28"				
20' long	EA.	1,910	95.50	2,006
25' long	"	2,370	102.00	2,472
30' long	"	2,890	109.00	2,999
32" x 32"				
20' long	EA.	2,480	102.00	2,582
25' long	"	3,100	109.00	3,209
30' long	"	3,700	117.00	3,817
36" x 36"				
20' long	EA.	3,120	109.00	3,229
25' long	"	3,900	117.00	4,017
30' long	"	4,680	127.00	4,807

03400.30 PRECAST SLABS

PRECAST SLABS	UNIT	MAT.	INST.	TOTAL
Prestressed flat slab				
6" thick, 4' wide				
20' span				
80 psf	S.F.	8.40	1.25	9.65
110 psf	"	8.40	1.25	9.65
25' span				
80 psf	S.F.	8.85	1.20	10.05
Cored slab				
6" thick, 4' wide				
20' span				
80 psf	S.F.	3.50	1.25	4.75
100 psf	"	3.50	1.25	4.75
130 psf	"	3.55	1.25	4.80
8" thick, 4' wide				
25' span				
70 psf	S.F.	3.85	1.20	5.05

03 CONCRETE

PRECAST CONCRETE	UNIT	MAT.	INST.	TOTAL
03400.30 PRECAST SLABS				
125 psf	S.F.	3.85	1.20	5.05
170 psf	"	3.90	1.20	5.10
30' span				
70 psf	S.F.	3.90	1.00	4.90
90 psf	"	4.15	1.00	5.15
35' span				
70 psf	S.F.	4.15	0.95	5.10
10" thick, 4' wide				
30' span				
75 psf	S.F.	4.20	1.00	5.20
100 psf	"	4.30	1.00	5.30
130 psf	"	4.40	1.00	5.40
35' span				
60 psf	S.F.	4.30	0.95	5.25
80 psf	"	4.40	0.95	5.35
120 psf	"	4.65	0.95	5.60
40' span				
65 psf	S.F.	4.65	0.76	5.41
Slabs, roof and floor members, 4' wide				
6" thick, 25' span	S.F.	3.55	1.20	4.75
8" thick, 30' span	"	4.20	0.92	5.12
10" thick, 40' span	"	5.20	0.82	6.02
Tee members				
Multiple tee, roof and floor				
Minimum	S.F.	4.75	0.76	5.51
Maximum	"	6.00	1.55	7.55
Double tee wall member				
Minimum	S.F.	4.40	0.87	5.27
Maximum	"	5.55	1.70	7.25
Single tee				
Short span, roof members				
Minimum	S.F.	5.00	0.92	5.92
Maximum	"	6.10	1.90	8.00
Long span, roof members				
Minimum	S.F.	6.35	0.76	7.11
Maximum	"	7.50	1.55	9.05
03400.40 PRECAST WALLS				
Wall panel, 8' x 20'				
Gray cement				
Liner finish				
4" wall	S.F.	6.35	0.87	7.22
5" wall	"	6.90	0.90	7.80
6" wall	"	7.60	0.92	8.52
8" wall	"	8.15	0.95	9.10
Sandblast finish				
4" wall	S.F.	7.25	0.87	8.12
5" wall	"	7.80	0.90	8.70
6" wall	"	8.50	0.92	9.42
8" wall	"	9.00	0.95	9.95
White cement				
Liner finish				

PRECAST CONCRETE	UNIT	MAT.	INST.	TOTAL
03400.40 PRECAST WALLS				
4" wall	S.F.	7.70	0.87	8.57
5" wall	"	8.25	0.90	9.15
6" wall	"	8.95	0.92	9.87
8" wall	"	9.45	0.95	10.40
Sandblast finish				
4" wall	S.F.	8.25	0.87	9.12
5" wall	"	8.85	0.90	9.75
6" wall	"	9.50	0.92	10.42
8" wall	"	10.05	0.95	11.00
Double tee wall panel, 24" deep				
Gray cement				
Liner finish	S.F.	4.10	1.00	5.10
Sandblast finish	"	5.20	1.00	6.20
White cement				
Form liner finish	S.F.	6.00	1.00	7.00
Sandblast finish	"	7.70	1.00	8.70
Partition panels				
4" wall	S.F.	6.40	1.00	7.40
5" wall	"	6.95	1.00	7.95
6" wall	"	7.65	1.00	8.65
8" wall	"	8.20	1.00	9.20
Cladding panels				
4" wall	S.F.	6.40	1.10	7.50
5" wall	"	6.95	1.10	8.05
6" wall	"	7.65	1.10	8.75
8" wall	"	8.20	1.10	9.30
Sandwich panel, 2.5" cladding panel, 2" insulation				
5" wall	S.F.	9.35	1.10	10.45
6" wall	"	9.90	1.10	11.00
8" wall	"	10.50	1.10	11.60
03400.90 PRECAST SPECIALTIES				
Precast concrete, coping, 4' to 8' long				
12" wide	L.F.	4.15	2.75	6.90
10" wide	"	3.70	3.15	6.85
Splash block, 30"x12"x4"	EA.	6.45	18.35	24.80
Stair unit, per riser	"	21.95	18.35	40.30
Sun screen and trellis, 8' long, 12" high				
4" thick blades	EA.	46.20	13.75	59.95
5" thick blades	"	55.50	13.75	69.25
6" thick blades	"	69.50	14.65	84.15
8" thick blades	"	92.50	14.65	107.15
Bearing pads for precast members, 2" wide strips				
1/8" thick	L.F.	0.12	0.08	0.20
1/4" thick	"	0.18	0.08	0.26
1/2" thick	"	0.19	0.08	0.27
3/4" thick	"	0.40	0.09	0.49
1" thick	"	0.42	0.10	0.52
1-1/2" thick	"	0.55	0.10	0.65

03 CONCRETE

CEMENTITOUS TOPPINGS

03550.10 CONCRETE TOPPINGS	UNIT	MAT.	INST.	TOTAL
Gypsum fill				
2" thick	S.F.	0.85	0.20	1.05
2-1/2" thick	"	0.98	0.20	1.18
3" thick	"	1.20	0.21	1.41
3-1/2" thick	"	1.35	0.21	1.56
4" thick	"	1.60	0.24	1.84
Formboard				
Mineral fiber board				
1" thick	S.F.	0.80	0.51	1.31
1-1/2" thick	"	2.15	0.58	2.73
Cement fiber board				
1" thick	S.F.	0.63	0.68	1.31
1-1/2" thick	"	0.81	0.78	1.59
Glass fiber board				
1" thick	S.F.	0.96	0.51	1.47
1-1/2" thick	"	1.30	0.58	1.88
Poured deck				
Vermiculite or perlite				
1 to 4 mix	C.Y.	109.00	32.10	141.10
1 to 6 mix	"	101.00	29.60	130.60
Vermiculite or perlite				
2" thick				
1 to 4 mix	S.F.	0.85	0.20	1.05
1 to 6 mix	"	0.63	0.18	0.81
3" thick				
1 to 4 mix	S.F.	1.20	0.30	1.50
1 to 6 mix	"	0.90	0.27	1.17
Concrete plank, lightweight				
2" thick	S.F.	4.35	1.55	5.90
2-1/2" thick	"	4.50	1.55	6.05
3-1/2" thick	"	4.75	1.70	6.45
4" thick	"	4.85	1.70	6.55
Channel slab, lightweight, straight				
2-3/4" thick	S.F.	3.35	1.55	4.90
3-1/2" thick	"	3.50	1.55	5.05
3-3/4" thick	"	3.75	1.55	5.30
4-3/4" thick	"	4.75	1.70	6.45
Gypsum plank				
2" thick	S.F.	1.55	1.55	3.10
3" thick	"	1.65	1.55	3.20
Cement fiber, T and G planks				
1" thick	S.F.	0.85	1.40	2.25
1-1/2" thick	"	0.90	1.40	2.30
2" thick	"	1.00	1.55	2.55
2-1/2" thick	"	1.10	1.55	2.65
3" thick	"	1.40	1.55	2.95
3-1/2" thick	"	1.65	1.70	3.35
4" thick	"	1.85	1.70	3.55
Bulb tees, sub-purlins and grout				
6' span	S.F.	0.34	1.70	2.04
8' span	"	0.40	1.55	1.95

03 CONCRETE

GROUT	UNIT	MAT.	INST.	TOTAL
03600.10 GROUTING				
Grouting for bases				
Nonshrink				
Metallic grout				
1" deep	S.F.	3.60	5.10	8.70
2" deep	"	7.20	5.65	12.85
Non-metallic grout				
1" deep	S.F.	2.35	5.10	7.45
2" deep	"	4.55	5.65	10.20
Fluid type				
Non-metallic				
1" deep	S.F.	2.80	5.10	7.90
2" deep	"	5.30	5.65	10.95
Grouting for joints				
Portland cement grout (1 cement to 3 sand, by volume)				
1/2" joint thickness				
6" wide joints	L.F.	0.07	0.85	0.92
8" wide joints	"	0.09	1.00	1.09
1" joint thickness				
4" wide joints	L.F.	0.09	0.79	0.88
6" wide joints	"	0.14	0.88	1.02
8" wide joints	"	0.18	1.05	1.23
Nonshrink, nonmetallic grout				
1/2" joint thickness				
4" wide joint	L.F.	0.42	0.73	1.15
6" wide joint	"	0.61	0.85	1.46
8" wide joint	"	0.79	1.00	1.79
1" joint thickness				
4" wide joint	L.F.	0.79	0.79	1.58
6" wide joint	"	1.20	0.88	2.08
8" wide joint	"	1.60	1.05	2.65

CONCRETE RESTORATION	UNIT	MAT.	INST.	TOTAL
03730.10 CONCRETE REPAIR				
Epoxy grout floor patch, 1/4" thick	S.F.	3.35	2.05	5.40
Grout, epoxy, 2 component system	C.F.			160.00
Epoxy sand	BAG			10.45
Epoxy modifier	GAL			69.50
Epoxy gel grout	S.F.	1.55	20.40	21.95
Injection valve, 1 way, threaded plastic	EA.	4.45	4.10	8.55
Grout crack seal, 2 component	C.F.	393.00	20.40	413.40
Grout, non shrink	"	37.00	20.40	57.40
Concrete, epoxy modified				
Sand mix	C.F.	59.50	8.15	67.65

CONCRETE RESTORATION

03730.10 CONCRETE REPAIR

CONCRETE RESTORATION	UNIT	MAT.	INST.	TOTAL
Gravel mix	C.F.	43.80	7.55	51.35
Concrete repair				
Soffit repair				
16" wide	L.F.	1.90	4.10	6.00
18" wide	"	2.00	4.25	6.25
24" wide	"	2.35	4.55	6.90
30" wide	"	2.65	4.85	7.50
32" wide	"	2.80	5.10	7.90
Edge repair				
2" spall	L.F.	0.90	5.10	6.00
3" spall	"	0.94	5.35	6.29
4" spall	"	0.97	5.50	6.47
6" spall	"	1.00	5.65	6.65
8" spall	"	1.05	6.00	7.05
9" spall	"	1.05	6.80	7.85
Crack repair, 1/8" crack	"	1.70	2.05	3.75
Reinforcing steel repair				
1 bar, 4 ft				
#4 bar	L.F.	0.22	3.65	3.87
#5 bar	"	0.34	3.65	3.99
#6 bar	"	0.45	3.90	4.35
#8 bar	"	0.78	3.90	4.68
#9 bar	"	1.00	4.20	5.20
#11 bar	"	1.55	4.20	5.75
Form fabric, nylon				
18" diameter	L.F.			6.85
20" diameter	"			7.00
24" diameter	"			11.55
30" diameter	"			12.00
36" diameter	"			13.75
Pile repairs				
Polyethylene wrap				
30 mil thick				
60" wide	S.F.	7.60	6.80	14.40
72" wide	"	8.15	8.15	16.30
60 mil thick				
60" wide	S.F.	9.20	6.80	16.00
80" wide	"	10.45	9.25	19.70
Pile spall, average repair 3'				
18" x 18"	EA.	22.45	17.00	39.45
20" x 20"	"	28.10	20.40	48.50

MORTAR AND GROUT

MORTAR AND GROUT	UNIT	MAT.	INST.	TOTAL
04100.10 MASONRY GROUT				
Grout, non shrink, non-metallic, trowelable	C.F.	6.75	0.77	7.52
Grout door frame, hollow metal				
Single	EA.	10.10	27.50	37.60
Double	"	14.60	28.90	43.50
Grout-filled concrete block (CMU)				
4" wide	S.F.	0.29	0.92	1.21
6" wide	"	0.61	1.00	1.61
8" wide	"	1.15	1.10	2.25
12" wide	"	1.35	1.15	2.50
Grout-filled individual cmu cells				
4" wide	L.F.	0.13	0.57	0.70
6" wide	"	0.29	0.55	0.84
8" wide	"	0.40	0.57	0.97
10" wide	"	0.54	0.63	1.17
12" wide	"	0.62	0.63	1.25
Bond beams or lintels, 8" deep				
6" thick	L.F.	0.62	0.87	1.49
8" thick	"	0.80	0.96	1.76
10" thick	"	1.00	1.05	2.05
12" thick	"	1.20	1.20	2.40
Cavity walls				
2" thick	S.F.	0.68	1.30	1.98
3" thick	"	1.00	1.30	2.30
4" thick	"	1.35	1.35	2.70
6" thick	"	2.05	1.60	3.65
04150.10 MASONRY ACCESSORIES				
Foundation vents	EA.	15.45	10.05	25.50
Bar reinforcing				
Horizontal				
#3 - #4	Lb.	0.33	1.00	1.33
#5 - #6	"	0.31	0.84	1.15
Vertical				
#3 - #4	Lb.	0.33	1.25	1.58
#5 - #6	"	0.31	1.00	1.31
Horizontal joint reinforcing				
Truss type				
4" wide, 6" wall	L.F.	0.09	0.10	0.19
6" wide, 8" wall	"	0.11	0.10	0.21
8" wide, 10" wall	"	0.12	0.11	0.23
10" wide, 12" wall	"	0.13	0.11	0.24
12" wide, 14" wall	"	0.17	0.12	0.29
Ladder type				
4" wide, 6" wall	L.F.	0.07	0.10	0.17
6" wide, 8" wall	"	0.08	0.10	0.18
8" wide, 10" wall	"	0.09	0.11	0.20
10" wide, 12" wall	"	0.10	0.11	0.21
Rectangular wall ties				
3/16" dia., galvanized				
2" x 6"	EA.	0.14	0.42	0.56
2" x 8"	"	0.16	0.42	0.58

MORTAR AND GROUT

04150.10 MASONRY ACCESSORIES

	UNIT	MAT.	INST.	TOTAL
2" x 10"	EA.	0.17	0.42	0.59
2" x 12"	"	0.20	0.42	0.62
4" x 6"	"	0.15	0.50	0.65
4" x 8"	"	0.18	0.50	0.68
4" x 10"	"	0.23	0.50	0.73
4" x 12"	"	0.26	0.50	0.76
1/4" dia., galvanized				
2" x 6"	EA.	0.25	0.42	0.67
2" x 8"	"	0.28	0.42	0.70
2" x 10"	"	0.32	0.42	0.74
2" x 12"	"	0.36	0.42	0.78
4" x 6"	"	0.29	0.50	0.79
4" x 8"	"	0.32	0.50	0.82
4" x 10"	"	0.36	0.50	0.86
4" x 12"	"	0.38	0.50	0.88
"Z" type wall ties, galvanized				
6" long				
1/8" dia.	EA.	0.14	0.42	0.56
3/16" dia.	"	0.14	0.42	0.56
1/4" dia.	"	0.15	0.42	0.57
8" long				
1/8" dia.	EA.	0.16	0.42	0.58
3/16" dia.	"	0.15	0.42	0.57
1/4" dia.	"	0.17	0.42	0.59
10" long				
1/8" dia.	EA.	0.15	0.42	0.57
3/16" dia.	"	0.18	0.42	0.60
1/4" dia.	"	0.21	0.42	0.63
Dovetail anchor slots				
Galvanized steel, filled				
24 ga.	L.F.	0.36	0.63	0.99
20 ga.	"	0.47	0.63	1.10
16 oz. copper, foam filled	"	0.96	0.63	1.59
Dovetail anchors				
16 ga.				
3-1/2" long	EA.	0.10	0.42	0.52
5-1/2" long	"	0.12	0.42	0.54
12 ga.				
3-1/2" long	EA.	0.13	0.42	0.55
5-1/2" long	"	0.28	0.42	0.70
Dovetail, triangular galvanized ties, 12 ga.				
3" x 3"	EA.	0.25	0.42	0.67
5" x 5"	"	0.28	0.42	0.70
7" x 7"	"	0.31	0.42	0.73
7" x 9"	"	0.33	0.42	0.75
Brick anchors				
Corrugated, 3-1/2" long				
16 ga.	EA.	0.13	0.42	0.55
12 ga.	"	0.20	0.42	0.62
Non-corrugated, 3-1/2" long				
16 ga.	EA.	0.15	0.42	0.57
12 ga.	"	0.26	0.42	0.68

04 MASONRY

MORTAR AND GROUT	UNIT	MAT.	INST.	TOTAL
04150.10 MASONRY ACCESSORIES				
Cavity wall anchors, corrugated, galvanized				
5" long				
16 ga.	EA.	0.31	0.42	0.73
12 ga.	"	0.47	0.42	0.89
7" long				
28 ga.	EA.	0.34	0.42	0.76
24 ga.	"	0.43	0.42	0.85
22 ga.	"	0.45	0.42	0.87
16 ga.	"	0.52	0.42	0.94
Mesh ties, 16 ga., 3" wide				
8" long	EA.	0.36	0.42	0.78
12" long	"	0.42	0.42	0.84
20" long	"	0.53	0.42	0.95
24" long	"	0.61	0.42	1.03
04150.20 MASONRY CONTROL JOINTS				
Control joint, cross shaped PVC	L.F.	2.20	0.63	2.83
Closed cell joint filler				
1/2"	L.F.	0.32	0.63	0.95
3/4"	"	0.62	0.63	1.25
Rubber, for				
4" wall	L.F.	1.95	0.63	2.58
6" wall	"	2.95	0.66	3.61
8" wall	"	3.20	0.70	3.90
PVC, for				
4" wall	L.F.	1.90	0.63	2.53
6" wall	"	2.30	0.66	2.96
8" wall	"	2.45	0.70	3.15
04150.50 MASONRY FLASHING				
Through-wall flashing				
5 oz. coated copper	S.F.	2.30	2.20	4.50
0.030" elastomeric	"	0.50	1.75	2.25

UNIT MASONRY	UNIT	MAT.	INST.	TOTAL
04210.10 BRICK MASONRY				
Standard size brick, running bond				
Face brick, red (6.4/sf)				
Veneer	S.F.	1.85	4.40	6.25
Cavity wall	"	1.85	3.75	5.60
9" solid wall	"	4.05	7.50	11.55
Common brick (6.4/sf)				

UNIT MASONRY

04210.10 BRICK MASONRY	UNIT	MAT.	INST.	TOTAL
Select common for veneers	S.F.	2.15	4.20	6.35
Back-up				
4" thick	S.F.	1.80	3.15	4.95
8" thick	"	3.55	5.05	8.60
Firewall				
12" thick	S.F.	5.35	8.40	13.75
16" thick	"	7.15	11.45	18.60
Glazed brick (7.4/sf)				
Veneer	S.F.	6.85	4.55	11.40
Buff or gray face brick (6.4/sf)				
Veneer	S.F.	1.80	4.20	6.00
Cavity wall	"	1.80	3.60	5.40
Jumbo or oversize brick (3/sf)				
4" veneer	S.F.	3.50	2.50	6.00
4" back-up	"	3.50	2.10	5.60
8" back-up	"	7.05	3.60	10.65
12" firewall	"	10.55	6.30	16.85
16" firewall	"	14.10	8.40	22.50
Norman brick, red face, (4.5/sf)				
4" veneer	S.F.	2.95	3.15	6.10
Cavity wall	"	2.55	2.80	5.35
Chimney, standard brick, including flue				
16" x 16"	L.F.	13.40	26.30	39.70
16" x 20"	"	16.35	26.30	42.65
16" x 24"	"	19.00	26.30	45.30
20" x 20"	"	19.15	32.90	52.05
20" x 24"	"	27.50	32.90	60.40
20" x 32"	"	34.70	37.60	72.30
Window sill, face brick on edge	"	2.10	6.30	8.40

04210.20 STRUCTURAL TILE	UNIT	MAT.	INST.	TOTAL
Structural glazed tile				
6T series, 5-1/2" x 12"				
Glazed on one side				
2" thick	S.F.	3.95	2.50	6.45
4" thick	"	4.95	2.50	7.45
6" thick	"	7.15	2.80	9.95
8" thick	"	9.80	3.15	12.95
Glazed on two sides				
4" thick	S.F.	7.70	3.15	10.85
6" thick	"	9.90	3.60	13.50
Special shapes				
Group 1	S.F.	3.95	5.05	9.00
Group 2	"	5.50	5.05	10.55
Group 3	"	6.85	5.05	11.90
Group 4	"	15.10	5.05	20.15
Group 5	"	18.15	5.05	23.20
Fire rated				
4" thick, 1 hr rating	S.F.	11.00	2.50	13.50
6" thick, 2 hr rating	"	14.30	2.80	17.10
8W series, 8" x 16"				

04 MASONRY

UNIT MASONRY	UNIT	MAT.	INST.	TOTAL
04210.20 STRUCTURAL TILE				
Glazed on one side				
2" thick	S.F.	4.30	1.70	6.00
4" thick	"	4.70	1.70	6.40
6" thick	"	6.65	1.95	8.60
8" thick	"	7.90	1.95	9.85
Glazed on two sides				
4" thick	S.F.	8.25	2.10	10.35
6" thick	"	10.75	2.50	13.25
8" thick	"	14.05	2.50	16.55
Special shapes				
Group 1	S.F.	6.40	3.60	10.00
Group 2	"	8.15	3.60	11.75
Group 3	"	9.00	3.60	12.60
Group 4	"	13.40	3.60	17.00
Group 5	"	36.40	3.60	40.00
Fire rated				
4" thick, 1 hr rating	S.F.	10.25	3.60	13.85
6" thick, 2 hr rating	"	14.20	3.60	17.80
04210.60 PAVERS, MASONRY				
Brick walk laid on sand, sand joints				
Laid flat, (4.5 per sf)	S.F.	1.20	2.80	4.00
Laid on edge, (7.2 per sf)	"	1.80	4.20	6.00
Precast concrete patio blocks				
2" thick				
Natural	S.F.	0.90	0.84	1.74
Colors	"	1.15	0.84	1.99
Exposed aggregates, local aggregate				
Natural	S.F.	0.97	0.84	1.81
Colors	"	1.10	0.84	1.94
Granite or limestone aggregate	"	0.86	0.84	1.70
White tumblestone aggregate	"	1.10	0.84	1.94
Stone pavers, set in mortar				
Bluestone				
1" thick				
Irregular	S.F.	1.70	6.30	8.00
Snapped rectangular	"	2.95	5.05	8.00
1-1/2" thick, random rectangular	"	2.85	6.30	9.15
2" thick, random rectangular	"	3.20	7.20	10.40
Slate				
Natural cleft				
Irregular, 3/4" thick	S.F.	1.55	7.20	8.75
Random rectangular				
1-1/4" thick	S.F.	3.85	6.30	10.15
1-1/2" thick	"	4.30	7.00	11.30
Granite blocks				
3" thick, 3" to 6" wide				
4" to 12" long	S.F.	3.10	8.40	11.50
6" to 15" long	"	2.10	7.20	9.30
Crushed stone, white marble, 3" thick	"	0.60	0.41	1.01

UNIT MASONRY

04220.10 CONCRETE MASONRY UNITS

	UNIT	MAT.	INST.	TOTAL
Hollow, load bearing				
4"	S.F.	1.05	1.95	3.00
6"	"	1.45	1.95	3.40
8"	"	1.85	2.20	4.05
10"	"	2.45	2.30	4.75
12"	"	2.85	2.55	5.40
Solid, load bearing				
4"	S.F.	1.55	1.95	3.50
6"	"	1.85	1.95	3.80
8"	"	2.10	2.20	4.30
10"	"	2.90	2.30	5.20
12"	"	3.30	2.55	5.85
Back-up block, 8" x 16"				
2"	S.F.	0.83	1.45	2.28
4"	"	1.10	1.55	2.65
6"	"	1.50	1.60	3.10
8"	"	1.60	1.75	3.35
10"	"	2.15	1.80	3.95
12"	"	2.25	1.95	4.20
Foundation wall, 8" x 16"				
6"	S.F.	1.45	1.80	3.25
8"	"	1.65	1.95	3.60
10"	"	2.40	2.10	4.50
12"	"	2.45	2.30	4.75
Solid				
6"	S.F.	2.05	1.95	4.00
8"	"	2.45	2.10	4.55
10"	"	2.70	2.30	5.00
12"	"	3.00	2.50	5.50
Exterior, styrofoam inserts, standard weight, 8" x 16"				
6"	S.F.	2.05	1.95	4.00
8"	"	2.30	2.10	4.40
10"	"	3.05	2.30	5.35
12"	"	3.15	2.50	5.65
Lightweight				
6"	S.F.	2.25	1.95	4.20
8"	"	2.70	2.10	4.80
10"	"	3.45	2.30	5.75
12"	"	3.65	2.50	6.15
Acoustical slotted block				
4"	S.F.	2.20	2.30	4.50
6"	"	2.80	2.30	5.10
8"	"	4.55	2.50	7.05
Filled cavities				
4"	S.F.	2.95	2.80	5.75
6"	"	3.65	2.95	6.60
8"	"	5.45	3.15	8.60
Hollow, split face				
4"	S.F.	3.40	1.85	5.25
6"	"	3.95	1.95	5.90
8"	"	4.95	2.10	7.05
10"	"	5.45	2.30	7.75

UNIT MASONRY

04220.10 CONCRETE MASONRY UNITS

	UNIT	MAT.	INST.	TOTAL
12"	S.F.	5.90	2.50	8.40
Split rib profile				
4"	S.F.	1.85	2.30	4.15
6"	"	2.35	2.30	4.65
8"	"	3.25	2.50	5.75
10"	"	3.50	2.50	6.00
12"	"	3.85	2.50	6.35
High strength block, 3500 psi				
2"	S.F.	0.99	1.85	2.84
4"	"	1.35	1.95	3.30
6"	"	1.65	1.95	3.60
8"	"	1.85	2.10	3.95
10"	"	2.15	2.30	4.45
12"	"	2.55	2.50	5.05
Solar screen concrete block				
4" thick				
6" x 6"	S.F.	3.70	5.60	9.30
8" x 8"	"	3.05	5.05	8.10
12" x 12"	"	2.25	3.85	6.10
8" thick				
8" x 16"	S.F.	2.50	3.60	6.10
Glazed block, cove base, glazed 1 side				
2"	L.F.	5.25	2.80	8.05
4"	"	5.55	2.80	8.35
6"	"	5.95	3.15	9.10
8"	"	6.45	3.15	9.60
Single face				
2"	S.F.	5.45	2.10	7.55
4"	"	5.90	2.10	8.00
6"	"	6.45	2.30	8.75
8"	"	7.05	2.50	9.55
10"	"	7.80	2.80	10.60
12"	"	8.60	2.95	11.55
Double face				
4"	S.F.	8.95	2.65	11.60
6"	"	9.65	2.80	12.45
8"	"	10.20	3.15	13.35
Corner or bullnose				
2"	EA.	6.80	3.15	9.95
4"	"	8.50	3.60	12.10
6"	"	10.20	3.60	13.80
8"	"	11.20	4.20	15.40
10"	"	12.45	4.55	17.00
12"	"	14.60	5.05	19.65
Gypsum unit masonry				
Partition blocks (12"x30")				
Solid				
2"	S.F.	0.62	1.00	1.62
Hollow				
3"	S.F.	0.65	1.00	1.65
4"	"	0.74	1.05	1.79
6"	"	0.78	1.15	1.93

04 MASONRY

UNIT MASONRY	UNIT	MAT.	INST.	TOTAL
04220.10 CONCRETE MASONRY UNITS				
Vertical reinforcing				
4' o.c., add 5% to labor				
2'8" o.c., add 15% to labor				
Interior partitions, add 10% to labor				
04220.90 BOND BEAMS & LINTELS				
Bond beam, no grout or reinforcement				
8" x 16" x				
4" thick	L.F.	1.35	1.95	3.30
6" thick	"	1.70	2.00	3.70
8" thick	"	2.05	2.10	4.15
10" thick	"	2.20	2.20	4.40
12" thick	"	2.20	2.30	4.50
Beam lintel, no grout or reinforcement				
8" x 16" x				
10" thick	L.F.	3.85	2.50	6.35
12" thick	"	3.85	2.80	6.65
Precast masonry lintel				
6 lf, 8" high x				
4" thick	L.F.	3.50	4.20	7.70
6" thick	"	4.70	4.20	8.90
8" thick	"	5.20	4.55	9.75
10" thick	"	6.35	4.55	10.90
10 lf, 8" high x				
4" thick	L.F.	4.60	2.65	7.25
6" thick	"	5.70	2.50	8.20
8" thick	"	6.35	2.80	9.15
10" thick	"	8.70	2.80	11.50
Steel angles and plates				
Minimum	Lb.	0.28	0.38	0.66
Maximum	"	0.31	0.66	0.97
Various size angle lintels				
1/4" stock				
3" x 3"	L.F.	1.50	1.65	3.15
3" x 3-1/2"	"	1.60	1.65	3.25
3/8" stock				
3" x 4"	L.F.	2.65	1.65	4.30
3-1/2" x 4"	"	2.80	1.55	4.35
4" x 4"	"	2.90	1.65	4.55
5" x 3-1/2"	"	3.20	1.55	4.75
6" x 3-1/2"	"	3.40	1.55	4.95
1/2" stock				
6" x 4"	L.F.	4.85	1.65	6.50
04240.10 CLAY TILE				
Hollow clay tile, for back-up, 12" x 12"				
Scored face				
Load bearing				
4" thick	S.F.	2.20	1.80	4.00
6" thick	"	2.65	1.85	4.50
8" thick	"	3.40	1.95	5.35

04 MASONRY

UNIT MASONRY

UNIT MASONRY	UNIT	MAT.	INST.	TOTAL
04240.10 CLAY TILE				
10" thick	S.F.	3.65	2.00	5.65
12" thick	"	4.95	2.10	7.05
Non-load bearing				
3" thick	S.F.	1.20	1.75	2.95
4" thick	"	2.00	1.80	3.80
6" thick	"	2.40	1.85	4.25
8" thick	"	3.20	1.95	5.15
12" thick	"	4.75	2.10	6.85
Partition, 12" x 12"				
In walls				
3" thick	S.F.	1.65	2.20	3.85
4" thick	"	2.15	2.10	4.25
6" thick	"	2.75	2.20	4.95
8" thick	"	3.75	2.30	6.05
10" thick	"	4.85	2.40	7.25
12" thick	"	5.35	2.50	7.85
Clay tile floors				
4" thick	S.F.	2.15	1.40	3.55
6" thick	"	2.75	1.50	4.25
8" thick	"	3.75	1.55	5.30
10" thick	"	4.85	1.70	6.55
12" thick	"	5.35	1.80	7.15
Terra cotta				
Coping, 10" or 12" wide, 3" thick	L.F.	15.40	5.05	20.45
04270.10 GLASS BLOCK				
Glass block, 4" thick				
6" x 6"	S.F.	24.20	8.40	32.60
8" x 8"	"	20.90	6.30	27.20
12" x 12"	"	17.60	5.05	22.65
Replacement glass blocks, 4" x 8" x 8"				
Minimum	S.F.	24.20	25.20	49.40
Maximum	"	28.60	50.50	79.10
04295.10 PARGING/MASONRY PLASTER				
Parging				
1/2" thick	S.F.	0.33	1.75	2.08
3/4" thick	"	0.44	2.20	2.64
1" thick	"	0.55	2.50	3.05

04 MASONRY

STONE	UNIT	MAT.	INST.	TOTAL
04400.10 STONE				
Rubble stone				
Walls set in mortar				
8" thick	S.F.	9.00	6.30	15.30
12" thick	"	9.90	10.05	19.95
18" thick	"	13.20	12.60	25.80
24" thick	"	18.00	16.85	34.85
Dry set wall				
8" thick	S.F.	10.20	4.20	14.40
12" thick	"	10.45	6.30	16.75
18" thick	"	14.30	8.40	22.70
24" thick	"	19.20	10.10	29.30
Cut stone				
Imported marble				
Facing panels				
3/4" thick	S.F.	24.00	10.10	34.10
1-1/2" thick	"	36.00	11.50	47.50
2-1/4" thick	"	38.50	14.00	52.50
Base				
1" thick				
4" high	L.F.	9.90	12.60	22.50
6" high	"	12.10	12.60	24.70
Columns, solid				
Plain faced	C.F.	55.00	503.00	558.00
Fluted	"	165.00	503.00	668.00
Flooring, travertine, minimum	S.F.	7.15	3.85	11.00
Average	"	9.35	5.05	14.40
Maximum	"	12.10	5.60	17.70
Domestic marble				
Facing panels				
7/8" thick	S.F.	19.80	10.05	29.85
1-1/2" thick	"	29.70	11.45	41.15
2-1/4" thick	"	36.30	14.00	50.30
Stairs				
12" treads	L.F.	17.60	12.60	30.20
6" risers	"	13.20	8.40	21.60
Thresholds, 7/8" thick, 3' long, 4" to 6" wide				
Plain	EA.	16.50	20.95	37.45
Beveled	"	18.70	20.95	39.65
Window sill				
6" wide, 2" thick	L.F.	8.60	10.05	18.65
Stools				
5" wide, 7/8" thick	L.F.	12.10	10.05	22.15
Limestone panels up to 12' x 5', smooth finish				
2" thick	S.F.	14.85	4.40	19.25
3" thick	"	19.00	4.40	23.40
4" thick	"	25.10	4.40	29.50
Miscellaneous limestone items				
Steps, 14" wide, 6" deep	L.F.	43.50	16.75	60.25
Coping, smooth finish	C.F.	60.50	8.40	68.90
Sills, lintels, jambs, smooth finish	"	60.50	10.05	70.55
Granite veneer facing panels, polished				
7/8" thick				

04 MASONRY

STONE	UNIT	MAT.	INST.	TOTAL
04400.10 STONE				
Black	S.F.	23.65	10.05	33.70
Gray	"	20.10	10.05	30.15
Base				
4" high	L.F.	11.00	5.05	16.05
6" high	"	13.20	5.60	18.80
Curbing, straight, 6" x 16"	"	12.10	18.35	30.45
Radius curbs, radius over 5'	"	14.85	24.45	39.30
Ashlar veneer				
4" thick, random	S.F.	33.00	10.05	43.05
Pavers, 4" x 4" split				
Gray	S.F.	19.80	5.25	25.05
Pink	"	19.60	5.05	24.65
Black	"	18.90	5.05	23.95
Slate, panels				
1" thick	S.F.	10.45	10.05	20.50
2" thick	"	17.35	11.45	28.80
Sills or stools				
1" thick				
6" wide	L.F.	6.60	10.05	16.65
10" wide	"	9.90	10.95	20.85
2" thick				
6" wide	L.F.	9.35	11.45	20.80
10" wide	"	17.60	12.60	30.20

MASONRY RESTORATION	UNIT	MAT.	INST.	TOTAL
04520.10 RESTORATION AND CLEANING				
Masonry cleaning				
Washing brick				
Smooth surface	S.F.	0.24	0.44	0.68
Rough surface	"	0.33	0.58	0.91
Steam clean masonry				
Smooth face				
Minimum	S.F.	0.00	0.32	0.32
Maximum	"	0.00	0.46	0.46
Rough face				
Minimum	S.F.	0.00	0.42	0.42
Maximum	"	0.00	0.63	0.63
Sandblast masonry				
Minimum	S.F.	0.29	0.51	0.80
Maximum	"	0.45	0.85	1.30
Pointing masonry				
Brick	S.F.	0.77	1.05	1.82

MASONRY RESTORATION

04520.10 RESTORATION AND CLEANING	UNIT	MAT.	INST.	TOTAL
Concrete block	S.F.	0.33	0.75	1.08
Cut and repoint				
Brick				
Minimum	S.F.	0.22	1.25	1.47
Maximum	"	0.36	2.50	2.86
Stone work	L.F.	1.00	1.95	2.95
Cut and recaulk				
Oil base caulks	L.F.	0.75	1.70	2.45
Butyl caulks	"	0.69	1.70	2.39
Polysulfides and acrylics	"	1.35	1.70	3.05
Silicones	"	1.60	1.70	3.30
Cement and sand grout on walls, to 1/8" thick				
Minimum	S.F.	0.32	1.00	1.32
Maximum	"	0.90	1.25	2.15
Brick removal and replacement				
Minimum	EA.	0.33	3.30	3.63
Average	"	0.44	4.40	4.84
Maximum	"	0.88	13.15	14.03
04550.10 REFRACTORIES				
Flue liners				
Rectangular				
8" x 12"	L.F.	4.20	4.40	8.60
12" x 12"	"	6.05	4.80	10.85
12" x 18"	"	9.35	5.25	14.60
16" x 16"	"	10.75	5.85	16.60
18" x 18"	"	13.40	6.25	19.65
20" x 20"	"	21.45	6.60	28.05
24" x 24"	"	28.60	7.50	36.10
Round				
18" dia.	L.F.	26.40	6.25	32.65
24" dia.	"	45.10	7.50	52.60

METAL FASTENING

	UNIT	MAT.	INST.	TOTAL
05050.10 STRUCTURAL WELDING				
Welding				
Single pass				
1/8"	L.F.	0.34	1.50	1.84
3/16"	"	0.68	2.00	2.68
1/4"	"	0.91	2.45	3.36
Miscellaneous steel shapes				
Plain	Lb.	0.68	0.06	0.74
Galvanized	"	1.05	0.10	1.15
Plates				
Plain	Lb.	0.63	0.07	0.70
Galvanized	"	0.94	0.12	1.06
05050.90 METAL ANCHORS				
Anchor bolts				
3/8" x				
8" long	EA.			2.75
10" long	"			2.95
12" long	"			3.20
1/2" x				
8" long	EA.			3.85
10" long	"			3.95
12" long	"			4.30
18" long	"			4.60
5/8" x				
8" long	EA.			3.50
10" long	"			3.85
12" long	"			4.05
18" long	"			4.30
24" long	"			4.60
3/4" x				
8" long	EA.			5.05
12" long	"			5.60
18" long	"			7.35
24" long	"			10.00
7/8" x				
8" long	EA.			4.95
12" long	"			5.40
18" long	"			8.25
24" long	"			10.35
1" x				
12" long	EA.			8.25
18" long	"			11.20
24" long	"			14.00
36" long	"			21.45
Expansion shield				
1/4"	EA.			0.44
3/8"	"			0.72
1/2"	"			1.15
5/8"	"			1.75
3/4"	"			2.85
1"	"			3.20

05 METALS

METAL FASTENING

METAL FASTENING	UNIT	MAT.	INST.	TOTAL
05050.90 METAL ANCHORS				
Non-drilling anchor				
1/4"	EA.			0.33
3/8"	"			0.45
1/2"	"			0.70
5/8"	"			1.20
3/4"	"			2.00
Self-drilling anchor				
1/4"	EA.			0.44
5/16"	"			0.61
3/8"	"			0.68
1/2"	"			1.00
5/8"	"			1.75
3/4"	"			3.20
7/8"	"			5.15
Add 25% for galvanized anchor bolts				
Channel door frame, with anchors	Lb.	1.00	0.32	1.32
Corner guard angle, with anchors	"	0.90	0.48	1.38
05050.95 METAL LINTELS				
Lintels, steel				
Plain	Lb.	0.68	0.74	1.42
Galvanized	"	0.99	0.74	1.73
05120.10 STRUCTURAL STEEL				
Beams and girders, A-36				
Welded	TON	1,130	305.00	1,435
Bolted	"	1,110	277.00	1,387
Columns				
Pipe				
6" dia.	Lb.	0.69	0.31	1.00
12" dia.	"	0.58	0.25	0.83
Purlins and girts				
Welded	TON	1,390	508.00	1,898
Bolted	"	1,360	436.00	1,796
Column base plates				
Up to 150 lb each	Lb.	0.76	0.19	0.95
Over 150 lb each	"	0.65	0.24	0.89
Structural pipe				
3" to 5" o.d.	TON	1,500	610.00	2,110
6" to 12" o.d.	"	1,390	436.00	1,826
Structural tube				
6" square				
Light sections	TON	1,790	610.00	2,400
Heavy sections	"	1,670	436.00	2,106
6" wide rectangular				
Light sections	TON	1,790	508.00	2,298
Heavy sections	"	1,670	381.00	2,051
Greater than 6" wide rectangular				
Light sections	TON	1,960	508.00	2,468
Heavy sections	"	1,850	381.00	2,231

METAL FASTENING

METAL FASTENING	UNIT	MAT.	INST.	TOTAL
05120.10 STRUCTURAL STEEL				
Miscellaneous structural shapes				
Steel angle	TON	1,270	763.00	2,033
Steel plate	"	1,440	508.00	1,948
Trusses, field welded				
60 lb/lf	TON	1,850	381.00	2,231
100 lb/lf	"	1,620	305.00	1,925
150 lb/lf	"	1,520	254.00	1,774
Bolted				
60 lb/lf	TON	1,820	339.00	2,159
100 lb/lf	"	1,590	277.00	1,867
150 lb/lf	"	1,510	235.00	1,745
Add for galvanizing	"			330.00
05200.10 METAL JOISTS				
Joist				
DLH series	TON	797.00	203.00	1,000
K series	"	774.00	203.00	977.00
LH series	"	785.00	203.00	988.00
05300.10 METAL DECKING				
Roof, 1-1/2" deep, non-composite				
16 ga.				
Primed	S.F.	1.60	0.51	2.11
Galvanized	"	1.65	0.51	2.16
18 ga.				
Primed	S.F.	1.15	0.51	1.66
Galvanized	"	1.30	0.51	1.81
20 ga.				
Primed	S.F.	0.94	0.51	1.45
Galvanized	"	1.05	0.51	1.56
22 ga.				
Primed	S.F.	0.71	0.51	1.22
Galvanized	"	0.88	0.51	1.39
Open type decking, galvanized				
1-1/2" deep				
18 ga.	S.F.	1.65	0.51	2.16
20 ga.	"	1.30	0.51	1.81
22 ga.	"	1.10	0.51	1.61
3" deep				
16 ga.	S.F.	3.85	0.55	4.40
18 ga.	"	3.30	0.55	3.85
20 ga.	"	2.80	0.55	3.35
22 ga.	"	1.85	0.55	2.40
4-1/2" deep				
16 ga.	S.F.	4.35	0.61	4.96
18 ga.	"	3.75	0.61	4.36
6" deep				
16 ga.	S.F.	4.60	0.68	5.28
18 ga.	"	3.85	0.68	4.53
7-1/2" deep				
16 ga.	S.F.	4.95	0.71	5.66
18 ga.	"	4.05	0.71	4.76

05 METALS

METAL FASTENING

05300.10 METAL DECKING	UNIT	MAT.	INST.	TOTAL
Cellular type				
1-1/2" deep, galvanized				
18-18 ga.	S.F.	4.25	0.61	4.86
22-18 ga.	"	3.90	0.61	4.51
3" deep, galvanized				
16-16 ga.	S.F.	5.50	0.68	6.18
18-16 ga.	"	5.50	0.68	6.18
18-18 ga.	"	4.60	0.68	5.28
20-18 ga.	"	4.20	0.68	4.88
4-1/2" deep, galvanized				
16-16 ga.	S.F.	8.60	0.73	9.33
18-16 ga.	"	8.60	0.73	9.33
18-18 ga.	"	7.15	0.73	7.88
20-18 ga.	"	5.95	0.73	6.68
Composite deck, non-cellular, galvanized				
1-1/2" deep				
18 ga.	S.F.	1.20	0.55	1.75
20 ga.	"	1.10	0.55	1.65
22 ga.	"	0.86	0.55	1.41
3" deep				
18 ga.	S.F.	1.65	0.59	2.24
20 ga.	"	1.00	0.59	1.59
22 ga.	"	0.86	0.59	1.45
Slab form floor deck				
9/16" deep				
28 ga.	S.F.	0.46	0.55	1.01
1-5/16" deep				
24 ga.	S.F.	0.73	0.58	1.31
22 ga.	"	0.92	0.58	1.50

COLD FORMED FRAMING

05410.10 METAL FRAMING	UNIT	MAT.	INST.	TOTAL
Furring channel, galvanized				
Beams and columns, 3/4"				
12" o.c.	S.F.	0.33	2.95	3.28
16" o.c.	"	0.28	2.70	2.98
Walls, 3/4"				
12" o.c.	S.F.	0.33	1.50	1.83
16" o.c.	"	0.28	1.25	1.53
24" o.c.	"	0.24	0.95	1.19
1-1/2"				
12" o.c.	S.F.	0.45	1.45	1.90

COLD FORMED FRAMING

05410.10 METAL FRAMING	UNIT	MAT.	INST.	TOTAL
16" o.c.	S.F.	0.36	1.20	1.56
24" o.c.	"	0.30	0.95	1.25
Stud, load bearing				
16" o.c.				
16 ga.				
2-1/2"	S.F.	1.05	1.30	2.35
3-5/8"	"	1.25	1.30	2.55
4"	"	1.25	1.25	2.50
6"	"	1.60	1.45	3.05
18 ga.				
2-1/2"	S.F.	0.78	1.25	2.03
3-5/8"	"	0.97	1.25	2.22
4"	"	1.05	1.25	2.30
6"	"	1.45	1.45	2.90
8"	"	1.60	1.45	3.05
20 ga.				
2-1/2"	S.F.	0.64	1.25	1.89
3-5/8"	"	0.71	1.25	1.96
4"	"	0.77	1.25	2.02
6"	"	0.99	1.45	2.44
8"	"	1.15	1.45	2.60
24" o.c.				
16 ga.				
2-1/2"	S.F.	0.77	1.15	1.92
3-5/8"	"	0.99	1.15	2.14
4"	"	0.82	1.10	1.92
6"	"	0.97	1.20	2.17
8"	"	1.00	1.20	2.20
18 ga.				
2-1/2"	S.F.	0.50	1.10	1.60
3-5/8"	"	0.58	1.10	1.68
4"	"	0.69	1.10	1.79
6"	"	0.80	1.20	2.00
8"	"	0.95	1.20	2.15
20 ga.				
2-1/2"	S.F.	0.42	1.10	1.52
3-5/8"	"	0.50	1.10	1.60
4"	"	0.62	1.10	1.72
6"	"	0.75	1.20	1.95
8"	"	0.84	1.20	2.04

05 METALS

METAL FABRICATIONS	UNIT	MAT.	INST.	TOTAL
05510.10 STAIRS				
Stock unit, steel, complete, per riser				
Tread				
3'-6" wide	EA.	76.50	35.70	112.20
4' wide	"	89.00	40.80	129.80
5' wide	"	103.00	47.60	150.60
Metal pan stair, cement filled, per riser				
3'-6" wide	EA.	87.50	28.60	116.10
4' wide	"	96.00	31.70	127.70
5' wide	"	107.00	35.70	142.70
Landing, steel pan	S.F.	31.20	7.15	38.35
Cast iron tread, steel stringers, stock units, per riser				
Tread				
3'-6" wide	EA.	133.00	35.70	168.70
4' wide	"	157.00	40.80	197.80
5' wide	"	184.00	47.60	231.60
Stair treads, abrasive, 12" x 3'-6"				
Cast iron				
3/8"	EA.	86.50	14.30	100.80
1/2"	"	98.00	14.30	112.30
Cast aluminum				
5/16"	EA.	98.00	14.30	112.30
3/8"	"	115.00	14.30	129.30
1/2"	"	144.00	14.30	158.30
05515.10 LADDERS				
Ladder, 18" wide		44.00	19.05	63.05
With cage	L.F.	50.50	19.80	70.30
Without cage	"	28.30	14.30	42.60
05520.10 RAILINGS				
Railing, pipe				
1-1/4" diameter, welded steel				
2-rail				
Primed	L.F.	16.20	5.70	21.90
Galvanized	"	19.50	5.70	25.20
3-rail				
Primed	L.F.	21.30	7.15	28.45
Galvanized	"	25.70	7.15	32.85
Wall mounted, single rail, welded steel				
Primed	L.F.	4.45	4.40	8.85
Galvanized	"	5.05	4.40	9.45
1-1/2" diameter, welded steel				
2-rail				
Primed	L.F.	19.75	5.70	25.45
Galvanized	"	23.65	5.70	29.35
3-rail				
Primed	L.F.	25.90	7.15	33.05
Galvanized	"	31.10	7.15	38.25
Wall mounted, single rail, welded steel				
Primed	L.F.	5.15	4.40	9.55

METAL FABRICATIONS	UNIT	MAT.	INST.	TOTAL
05520.10 RAILINGS				
Galvanized	L.F.	6.05	4.40	10.45
2" diameter, welded steel				
2-rail				
Primed	L.F.	22.75	6.35	29.10
Galvanized	"	25.80	6.35	32.15
3-rail				
Primed	L.F.	26.80	8.15	34.95
Galvanized	"	30.90	8.15	39.05
Wall mounted, single rail, welded steel				
Primed	L.F.	6.15	4.75	10.90
Galvanized	"	7.05	4.75	11.80
05530.10 METAL GRATING				
Floor plate, checkered, steel				
1/4"				
Primed	S.F.	4.05	0.41	4.46
Galvanized	"	5.80	0.41	6.21
3/8"				
Primed	S.F.	5.70	0.44	6.14
Galvanized	"	8.25	0.44	8.69
Aluminum grating, pressure-locked bearing bars				
3/4" x 1/8"	S.F.	10.75	0.71	11.46
1" x 1/8"	"	10.75	0.71	11.46
1-1/4" x 1/8"	"	11.55	0.71	12.26
1-1/4" x 3/16"	"	14.85	0.71	15.56
1-1/2" x 1/8"	"	12.65	0.71	13.36
1-3/4" x 3/16"	"	17.35	0.71	18.06
Miscellaneous expenses				
Cutting				
Minimum	L.F.	0.00	1.90	1.90
Maximum	"	0.00	2.85	2.85
Banding				
Minimum	L.F.	0.00	4.75	4.75
Maximum	"	0.00	5.70	5.70
Toe plates				
Minimum	L.F.	0.00	5.70	5.70
Maximum	"	0.00	7.15	7.15
Steel grating, primed				
3/4" x 1/8"	S.F.	4.70	0.95	5.65
1" x 1/8"	"	4.70	0.95	5.65
1-1/4" x 1/8"	"	5.00	0.95	5.95
1-1/4" x 3/16"	"	6.35	0.95	7.30
1-1/2" x 1/8"	"	5.80	0.95	6.75
1-3/4" x 3/16"	"	8.25	0.95	9.20
Galvanized				
3/4" x 1/8"	S.F.	5.80	0.95	6.75
1" x 1/8"	"	5.80	0.95	6.75
1-1/4" x 1/8"	"	6.25	0.95	7.20
1-1/4" x 3/16"	"	8.00	0.95	8.95
1-1/2" x 1/8"	"	7.15	0.95	8.10
1-3/4" x 3/16"	"	10.45	0.95	11.40

METAL FABRICATIONS

	UNIT	MAT.	INST.	TOTAL
05530.10 METAL GRATING				
Miscellaneous expenses				
Cutting				
Minimum	L.F.	0.00	2.05	2.05
Maximum	"	0.00	3.15	3.15
Banding				
Minimum	L.F.	0.00	5.20	5.20
Maximum	"	0.00	6.35	6.35
Toe plates				
Minimum	L.F.	0.00	6.35	6.35
Maximum	"	0.00	8.15	8.15
05540.10 CASTINGS				
Miscellaneous castings				
Light sections	Lb.	0.87	0.57	1.44
Heavy sections	"	0.76	0.41	1.17
Manhole covers and frames				
Regular, city type				
18" dia.				
100 lb	EA.	248.00	57.00	305.00
24" dia.				
200 lb	EA.	258.00	57.00	315.00
300 lb	"	264.00	63.50	327.50
400 lb	"	286.00	63.50	349.50
26" dia., 475 lb	"	308.00	71.50	379.50
30" dia., 600 lb	"	352.00	81.50	433.50
8" square, 75 lb	"	98.50	11.40	109.90
24" square				
126 lb	EA.	209.00	57.00	266.00
500 lb	"	352.00	71.50	423.50
Watertight type				
20" dia., 200 lb	EA.	176.00	71.50	247.50
24" dia., 350 lb	"	198.00	95.00	293.00
Steps, cast iron				
7" x 9"	EA.	8.25	4.75	13.00
8" x 9"	"	8.25	4.75	13.00
Manhole covers and frames, aluminum				
12" x 12"	EA.	40.20	11.40	51.60
18" x 18"	"	42.40	11.40	53.80
24" x 24"	"	45.70	14.30	60.00
Corner protection				
Steel angle guard with anchors				
2" x 2" x 3/16"	L.F.	12.10	4.10	16.20
2" x 3" x 1/4"	"	14.30	4.10	18.40
3" x 3" x 5/16"	"	18.70	4.10	22.80
3" x 4" x 5/16"	"	23.10	4.40	27.50
4" x 4" x 5/16"	"	23.10	4.40	27.50

MISC. FABRICATIONS

MISC. FABRICATIONS	UNIT	MAT.	INST.	TOTAL
05580.10 METAL SPECIALTIES				
Kick plate				
4" high x 1/4" thick				
Primed	L.F.	4.05	5.70	9.75
Galvanized	"	4.60	5.70	10.30
6" high x 1/4" thick				
Primed	L.F.	4.40	6.35	10.75
Galvanized	"	5.20	6.35	11.55
05700.10 ORNAMENTAL METAL				
Railings, vertical square bars, 6" o.c., with shaped top rails				
Steel	L.F.	34.00	14.30	48.30
Aluminum	"	53.00	14.30	67.30
Bronze	"	97.00	19.10	116.10
Stainless steel	"	80.50	19.10	99.60
Laminated metal or wood handrails with metal supports				
2-1/2" round or oval shape	L.F.	84.50	14.30	98.80
Grilles and louvers				
Fixed type louvers				
4 through 10 sf	S.F.	14.45	4.75	19.20
Over 10 sf	"	18.50	3.55	22.05
Movable type louvers				
4 through 10 sf	S.F.	17.90	4.75	22.65
Over 10 sf	"	19.90	3.55	23.45
Aluminum louvers				
Residential use, fixed type, with screen				
8" x 8"	EA.	15.30	14.30	29.60
12" x 12"	"	17.35	14.30	31.65
12" x 18"	"	19.65	14.30	33.95
14" x 24"	"	25.40	14.30	39.70
18" x 24"	"	31.20	14.30	45.50
30" x 24"	"	37.00	15.85	52.85
05800.10 EXPANSION CONTROL				
Expansion joints with covers, floor assembly type				
With 1" space				
Aluminum	L.F.	15.00	4.75	19.75
Bronze	"	30.00	4.75	34.75
Stainless steel	"	23.10	4.75	27.85
With 2" space				
Aluminum	L.F.	18.50	4.75	23.25
Bronze	"	32.30	4.75	37.05
Stainless steel	"	27.70	4.75	32.45
Ceiling and wall assembly type				
With 1" space				
Aluminum	L.F.	9.25	5.70	14.95
Bronze	"	32.30	5.70	38.00
Stainless steel	"	28.90	5.70	34.60
With 2" space				
Aluminum	L.F.	10.40	5.70	16.10
Bronze	"	34.70	5.70	40.40
Stainless steel	"	30.00	5.70	35.70

MISC. FABRICATIONS

05800.10 EXPANSION CONTROL

	UNIT	MAT.	INST.	TOTAL
Exterior roof and wall, aluminum				
Roof to roof				
With 1" space	L.F.	27.70	4.75	32.45
With 2" space	"	30.00	4.75	34.75
Roof to wall				
With 1" space	L.F.	21.95	5.20	27.15
With 2" space	"	25.40	5.20	30.60
Flat wall to wall				
With 1" space	L.F.	10.40	4.75	15.15
With 2" space	"	11.55	4.75	16.30
Corner to flat wall				
With 1" space	L.F.	12.70	5.70	18.40
With 2" in space	"	12.70	5.70	18.40

FASTENERS AND ADHESIVES

06050.10 ACCESSORIES

FASTENERS AND ADHESIVES	UNIT	MAT.	INST.	TOTAL
Column/post base, cast aluminum				
4" x 4"	EA.	1.80	6.50	8.30
6" x 6"	"	4.10	6.50	10.60
Bridging, metal, per pair				
12" o.c.	EA.	0.14	2.55	2.69
16" o.c.	"	0.14	2.35	2.49
Anchors				
Bolts, threaded two ends, with nuts and washers				
1/2" dia.				
4" long	EA.	0.83	1.70	2.53
7-1/2" long	"	1.05	1.70	2.75
3/4" dia.				
7-1/2" long	EA.	1.65	1.70	3.35
15" long	"	2.15	1.70	3.85
Framing anchors				
10 gauge	EA.	0.46	2.15	2.61
Bolts, carriage				
1/4 x 4	EA.	0.87	2.70	3.57
5/16 x 6	"	0.92	2.85	3.77
3/8 x 6	"	1.05	2.85	3.90
1/2 x 6	"	1.20	2.85	4.05
Joist and beam hangers				
18 ga.				
2 x 4	EA.	0.41	2.70	3.11
2 x 6	"	0.46	2.70	3.16
2 x 8	"	0.55	2.70	3.25
2 x 10	"	0.66	3.00	3.66
2 x 12	"	0.71	3.40	4.11
16 ga.				
3 x 6	EA.	0.78	2.85	3.63
3 x 8	"	1.30	2.85	4.15
3 x 10	"	1.45	3.05	4.50
3 x 12	"	1.55	3.45	5.00
3 x 14	"	1.80	3.65	5.45
4 x 6	"	1.60	2.85	4.45
4 x 8	"	1.75	2.85	4.60
4 x 10	"	1.95	3.05	5.00
4 x 12	"	2.20	3.45	5.65
4 x 14	"	2.30	3.65	5.95
Rafter anchors, 18 ga., 1-1/2" wide				
5-1/4" long	EA.	0.41	2.15	2.56
10-3/4" long	"	0.56	2.15	2.71
Shear plates				
2-5/8" dia.	EA.	1.30	2.00	3.30
4" dia.	"	2.75	2.15	4.90
Sill anchors				
Embedded in concrete	EA.	5.80	2.60	8.40
Split rings				
2-1/2" dia.	EA.	0.78	2.85	3.63
4" dia.	"	1.50	3.20	4.70
Strap ties, 14 ga., 1-3/8" wide				
12" long	EA.	0.88	2.25	3.13

06 WOOD AND PLASTICS

FASTENERS AND ADHESIVES

06050.10 ACCESSORIES

	UNIT	MAT.	INST.	TOTAL
18" long	EA.	1.05	2.45	3.50
24" long	"	1.25	2.70	3.95
36" long	"	1.80	3.00	4.80
Toothed rings				
2-5/8" dia.	EA.	0.65	4.30	4.95
4" dia.	"	0.74	5.15	5.89

ROUGH CARPENTRY

06110.10 BLOCKING

	UNIT	MAT.	INST.	TOTAL
Steel construction				
Walls				
2x4	L.F.	0.29	1.80	2.09
2x6	"	0.42	2.10	2.52
2x8	"	0.58	2.25	2.83
2x10	"	0.94	2.45	3.39
2x12	"	1.20	2.70	3.90
Ceilings				
2x4	L.F.	0.29	2.10	2.39
2x6	"	0.42	2.45	2.87
2x8	"	0.58	2.70	3.28
2x10	"	0.94	3.00	3.94
2x12	"	1.20	3.40	4.60
Wood construction				
Walls				
2x4	L.F.	0.29	1.50	1.79
2x6	"	0.42	1.70	2.12
2x8	"	0.58	1.80	2.38
2x10	"	0.94	1.95	2.89
2x12	"	1.20	2.10	3.30
Ceilings				
2x4	L.F.	0.29	1.70	1.99
2x6	"	0.42	1.95	2.37
2x8	"	0.58	2.10	2.68
2x10	"	0.94	2.25	3.19
2x12	"	1.20	2.45	3.65

06110.20 CEILING FRAMING

	UNIT	MAT.	INST.	TOTAL
Ceiling joists				
12" o.c.				
2x4	S.F.	0.34	0.65	0.99
2x6	"	0.50	0.64	1.14
2x8	"	0.69	0.71	1.40
2x10	"	1.00	0.71	1.71

ROUGH CARPENTRY	UNIT	MAT.	INST.	TOTAL
06110.20 CEILING FRAMING				
2x12	S.F.	1.30	0.76	2.06
16" o.c.				
2x4	S.F.	0.29	0.52	0.81
2x6	"	0.42	0.51	0.93
2x8	"	0.58	0.56	1.14
2x10	"	0.85	0.56	1.41
2x12	"	1.10	0.58	1.68
24" o.c.				
2x4	S.F.	0.22	0.41	0.63
2x6	"	0.33	0.43	0.76
2x8	"	0.46	0.45	0.91
2x10	"	0.68	0.48	1.16
2x12	"	0.88	0.50	1.38
Headers and nailers				
2x4	L.F.	0.28	0.83	1.11
2x6	"	0.42	0.86	1.28
2x8	"	0.58	0.92	1.50
2x10	"	0.85	0.99	1.84
2x12	"	1.10	1.05	2.15
Sister joists for ceilings				
2x4	L.F.	0.29	1.95	2.24
2x6	"	0.42	2.25	2.67
2x8	"	0.58	2.70	3.28
2x10	"	0.94	3.40	4.34
2x12	"	1.20	4.50	5.70
06110.30 FLOOR FRAMING				
Floor joists				
12" o.c.				
2x6	S.F.	0.50	0.54	1.04
2x8	"	0.69	0.55	1.24
2x10	"	1.10	0.56	1.66
2x12	"	1.45	0.59	2.04
2x14	"	1.65	0.54	2.19
3x6	"	1.30	0.55	1.85
3x8	"	1.65	0.56	2.21
3x10	"	2.15	0.59	2.74
3x12	"	2.40	0.61	3.01
3x14	"	2.75	0.64	3.39
4x6	"	1.95	0.54	2.49
4x8	"	2.10	0.56	2.66
4x10	"	2.65	0.58	3.23
4x12	"	3.20	0.61	3.81
4x14	"	3.75	0.64	4.39
16" o.c.				
2x6	S.F.	0.42	0.45	0.87
2x8	"	0.58	0.46	1.04
2x10	"	0.94	0.47	1.41
2x12	"	1.20	0.48	1.68
2x14	"	1.40	0.48	1.88
3x6	"	1.10	0.45	1.55

06 WOOD AND PLASTICS

ROUGH CARPENTRY	UNIT	MAT.	INST.	TOTAL
06110.30 FLOOR FRAMING				
3x8	S.F.	1.40	0.46	1.86
3x10	"	1.80	0.48	2.28
3x12	"	2.00	0.49	2.49
3x14	"	2.30	0.51	2.81
4x6	"	1.60	0.44	2.04
4x8	"	1.75	0.46	2.21
4x10	"	2.20	0.48	2.68
4x12	"	2.65	0.49	3.14
4x14	"	3.10	0.51	3.61
Sister joists for floors				
2x4	L.F.	0.29	1.70	1.99
2x6	"	0.38	1.95	2.33
2x8	"	0.58	2.25	2.83
2x10	"	0.94	2.70	3.64
2x12	"	1.20	3.40	4.60
3x6	"	0.99	2.70	3.69
3x8	"	1.30	3.00	4.30
3x10	"	1.65	3.40	5.05
3x12	"	2.00	3.85	5.85
4x6	"	1.60	2.70	4.30
4x8	"	1.75	3.00	4.75
4x10	"	2.20	3.40	5.60
4x12	"	2.65	3.85	6.50
06110.40 FURRING				
Furring, wood strips				
Walls				
On masonry or concrete walls				
1x2 furring				
12" o.c.	S.F.	0.14	0.85	0.99
16" o.c.	"	0.12	0.77	0.89
24" o.c.	"	0.10	0.68	0.78
1x3 furring				
12" o.c.	S.F.	0.19	0.85	1.04
16" o.c.	"	0.17	0.77	0.94
24" o.c.	"	0.14	0.68	0.82
On wood walls				
1x2 furring				
12" o.c.	S.F.	0.14	0.60	0.74
16" o.c.	"	0.12	0.54	0.66
24" o.c.	"	0.10	0.47	0.57
1x3 furring				
12" o.c.	S.F.	0.19	0.60	0.79
16" o.c.	"	0.17	0.54	0.71
24" o.c.	"	0.14	0.47	0.61
Ceilings				
On masonry or concrete ceilings				
1x2 furring				
12" o.c.	S.F.	0.14	1.50	1.64
16" o.c.	"	0.12	1.35	1.47
24" o.c.	"	0.10	1.20	1.30

ROUGH CARPENTRY

06110.40 FURRING

	UNIT	MAT.	INST.	TOTAL
1x3 furring				
12" o.c.	S.F.	0.19	1.50	1.69
16" o.c.	"	0.17	1.35	1.52
24" o.c.	"	0.14	1.15	1.29
On wood ceilings				
1x2 furring				
12" o.c.	S.F.	0.14	1.00	1.14
16" o.c.	"	0.12	0.90	1.02
24" o.c.	"	0.10	0.79	0.89
1x3				
12" o.c.	S.F.	0.19	1.00	1.19
16" o.c.	"	0.17	0.90	1.07
24" o.c.	"	0.14	0.78	0.92

06110.50 ROOF FRAMING

	UNIT	MAT.	INST.	TOTAL
Roof framing				
Rafters, gable end				
0-2 pitch (flat to 2-in-12)				
12" o.c.				
2x4	S.F.	0.34	0.56	0.90
2x6	"	0.50	0.59	1.09
2x8	"	0.69	0.62	1.31
2x10	"	1.10	0.62	1.72
2x12	"	1.40	0.65	2.05
16" o.c.				
2x6	S.F.	0.42	0.48	0.90
2x8	"	0.58	0.50	1.08
2x10	"	0.92	0.50	1.42
2x12	"	1.20	0.52	1.72
24" o.c.				
2x6	S.F.	0.36	0.39	0.75
2x8	"	0.50	0.41	0.91
2x10	"	0.74	0.42	1.16
2x12	"	0.96	0.43	1.39
4-6 pitch (4-in-12 to 6-in-12)				
12" o.c.				
2x4	S.F.	0.37	0.56	0.93
2x6	"	0.54	0.58	1.12
2x8	"	0.73	0.61	1.34
2x10	"	1.05	0.64	1.69
2x12	"	1.35	0.68	2.03
16" o.c.				
2x6	S.F.	0.43	0.48	0.91
2x8	"	0.61	0.49	1.10
2x10	"	0.88	0.51	1.39
2x12	"	1.10	0.54	1.64
24" o.c.				
2x6	S.F.	0.34	0.40	0.74
2x8	"	0.50	0.41	0.91
2x10	"	0.70	0.44	1.14
2x12	"	0.90	0.49	1.39

ROUGH CARPENTRY

06110.50 ROOF FRAMING

ROUGH CARPENTRY	UNIT	MAT.	INST.	TOTAL
8-12 pitch (8-in-12 to 12-in-12)				
12" o.c.				
2x4	S.F.	0.38	0.58	0.96
2x6	"	0.56	0.61	1.17
2x8	"	0.80	0.64	1.44
2x10	"	1.15	0.68	1.83
2x12	"	1.45	0.71	2.16
16" o.c.				
2x6	S.F.	0.47	0.49	0.96
2x8	"	0.67	0.51	1.18
2x10	"	0.97	0.54	1.51
2x12	"	1.25	0.56	1.81
24" o.c.				
2x6	S.F.	0.37	0.41	0.78
2x8	"	0.53	0.43	0.96
2x10	"	0.79	0.44	1.23
2x12	"	1.00	0.46	1.46
Ridge boards				
2x6	L.F.	0.46	1.30	1.76
2x8	"	0.64	1.45	2.09
2x10	"	0.92	1.60	2.52
2x12	"	1.20	1.85	3.05
Hip rafters				
2x6	L.F.	0.46	0.93	1.39
2x8	"	0.64	0.96	1.60
2x10	"	0.92	1.00	1.92
2x12	"	1.20	1.05	2.25
Jack rafters				
4-6 pitch (4-in-12 to 6-in-12)				
16" o.c.				
2x6	S.F.	0.45	0.76	1.21
2x8	"	0.64	0.78	1.42
2x10	"	0.92	0.83	1.75
2x12	"	1.20	0.86	2.06
24" o.c.				
2x6	S.F.	0.36	0.58	0.94
2x8	"	0.52	0.60	1.12
2x10	"	0.74	0.63	1.37
2x12	"	0.95	0.64	1.59
8-12 pitch (8-in-12 to 12-in-12)				
16" o.c.				0.47
2x6	S.F.	0.70	0.80	1.50
2x8	"	1.00	0.83	1.83
2x10	"	1.30	0.86	2.16
2x12	"	1.55	0.89	2.44
24" o.c.				0.39
2x6	S.F.	0.58	0.61	1.19
2x8	"	0.84	0.63	1.47
2x10	"	1.05	0.64	1.69
2x12	"	1.25	0.66	1.91
Sister rafters				
2x4	L.F.	0.29	1.95	2.24

06 WOOD AND PLASTICS

ROUGH CARPENTRY	UNIT	MAT.	INST.	TOTAL
06110.50 ROOF FRAMING				
2x6	L.F.	0.42	2.25	2.67
2x8	"	0.58	2.70	3.28
2x10	"	0.94	3.40	4.34
2x12	"	1.20	4.50	5.70
Fascia boards				
2x4	L.F.	0.30	1.30	1.60
2x6	"	0.46	1.30	1.76
2x8	"	0.64	1.45	2.09
2x10	"	0.92	1.45	2.37
2x12	"	1.20	1.60	2.80
Cant strips				
Fiber				
3x3	L.F.	0.19	0.74	0.93
4x4	"	0.25	0.79	1.04
Wood				
3x3	L.F.	0.87	0.78	1.65
06110.60 SLEEPERS				
Sleepers, over concrete				
12" o.c.				
1x2	S.F.	0.15	0.62	0.77
1x3	"	0.20	0.65	0.85
2x4	"	0.34	0.75	1.09
2x6	"	0.50	0.80	1.30
16" o.c.				
1x2	S.F.	0.13	0.54	0.67
1x3	"	0.18	0.54	0.72
2x4	"	0.29	0.65	0.94
2x6	"	0.42	0.68	1.10
06110.65 SOFFITS				
Soffit framing				
2x3	L.F.	0.22	1.95	2.17
2x4	"	0.29	2.10	2.39
2x6	"	0.42	2.25	2.67
2x8	"	0.58	2.45	3.03
06110.70 WALL FRAMING				
Framing wall, studs				
12" o.c.				
2x3	S.F.	0.25	0.50	0.75
2x4	"	0.34	0.50	0.84
2x6	"	0.54	0.52	1.06
2x8	"	0.76	0.54	1.30
16" o.c.				
2x3	S.F.	0.22	0.42	0.64
2x4	"	0.29	0.42	0.71
2x6	"	0.46	0.43	0.89
2x8	"	0.64	0.45	1.09
24" o.c.				

06 WOOD AND PLASTICS

ROUGH CARPENTRY

ROUGH CARPENTRY	UNIT	MAT.	INST.	TOTAL
06110.70 WALL FRAMING				
2x3	S.F.	0.19	0.37	0.56
2x4	"	0.23	0.37	0.60
2x6	"	0.36	0.37	0.73
2x8	"	0.50	0.38	0.88
Plates, top or bottom				
2x3	L.F.	0.22	0.80	1.02
2x4	"	0.29	0.85	1.14
2x6	"	0.46	0.86	1.32
2x8	"	0.64	0.93	1.57
Headers, door or window				
2x6				
Single				
3' long	EA.	1.25	13.55	14.80
6' long	"	2.50	16.95	19.45
Double				
3' long	EA.	2.50	15.05	17.55
6' long	"	5.00	19.35	24.35
2x8				
Single				
4' long	EA.	2.30	16.95	19.25
8' long	"	4.60	20.85	25.45
Double				
4' long	EA.	4.60	19.35	23.95
8' long	"	9.25	24.65	33.90
2x10				
Single				
5' long	EA.	4.65	20.85	25.50
10' long	"	9.30	27.10	36.40
Double				
5' long	EA.	9.30	22.60	31.90
10' long	"	18.55	27.10	45.65
2x12				
Single				
6' long	EA.	7.10	19.95	27.05
12' long	"	14.20	25.90	40.10
Double				
6' long	EA.	13.00	23.40	36.40
12' long	"	26.10	28.60	54.70
06115.10 FLOOR SHEATHING				
Sub-flooring, plywood, CDX				
1/2" thick	S.F.	0.44	0.34	0.78
5/8" thick	"	0.53	0.39	0.92
3/4" thick	"	0.63	0.45	1.08
Structural plywood				
1/2" thick	S.F.	0.48	0.34	0.82
5/8" thick	"	0.50	0.37	0.87
3/4" thick	"	0.65	0.40	1.05
Board type subflooring				
1x6				
Minimum	S.F.	0.64	0.60	1.24
Maximum	"	0.77	0.68	1.45

06 WOOD AND PLASTICS

ROUGH CARPENTRY	UNIT	MAT.	INST.	TOTAL
06115.10 FLOOR SHEATHING				
1x8				
Minimum	S.F.	0.75	0.57	1.32
Maximum	"	0.83	0.64	1.47
1x10				
Minimum	S.F.	0.87	0.54	1.41
Maximum	"	0.94	0.60	1.54
Underlayment				
Hardboard, 1/4" tempered	S.F.	0.33	0.34	0.67
Plywood				
3/8" thick	S.F.	0.38	0.34	0.72
1/2" thick	"	0.44	0.36	0.80
5/8" thick	"	0.53	0.39	0.92
3/4" thick	"	0.63	0.42	1.05
06115.20 ROOF SHEATHING				
Sheathing				
3/8" thick	S.F.	0.38	0.35	0.73
1/2" thick	"	0.44	0.36	0.80
5/8" thick	"	0.53	0.39	0.92
3/4" thick	"	0.64	0.40	1.04
Structural plywood				
3/8" thick	S.F.	0.38	0.33	0.71
1/2" thick	"	0.41	0.34	0.75
5/8" thick	"	0.50	0.37	0.87
3/4" thick	"	0.65	0.40	1.05
06115.30 WALL SHEATHING				
Sheathing				
Plywood, CDX				
3/8" thick	S.F.	0.38	0.40	0.78
1/2" thick	"	0.44	0.42	0.86
5/8" thick	"	0.53	0.45	0.98
3/4" thick	"	0.63	0.49	1.12
Waferboard				
3/8" thick	S.F.	0.33	0.40	0.73
1/2" thick	"	0.41	0.42	0.83
5/8" thick	"	0.50	0.45	0.95
3/4" thick	"	0.55	0.49	1.04
Structural plywood				
3/8" thick	S.F.	0.38	0.38	0.76
1/2" thick	"	0.41	0.40	0.81
5/8" thick	"	0.50	0.43	0.93
3/4" thick	"	0.65	0.47	1.12
Gypsum, 1/2" thick	"	0.30	0.40	0.70
Asphalt impregnated fiberboard, 1/2" thick	"	0.36	0.40	0.76
06125.10 WOOD DECKING				
Decking, T&G solid				
Cedar				
3" thick	S.F.	4.50	0.64	5.14
4" thick	"	5.55	0.69	6.24

06 WOOD AND PLASTICS

ROUGH CARPENTRY

	UNIT	MAT.	INST.	TOTAL
06125.10 WOOD DECKING				
Fir				
3" thick	S.F.	2.00	0.65	2.65
4" thick	"	2.20	0.69	2.89
Southern yellow pine				
3" thick	S.F.	2.10	0.74	2.84
4" thick	"	2.25	0.80	3.05
White pine				
3" thick	S.F.	2.70	0.65	3.35
4" thick	"	3.55	0.69	4.24
06130.10 HEAVY TIMBER				
Mill framed structures				
Beams to 20' long				
Douglas fir				
6x8	L.F.	4.40	3.20	7.60
6x10	"	5.30	3.30	8.60
6x12	"	5.65	3.55	9.20
6x14	"	6.45	3.70	10.15
6x16	"	8.10	3.85	11.95
8x10	"	6.75	3.30	10.05
8x12	"	7.80	3.55	11.35
8x14	"	8.80	3.70	12.50
8x16	"	10.80	3.85	14.65
Southern yellow pine				
6x8	L.F.	4.15	3.20	7.35
6x10	"	5.20	3.30	8.50
6x12	"	5.80	3.55	9.35
6x14	"	6.75	3.70	10.45
6x16	"	7.60	3.85	11.45
8x10	"	6.35	3.30	9.65
8x12	"	8.00	3.55	11.55
8x14	"	9.35	3.70	13.05
8x16	"	10.15	3.85	14.00
Columns to 12' high				
Douglas fir				
6x6	L.F.	3.80	4.80	8.60
8x8	"	5.90	4.80	10.70
10x10	"	7.80	5.35	13.15
12x12	"	8.65	5.35	14.00
Southern yellow pine				
6x6	L.F.	3.35	4.80	8.15
8x8	"	5.60	4.80	10.40
10x10	"	6.95	5.35	12.30
12x12	"	8.20	5.35	13.55
Posts, treated				
4x4	L.F.	0.92	1.05	1.97
6x6	"	2.20	1.30	3.50
06190.20 WOOD TRUSSES				
Truss, fink, 2x4 members				
3-in-12 slope				

ROUGH CARPENTRY

06190.20 WOOD TRUSSES	UNIT	MAT.	INST.	TOTAL
24' span	EA.	38.70	27.50	66.20
26' span	"	39.30	27.50	66.80
28' span	"	41.60	29.20	70.80
30' span	"	53.00	29.20	82.20
34' span	"	58.00	31.00	89.00
38' span	"	71.50	31.00	102.50
5-in-12 slope				
24' span	EA.	39.30	28.30	67.60
28' span	"	48.50	29.20	77.70
30' span	"	55.00	30.10	85.10
32' span	"	75.00	30.10	105.10
40' span	"	88.00	32.10	120.10
Gable, 2x4 members				
5-in-12 slope				
24' span	EA.	54.50	28.30	82.80
26' span	"	59.00	28.30	87.30
28' span	"	68.00	29.20	97.20
30' span	"	73.00	30.10	103.10
32' span	"	75.00	30.10	105.10
36' span	"	81.00	31.00	112.00
40' span	"	92.50	32.10	124.60
King post type, 2x4 members				
4-in-12 slope				
16' span	EA.	29.50	26.00	55.50
18' span	"	30.30	26.70	57.00
24' span	"	34.70	28.30	63.00
26' span	"	37.00	28.30	65.30
30' span	"	51.00	30.10	81.10
34' span	"	55.50	30.10	85.60
38' span	"	69.50	31.00	100.50
42' span	"	86.50	33.20	119.70

FINISH CARPENTRY

06200.10 FINISH CARPENTRY	UNIT	MAT.	INST.	TOTAL
Mouldings and trim				
Apron, flat				
9/16 x 2	L.F.	0.53	1.30	1.83
9/16 x 3-1/2	"	0.77	1.35	2.12
Base				
Colonial				
7/16 x 2-1/4	L.F.	0.71	1.30	2.01
7/16 x 3	"	0.86	1.30	2.16

FINISH CARPENTRY

06200.10 FINISH CARPENTRY

FINISH CARPENTRY	UNIT	MAT.	INST.	TOTAL
7/16 x 3-1/4	L.F.	0.95	1.30	2.25
9/16 x 3	"	0.90	1.35	2.25
9/16 x 3-1/4	"	0.98	1.35	2.33
11/16 x 2-1/4	"	0.81	1.45	2.26
Ranch				
7/16 x 2-1/4	L.F.	0.71	1.30	2.01
7/16 x 3-1/4	"	0.95	1.30	2.25
9/16 x 2-1/4	"	0.81	1.35	2.16
9/16 x 3	"	0.92	1.35	2.27
9/16 x 3-1/4	"	0.97	1.35	2.32
Casing				
11/16 x 2-1/2	L.F.	0.89	1.20	2.09
11/16 x 3-1/2	"	1.15	1.25	2.40
Chair rail				
9/16 x 2-1/2	L.F.	0.67	1.30	1.97
9/16 x 3-1/2	"	1.00	1.30	2.30
Closet pole				
1-1/8" dia.	L.F.	0.87	1.70	2.57
1-5/8" dia.	"	0.98	1.70	2.68
Cove				
9/16 x 1-3/4	L.F.	0.54	1.30	1.84
11/16 x 2-3/4	"	0.78	1.30	2.08
Crown				
9/16 x 1-5/8	L.F.	0.87	1.70	2.57
9/16 x 2-5/8	"	0.98	2.00	2.98
11/16 x 3-5/8	"	1.15	2.15	3.30
11/16 x 4-1/4	"	1.55	2.35	3.90
11/16 x 5-1/4	"	2.10	2.55	4.65
Drip cap				
1-1/16 x 1-5/8	L.F.	0.79	1.30	2.09
Glass bead				
3/8 x 3/8	L.F.	0.20	1.60	1.80
1/2 x 9/16	"	0.25	1.60	1.85
5/8 x 5/8	"	0.28	1.60	1.88
3/4 x 3/4	"	0.35	1.60	1.95
Half round				
1/2	L.F.	0.23	1.05	1.28
5/8	"	0.28	1.05	1.33
3/4	"	0.29	1.05	1.34
Lattice				
1/4 x 7/8	L.F.	0.21	1.05	1.26
1/4 x 1-1/8	"	0.23	1.05	1.28
1/4 x 1-3/8	"	0.25	1.05	1.30
1/4 x 1-3/4	"	0.29	1.05	1.34
1/4 x 2	"	0.35	1.05	1.40
Ogee molding				
5/8 x 3/4	L.F.	0.46	1.30	1.76
11/16 x 1-1/8	"	0.69	1.30	1.99
11/16 x 1-3/8	"	0.84	1.30	2.14
Parting bead				
3/8 x 7/8	L.F.	0.35	1.60	1.95
Quarter round				

FINISH CARPENTRY	UNIT	MAT.	INST.	TOTAL
06200.10 FINISH CARPENTRY				
1/4 x 1/4	L.F.	0.12	1.05	1.17
3/8 x 3/8	"	0.18	1.05	1.23
1/2 x 1/2	"	0.23	1.05	1.28
11/16 x 11/16	"	0.25	1.10	1.35
3/4 x 3/4	"	0.29	1.10	1.39
1-1/16 x 1-1/16	"	0.35	1.15	1.50
Railings, balusters				
1-1/8 x 1-1/8	L.F.	1.50	2.60	4.10
1-1/2 x 1-1/2	"	1.30	2.35	3.65
Screen moldings				
1/4 x 3/4	L.F.	0.30	2.15	2.45
5/8 x 5/16	"	0.30	2.15	2.45
Shoe				
7/16 x 11/16	L.F.	0.32	1.05	1.37
Sash beads				
1/2 x 3/4	L.F.	0.98	2.15	3.13
1/2 x 7/8	"	0.35	2.15	2.50
1/2 x 1-1/8	"	0.44	2.35	2.79
5/8 x 7/8	"	0.46	2.35	2.81
Stop				
5/8 x 1-5/8				
Colonial	L.F.	0.31	1.60	1.91
Ranch	"	0.29	1.60	1.89
Stools				
11/16 x 2-1/4	L.F.	1.40	2.85	4.25
11/16 x 2-1/2	"	1.50	2.85	4.35
11/16 x 5-1/4	"	1.60	3.20	4.80
Exterior trim, casing, select pine, 1x3	"	0.95	1.30	2.25
Douglas fir				
1x3	L.F.	0.46	1.30	1.76
1x4	"	0.64	1.30	1.94
1x6	"	0.81	1.45	2.26
1x8	"	1.15	1.60	2.75
Cornices, white pine, #2 or better				
1x2	L.F.	0.29	1.30	1.59
1x4	"	0.48	1.30	1.78
1x6	"	0.64	1.45	2.09
1x8	"	0.89	1.50	2.39
1x10	"	1.05	1.60	2.65
1x12	"	1.45	1.70	3.15
Shelving, pine				
1x8	L.F.	0.89	2.00	2.89
1x10	"	1.05	2.05	3.10
1x12	"	1.55	2.15	3.70
Plywood shelf, 3/4", with edge band, 12" wide	"	1.00	2.55	3.55
Adjustable shelf, and rod, 12" wide				
3' to 4' long	EA.	8.10	6.45	14.55
5' to 8' long	"	13.85	8.55	22.40
Prefinished wood shelves with brackets and supports				
8" wide				
3' long	EA.	23.10	6.45	29.55
4' long	"	28.90	6.45	35.35

06 WOOD AND PLASTICS

FINISH CARPENTRY

FINISH CARPENTRY	UNIT	MAT.	INST.	TOTAL
06200.10 FINISH CARPENTRY				
6' long	EA.	20.80	6.45	27.25
10" wide				
3' long	EA.	27.70	6.80	34.50
4' long	"	41.60	6.45	48.05
6' long	"	48.50	6.45	54.95
06220.10 MILLWORK				
Countertop, laminated plastic				
25" x 7/8" thick				
Minimum	L.F.	10.45	6.80	17.25
Average	"	17.25	9.05	26.30
Maximum	"	27.50	10.85	38.35
25" x 1-1/4" thick				
Minimum	L.F.	11.55	9.05	20.60
Average	"	18.50	10.85	29.35
Maximum	"	31.40	13.55	44.95
Add for cutouts	EA.	0.00	16.95	16.95
Backsplash, 4" high, 7/8" thick	L.F.	9.00	5.40	14.40
Plywood, sanded, A-C				
1/4" thick	S.F.	0.46	0.86	1.32
3/8" thick	"	0.53	0.92	1.45
1/2" thick	"	0.62	0.99	1.61
A-D				
1/4" thick	S.F.	0.40	0.86	1.26
3/8" thick	"	0.48	0.92	1.40
1/2" thick	"	0.58	0.99	1.57
Base cabinets, 34-1/2" high, 24" deep, hardwood, no tops				
Minimum	L.F.	38.50	10.85	49.35
Average	"	54.00	13.55	67.55
Maximum	"	88.00	18.10	106.10
Wall cabinets				
Minimum	L.F.	27.50	9.05	36.55
Average	"	41.80	10.85	52.65
Maximum	"	77.00	13.55	90.55

WOOD TREATMENT

WOOD TREATMENT	UNIT	MAT.	INST.	TOTAL
06300.10 WOOD TREATMENT				
Creosote preservative treatment				
8 lb/cf	B.F.			0.20
10 lb/cf	"			0.28
Salt preservative treatment				

06 WOOD AND PLASTICS

WOOD TREATMENT

06300.10 WOOD TREATMENT	UNIT	MAT.	INST.	TOTAL
Oil borne				
Minimum	B.F.			0.17
Maximum	"			0.24
Water borne				
Minimum	B.F.			0.13
Maximum	"			0.20
Fire retardant treated lumber				
Minimum	B.F.			0.28
Maximum	"			0.33
Kiln dried, softwood, add to framing costs				
1" thick	B.F.			0.09
2" thick	"			0.13
3" thick	"			0.18
4" thick	"			0.22

ARCHITECTURAL WOODWORK

06420.10 PANEL WORK	UNIT	MAT.	INST.	TOTAL
Hardboard, tempered, 1/4" thick				
Natural faced	S.F.	0.34	0.64	0.98
Plastic faced	"	0.68	0.73	1.41
Pegboard, natural	"	0.40	0.64	1.04
Plastic faced	"	0.57	0.73	1.30
Untempered, 1/4" thick				
Natural faced	S.F.	0.32	0.64	0.96
Plastic faced	"	0.65	0.73	1.38
Pegboard, natural	"	0.36	0.64	1.00
Plastic faced	"	0.54	0.73	1.27
Plywood unfinished, 1/4" thick				
Birch				
Natural	S.F.	1.00	0.86	1.86
Select	"	1.30	0.86	2.16
Knotty pine	"	1.30	0.86	2.16
Cedar (closet lining)				
Standard boards T&G	S.F.	1.10	0.86	1.96
Particle board	"	0.87	0.86	1.73
Plywood, prefinished, 1/4" thick, premium grade				
Birch veneer	S.F.	1.60	1.05	2.65
Cherry veneer	"	1.75	1.05	2.80
Chestnut veneer	"	0.80	1.05	1.85
Lauan veneer	"	0.46	1.05	1.51
Mahogany veneer	"	1.60	1.05	2.65
Oak veneer (red)	"	1.30	1.05	2.35
Pecan veneer	"	1.55	1.05	2.60

06 WOOD AND PLASTICS

ARCHITECTURAL WOODWORK	UNIT	MAT.	INST.	TOTAL
06420.10 PANEL WORK				
Rosewood veneer	S.F.	1.65	1.05	2.70
Teak veneer	"	3.20	1.05	4.25
Walnut veneer	"	1.55	1.05	2.60
06430.10 STAIRWORK				
Risers, 1x8, 42" wide				
White oak	EA.	12.60	12.95	25.55
Pine	"	8.95	12.95	21.90
Treads, 1-1/16" x 9-1/2" x 42"				
White oak	EA.	16.40	16.20	32.60
06440.10 COLUMNS				
Column, hollow, round				
12" diameter				
10' high	EA.	378.00	36.80	414.80
12' high	"	454.00	39.40	493.40
14' high	"	485.00	44.00	529.00
16' high	"	554.00	55.00	609.00
24" diameter				
16' high	EA.	2,020	55.00	2,075
18' high	"	2,380	58.00	2,438
20' high	"	2,430	58.00	2,488
22' high	"	2,600	61.00	2,661
24' high	"	2,880	61.00	2,941

07 THERMAL AND MOISTURE

MOISTURE PROTECTION

07100.10 WATERPROOFING	UNIT	MAT.	INST.	TOTAL
Membrane waterproofing, elastomeric				
Butyl				
1/32" thick	S.F.	0.45	0.82	1.27
1/16" thick	"	0.61	0.85	1.46
Butyl with nylon				
1/32" thick	S.F.	0.55	0.82	1.37
1/16" thick	"	0.71	0.85	1.56
Neoprene				
1/32" thick	S.F.	0.92	0.82	1.74
1/16" thick	"	1.50	0.85	2.35
Neoprene with nylon				
1/32" thick	S.F.	0.94	0.82	1.76
1/16" thick	"	1.50	0.85	2.35
Bituminous membrane waterproofing, asphalt felt, 15 lb.				
One ply	S.F.	0.30	0.51	0.81
Two ply	"	0.36	0.62	0.98
Three ply	"	0.48	0.73	1.21
Four ply	"	0.50	0.85	1.35
Five ply	"	0.55	1.05	1.60
Modified asphalt membrane waterproofing, fibrous asphalt				
One ply	S.F.	0.71	0.85	1.56
Two ply	"	1.05	1.00	2.05
Three ply	"	1.20	1.15	2.35
Four ply	"	1.45	1.35	2.80
Five ply	"	1.65	1.65	3.30
Asphalt coated protective board				
1/8" thick	S.F.	0.37	0.51	0.88
1/4" thick	"	0.55	0.51	1.06
3/8" thick	"	0.62	0.51	1.13
1/2" thick	"	0.75	0.54	1.29
Cement protective board				
3/8" thick	S.F.	1.35	0.68	2.03
1/2" thick	"	1.95	0.68	2.63
Fluid applied, neoprene		0.00	0.51	0.51
50 mil	S.F.	0.73	0.68	1.41
90 mil	"	1.10	0.68	1.78
Tab extended polyurethane				
.050" thick	S.F.	0.73	0.51	1.24
Fluid applied rubber based polyurethane				
6 mil	S.F.	0.36	0.64	1.00
15 mil	"	0.88	0.51	1.39
Bentonite waterproofing, panels				
3/16" thick	S.F.	0.17	0.51	0.68
1/4" thick	"	0.98	0.51	1.49
5/8" thick	"	1.30	0.54	1.84
Granular admixtures, trowel on, 3/8" thick	"	1.10	0.51	1.61
Metallic oxide waterproofing, iron compound, troweled				
5/8" thick	S.F.	0.83	0.51	1.34
3/4" thick	"	0.99	0.58	1.57

07 THERMAL AND MOISTURE

MOISTURE PROTECTION

07150.10 DAMPROOFING

MOISTURE PROTECTION	UNIT	MAT.	INST.	TOTAL
Silicone dampproofing, sprayed on				
Concrete surface				
1 coat	S.F.	0.30	0.12	0.42
2 coats	"	0.52	0.16	0.68
Concrete block				
1 coat	S.F.	0.32	0.14	0.46
2 coats	"	0.54	0.19	0.73
Brick				
1 coat	S.F.	0.35	0.16	0.51
2 coats	"	0.57	0.21	0.78

07160.10 BITUMINOUS DAMPPROOFING

Item	UNIT	MAT.	INST.	TOTAL
Building paper, asphalt felt				
15 lb	S.F.	0.07	0.85	0.92
30 lb	"	0.11	0.88	0.99
Asphalt dampproofing, troweled, cold, primer plus				
1 coat	S.F.	0.48	0.68	1.16
2 coats	"	0.74	1.00	1.74
3 coats	"	1.10	1.25	2.35
Fibrous asphalt dampproofing, hot troweled, primer plus				
1 coat	S.F.	0.71	0.82	1.53
2 coats	"	0.89	1.15	2.04
3 coats	"	0.95	1.45	2.40
Asphaltic paint dampproofing, per coat				
Brush on	S.F.	0.13	0.29	0.42
Spray on	"	0.22	0.23	0.45

07190.10 VAPOR BARRIERS

Item	UNIT	MAT.	INST.	TOTAL
Vapor barrier, polyethylene				
2 mil	S.F.	0.01	0.11	0.12
6 mil	"	0.02	0.11	0.13
8 mil	"	0.03	0.12	0.15
10 mil	"	0.04	0.12	0.16

INSULATION

07210.10 BATT INSULATION

INSULATION	UNIT	MAT.	INST.	TOTAL
Ceiling, fiberglass, unfaced				
3-1/2" thick, R11	S.F.	0.24	0.25	0.49
6" thick, R19	"	0.41	0.28	0.69
9" thick, R30	"	0.63	0.33	0.96
Suspended ceiling, unfaced				
3-1/2" thick, R11	S.F.	0.24	0.24	0.48

INSULATION	UNIT	MAT.	INST.	TOTAL
07210.10 BATT INSULATION				
6" thick, R19	S.F.	0.41	0.26	0.67
9" thick, R30	"	0.63	0.30	0.93
Crawl space, unfaced				
3-1/2" thick, R11	S.F.	0.24	0.33	0.57
6" thick, R19	"	0.41	0.35	0.76
9" thick, R30	"	0.63	0.39	1.02
Wall, fiberglass				
Paper backed				
2" thick, R7	S.F.	0.17	0.22	0.39
3" thick, R8	"	0.18	0.24	0.42
4" thick, R11	"	0.28	0.25	0.53
6" thick, R19	"	0.42	0.26	0.68
Foil backed, 1 side				
2" thick, R7	S.F.	0.35	0.22	0.57
3" thick, R11	"	0.37	0.24	0.61
4" thick, R14	"	0.42	0.25	0.67
6" thick, R21	"	0.44	0.26	0.70
Foil backed, 2 sides				
2" thick, R7	S.F.	0.42	0.25	0.67
3" thick, R11	"	0.52	0.26	0.78
4" thick, R14	"	0.61	0.28	0.89
6" thick, R21	"	0.65	0.30	0.95
Unfaced				
2" thick, R7	S.F.	0.18	0.22	0.40
3" thick, R9	"	0.20	0.24	0.44
4" thick, R11	"	0.24	0.25	0.49
6" thick, R19	"	0.41	0.26	0.67
Mineral wool batts				
Paper backed				
2" thick, R6	S.F.	0.34	0.22	0.56
4" thick, R12	"	0.42	0.24	0.66
6" thick, R19	"	0.48	0.26	0.74
Fasteners, self adhering, attached to ceiling deck				
2-1/2" long	EA.	0.11	0.35	0.46
4-1/2" long	"	0.13	0.39	0.52
Capped, self-locking washers for fastening insulation	"	0.11	0.21	0.32
07210.20 BOARD INSULATION				
Insulation, rigid				
Fiberglass, roof				
0.75" thick, R2.78	S.F.	0.17	0.19	0.36
1.06" thick, R4.17	"	0.38	0.20	0.58
1.31" thick, R5.26	"	0.55	0.21	0.76
1.63" thick, R6.67	"	0.61	0.22	0.83
2.25" thick, R8.33	"	0.66	0.24	0.90
Composite board, roof				
1-1/2" thick, R6.67	S.F.	0.66	0.20	0.86
1-5/8" thick, R7.69	"	0.71	0.21	0.92
2" thick, R10.0	"	1.25	0.23	1.48
2-1/4" thick, R12.50	"	1.45	0.24	1.69
2-1/2" thick, R14.29	"	1.55	0.25	1.80
2-3/4" thick, R16.67	"	1.70	0.27	1.97

INSULATION	UNIT	MAT.	INST.	TOTAL
07210.20 BOARD INSULATION				
3-1/4" thick, R20.00	S.F.	2.15	0.29	2.44
Perlite board, roof				
1.00" thick, R2.78	S.F.	0.44	0.18	0.62
1.50" thick, R4.17	"	0.68	0.18	0.86
2.00" thick, R5.92	"	0.84	0.19	1.03
2.50" thick, R6.67	"	1.00	0.20	1.20
3.00" thick, R8.33	"	1.30	0.21	1.51
4.00" thick, R10.00	"	1.45	0.22	1.67
5.25" thick, R14.29	"	1.60	0.24	1.84
Rigid urethane				
Roof				
1" thick, R6.67	S.F.	0.58	0.17	0.75
1.20" thick, R8.33	"	0.73	0.17	0.90
1.50" thick, R11.11	"	0.81	0.18	0.99
2" thick, R14.29	"	1.10	0.19	1.29
2.25" thick, R16.67	"	1.35	0.19	1.54
Wall				
1" thick, R6.67	S.F.	0.53	0.21	0.74
1.5" thick, R11.11	"	0.81	0.23	1.04
2" thick, R14.29	"	1.10	0.24	1.34
Polystyrene				
Roof				
1.0" thick, R4.17	S.F.	0.61	0.17	0.78
1.5" thick, R6.26	"	1.00	0.18	1.18
2.0" thick, R8.33	"	1.05	0.19	1.24
Wall				
1.0" thick, R4.17	S.F.	0.56	0.21	0.77
1.5" thick, R6.26	"	0.92	0.23	1.15
2.0" thick, R8.33	"	1.05	0.24	1.29
Rigid board insulation, deck				
Mineral fiberboard				
1" thick, R3.0	S.F.	0.33	0.18	0.51
2" thick, R5.26	"	0.76	0.19	0.95
Fiberglass				
1" thick, R4.3	S.F.	0.58	0.17	0.75
2" thick, R8.5	"	0.87	0.19	1.06
Polystyrene				
1" thick, R5.4	S.F.	0.56	0.17	0.73
2" thick, R10.8	"	1.05	0.19	1.24
Urethane				
.75" thick, R5.4	S.F.	0.54	0.17	0.71
1" thick, R6.4	"	0.65	0.17	0.82
1.5" thick, R10.7	"	0.78	0.18	0.96
2" thick, R14.3	"	0.89	0.19	1.08
Foamglass				
1" thick, R1.8	S.F.	1.75	0.17	1.92
2" thick, R5.26	"	2.25	0.19	2.44
Wood fiber				
1" thick, R3.85	S.F.	2.00	0.17	2.17
2" thick, R7.7	"	2.40	0.19	2.59
Particle board				
3/4" thick, R2.08	S.F.	0.53	0.17	0.70

07 THERMAL AND MOISTURE

INSULATION	UNIT	MAT.	INST.	TOTAL
07210.20 BOARD INSULATION				
1" thick, R2.77	S.F.	0.58	0.17	0.75
2" thick, R5.50	"	0.61	0.19	0.80
07210.60 LOOSE FILL INSULATION				
Blown-in type				
Fiberglass				
5" thick, R11	S.F.	0.20	0.18	0.38
6" thick, R13	"	0.24	0.21	0.45
9" thick, R19	"	0.28	0.30	0.58
Rockwool, attic application				
6" thick, R13	S.F.	0.35	0.21	0.56
8" thick, R19	"	0.43	0.26	0.69
10" thick, R22	"	0.52	0.33	0.85
12" thick, R26	"	0.67	0.35	1.02
15" thick, R30	"	0.81	0.42	1.23
Poured type				
Fiberglass				
1" thick, R4	S.F.	0.12	0.13	0.25
2" thick, R8	"	0.23	0.15	0.38
3" thick, R12	"	0.35	0.18	0.53
4" thick, R16	"	0.46	0.21	0.67
Mineral wool				
1" thick, R3	S.F.	0.15	0.13	0.28
2" thick, R6	"	0.29	0.15	0.44
3" thick, R9	"	0.46	0.18	0.64
4" thick, R12	"	0.58	0.21	0.79
Vermiculite or perlite				
2" thick, R4.8	S.F.	0.28	0.15	0.43
3" thick, R7.2	"	0.44	0.18	0.62
4" thick, R9.6	"	0.57	0.21	0.78
Masonry, poured vermiculite or perlite				
4" block	S.F.	0.21	0.11	0.32
6" block	"	0.28	0.13	0.41
8" block	"	0.43	0.15	0.58
10" block	"	0.61	0.16	0.77
12" block	"	0.66	0.18	0.84
07210.70 SPRAYED INSULATION				
Foam, sprayed on				
Polystyrene				
1" thick, R4	S.F.	0.28	0.21	0.49
2" thick, R8	"	0.55	0.28	0.83
Urethane				
1" thick, R7.7	S.F.	0.33	0.21	0.54
2" thick, R15.4	"	0.55	0.28	0.83
07250.10 FIREPROOFING				
Sprayed on				
1" thick				
On beams	S.F.	0.36	0.47	0.83
On columns	"	0.38	0.42	0.80

07 THERMAL AND MOISTURE

INSULATION

07250.10 FIREPROOFING

	UNIT	MAT.	INST.	TOTAL
On decks				
Flat surface	S.F.	0.36	0.21	0.57
Fluted surface	"	0.40	0.26	0.66
1-1/2" thick				
On beams	S.F.	0.55	0.61	1.16
On columns	"	0.61	0.53	1.14
On decks				
Flat surface	S.F.	0.55	0.26	0.81
Fluted surface	"	0.58	0.35	0.93

SHINGLES AND TILES

07310.10 ASPHALT SHINGLES

	UNIT	MAT.	INST.	TOTAL
Standard asphalt shingles, strip shingles				
210 lb/square	SQ.	36.30	26.20	62.50
235 lb/square	"	38.50	29.10	67.60
240 lb/square	"	39.60	32.70	72.30
260 lb/square	"	61.00	37.40	98.40
300 lb/square	"	63.50	43.60	107.10
385 lb/square	"	88.00	52.50	140.50
Roll roofing, mineral surface				
90 lb	SQ.	18.50	18.70	37.20
110 lb	"	30.00	21.80	51.80
140 lb	"	31.20	26.20	57.40

07310.30 METAL SHINGLES

	UNIT	MAT.	INST.	TOTAL
Aluminum, .020" thick				
Plain	SQ.	116.00	52.50	168.50
Colors	"	150.00	52.50	202.50
Steel, galvanized				
26 ga.				
Plain	SQ.	154.00	52.50	206.50
Colors	"	198.00	52.50	250.50
24 ga.				
Plain	SQ.	187.00	50.00	237.00
Colors	"	242.00	50.00	292.00
Porcelain enamel, 22 ga.				
Minimum	SQ.	473.00	65.50	538.50
Average	"	539.00	65.50	604.50
Maximum	"	605.00	65.50	670.50

07 THERMAL AND MOISTURE

SHINGLES AND TILES	UNIT	MAT.	INST.	TOTAL
07310.60 SLATE SHINGLES				
Slate shingles				
Pennsylvania				
Ribbon	SQ.	346.00	131.00	477.00
Clear	"	462.00	131.00	593.00
Vermont				
Black	SQ.	404.00	131.00	535.00
Grey	"	439.00	125.00	564.00
Green	"	444.00	125.00	569.00
Red	"	462.00	125.00	587.00
Replacement shingles				
Small jobs	EA.	5.80	8.70	14.50
Large jobs	S.F.	3.85	4.35	8.20
07310.70 WOOD SHINGLES				
Wood shingles, on roofs				
White cedar, #1 shingles				
4" exposure	SQ.	154.00	87.00	241.00
5" exposure	"	138.00	65.50	203.50
#2 shingles				
4" exposure	SQ.	120.00	84.00	204.00
5" exposure	"	102.00	63.00	165.00
Resquared and rebutted				
4" exposure	SQ.	150.00	84.00	234.00
5" exposure	"	126.00	63.00	189.00
On walls				
White cedar, #1 shingles				
4" exposure	SQ.	154.00	131.00	285.00
5" exposure	"	138.00	105.00	243.00
6" exposure	"	116.00	83.50	199.50
#2 shingles				
4" exposure	SQ.	110.00	125.00	235.00
5" exposure	"	93.50	100.00	193.50
6" exposure	"	77.00	83.50	160.50
Add for fire retarding	"			110.00
07310.80 WOOD SHAKES				
Shakes, hand split, 24" red cedar, on roofs				
5" exposure	SQ.	107.00	131.00	238.00
7" exposure	"	108.00	101.00	209.00
9" exposure	"	88.00	87.00	175.00
On walls				
6" exposure	SQ.	102.00	125.00	227.00
8" exposure	"	95.50	100.00	195.50
10" exposure	"	86.00	83.50	169.50
Add for fire retarding	"			110.00

07 THERMAL AND MOISTURE

ROOFING AND SIDING	UNIT	MAT.	INST.	TOTAL
07410.10 MANUFACTURED ROOFS				
Aluminum roof panels, for structural steel framing				
Corrugated				
Unpainted finish				
.024"	S.F.	1.20	0.63	1.83
.030"	"	1.45	0.63	2.08
Painted finish				
.024"	S.F.	1.70	0.63	2.33
.030"	"	1.95	0.63	2.58
V-beam				
Unpainted finish				
.032"	S.F.	1.40	0.63	2.03
.040"	"	1.75	0.63	2.38
.050"	"	2.25	0.63	2.88
Painted finish				
.032"	S.F.	1.85	0.63	2.48
.040"	"	2.25	0.63	2.88
.050"	"	2.75	0.63	3.38
Steel roof panels, for structural steel framing				
Corrugated, painted				
18 ga.	S.F.	3.65	0.63	4.28
20 ga.	"	3.40	0.63	4.03
22 ga.	"	2.75	0.63	3.38
Box rib, painted				
18 ga.	S.F.	3.95	0.66	4.61
20 ga.	"	3.25	0.66	3.91
22 ga.	"	2.90	0.66	3.56
4" rib, painted				
18 ga.	S.F.	4.50	0.70	5.20
20 ga.	"	3.80	0.70	4.50
22 ga.	"	3.45	0.70	4.15
Standing seam roof				
2" high seam, painted				
22 ga.	S.F.	4.40	1.00	5.40
24 ga.	"	4.30	1.00	5.30
26 ga.	"	4.15	1.00	5.15
07410.30 MANUFACTURED WALLS				
Sandwich panels with 1-1/2" fiberglass insulation				
Galvanized 18 ga. steel interior panels				
Exterior panels				
16 ga. aluminum	S.F.	4.15	3.80	7.95
18 ga. galvanized steel	"	5.90	3.80	9.70
20 ga. painted steel	"	6.00	3.80	9.80
20 ga. stainless steel	"	6.15	3.80	9.95
Metal liner panels, 1-3/8" thick, 24" wide				
Galvanized				
22 ga.	S.F.	2.10	0.95	3.05
20 ga.	"	2.35	0.95	3.30
18 ga.	"	2.90	0.95	3.85
Primed				
22 ga.	S.F.	1.60	0.95	2.55

07 THERMAL AND MOISTURE

ROOFING AND SIDING	UNIT	MAT.	INST.	TOTAL
07410.30 MANUFACTURED WALLS				
20 ga.	S.F.	1.95	0.95	2.90
18 ga.	"	2.35	0.95	3.30
07440.10 AGGREGATE COATED PANELS				
Dryvit type system				
1" thick	S.F.	1.40	0.95	2.35
1-1/2" thick	"	1.50	1.00	2.50
2" thick	"	1.60	1.20	2.80
07460.10 METAL SIDING PANELS				
Aluminum siding panels				
Corrugated				
Plain finish				
.024"	S.F.	1.10	1.15	2.25
.032"	"	1.30	1.15	2.45
Painted finish				
.024"	S.F.	1.35	1.15	2.50
.032"	"	1.60	1.15	2.75
V. beam				
Plain finish				
.032"	S.F.	1.25	1.15	2.40
.040"	"	1.40	1.15	2.55
.050"	"	1.65	1.15	2.80
Painted finish				
.032"	S.F.	1.50	1.15	2.65
.040"	"	1.80	1.15	2.95
.050"	"	2.60	1.15	3.75
4" rib				
Plain finish				
.032"	S.F.	1.85	1.30	3.15
.040"	"	1.65	1.30	2.95
.050"	"	1.90	1.30	3.20
Painted finish				
.032"	S.F.	1.65	1.30	2.95
.040"	"	1.95	1.30	3.25
.050"	"	2.15	1.30	3.45
Steel siding panels				
Corrugated				
22 ga.	S.F.	2.90	1.90	4.80
24 ga.	"	2.65	1.90	4.55
26 ga.	"	2.20	1.90	4.10
Box rib				
20 ga.	S.F.	3.25	1.90	5.15
22 ga.	"	2.90	1.90	4.80
24 ga.	"	2.65	1.90	4.55
26 ga.	"	2.15	1.90	4.05
07460.50 PLASTIC SIDING				
Horizontal vinyl siding, solid				
8" wide				
Standard	S.F.	0.65	1.05	1.70
Insulated	"	1.05	1.00	2.05

07 THERMAL AND MOISTURE

ROOFING AND SIDING	UNIT	MAT.	INST.	TOTAL
07460.50 PLASTIC SIDING				
10" wide				
Standard	S.F.	0.71	0.97	1.68
Insulated	"	1.15	0.93	2.08
Vinyl moldings for doors and windows	L.F.	0.41	1.05	1.46
07460.60 PLYWOOD SIDING				
Rough sawn cedar, 3/8" thick	S.F.	0.87	0.86	1.73
Fir, 3/8" thick	"	0.46	0.86	1.32
Texture 1-11, 5/8" thick				
Cedar	S.F.	1.20	0.93	2.13
Fir	"	0.64	0.93	1.57
Redwood	"	1.20	0.90	2.10
Southern Yellow Pine	"	0.71	0.93	1.64
07460.70 STEEL SIDING				
Ribbed, sheets, galvanized				
22 ga.	S.F.	1.40	1.15	2.55
24 ga.	"	1.20	1.15	2.35
26 ga.	"	0.81	1.15	1.96
28 ga.	"	0.58	1.15	1.73
Primed				
24 ga.	S.F.	1.65	1.15	2.80
26 ga.	"	1.05	1.15	2.20
28 ga.	"	0.71	1.15	1.86
07460.80 WOOD SIDING				
Beveled siding, cedar				
A grade				
1/2 x 6	S.F.	1.85	1.30	3.15
1/2 x 8	"	2.20	1.05	3.25
3/4 x 10	"	2.00	0.86	2.86
Clear				
1/2 x 6	S.F.	1.85	1.30	3.15
1/2 x 8	"	2.15	1.05	3.20
3/4 x 10	"	1.95	0.86	2.81
B grade				
1/2 x 6	S.F.	1.60	1.30	2.90
1/2 x 8	"	1.85	10.30	12.15
3/4 x 10	"	1.75	0.86	2.61
Board and batten				
Cedar				
1x6	S.F.	2.10	1.30	3.40
1x8	"	2.05	1.05	3.10
1x10	"	1.65	0.92	2.57
1x12	"	1.55	0.84	2.39
Pine				
1x6	S.F.	0.61	1.30	1.91
1x8	"	0.60	1.05	1.65
1x10	"	0.50	0.92	1.42
1x12	"	0.52	0.84	1.36

ROOFING AND SIDING

07460.80 WOOD SIDING

	UNIT	MAT.	INST.	TOTAL
Redwood				
1x6	S.F.	2.75	1.30	4.05
1x8	"	2.75	1.05	3.80
1x10	"	2.30	0.92	3.22
1x12	"	2.30	0.84	3.14
Tongue and groove				
Cedar				
1x4	S.F.	2.40	1.45	3.85
1x6	"	2.55	1.35	3.90
1x8	"	2.15	1.30	3.45
1x10	"	2.30	1.25	3.55
Pine				
1x4	S.F.	0.71	1.45	2.16
1x6	"	0.72	1.35	2.07
1x8	"	0.61	1.30	1.91
1x10	"	0.60	1.25	1.85
Redwood				
1x4	S.F.	2.60	1.45	4.05
1x6	"	2.65	1.35	4.00
1x8	"	2.35	1.30	3.65
1x10	"	2.40	1.25	3.65

MEMBRANE ROOFING

07510.10 BUILT-UP ASPHALT ROOFING

	UNIT	MAT.	INST.	TOTAL
Built-up roofing, asphalt felt, including gravel				
2 ply	SQ.	25.40	65.50	90.90
3 ply	"	35.80	87.00	122.80
4 ply	"	47.40	105.00	152.40
Walkway, for built-up roofs				
3' x 3' x				
1/2" thick	S.F.	2.30	0.83	3.13
3/4" thick	"	3.00	0.83	3.83
1" thick	"	3.35	0.83	4.18
Roof bonds				
Asphalt felt				
10 yrs	SQ.			17.60
20 yrs	"			18.70
Cant strip, 4" x 4"				
Treated wood	L.F.	1.25	0.72	1.97
Foamglass	"	0.35	0.63	0.98
Mineral fiber	"	0.23	0.63	0.86
New gravel for built-up roofing, 400 lb/sq	SQ.	18.90	50.50	69.40

MEMBRANE ROOFING	UNIT	MAT.	INST.	TOTAL
07510.10 BUILT-UP ASPHALT ROOFING				
Roof gravel (ballast)	C.Y.	11.55	125.00	136.55
Aluminum coating, top surfacing, for built-up roofing	SQ.	18.50	41.70	60.20
Remove & replace gravel, includes flood coat	"	23.10	83.50	106.60
07530.10 SINGLE-PLY ROOFING				
Elastic sheet roofing				
Neoprene, 1/16" thick	S.F.	1.15	0.33	1.48
EPDM rubber				
45 mil	S.F.	0.76	0.33	1.09
60 mil	"	1.05	0.33	1.38
PVC				
45 mil	S.F.	1.05	0.33	1.38
60 mil	"	1.75	0.33	2.08
Flashing				
Pipe flashing, 90 mil thick				
1" pipe	EA.	13.20	6.55	19.75
2" pipe	"	14.85	6.55	21.40
3" pipe	"	15.05	6.90	21.95
4" pipe	"	16.15	6.90	23.05
5" pipe	"	17.50	7.25	24.75
6" pipe	"	18.90	7.25	26.15
8" pipe	"	20.90	7.70	28.60
10" pipe	"	24.75	8.70	33.45
12" pipe	"	29.70	8.70	38.40
Neoprene flashing, 60 mil thick strip				
6" wide	L.F.	0.85	2.20	3.05
12" wide	"	1.65	3.25	4.90
18" wide	"	2.40	4.35	6.75
24" wide	"	3.15	6.55	9.70
Adhesives				
Mastic sealer, applied at joints only				
1/4" bead	L.F.	0.04	0.13	0.17
Fluid applied roofing				
Urethane, 2 components, elastomeric top membrane				
1" thick	S.F.	1.50	0.42	1.92
Vinyl liquid roofing, 2 coats, 2 mils per coat	"	2.30	0.36	2.66
Silicone roofing, 2 coats sprayed, 16 mil per coat	"	1.75	0.42	2.17
Inverted roof system				
Insulated membrane with coarse gravel ballast				
3 ply with 2" polystyrene	S.F.	3.25	0.42	3.67
Ballast, 3/4" through 1-1/2" dia. river gravel, 100lb/sf	"	0.19	26.20	26.39
Walkway for membrane roofs, 1/2" thick	"	1.15	0.87	2.02

07 THERMAL AND MOISTURE

FLASHING AND SHEET METAL

	UNIT	MAT.	INST.	TOTAL
07610.10 METAL ROOFING				
Sheet metal roofing, copper, 16 oz, batten seam	SQ.	440.00	174.00	614.00
Standing seam	"	429.00	164.00	593.00
Aluminum roofing, natural finish				
Corrugated, on steel frame				
.0175" thick	SQ.	66.00	74.50	140.50
.0215" thick	"	1.40	71.50	72.90
.024" thick	"	1.85	71.50	73.35
.032" thick	"	154.00	74.50	228.50
V-beam, on steel frame				
.032" thick	SQ.	2.15	71.50	73.65
.040" thick	"	2.95	71.50	74.45
.050" thick	"	2.95	71.50	74.45
Ridge cap				
.019" thick	L.F.	2.60	0.83	3.43
Corrugated galvanized steel roofing, on steel frame				
28 ga.	SQ.	77.00	74.50	151.50
26 ga.	"	2.20	71.50	73.70
24 ga.	"	2.30	71.50	73.80
22 ga.	"	110.00	74.50	184.50
26 ga., factory insulated with 1" polystyrene	"	3.85	100.00	103.85
Ridge roll				
10" wide	L.F.	1.20	0.83	2.03
20" wide	"	2.45	1.00	3.45
07620.10 FLASHING AND TRIM				
Counter flashing				
Aluminum, .032"	S.F.	0.88	2.60	3.48
Stainless steel, .015"	"	2.65	2.60	5.25
16 oz.	"	2.65	2.60	5.25
20 oz.	"	2.85	2.50	5.35
24 oz.	"	3.30	2.50	5.80
32 oz.	"	4.40	2.50	6.90
Valley flashing				
Aluminum, .032"	S.F.	0.96	1.55	2.51
Stainless steel, .015	"	2.90	1.55	4.45
Copper				
16 oz.	S.F.	2.65	1.55	4.20
20 oz.	"	2.85	2.10	4.95
24 oz.	"	3.30	1.55	4.85
32 oz.	"	4.40	1.55	5.95
Base flashing				
Aluminum, .040"	S.F.	1.20	2.20	3.40
Stainless steel, .018"	"	2.85	2.20	5.05
16 oz.	"	2.65	2.20	4.85
20 oz.	"	2.85	1.55	4.40
24 oz.	"	3.95	2.20	6.15
32 oz.	"	4.40	2.10	6.50
Waterstop, "T" section, 22 ga.				2.55
1-1/2" x 3"	L.F.	1.55	1.25	2.80
2" x 2"	"	1.70	1.25	2.95
4" x 3"	"	1.95	1.25	3.20

FLASHING AND SHEET METAL

07620.10 FLASHING AND TRIM

	UNIT	MAT.	INST.	TOTAL
6" x 4"	L.F.	2.05	1.25	3.30
8" x 4"	"	2.55	1.25	3.80
Scupper outlets				
10" x 10" x 4"	EA.	15.75	6.55	22.30
22" x 4" x 4"	"	19.35	6.55	25.90
8" x 8" x 5"	"	15.75	6.55	22.30
Flashing and trim, aluminum				
.019" thick	S.F.	0.83	1.80	2.63
.032" thick	"	1.00	1.80	2.80
.040" thick	"	0.97	1.95	2.92
Neoprene sheet flashing, .060" thick	"	0.79	1.55	2.34
Copper, paper backed				
2 oz.	S.F.	0.99	2.50	3.49
Drainage boots, roof, cast iron				
2 x 3	L.F.	29.00	3.25	32.25
3 x 4	"	36.30	3.25	39.55
4 x 5	"	52.00	3.50	55.50
4 x 6	"	49.60	3.50	53.10
5 x 7	"	59.50	3.75	63.25
6 x 6	EA.	56.00	6.55	62.55
8 x 8	"	59.50	6.55	66.05
8 x 10	"	64.00	6.55	70.55
8 x 12	"	77.00	6.55	83.55
Reglets, copper 10 oz.	L.F.	2.00	1.75	3.75
Stainless steel, .020"	"	1.40	1.75	3.15
Gravel stop				
Aluminum, .032"				
4"	L.F.	3.05	0.83	3.88
6"	"	3.60	0.83	4.43
8"	"	5.45	0.96	6.41
10"	"	6.35	0.96	7.31
Copper, 16 oz.				
4"	L.F.	4.70	0.83	5.53
6"	"	5.95	0.83	6.78
8"	"	6.15	0.96	7.11
10"	"	7.40	0.96	8.36

07620.20 GUTTERS AND DOWNSPOUTS

	UNIT	MAT.	INST.	TOTAL
Copper gutter and downspout				
Downspouts, 16 oz. copper				
Round				
3" dia.	L.F.	4.45	1.75	6.20
4" dia.	"	5.70	1.75	7.45
Rectangular, corrugated				
2" x 3"	L.F.	4.30	1.65	5.95
3" x 4"	"	5.60	1.65	7.25
Rectangular, flat surface				
2" x 3"	L.F.	4.70	1.75	6.45
3" x 4"	"	6.65	1.75	8.40
Lead-coated copper downspouts				
Round				
3" dia.	L.F.	5.50	1.65	7.15

FLASHING AND SHEET METAL

07620.20 GUTTERS AND DOWNSPOUTS

FLASHING AND SHEET METAL	UNIT	MAT.	INST.	TOTAL
4" dia.	L.F.	6.60	1.85	8.45
Rectangular, corrugated				
2" x 3"	L.F.	4.05	1.75	5.80
3" x 4"	"	4.85	1.75	6.60
Rectangular, plain				
2" x 3"	L.F.	3.50	1.75	5.25
3" x 4"	"	4.20	1.75	5.95
Gutters, 16 oz. copper				
Half round				
4" wide	L.F.	4.10	2.60	6.70
5" wide	"	4.95	2.90	7.85
Type K				
4" wide	L.F.	4.50	2.60	7.10
5" wide	"	4.95	2.90	7.85
Lead-coated copper gutters				
Half round				
4" wide	L.F.	3.85	2.60	6.45
6" wide	"	6.05	2.90	8.95
Type K				
4" wide	L.F.	5.30	2.60	7.90
5" wide	"	5.70	2.90	8.60
Aluminum gutter and downspout				
Downspouts				
2" x 3"	L.F.	0.63	1.75	2.38
3" x 4"	"	0.84	1.85	2.69
4" x 5"	"	0.94	2.00	2.94
Round				
3" dia.	L.F.	1.00	1.75	2.75
4" dia.	"	1.30	1.85	3.15
Gutters, stock units				
4" wide	L.F.	1.05	2.75	3.80
5" wide	"	1.15	2.90	4.05
Galvanized steel gutter and downspout				
Downspouts, round corrugated				
3" dia.	L.F.	0.88	1.75	2.63
4" dia.	"	0.99	1.75	2.74
5" dia.	"	1.20	1.85	3.05
6" dia.	"	1.55	1.85	3.40
Rectangular				
2" x 3"	L.F.	0.88	1.75	2.63
3" x 4"	"	1.40	1.65	3.05
4" x 4"	"	1.75	1.65	3.40
Gutters, stock units				
5" wide				
Plain	L.F.	0.99	2.90	3.89
Painted	"	1.10	2.90	4.00
6" wide				
Plain	L.F.	1.30	3.10	4.40
Painted	"	1.55	3.10	4.65

ROOFING SPECIALITES

07700.10 MANUFACTURED SPECIALTIES

ROOFING SPECIALITES	UNIT	MAT.	INST.	TOTAL
Moisture relief vent				
Aluminum	EA.	11.20	3.60	14.80
Copper	"	20.90	3.60	24.50
Expansion joint				
Aluminum				
Opening to 2.5"	L.F.	6.05	1.80	7.85
Opening to 3.5"	"	7.25	1.95	9.20
Copper, 16 oz.				
Opening to 2.5"	L.F.	9.55	1.80	11.35
Opening to 3.5"	"	11.50	1.95	13.45
Butyl or neoprene				
4" wide				
16 oz. copper bellows	L.F.	10.90	2.10	13.00
28 ga. stainless steel bellows	"	9.45	2.10	11.55
6" wide				
Copper bellows	L.F.	10.65	2.30	12.95
Stainless steel				
Opening to 2.5"	L.F.	11.75	1.80	13.55
Opening to 3.5"	"	13.90	1.95	15.85
Smoke vent, 48" x 48"				
Aluminum	EA.	470.00	63.00	533.00
Galvanized steel	"	455.00	63.00	518.00
Heat/smoke vent, 48" x 96"				
Aluminum	EA.	1,810	84.00	1,894
Galvanized steel	"	1,770	84.00	1,854
Ridge vent strips				
Mill finish	L.F.	2.50	1.70	4.20
Connectors	EA.	10.90	6.25	17.15
End cap	"	21.80	7.15	28.95
Soffit vents				
Mill finish				
2-1/2" wide	L.F.	0.28	1.00	1.28
3" wide	"	0.31	1.00	1.31
6" wide	"	0.58	1.00	1.58
Roof hatches				
Steel, plain, primed				
2'6" x 3'0"	EA.	396.00	63.00	459.00
2'6" x 4'6"	"	544.00	83.50	627.50
2'6" x 8'0"	"	847.00	125.00	972.00
Galvanized steel				
2'6" x 3'0"	EA.	409.00	63.00	472.00
2'6" x 4'6"	"	575.00	83.50	658.50
2'6" x 8'0"	"	901.00	125.00	1,026
Aluminum				
2'6" x 3'0"	EA.	436.00	63.00	499.00
2'6" x 4'6"	"	605.00	83.50	688.50
2'6" x 8'0"	"	1,080	125.00	1,205
Ceiling access doors				
Swing up model, metal frame				
Steel door				
2'6" x 2'6"	EA.	275.00	26.20	301.20
2'6" x 3'0"	"	297.00	26.20	323.20

ROOFING SPECIALITES

07700.10 MANUFACTURED SPECIALTIES

	UNIT	MAT.	INST.	TOTAL
Aluminum door				
2'6" x 2'6"	EA.	297.00	26.20	323.20
2'6" x 3'0"	"	319.00	26.20	345.20
Swing down model, metal frame				
Steel door				
2'6" x 2'6"	EA.	275.00	26.20	301.20
2'6" x 3'0"	"	297.00	26.20	323.20
Aluminum door				
2'6" x 2'6"	EA.	286.00	26.20	312.20
2'6" x 3'0"	"	308.00	26.20	334.20
Gravity ventilators, with curb, base, damper and screen				
Stationary siphon				
6" dia.	EA.	19.80	17.45	37.25
12" dia.	"	30.80	17.45	48.25
24" dia.	"	132.00	26.20	158.20
36" dia.	"	242.00	26.20	268.20
Wind driven spinner				
6" dia.	EA.	30.80	17.45	48.25
12" dia.	"	41.80	17.45	59.25
24" dia.	"	154.00	26.20	180.20
36" dia.	"	330.00	26.20	356.20
Stationary mushroom				
16" dia.	EA.	198.00	26.20	224.20
30" dia.	"	462.00	32.70	494.70
36" dia.	"	594.00	43.60	637.60
42" dia.	"	880.00	52.50	932.50

SKYLIGHTS

07810.10 PLASTIC SKYLIGHTS

	UNIT	MAT.	INST.	TOTAL
Single thickness, not including mounting curb				
2' x 4'	EA.	158.00	31.50	189.50
4' x 4'	"	253.00	41.90	294.90
5' x 5'	"	333.00	62.50	395.50
6' x 8'	"	641.00	83.50	724.50
Double thickness, not including mounting curb				
2' x 4'	EA.	201.00	31.50	232.50
4' x 4'	"	317.00	41.90	358.90
5' x 5'	"	438.00	62.50	500.50
6' x 8'	"	786.00	83.50	869.50
Metal framed skylights				
Translucent panels, 2-1/2" thick	S.F.	21.80	2.50	24.30
Continuous vaults, 8' wide				
Single glazed	S.F.	30.20	3.15	33.35

07 THERMAL AND MOISTURE

SKYLIGHTS

07810.10 PLASTIC SKYLIGHTS	UNIT	MAT.	INST.	TOTAL
Double glazed	S.F.	48.40	3.60	52.00

JOINT SEALERS

07920.10 CAULKING	UNIT	MAT.	INST.	TOTAL
Caulk exterior, two component				
1/4 x 1/2	L.F.	0.24	1.35	1.59
3/8 x 1/2	"	0.36	1.50	1.86
1/2 x 1/2	"	0.47	1.70	2.17
Caulk interior, single component				
1/4 x 1/2	L.F.	0.15	1.30	1.45
3/8 x 1/2	"	0.23	1.45	1.68
1/2 x 1/2	"	0.30	1.60	1.90
Butyl rubber fillers				
1/4" x 1/4"	L.F.	0.52	0.51	1.03
1/2" x 1/2"	"	0.69	0.86	1.55
1/2" x 3/4"	"	1.10	1.05	2.15
3/4" x 3/4"	"	1.40	1.05	2.45
1" x 1"	"	1.65	1.15	2.80
Neoprene, closed cell				
1/8" x 3"	L.F.	0.03	0.64	0.67
1/8" x 6"	"	0.07	0.71	0.78
1/4" x 3"	"	0.07	0.79	0.86
1/4" x 6"	"	0.13	0.94	1.07
1/2" x 6"	"	0.25	1.15	1.40
1/2" x 9"	"	0.38	1.30	1.68
1/2" x 12"	"	0.50	1.30	1.80
Seals, "O" ring type cord				
1/4" dia.	L.F.	0.46	0.64	1.10
1/2" dia.	"	1.40	0.68	2.08
1" dia.	"	4.90	0.71	5.61
1-1/4" dia.	"	6.25	0.76	7.01
1-1/2" dia.	"	8.30	0.80	9.10
1-3/4" dia.	"	12.25	0.83	13.08
2" dia.	"	15.85	0.86	16.71
Polyvinyl chloride, closed cell				
1/4" x 2"	L.F.	0.32	0.92	1.24
1/4" x 6"	"	0.98	1.15	2.13
Silicon foam penetration seal				
1/4" x 1/2"	L.F.	0.12	0.32	0.44
1/2" x 1/2"	"	0.21	0.43	0.64
1/2" x 3/4"	"	0.31	0.51	0.82

JOINT SEALERS

07920.10 CAULKING	UNIT	MAT.	INST.	TOTAL
3/4" x 3/4"	L.F.	0.46	0.64	1.10
1/8" x 1"	"	0.12	0.32	0.44
1/8" x 3"	"	0.31	0.51	0.82
1/4" x 3"	"	0.63	0.64	1.27
1/4" x 6"	"	1.30	0.86	2.16
1/2" x 6"	"	2.50	2.00	4.50
1/2" x 9"	"	3.75	3.20	6.95
1/2" x 12"	"	4.95	4.70	9.65
Oil base sealants and caulking				
1/4" x 1/4"	L.F.	0.03	0.64	0.67
1/4" x 3/8"	"	0.06	0.67	0.73
1/4" x 1/2"	"	0.07	0.70	0.77
3/8" x 3/8"	"	0.08	0.73	0.81
3/8" x 1/2"	"	0.10	0.78	0.88
3/8" x 5/8"	"	0.14	0.83	0.97
3/8" x 3/4"	"	0.17	0.89	1.06
1/2" x 1/2"	"	0.14	0.99	1.13
1/2" x 5/8"	"	0.19	1.10	1.29
1/2" x 3/4"	"	0.22	1.30	1.52
1/2" x 7/8"	"	0.25	1.30	1.55
1/2" x 1"	"	0.29	1.35	1.64
3/4" x 3/4"	"	0.33	1.40	1.73
1" x 1"	"	0.58	1.45	2.03
Polyurethane compounds				
1/4" x 1/4"	L.F.	0.09	0.64	0.73
1/4" x 3/8"	"	0.14	0.67	0.81
1/4" x 1/2"	"	0.19	0.70	0.89
3/8" x 3/8"	"	0.21	0.73	0.94
3/8" x 1/2"	"	0.28	0.78	1.06
3/8" x 5/8"	"	0.35	0.83	1.18
3/8" x 3/4"	"	0.42	0.89	1.31
1/2" x 1/2"	"	0.37	0.99	1.36
1/2" x 5/8"	"	0.46	1.10	1.56
1/2" x 3/4"	"	0.55	1.30	1.85
1/2" x 7/8"	"	0.65	1.30	1.95
1/2" x 1"	"	0.75	1.45	2.20
3/4" x 3/4"	"	0.85	1.40	2.25
3/4" x 1"	"	1.15	1.45	2.60
Backer rod, polyethylene				
1/4"	L.F.	0.04	0.64	0.68
1/2"	"	0.06	0.68	0.74
3/4"	"	0.06	0.71	0.77
1"	"	0.09	0.76	0.85

08 DOORS AND WINDOWS

METAL	UNIT	MAT.	INST.	TOTAL
08110.10 METAL DOORS				
Flush hollow metal, standard duty, 20 ga., 1-3/8"				
2-6 x 6-8	EA.	145.00	28.80	173.80
2-8 x 6-8	"	149.00	28.80	177.80
3-0 x 6-8	"	164.00	28.80	192.80
1-3/4"				
2-6 x 6-8	EA.	173.00	28.80	201.80
2-8 x 6-8	"	183.00	28.80	211.80
3-0 x 6-8	"	195.00	28.80	223.80
2-6 x 7-0	"	173.00	28.60	201.60
2-8 x 7-0	"	178.00	28.60	206.60
3-0 x 7-0	"	189.00	28.60	217.60
Heavy duty, 20 ga., unrated, 1-3/4"				
2-8 x 6-8	EA.	151.00	28.80	179.80
3-0 x 6-8	"	164.00	28.80	192.80
2-8 x 7-0	"	165.00	28.60	193.60
3-0 x 7-0	"	195.00	28.60	223.60
3-4 x 7-0	"	231.00	28.60	259.60
18 ga., 1-3/4", unrated door				
2-0 x 7-0	EA.	172.00	28.60	200.60
2-4 x 7-0	"	184.00	28.60	212.60
2-6 x 7-0	"	188.00	28.60	216.60
2-8 x 7-0	"	195.00	28.60	223.60
3-0 x 7-0	"	210.00	28.60	238.60
3-4 x 7-0	"	254.00	28.60	282.60
2", unrated door				
2-0 x 7-0	EA.	191.00	32.20	223.20
2-4 x 7-0	"	202.00	32.20	234.20
2-6 x 7-0	"	208.00	32.20	240.20
2-8 x 7-0	"	217.00	32.20	249.20
3-0 x 7-0	"	233.00	32.20	265.20
3-4 x 7-0	"	283.00	32.20	315.20
Galvanized metal door				
3-0 x 7-0	EA.	271.00	32.20	303.20
For lead lining in doors	"			550.00
For sound attenuation	"			22.00
Vision glass				
8" x 8"	EA.	69.50	32.20	101.70
8" x 48"	"	103.00	32.20	135.20
Fixed metal louver	"	81.00	25.70	106.70
For fire rating, add				275.00
3 hr door	EA.			44.00
1-1/2 hr door	"			22.00
3/4 hr door	"			11.00
1' extra height, add to material, 20%				
1'6" extra height, add to material, 60%				
For dutch doors with shelf, add to material, 100%				
08110.40 METAL DOOR FRAMES				
Hollow metal, stock, 18 ga., 4-3/4" x 1-3/4"				
2-0 x 7-0	EA.	61.50	32.40	93.90
2-4 x 7-0	"	61.50	32.40	93.90
2-6 x 7-0	"	63.00	32.40	95.40

METAL	UNIT	MAT.	INST.	TOTAL
08110.40 METAL DOOR FRAMES				
2-8 x 7-0	EA.	58.00	32.20	90.20
3-0 x 7-0	"	64.50	32.40	96.90
4-0 x 7-0	"	63.50	42.90	106.40
5-0 x 7-0	"	64.50	42.90	107.40
6-0 x 7-0	"	67.00	42.90	109.90
16 ga., 6-3/4" x 1-3/4"		0.00	35.70	35.70
2-0 x 7-0	EA.	60.00	35.70	95.70
2-4 x 7-0	"	63.00	35.70	98.70
2-6 x 7-0	"	63.00	35.70	98.70
2-8 x 7-0	"	64.50	35.70	100.20
3-0 x 7-0	"	69.50	35.70	105.20
4-0 x 7-0	"	78.50	47.60	126.10
6-0 x 7-0	"	83.00	47.60	130.60
Transom frame				
3-4 x 1-6	EA.	243.00	35.70	278.70
3-8 x 1-6	"	243.00	35.70	278.70
6-4 x 1-6	"	289.00	35.70	324.70
Transom sash				
3-0 x 1-4	EA.	375.00	35.70	410.70
3-4 x 1-4	"	375.00	35.70	410.70
6-0 x 1-4	"	422.00	35.70	457.70
1' extension of frame, add	"			11.00
14 ga. frame, add	"			11.00
For fire rating, add				
3 hour	EA.			22.00
1-1/2 hour	"			22.00
3/4 hour	"			22.00
Lead lining in frame, add	"			66.00
Sidelights, complete				
1-0 x 7-2	EA.	225.00	35.70	260.70
1-4 x 7-2	"	243.00	35.70	278.70
1-0 x 8-8	"	254.00	35.70	289.70
1-6 x 8-8	"	266.00	35.70	301.70
16 ga., 4-3/4" x 1-3/4"				
2-0 x 7-0	EA.	56.50	35.70	92.20
2-4 x 7-0	"	56.50	35.70	92.20
2-6 x 7-0	"	58.00	35.70	93.70
2-8 x 7-0	"	58.00	35.70	93.70
3-0 x 7-0	"	60.00	35.70	95.70
4-0 x 7-0	"	64.50	47.60	112.10
6-0 x 7-0	"	69.50	47.60	117.10
Transom frame				
3-4 x 1-6	EA.	225.00	35.70	260.70
3-8 x 1-6	"	225.00	35.70	260.70
6-4 x 1-6	"	277.00	35.70	312.70
Transom sash				
3-0 x 1-4	EA.	364.00	35.70	399.70
3-4 x 1-4	"	364.00	35.70	399.70
6-0 x 1-4	"	416.00	35.70	451.70
1' extension of door frame, add	"			11.00
14 ga., metal frame, add	"			11.00
For fire rating, add				

METAL	UNIT	MAT.	INST.	TOTAL
08110.40 METAL DOOR FRAMES				
3 hour	EA.			22.00
1-1/2 hour	"			22.00
3/4 hour	"			22.00
Lead lining in frame, add	"			66.00
Sidelights, complete				
1-0 x 7-2	EA.	214.00	35.70	249.70
1-4 x 7-2	"	222.00	35.70	257.70
1-0 x 8-8	"	243.00	35.70	278.70
1-4 x 8-8	"	248.00	35.70	283.70
16 ga., 5-3/4" x 1-3/4"				
2-0 x 7-0	EA.	63.50	32.20	95.70
2-4 x 7-0	"	63.50	32.20	95.70
2-6 x 7-0	"	66.00	32.20	98.20
2-8 x 7-0	"	66.00	32.20	98.20
3-0 x 7-0	"	67.00	32.20	99.20
4-0 x 7-0	"	73.00	42.90	115.90
5-0 x 7-0	"	76.00	42.90	118.90
6-0 x 7-0	"	78.50	42.90	121.40
Mullions, vertical				
5-1/4" x 1-3/4"	L.F.	7.15	3.20	10.35
5-1/4" x 2"	"	9.00	3.20	12.20
Horizontal				
5-1/4" x 1-3/4"	L.F.	7.15	3.20	10.35
5-1/4" x 2"	"	9.00	3.20	12.20
08120.10 ALUMINUM DOORS				
Aluminum doors, commercial				
Narrow stile				
2-6 x 7-0	EA.	374.00	143.00	517.00
3-0 x 7-0	"	377.00	143.00	520.00
3-6 x 7-0	"	392.00	143.00	535.00
Pair				
5-0 x 7-0	EA.	687.00	286.00	973.00
6-0 x 7-0	"	687.00	286.00	973.00
7-0 x 7-0	"	716.00	286.00	1,002
Wide stile				
2-6 x 7-0	EA.	642.00	143.00	785.00
3-0 x 7-0	"	690.00	143.00	833.00
3-6 x 7-0	"	744.00	143.00	887.00
Pair				
5-0 x 7-0	EA.	1,500	286.00	1,786
6-0 x 7-0	"	1,500	286.00	1,786
7-0 x 7-0	"	1,560	286.00	1,846

08 DOORS AND WINDOWS

WOOD AND PLASTIC	UNIT	MAT.	INST.	TOTAL
08210.10 WOOD DOORS				
Solid core, 1-3/8" thick				
Birch faced				
2-4 x 7-0	EA.	101.00	32.20	133.20
2-8 x 7-0	"	107.00	32.20	139.20
3-0 x 7-0	"	112.00	32.20	144.20
3-4 x 7-0	"	119.00	32.20	151.20
2-4 x 6-8	"	108.00	32.40	140.40
2-6 x 6-8	"	114.00	32.40	146.40
2-8 x 6-8	"	120.00	32.40	152.40
3-0 x 6-8	"	125.00	32.40	157.40
Lauan faced				
2-4 x 6-8	EA.	87.50	32.20	119.70
2-8 x 6-8	"	92.00	32.20	124.20
3-0 x 6-8	"	98.00	32.20	130.20
3-4 x 6-8	"	95.50	32.20	127.70
Tempered hardboard faced				
2-4 x 7-0	EA.	92.00	32.20	124.20
2-8 x 7-0	"	98.00	32.20	130.20
3-0 x 7-0	"	105.00	32.20	137.20
3-4 x 7-0	"	113.00	32.20	145.20
Hollow core, 1-3/8" thick				
Birch faced				
2-4 x 7-0	EA.	52.50	32.20	84.70
2-8 x 7-0	"	55.50	32.20	87.70
3-0 x 7-0	"	58.00	32.20	90.20
3-4 x 7-0	"	62.50	32.20	94.70
Lauan faced				
2-4 x 6-8	EA.	35.60	32.40	68.00
2-6 x 6-8	"	38.20	32.40	70.60
2-8 x 6-8	"	40.70	32.40	73.10
3-0 x 6-8	"	43.20	32.40	75.60
3-4 x 6-8	"	43.10	32.20	75.30
Tempered hardboard faced				
2-4 x 7-0	EA.	35.00	32.20	67.20
2-6 x 7-0	"	37.30	32.20	69.50
2-8 x 7-0	"	39.60	32.20	71.80
3-0 x 7-0	"	42.00	32.20	74.20
3-4 x 7-0	"	44.30	32.20	76.50
Solid core, 1-3/4" thick				
Birch faced				
2-4 x 7-0	EA.	119.00	32.20	151.20
2-6 x 7-0	"	126.00	32.20	158.20
2-8 x 7-0	"	131.00	32.20	163.20
3-0 x 7-0	"	136.00	32.20	168.20
3-4 x 7-0	"	142.00	32.20	174.20
Lauan faced				
2-4 x 7-0	EA.	105.00	32.20	137.20
2-6 x 7-0	"	119.00	32.20	151.20
2-8 x 7-0	"	126.00	32.20	158.20
3-0 x 7-0	"	131.00	32.20	163.20
3-0 x 7-0	"	138.00	32.20	170.20
Tempered hardboard faced				

WOOD AND PLASTIC	UNIT	MAT.	INST.	TOTAL
08210.10 WOOD DOORS				
2-4 x 7-0	EA.	105.00	32.20	137.20
2-6 x 7-0	"	117.00	32.20	149.20
2-8 x 7-0	"	128.00	32.20	160.20
3-0 x 7-0	"	134.00	32.20	166.20
3-4 x 7-0	"	142.00	32.20	174.20
Hollow core, 1-3/4" thick				
Birch faced				
2-4 x 7-0	EA.	53.50	32.20	85.70
2-6 x 7-0	"	56.00	32.20	88.20
2-8 x 7-0	"	58.50	32.20	90.70
3-0 x 7-0	"	60.50	32.20	92.70
3-4 x 7-0	"	64.00	32.20	96.20
Lauan faced				
2-4 x 6-8	EA.	39.60	32.20	71.80
2-6 x 6-8	"	42.00	32.20	74.20
2-8 x 6-8	"	44.30	32.20	76.50
3-0 x 6-8	"	46.60	32.20	78.80
3-4 x 6-8	"	50.00	32.20	82.20
Tempered hardboard				
2-4 x 7-0	EA.	39.60	32.20	71.80
2-6 x 7-0	"	42.00	32.20	74.20
2-8 x 7-0	"	44.30	32.20	76.50
3-0 x 7-0	"	46.60	32.20	78.80
3-4 x 7-0	"	50.00	32.20	82.20
Add-on, louver	"	11.65	25.70	37.35
Glass	"	46.60	25.70	72.30
Exterior doors, 3-0 x 7-0 x 2-1/2", solid core				
Carved				
One face	EA.	606.00	64.50	670.50
Two faces	"	921.00	64.50	985.50
Closet doors, 1-3/4" thick				
Bi-fold or bi-passing, includes frame and trim				
Paneled				
4-0 x 6-8	EA.	240.00	43.20	283.20
6-0 x 6-8	"	293.00	43.20	336.20
Louvered				
4-0 x 6-8	EA.	165.00	43.20	208.20
6-0 x 6-8	"	211.00	43.20	254.20
Flush				
4-0 x 6-8	EA.	146.00	43.20	189.20
6-0 x 6-8	"	178.00	43.20	221.20
Primed				
4-0 x 6-8	EA.	153.00	43.20	196.20
6-0 x 6-8	"	184.00	43.20	227.20
08210.90 WOOD FRAMES				
Frame, interior, pine				
2-6 x 6-8	EA.	26.50	37.00	63.50
2-8 x 6-8	"	27.70	37.00	64.70
3-0 x 6-8	"	27.70	37.00	64.70
5-0 x 6-8	"	30.10	37.00	67.10
6-0 x 6-8	"	31.50	37.00	68.50

08 DOORS AND WINDOWS

WOOD AND PLASTIC

WOOD AND PLASTIC	UNIT	MAT.	INST.	TOTAL
08210.90 WOOD FRAMES				
2-6 x 7-0	EA.	26.60	36.70	63.30
2-8 x 7-0	"	28.90	36.70	65.60
3-0 x 7-0	"	30.00	36.70	66.70
5-0 x 7-0	"	32.30	51.50	83.80
6-0 x 7-0	"	33.50	51.50	85.00
Exterior, custom, with threshold, including trim				
Walnut				
3-0 x 7-0	EA.	289.00	64.50	353.50
6-0 x 7-0	"	346.00	64.50	410.50
Oak				
3-0 x 7-0	EA.	266.00	64.50	330.50
6-0 x 7-0	"	323.00	64.50	387.50
Pine				
2-4 x 7-0	EA.	52.00	51.50	103.50
2-6 x 7-0	"	54.50	51.50	106.00
2-8 x 7-0	"	54.50	51.50	106.00
3-0 x 7-0	"	63.50	51.50	115.00
3-4 x 7-0	"	69.50	51.50	121.00
6-0 x 7-0	"	125.00	85.50	210.50
08300.10 SPECIAL DOORS				
Vault door and frame, class 5, steel	EA.	2,860	257.00	3,117
Overhead door, coiling insulated				
Chain gear, no frame, 12' x 12'	EA.	1,490	322.00	1,812
Aluminum, bronze glass panels, 12-9 x 13-0	"	1,600	257.00	1,857
Garage door, flush insulated metal, primed, 9-0 x 7-0	"	624.00	86.50	710.50
Sliding metal fire doors, motorized, fusible link, 3 hr.				
3-0 x 6-8	EA.	2,060	514.00	2,574
3-8 x 6-8	"	2,060	514.00	2,574
4-0 x 8-0	"	2,120	514.00	2,634
5-0 x 8-0	"	2,170	514.00	2,684
Metal clad doors, including electric motor				
Light duty				
Minimum	S.F.	22.90	4.30	27.20
Maximum	"	37.80	10.30	48.10
Heavy duty				
Minimum	S.F.	44.60	12.85	57.45
Maximum	"	75.50	16.10	91.60
Hangar doors, based on 150' openings				
To 20' high	S.F.	25.20	3.85	29.05
20' to 40' high	"	22.90	2.40	25.30
40' to 60' high	"	25.20	1.60	26.80
60' to 80' high	"	28.60	0.96	29.56
Over 80' high	"	40.00	0.77	40.77
Counter doors, (roll-up shutters), standard, manual				
Opening, 4' high				
4' wide	EA.	758.00	214.00	972.00
6' wide	"	1,040	214.00	1,254
8' wide	"	1,110	234.00	1,344
10' wide	"	1,220	322.00	1,542
14' wide	"	1,570	322.00	1,892

WOOD AND PLASTIC

08300.10 SPECIAL DOORS

Item	UNIT	MAT.	INST.	TOTAL
6' high				
4' wide	EA.	770.00	214.00	984.00
6' wide	"	1,160	234.00	1,394
8' wide	"	1,340	257.00	1,597
10' wide	"	1,600	322.00	1,922
14' wide	"	1,750	367.00	2,117
For stainless steel, add to material, 80%				
For motor operator, add	EA.			770.00
Service doors, (roll up shutters), standard, manual				
Opening				
8' high x 8' wide	EA.	968.00	143.00	1,111
10' high x 10' wide	"	1,280	214.00	1,494
12' high x 12' wide	"	1,710	322.00	2,032
14' high x 14' wide	"	2,240	429.00	2,669
16' high x 14' wide	"	2,540	429.00	2,969
20' high x 14' wide	"	3,240	643.00	3,883
24' high x 16' wide	"	4,590	572.00	5,162
For motor operator				
Up to 12-0 x 12-0, add	EA.			880.00
Over 12-0 x 12-0, add	"			1,100
Roll-up doors				
13-0 high x 14-0 wide	EA.	763.00	370.00	1,133
12-0 high x 14-0 wide	"	979.00	370.00	1,349
Top coiling grilles, manually operated, steel or aluminum				
Opening, 4' high x				
4' wide	EA.	1,150	103.00	1,253
6' wide	"	1,150	103.00	1,253
8' wide	"	1,150	143.00	1,293
12' wide	"	1,360	143.00	1,503
16' wide	"	1,710	214.00	1,924
6' high x				
4' wide	EA.	1,150	214.00	1,364
6' wide	"	1,150	234.00	1,384
8' wide	"	1,150	257.00	1,407
12' wide	"	1,460	286.00	1,746
16' wide	"	1,740	367.00	2,107
Side coiling grilles, manually operated, aluminum				
Opening, 8' high x				
18' wide	EA.	6,410	2,400	8,810
24' wide	"	7,580	2,750	10,330
12' high x				
12' wide	EA.	5,480	2,400	7,880
18' wide	"	8,160	2,750	10,910
24' wide	"	17,840	3,210	21,050
Accordion folding doors, tracks and fittings included				
Vinyl covered, 2 layers	S.F.	8.00	10.35	18.35
Woven mahogany and vinyl	"	10.00	10.35	20.35
Economy vinyl	"	6.70	10.35	17.05
Rigid polyvinyl chloride	"	10.70	10.35	21.05
Sectional wood overhead doors, frames not included				
Commercial grade, heavy duty, 1-3/4" thick, manual				
8' x 8'	EA.	541.00	216.00	757.00

WOOD AND PLASTIC

08300.10 SPECIAL DOORS

Description	UNIT	MAT.	INST.	TOTAL
10' x 10'	EA.	840.00	236.00	1,076
12' x 12'	"	1,130	259.00	1,389
Chain hoist				
12' x 16' high	EA.	1,750	322.00	2,072
14' x 14' high	"	1,630	429.00	2,059
20' x 8' high	"	1,520	514.00	2,034
16' high	"	3,500	643.00	4,143
Sectional metal overhead doors, complete				
Residential grade, manual				
9' x 7'	EA.	305.00	104.00	409.00
16' x 7'	"	687.00	130.00	817.00
Commercial grade				
8' x 8'	EA.	466.00	214.00	680.00
10' x 10'	"	700.00	234.00	934.00
12' x 12'	"	933.00	257.00	1,190
20' x 14', with chain hoist	"	2,330	514.00	2,844
Sliding glass doors				
Tempered plate glass, 1/4" thick				
6' wide				
Economy grade	EA.	636.00	86.50	722.50
Premium grade	"	738.00	86.50	824.50
12' wide				
Economy grade	EA.	933.00	129.00	1,062
Premium grade	"	1,280	129.00	1,409
Insulating glass, 5/8" thick				
6' wide				
Economy grade	EA.	890.00	86.50	976.50
Premium grade	"	1,650	86.50	1,737
12' wide				
Economy grade	EA.	700.00	129.00	829.00
Premium grade	"	1,280	129.00	1,409
1" thick				
6' wide				
Economy grade	EA.	909.00	85.50	994.50
Premium grade	"	1,050	85.50	1,136
12' wide				
Economy grade	EA.	1,400	129.00	1,529
Premium grade	"	1,750	129.00	1,879
Added costs				
Custom quality, add to material, 30%				
Vertical lift doors, channel frame construction				
20' high x				
10' wide	EA.	11,660	440.00	12,100
15' wide	"	12,830	440.00	13,270
20' wide	"	13,990	786.00	14,776
25' wide	"	15,160	786.00	15,946
25' high				
20' wide	EA.	17,490	786.00	18,276
25' wide	"	18,660	917.00	19,577
30' high				
25' wide	EA.	19,820	917.00	20,737
30' wide	"	22,150	917.00	23,067

08 DOORS AND WINDOWS

WOOD AND PLASTIC

08300.10 SPECIAL DOORS	UNIT	MAT.	INST.	TOTAL
35' wide	EA.	24,490	917.00	25,407
Residential storm door				
Minimum	EA.	93.50	42.90	136.40
Average	"	128.00	42.90	170.90
Maximum	"	280.00	64.50	344.50

STOREFRONTS

08410.10 STOREFRONTS	UNIT	MAT.	INST.	TOTAL
Storefront, aluminum and glass				
Minimum	S.F.	15.10	3.60	18.70
Average	"	20.15	4.10	24.25
Maximum	"	27.70	4.75	32.45
Entrance doors				
1/2" thick glass				
3' x 7'	EA.	2,430	238.00	2,668
6' x 7'	"	4,970	357.00	5,327
3/4" thick glass				
3' x 7'	EA.	2,710	238.00	2,948
6' x 7'	"	5,430	357.00	5,787
1" thick glass				
3' x 7'	EA.	3,000	238.00	3,238
6' x 7'	"	6,060	357.00	6,417
For anodized color, add to material, 25%				
Revolving doors				
7'diameter, 7' high				
Minimum	EA.	8,430	2,750	11,180
Average	"	10,050	4,400	14,450
Maximum	"	17,900	5,500	23,400

METAL WINDOWS

08510.10 STEEL WINDOWS	UNIT	MAT.	INST.	TOTAL
Steel windows, primed				
Casements				
Operable				

METAL WINDOWS

METAL WINDOWS	UNIT	MAT.	INST.	TOTAL
08510.10 STEEL WINDOWS				
Minimum	S.F.	13.85	1.90	15.75
Maximum	"	11.55	1.80	13.35
Fixed sash	"	6.95	1.45	8.40
Double hung	"	23.10	1.60	24.70
Industrial windows				
Horizontally pivoted sash	S.F.	12.70	1.90	14.60
Fixed sash	"	11.55	1.60	13.15
Security sash				
Operable	S.F.	17.35	1.90	19.25
Fixed	"	11.55	1.60	13.15
Picture window	"	13.85	1.60	15.45
Projecting sash				
Minimum	S.F.	13.85	1.80	15.65
Maximum	"	15.00	1.80	16.80
Mullions	L.F.	6.95	1.45	8.40
08520.10 ALUMINUM WINDOWS				
Jalousie				
3-0 x 4-0	EA.	156.00	35.70	191.70
3-0 x 5-0	"	185.00	35.70	220.70
Fixed window				
6 sf to 8 sf	S.F.	8.80	4.10	12.90
12 sf to 16 sf	"	8.20	3.20	11.40
Projecting window				
6 sf to 8 sf	S.F.	20.15	7.15	27.30
12 sf to 16 sf	"	17.65	4.75	22.40
Horizontal sliding				
6 sf to 8 sf	S.F.	13.85	3.60	17.45
12 sf to 16 sf	"	12.60	2.85	15.45
Double hung				
6 sf to 8 sf	S.F.	12.60	5.70	18.30
10 sf to 12 sf	"	12.00	4.75	16.75
Storm window, 0.5 cfm, up to				
60 u.i. (united inches)	EA.	45.40	14.30	59.70
70 u.i.	"	46.60	14.30	60.90
80 u.i.	"	50.50	14.30	64.80
90 u.i.	"	47.40	15.85	63.25
100 u.i.	"	48.50	15.85	64.35
2.0 cfm, up to				
60 u.i.	EA.	52.00	14.30	66.30
70 u.i.	"	53.00	14.30	67.30
80 u.i.	"	55.50	14.30	69.80
90 u.i.	"	60.00	15.85	75.85
100 u.i.	"	60.00	15.85	75.85

WOOD AND PLASTIC	UNIT	MAT.	INST.	TOTAL
08600.10 WOOD WINDOWS				
Double hung				
24" x 36"				
Minimum	EA.	142.00	25.90	167.90
Average	"	204.00	32.40	236.40
Maximum	"	254.00	43.20	297.20
24" x 48"				
Minimum	EA.	163.00	25.70	188.70
Average	"	212.00	32.20	244.20
Maximum	"	257.00	42.90	299.90
30" x 48"				
Minimum	EA.	204.00	28.80	232.80
Average	"	254.00	37.00	291.00
Maximum	"	305.00	52.00	357.00
30" x 60"				
Minimum	EA.	210.00	28.60	238.60
Average	"	257.00	36.70	293.70
Maximum	"	303.00	51.50	354.50
Casement				
1 leaf, 22" x 38" high				
Minimum	EA.	204.00	25.90	229.90
Average	"	229.00	32.40	261.40
Maximum	"	254.00	43.20	297.20
2 leaf, 50" x 50" high				
Minimum	EA.	509.00	32.40	541.40
Average	"	611.00	43.20	654.20
Maximum	"	661.00	65.00	726.00
3 leaf, 71" x 62" high				
Minimum	EA.	816.00	32.20	848.20
Average	"	898.00	42.90	940.90
Maximum	"	979.00	64.50	1,044
4 leaf, 95" x 75" high				
Minimum	EA.	902.00	38.70	940.70
Average	"	1,190	51.50	1,242
Maximum	"	1,500	85.50	1,586
5 leaf, 119" x 75" high				
Minimum	EA.	1,170	36.70	1,207
Average	"	1,590	51.50	1,642
Maximum	"	1,850	85.50	1,936
Picture window, fixed glass, 54" x 54" high				
Minimum	EA.	382.00	32.40	414.40
Average	"	477.00	37.00	514.00
Maximum	"	649.00	43.20	692.20
68" x 55" high, minimum	"	554.00	32.20	586.20
Average	"	641.00	36.70	677.70
Maximum	"	793.00	42.90	835.90
Sliding, 40" x 31" high				
Minimum	EA.	178.00	25.90	203.90
Average	"	280.00	32.40	312.40
Maximum	"	331.00	43.20	374.20
52" x 39" high				
Minimum	EA.	210.00	32.20	242.20
Average	"	326.00	36.70	362.70

WOOD AND PLASTIC

08600.10 WOOD WINDOWS

Description	UNIT	MAT.	INST.	TOTAL
Maximum	EA.	361.00	42.90	403.90
64" x 72" high				
Minimum	EA.	350.00	32.20	382.20
Average	"	548.00	42.90	590.90
Maximum	"	612.00	51.50	663.50
Awning windows				
34" x 21" high				
Minimum	EA.	188.00	25.90	213.90
Average	"	216.00	32.40	248.40
Maximum	"	267.00	43.20	310.20
40" x 21" high				
Minimum	EA.	223.00	28.80	251.80
Average	"	242.00	37.00	279.00
Maximum	"	293.00	52.00	345.00
48" x 27" high				
Minimum	EA.	210.00	28.60	238.60
Average	"	268.00	36.70	304.70
Maximum	"	326.00	51.50	377.50
60" x 36" high				
Minimum	EA.	222.00	32.20	254.20
Average	"	431.00	42.90	473.90
Maximum	"	501.00	51.50	552.50
Window frame, milled		2.35	6.45	8.80
Minimum	L.F.	2.00	5.40	7.40
Average	"	2.40	6.80	9.20
Maximum	"	3.85	9.05	12.90

HARDWARE

08710.10 HINGES

Description	UNIT	MAT.	INST.	TOTAL
Hinges				
3 x 3 butts, steel, interior, plain bearing	PAIR			11.00
4 x 4 butts, steel, standard	"			16.50
5 x 4-1/2 butts, bronze/s. steel, heavy duty	"			44.00
Pivot hinges				
Top pivot	EA.			39.60
Intermediate pivot	"			42.90
Bottom pivot	"			88.00
BHMA specifications				
3-1/2 x 3-1/2, full mortise butts				
Plain bearing	PAIR			13.20
Ball bearing	"			16.50
Half surface butts	"			24.20
4 x 4				

08 DOORS AND WINDOWS

HARDWARE	UNIT	MAT.	INST.	TOTAL
08710.10 HINGES				
Full mortise butts, plain bearing, standard duty	PAIR			14.30
Full mortise butts, ball bearing	"			17.60
Half surface butts				
Standard duty	PAIR			24.20
Ball bearing	"			24.20
4-1/2 x 4-1/2				
Full mortise butts, plain bearing	PAIR			15.40
Ball bearing, heavy duty	"			38.50
Half mortise and half surface butts				
Plain bearing	PAIR			24.20
Full surface and half surface butts				
Standard duty	PAIR			37.40
Heavy duty	"			103.00
Full mortise and full slide-in butts, ball bearing	"			17.60
Half mortise butts, ball bearing				
Standard duty	PAIR			82.50
Heavy duty	"			97.00
5 x 5, ball bearing				
Full mortise butts	PAIR			38.50
Half mortise, full & half surface butts	"			86.00
Full mortise, full surface and half surface butts	"			117.00
4 x 4				
Full mortise butts, plain bearing, standard duty	PAIR			7.70
5 x 4-1/2				
Full mortise butts, ball bearing, heavy duty	PAIR			44.00
08710.20 LOCKSETS				
Latchset, heavy duty				
Cylindrical	EA.	99.00	16.20	115.20
Mortise	"	89.00	25.90	114.90
Lockset, heavy duty				
Cylindrical	EA.	124.00	16.20	140.20
Mortise	"	143.00	25.90	168.90
Mortise locks and latchsets, chrome				
Latchset passage or closet latch	EA.	90.50	21.45	111.95
Privacy (bath or bedroom)	"	113.00	21.45	134.45
Entry lockset	"	136.00	21.45	157.45
Classroom lockset (outside key operated)	"	136.00	21.45	157.45
Storeroom lock	"	136.00	21.45	157.45
Front door lock	"	136.00	21.45	157.45
Dormitory or exit lock	"	136.00	21.45	157.45
Preassembled locks and latches, brass				
Latchset, passage or closet latch	EA.	147.00	21.45	168.45
Lockset				
Privacy (bath or bathroom)	EA.	181.00	21.45	202.45
Entry lock	"	215.00	21.45	236.45
Classroom lock (outside key, operated)	"	215.00	21.45	236.45
Storeroom lock	"	261.00	21.45	282.45
Bored locks and latches, satin chrome plated				
Latchset passage or closet latch	EA.	79.50	21.45	100.95
Lockset				

HARDWARE	UNIT	MAT.	INST.	TOTAL
08710.20 LOCKSETS				
Privacy (bath or bedroom)	EA.	102.00	21.45	123.45
Entry lock	"	119.00	21.45	140.45
Classroom lock	"	119.00	21.45	140.45
Corridor lock	"	119.00	21.45	140.45
Miscellaneous locks				
Exit lock with alarm, single door	EA.	374.00	103.00	477.00
Electric strike				
Rim mounted wrought steel	EA.	312.00	64.50	376.50
Mortised, wrought steel with bronze plating	"	102.00	103.00	205.00
Dead bolt				
Bored, wrought brass, keyed both sides	EA.	62.50	42.90	105.40
Mortised, cast brass	"	68.00	42.90	110.90
Lockset, cipher, mechanical	"	1,080	25.70	1,106
08710.30 CLOSERS				
Door closers				
Surface mounted, traditional type, parallel arm				
Standard	EA.	110.00	32.40	142.40
Heavy duty	"	121.00	32.40	153.40
Modern type, parallel arm, standard duty	"	110.00	32.20	142.20
Overhead, concealed, pivot hung, single acting				
Interior	EA.	185.00	32.20	217.20
Exterior	"	381.00	32.20	413.20
Floor concealed, single acting, offset, pivoted				
Interior	EA.	335.00	85.50	420.50
Exterior	"	427.00	85.50	512.50
08710.40 DOOR TRIM				
Door bumper, bronze, wall type	EA.	2.05	5.15	7.20
Wall type, 4" dia. with convex rubber pad, aluminum	"	18.65	5.15	23.80
Floor type				
Aluminum	EA.	2.90	5.15	8.05
Brass	"	2.90	5.15	8.05
Door holders				
Wall type, bronze	EA.	2.90	5.15	8.05
Overhead	"	29.20	12.85	42.05
Floor type	"	29.20	12.85	42.05
Plunger type	"	15.75	12.85	28.60
Wall type, aluminum	"	18.65	12.85	31.50
Surface bolt	"	14.00	5.15	19.15
Panic device		373.00	64.50	437.50
Rim type with thumb piece	EA.			
Mortise	"	515.00	65.00	580.00
Vertical rod	"	598.00	65.00	663.00
Labeled, rim type	"	534.00	65.00	599.00
Mortise	"	642.00	65.00	707.00
Vertical rod	"	725.00	65.00	790.00
Silencers, rubber type	"	0.12	5.15	5.27
Dust proof strike with plate, brass	"	10.50	8.55	19.05
Flush bolt, lever extension, brass, rated	"	18.65	5.15	23.80

HARDWARE

	UNIT	MAT.	INST.	TOTAL
08710.40 DOOR TRIM				
Surface bolt with strike, brass, 6" long	EA.	14.00	5.15	19.15
Door coordinator, labeled, brass, satin chrome	"	67.50	18.35	85.85
Door plates				
Kick plate, aluminum, 3 beveled edges				
10" x 28"	EA.	10.50	12.85	23.35
10" x 30"	"	12.70	12.95	25.65
10" x 34"	"	14.00	12.95	26.95
10" x 38"	"	15.25	12.95	28.20
Push plate, 4" x 16"				
Aluminum	EA.	3.80	5.20	9.00
Bronze	"	7.65	5.20	12.85
Stainless steel	"	5.70	5.20	10.90
Armor plate, 40" x 34"	"	51.50	10.30	61.80
Pull handle, 4" x 16"				
Aluminum	EA.	10.50	5.15	15.65
Bronze	"	18.65	5.15	23.80
Stainless steel	"	18.65	5.15	23.80
Hasp assembly				
3"	EA.	0.81	4.30	5.11
4-1/2"	"	1.10	5.70	6.80
6"	"	1.75	7.35	9.10
Electro-magnetic door holder				
Wall mounted	EA.	105.00	85.50	190.50
Floor mounted	"	163.00	85.50	248.50
Smoke detector door holder				
Photo electric type	EA.	169.00	85.50	254.50
Ionization type	"	163.00	85.50	248.50
Pneumatic operators, activated by rubber mats				
Swing				
Single	EA.	787.00	214.00	1,001
Double	"	1,040	322.00	1,362
Sliding				
Single	EA.	711.00	214.00	925.00
Double	"	1,280	322.00	1,602
08710.60 WEATHERSTRIPPING				
Weatherstrip, head and jamb, metal strip, neoprene bulb				
Standard duty	L.F.	2.50	1.45	3.95
Heavy duty	"	3.30	1.60	4.90
Spring type				
Metal doors	EA.	31.80	64.50	96.30
Wood doors	"	31.80	85.50	117.30
Sponge type with adhesive backing	"	28.90	25.70	54.60
Astragal				
1-3/4" x 13 ga., aluminum	L.F.	2.90	2.15	5.05
1-3/8" x 5/8", oak	"	2.90	1.70	4.60
Thresholds				
Bronze	L.F.	20.15	6.50	26.65
Aluminum				
Plain	L.F.	7.55	6.50	14.05
Vinyl insert	"	15.10	6.50	21.60
Aluminum with grit	"	13.25	6.50	19.75

HARDWARE

08710.60 WEATHERSTRIPPING	UNIT	MAT.	INST.	TOTAL
Steel				
Plain	L.F.	10.10	6.50	16.60
Interlocking	"	15.75	21.60	37.35

GLAZING

08810.10 GLAZING	UNIT	MAT.	INST.	TOTAL
Sheet glass, 1/8" thick	S.F.	4.75	1.60	6.35
Plate glass, bronze or grey, 1/4" thick	"	7.05	2.60	9.65
Clear	"	5.40	2.60	8.00
Polished	"	6.00	2.60	8.60
Plexiglass				
1/8" thick	S.F.	1.90	2.60	4.50
1/4" thick	"	3.80	1.60	5.40
Float glass, clear				
3/16" thick	S.F.	3.65	2.40	6.05
1/4" thick	"	4.65	2.60	7.25
5/16" thick	"	6.60	2.85	9.45
3/8" thick	"	8.45	3.60	12.05
1/2" thick	"	12.45	4.75	17.20
5/8" thick	"	14.30	5.70	20.00
3/4" thick	"	16.75	7.15	23.90
1" thick	"	31.20	9.50	40.70
Tinted glass, polished plate, twin ground				
3/16" thick	S.F.	3.90	2.40	6.30
1/4" thick	"	4.90	2.60	7.50
3/8" thick	"	8.70	3.60	12.30
1/2" thick	"	12.95	4.75	17.70
Total, full vision, all glass window system				
To 10' high				
Minimum	S.F.	19.65	7.15	26.80
Average	"	20.80	7.15	27.95
Maximum	"	21.95	7.15	29.10
10' to 20' high				
Minimum	S.F.	24.25	7.15	31.40
Average	"	26.60	7.15	33.75
Maximum	"	28.90	7.15	36.05
Insulated glass, bronze or gray				
1/2" thick	S.F.	9.25	4.75	14.00
1" thick	"	11.00	7.15	18.15
Clear				
1/2" thick	S.F.	7.30	4.75	12.05
1" thick	"	10.30	7.15	17.45

GLAZING	UNIT	MAT.	INST.	TOTAL
08810.10 GLAZING				
Spandrel glass, bronze or grey, polished, 1 side, 1/4" thick	S.F.	6.35	2.60	8.95
Tempered glass (safety)				
Clear sheet glass				
1/8" thick	S.F.	5.65	1.60	7.25
3/16" thick	"	6.35	2.20	8.55
Clear float glass				
1/4" thick	S.F.	5.90	2.40	8.30
5/16" thick	"	10.80	2.85	13.65
3/8" thick	"	15.60	3.55	19.15
1/2" thick	"	25.40	4.75	30.15
5/8" thick	"	27.80	5.70	33.50
3/4" thick	"	33.40	9.50	42.90
Tinted float glass				
3/16" thick	S.F.	7.30	2.20	9.50
1/4" thick	"	7.50	2.40	9.90
3/8" thick	"	18.50	3.55	22.05
1/2" thick	"	29.10	4.75	33.85
Laminated glass				
Float safety glass with polyvinyl plastic interlayer				
1/4", sheet or float				
Two lites, 1/8" thick, clear glass	S.F.	7.40	2.40	9.80
1/2" thick, float glass				
Two lites, 1/4" thick, clear glass	S.F.	12.85	4.75	17.60
Tinted glass	"	13.15	4.75	17.90
Insulating glass, two lites, clear float glass				
1/2" thick	S.F.	7.85	4.75	12.60
5/8" thick	"	9.05	5.70	14.75
3/4" thick	"	9.95	7.15	17.10
7/8" thick	"	10.45	8.20	18.65
1" thick	"	11.20	9.55	20.75
Glass seal edge				
3/8" thick	S.F.	6.05	4.75	10.80
Tinted glass				
1/2" thick	S.F.	12.35	4.75	17.10
1" thick	"	13.30	9.50	22.80
Tempered, clear				
1" thick	S.F.	39.30	9.50	48.80
Wire reinforced	"	42.70	9.50	52.20
Plate mirror glass				
1/4" thick				
15 sf	S.F.	7.95	2.85	10.80
Over 15 sf	"	7.30	2.60	9.90
Transparent, one way vision, 1/4" thick	"	14.10	2.85	16.95
Sheet mirror glass				
3/16" thick	S.F.	6.60	2.85	9.45
1/4" thick	"	7.05	2.40	9.45
Wall tiles, 12" x 12"				
Clear glass	S.F.	2.10	1.60	3.70
Veined glass	"	2.45	1.60	4.05
Wire glass, 1/4" thick				
Clear	S.F.	6.95	9.50	16.45
Hammered	"	8.10	9.50	17.60

GLAZING	UNIT	MAT.	INST.	TOTAL
08810.10 GLAZING				
Obscure	S.F.	10.40	9.50	19.90
Bullet resistant glass, plate glass with inter-leaved vinyl				
1-3/16" thick				
To 15 sf	S.F.	22.05	14.30	36.35
Over 15 sf	"	25.80	14.30	40.10
2" thick				
To 15 sf	S.F.	43.90	23.80	67.70
Over 15 sf	"	54.50	23.80	78.30
Glazing accessories				
Neoprene glazing gaskets				
1/4" glass	L.F.	0.81	1.15	1.96
3/8" glass	"	1.00	1.20	2.20
1/2" glass	"	1.30	1.25	2.55
3/4" glass	"	1.75	1.30	3.05
1" glass	"	1.95	1.45	3.40
Mullion section				
1/4" glass	L.F.	0.35	0.57	0.92
3/8" glass	"	0.46	0.71	1.17
1/2" glass	"	0.69	0.82	1.51
3/4" glass	"	1.05	0.95	2.00
1" glass	"	1.40	1.15	2.55
Moulded corners	EA.	1.50	19.05	20.55

GLAZED CURTAIN WALLS	UNIT	MAT.	INST.	TOTAL
08910.10 GLAZED CURTAIN WALLS				
Curtain wall, aluminum system, framing sections				
2" x 3"				
Jamb	L.F.	6.15	2.40	8.55
Horizontal	"	6.25	2.40	8.65
Mullion	"	8.30	2.40	10.70
2" x 4"				
Jamb	L.F.	8.30	3.55	11.85
Horizontal	"	8.55	3.55	12.10
Mullion	"	8.30	3.55	11.85
3" x 5-1/2"				
Jamb	L.F.	11.00	3.55	14.55
Horizontal	"	12.25	3.55	15.80
Mullion	"	11.10	3.55	14.65
4" corner mullion	"	14.65	4.75	19.40
Coping sections				
1/8" x 8"	L.F.	15.35	4.75	20.10
1/8" x 9"	"	15.50	4.75	20.25
1/8" x 12-1/2"	"	15.85	5.70	21.55

GLAZED CURTAIN WALLS

08910.10 GLAZED CURTAIN WALLS	UNIT	MAT.	INST.	TOTAL
Sill section				
1/8" x 6"	L.F.	15.90	2.85	18.75
1/8" x 7"	"	15.30	2.85	18.15
1/8" x 8-1/2"	"	15.60	2.85	18.45
Column covers, aluminum				
1/8" x 26"	L.F.	15.30	7.15	22.45
1/8" x 34"	"	15.30	7.50	22.80
1/8" x 38"	"	15.50	7.50	23.00
Doors				
Aluminum framed, standard hardware				
Narrow stile				
2-6 x 7-0	EA.			
3-0 x 7-0	"			
3-6 x 7-0	"			
Wide stile				
2-6 x 7-0	EA.			
3-0 x 7-0	"			
3-6 x 7-0	"			
Flush panel doors, to match adjacent wall panels				
2-6 x 7-0	EA.	497.00	178.00	675.00
3-0 x 7-0	"	523.00	178.00	701.00
3-6 x 7-0	"	543.00	178.00	721.00
Wall panel, insulated				
"U"=.08	S.F.	6.25	2.40	8.65
"U"=.10	"	5.90	2.40	8.30
"U"=.15	"	5.30	2.40	7.70
Window wall system, complete				
Minimum	S.F.			
Average	"			
Maximum	"			
Added costs				
For bronze, add 20% to material				
For stainless steel, add 50% to material				

SUPPORT SYSTEMS	UNIT	MAT.	INST.	TOTAL
09110.10 METAL STUDS				
Studs, non load bearing, galvanized				
2-1/2", 20 ga.				
12" o.c.	S.F.	0.53	0.54	1.07
16" o.c.	"	0.44	0.43	0.87
25 ga.				
12" o.c.	S.F.	0.38	0.54	0.92
16" o.c.	"	0.28	0.43	0.71
24" o.c.	"	0.20	0.36	0.56
3-5/8", 20 ga.				
12" o.c.	S.F.	0.55	0.65	1.20
16" o.c.	"	0.53	0.52	1.05
24" o.c.	"	0.47	0.43	0.90
25 ga.				
12" o.c.	S.F.	0.34	0.65	0.99
16" o.c.	"	0.29	0.52	0.81
24" o.c.	"	0.26	0.43	0.69
4", 20 ga.				
12" o.c.	S.F.	0.73	0.65	1.38
16" o.c.	"	0.62	0.52	1.14
24" o.c.	"	0.53	0.43	0.96
25 ga.				
12" o.c.	S.F.	0.40	0.65	1.05
16" o.c.	"	0.37	0.52	0.89
24" o.c.	"	0.30	0.43	0.73
6", 20 ga.				
12" o.c.	S.F.	0.89	0.81	1.70
16" o.c.	"	0.76	0.65	1.41
24" o.c.	"	0.62	0.54	1.16
25 ga.				
12" o.c.	S.F.	0.49	0.81	1.30
16" o.c.	"	0.44	0.65	1.09
24" o.c.	"	0.40	0.54	0.94
Load bearing studs, galvanized				
3-5/8", 16 ga.				
12" o.c.	S.F.	1.50	0.65	2.15
16" o.c.	"	1.35	0.52	1.87
18 ga.				
12" o.c.	S.F.	1.25	0.43	1.68
16" o.c.	"	1.15	0.52	1.67
4", 16 ga.				
12" o.c.	S.F.	1.60	0.65	2.25
16" o.c.	"	1.30	0.52	1.82
6", 16 ga.				
12" o.c.	S.F.	1.75	0.81	2.56
16" o.c.	"	1.65	0.65	2.30
Furring				
On beams and columns				
7/8" channel	L.F.	0.34	1.75	2.09
1-1/2" channel	"	0.56	1.75	2.31
On ceilings				
3/4" furring channels				
12" o.c.	S.F.	0.38	1.85	2.23

09 FINISHES

SUPPORT SYSTEMS

09110.10 METAL STUDS	UNIT	MAT.	INST.	TOTAL
16" o.c.	S.F.	0.31	1.75	2.06
24" o.c.	"	0.22	0.97	1.19
1-1/2" furring channels		0.28	1.45	1.73
12" o.c.	S.F.	0.40	1.25	1.65
16" o.c.	"	0.33	1.15	1.48
24" o.c.	"	0.28	1.05	1.33
On walls				
3/4" furring channels				
12" o.c.	S.F.	0.31	0.90	1.21
16" o.c.	"	0.26	0.85	1.11
24" o.c.	"	0.22	0.80	1.02
1-1/2" furring channels				
12" o.c.	S.F.	0.40	0.97	1.37
16" o.c.	"			
24" o.c.	"	0.28	0.85	1.13

LATH AND PLASTER

09205.10 GYPSUM LATH	UNIT	MAT.	INST.	TOTAL
Gypsum lath, 1/2" thick				
Clipped	S.Y.	4.80	1.45	6.25
Nailed	"	4.40	1.60	6.00

09205.20 METAL LATH	UNIT	MAT.	INST.	TOTAL
Diamond expanded, galvanized				
2.5 lb., on walls				
Nailed	S.Y.	1.55	3.20	4.75
Wired	"	1.60	3.65	5.25
On ceilings				
Nailed	S.Y.	1.80	3.65	5.45
Wired	"	1.85	4.30	6.15
3.4 lb., on walls				
Nailed	S.Y.	2.00	3.20	5.20
Wired	"	2.05	3.65	5.70
On ceilings				
Nailed	S.Y.	2.05	3.65	5.70
Wired	"	2.10	4.30	6.40
Flat rib				
2.75 lb., on walls				
Nailed	S.Y.	1.95	3.20	5.15
Wired	"	2.00	3.65	5.65
On ceilings				

LATH AND PLASTER	UNIT	MAT.	INST.	TOTAL
09205.20 METAL LATH				
Nailed	S.Y.	2.00	3.65	5.65
Wired	"	2.05	4.30	6.35
3.41 lb., on walls				
Nailed	S.Y.	2.25	3.20	5.45
Wired	"	2.30	3.65	5.95
On ceilings				
Nailed	S.Y.	2.35	3.65	6.00
Wired	"	2.40	4.30	6.70
Stucco lath				
1.8 lb.	S.Y.	3.15	3.25	6.40
3.6 lb.	"	3.60	3.25	6.85
Paper backed				
Minimum	S.Y.	3.70	2.60	6.30
Maximum	"	4.20	3.70	7.90
09205.60 PLASTER ACCESSORIES				
Expansion joint, 3/4", 26 ga., galvanized, one piece	L.F.	0.77	0.64	1.41
Plaster corner beads, 3/4", galvanized	"	0.24	0.73	0.97
Casing bead, expanded flange, galvanized	"	0.24	0.64	0.88
Expanded wing, 1-1/4" wide, galvanized	"	0.18	0.64	0.82
Joint clips for lath	EA.	0.21	0.13	0.34
Metal base, galvanized, 2-1/2" high	L.F.	0.46	0.86	1.32
Stud clips for gypsum lath	EA.	0.21	0.13	0.34
Tie wire galvanized, 18 ga., 25 lb. hank	"			27.50
Sound deadening board, 1/4", nailed or clipped	S.F.	0.24	0.43	0.67
09210.10 PLASTER				
Gypsum plaster, trowel finish, 2 coats				
Ceilings	S.Y.	3.95	6.10	10.05
Walls	"	3.95	5.70	9.65
3 coats				
Ceilings	S.Y.	5.50	8.55	14.05
Walls	"	5.50	7.30	12.80
Vermiculite plaster				
2 coats				
Ceilings	S.Y.	4.15	9.10	13.25
Walls	"	4.15	8.50	12.65
3 coats				
Ceilings	S.Y.	6.50	11.60	18.10
Walls	"	6.50	10.20	16.70
Keenes cement plaster				
2 coats				
Ceilings	S.Y.	1.95	7.30	9.25
Walls	"	1.95	6.40	8.35
3 coats				
Ceilings	S.Y.	2.10	8.50	10.60
Walls	"	2.10	7.30	9.40
On columns, add to installation, 50%	"			
Chases, fascia, and soffits, add to installation, 50%	"			
Beams, add to installation, 50%	"			
Patch cracks		0.85	1.30	2.15

09 FINISHES

LATH AND PLASTER

LATH AND PLASTER	UNIT	MAT.	INST.	TOTAL
09220.10 PORTLAND CEMENT PLASTER				
Stucco, portland, gray, 3 coat, 1" thick				
Sand finish	S.Y.	4.15	11.15	15.30
Trowel finish	"	4.15	11.65	15.80
White cement				
Sand finish	S.Y.	4.75	11.65	16.40
Trowel finish	"	4.75	12.80	17.55
Scratch coat				
For ceramic tile	S.Y.	1.45	2.55	4.00
For quarry tile	"	1.45	2.55	4.00
Portland cement plaster				
2 coats, 1/2"	S.Y.	0.29	0.57	0.86
3 coats, 7/8"	"	0.40	0.73	1.13
09250.10 GYPSUM BOARD				
Drywall, plasterboard, 3/8" clipped to				
Metal furred ceiling	S.F.	0.28	0.29	0.57
Columns and beams	"	0.28	0.65	0.93
Walls	"	0.28	0.26	0.54
Nailed or screwed to				
Wood framed ceiling	S.F.	0.28	0.26	0.54
Columns and beams	"	0.28	0.58	0.86
Walls	"	0.28	0.24	0.52
1/2", clipped to				
Metal furred ceiling	S.F.	0.34	0.29	0.63
Columns and beams	"	0.34	0.65	0.99
Walls	"	0.34	0.26	0.60
Nailed or screwed to				
Wood framed ceiling	S.F.	0.34	0.26	0.60
Columns and beams	"	0.34	0.58	0.92
Walls	"	0.34	0.24	0.58
5/8", clipped to				
Metal furred ceiling	S.F.	0.37	0.32	0.69
Columns and beams	"	0.37	0.72	1.09
Walls	"	0.37	0.29	0.66
Nailed or screwed to				
Wood framed ceiling	S.F.	0.37	0.32	0.69
Columns and beams	"	0.37	0.72	1.09
Walls	"	0.37	0.29	0.66
Vinyl faced, clipped to metal studs				
1/2"	S.F.	1.10	0.32	1.42
5/8"	"	1.15	0.32	1.47
Add for				
Fire resistant	S.F.			0.06
Water resistant	"			0.10
Water and fire resistant	"			0.13
Taping and finishing joints		0.04	0.17	0.21
Minimum	S.F.	0.03	0.18	0.21
Average	"	0.04	0.23	0.27
Maximum	"	0.06	0.27	0.33
Casing bead		0.10	0.86	0.96
Minimum	L.F.	0.09	0.77	0.86
Average	"	0.10	0.90	1.00

09 FINISHES

LATH AND PLASTER

09250.10 GYPSUM BOARD	UNIT	MAT.	INST.	TOTAL
Maximum	L.F.	0.13	1.35	1.48
Corner bead		0.12	0.86	0.98
Minimum	L.F.	0.10	0.77	0.87
Average	"	0.12	0.90	1.02
Maximum	"	0.15	1.35	1.50

TILE

09310.10 CERAMIC TILE	UNIT	MAT.	INST.	TOTAL
Glazed wall tile, 4-1/4" x 4-1/4"				
Minimum	S.F.	1.65	1.80	3.45
Average	"	2.55	2.10	4.65
Maximum	"	6.70	2.55	9.25
Base, 4-1/4" high				
Minimum	L.F.	2.35	3.15	5.50
Average	"	5.85	3.15	9.00
Maximum	"	4.10	3.15	7.25
Unglazed floor tile				
Portland cement bed, cushion edge, face mounted				
1" x 1"	S.F.	4.95	2.30	7.25
1" x 2"	"	7.95	2.20	10.15
2" x 2"	"	5.20	2.10	7.30
Adhesive bed, with white grout				
1" x 1"	S.F.	4.40	2.30	6.70
1" x 2"	"	4.50	2.20	6.70
2" x 2"	"	4.60	2.10	6.70
Organic adhesive bed, thin set, back mounted				
1" x 1"	S.F.	4.50	2.30	6.80
1" x 2"	"	4.60	2.20	6.80
2" x 2"	"	4.80	2.10	6.90
For group 2 colors, add to material, 10%				
For group 3 colors, add to material, 20%				
For abrasive surface, add to material, 25%				
Unglazed wall tile				
Organic adhesive, face mounted cushion edge				
1" x 1"				
Minimum	S.F.	1.50	2.10	3.60
Average	"	1.55	2.30	3.85
Maximum	"	1.65	2.50	4.15
1" x 2"				
Minimum	S.F.	1.50	2.00	3.50
Average	"	1.65	2.20	3.85
Maximum	"	2.10	2.40	4.50

09 FINISHES

TILE	UNIT	MAT.	INST.	TOTAL
09310.10 CERAMIC TILE				
2" x 2"				
Minimum	S.F.	1.65	1.95	3.60
Average	"	1.75	2.10	3.85
Maximum	"	2.20	2.30	4.50
Back mounted				
1" x 1"				
Minimum	S.F.	1.50	2.10	3.60
Average	"	1.55	2.30	3.85
Maximum	"	1.65	2.50	4.15
1" x 2"				
Minimum	S.F.	1.50	2.00	3.50
Average	"	1.65	2.20	3.85
Maximum	"	2.10	2.40	4.50
2" x 2"				
Minimum	S.F.	1.65	1.95	3.60
Average	"	1.75	2.10	3.85
Maximum	"	2.20	2.30	4.50
For glazed finish, add to material, 25%				
For glazed mosaic, add to material, 100%				
For metallic colors, add to material, 125%				
For exterior wall use, add to total, 25%				
For exterior soffit, add to total, 25%				
For portland cement bed, add to total, 25%				
For dry set portland cement bed, add to total, 10%				
Conductive floor tile, unglazed square edged				
Portland cement bed				
1 x 1	S.F.	4.30	3.15	7.45
1-9/16 x 1-9/16	"	4.10	3.15	7.25
Dry set				
1 x 1	S.F.	4.55	3.15	7.70
1-9/16 x 1-9/16	"	4.20	3.15	7.35
Epoxy bed with epoxy joints				
1 x 1	S.F.	4.80	3.15	7.95
1-9/16 x 1-9/16	"	4.55	3.15	7.70
For WWF in bed add to total, 15%				
For abrasive surface, add to material, 40%				
Ceramic accessories				
Towel bar, 24" long				
Minimum	EA.	17.60	10.55	28.15
Average	"	22.00	13.15	35.15
Maximum	"	33.00	17.55	50.55
Soap dish		17.50	20.95	38.45
Minimum	EA.	15.40	17.55	32.95
Average	"	17.60	21.95	39.55
Maximum	"	24.20	26.30	50.50
09330.10 QUARRY TILE				
Floor				
4 x 4 x 1/2"	S.F.	3.70	3.35	7.05
6 x 6 x 1/2"	"	3.75	3.15	6.90
6 x 6 x 3/4"	"	4.30	3.15	7.45

TILE	UNIT	MAT.	INST.	TOTAL
09330.10 QUARRY TILE				
Wall, applied to 3/4" portland cement bed				
4 x 4 x 1/2"	S.F.	3.30	5.05	8.35
6 x 6 x 3/4"	"	4.15	4.20	8.35
Cove base				
5 x 6 x 1/2" straight top	L.F.	3.25	4.20	7.45
6 x 6 x 3/4" round top	"	3.45	4.20	7.65
Stair treads 6 x 6 x 3/4"	"	5.25	6.30	11.55
Window sill 6 x 8 x 3/4"	"	4.20	5.05	9.25
For abrasive surface, add to material, 25%				
09410.10 TERRAZZO				
Floors on concrete, 1-3/4" thick, 5/8" topping				
Gray cement	S.F.	2.85	3.65	6.50
White cement	"	3.10	3.65	6.75
Sand cushion, 3" thick, 5/8" top, 1/4"				
Gray cement	S.F.	3.35	4.25	7.60
White cement	"	3.75	4.25	8.00
Monolithic terrazzo, 3-1/2" base slab, 5/8" topping	"	2.40	3.20	5.60
Terrazzo wainscot, cast-in-place, 1/2" thick	"	6.55	6.40	12.95
Base, cast in place, terrazzo cove type, 6" high	L.F.	7.15	3.65	10.80
Curb, cast in place, 6" wide x 6" high, polished top	"	14.85	12.75	27.60
For venetian type terrazzo, add to material, 10%				
For abrasive heavy duty terrazzo, add to material, 15%				
Divider strips				
Zinc	L.F.			0.88
Brass	"			1.65
Stairs, cast-in-place, topping on concrete or metal				
1-1/2" thick treads, 12" wide	L.F.	25.50	4.25	29.75
Combined tread and riser	"	36.00	12.75	48.75
Precast terrazzo, thin set				
Terrazzo tiles, non-slip surface				
9" x 9" x 1" thick	S.F.	10.65	3.65	14.30
12" x 12"				
1" thick	S.F.	11.65	3.40	15.05
1-1/2" thick	"	12.20	3.65	15.85
18" x 18" x 1-1/2" thick	"	15.80	3.65	19.45
24" x 24" x 1-1/2" thick	"	20.60	3.00	23.60
For white cement, add to material, 10%				
For venetian type terrazzo, add to material, 25%				
Terrazzo wainscot				
12" x 12" x 1" thick	S.F.	10.75	6.40	17.15
18" x 18" x 1-1/2" thick	"	17.25	7.30	24.55
Base				
6" high				
Straight	L.F.	7.15	1.95	9.10
Coved	"	8.65	1.95	10.60
8" high				
Straight	L.F.	7.95	2.15	10.10
Coved	"	9.40	2.15	11.55
Terrazzo curbs				
8" wide x 8" high	L.F.	28.00	4.65	32.65

09 FINISHES

TILE

09410.10 TERRAZZO

	UNIT	MAT.	INST.	TOTAL
6" wide x 6" high	L.F.	22.55	4.65	27.20
Precast terrazzo stair treads, 12" wide				
1-1/2" thick				
Diamond pattern	L.F.	15.45	4.65	20.10
Non-slip surface	"	17.80	4.65	22.45
2" thick				
Diamond pattern	L.F.	18.30	4.65	22.95
Non-slip surface	"	22.90	5.10	28.00
Stair risers, 1" thick to 6" high				
Straight sections	L.F.	7.70	2.55	10.25
Cove sections	"	9.15	2.55	11.70
Combined tread and riser				
Straight sections				
1-1/2" tread, 3/4" riser	L.F.	37.10	7.30	44.40
3" tread, 1" riser	"	39.30	7.30	46.60
Curved sections				
2" tread, 1" riser	L.F.	42.00	8.50	50.50
3" tread, 1" riser	"	43.40	8.50	51.90
Stair stringers, notched for treads and risers				
1" thick	L.F.	19.25	6.40	25.65
2" thick	"	19.70	8.50	28.20
Landings, structural, nonslip				
1-1/2" thick	S.F.	14.85	4.25	19.10
3" thick	"	15.90	5.10	21.00
Conductive terrazzo, spark proof industrial floor				
Epoxy terrazo				
Floor	S.F.	7.00	1.60	8.60
Base	"	7.00	2.15	9.15
Polyacrylate				
Floor	S.F.	5.70	1.60	7.30
Base	"	5.70	2.15	7.85
Polyester				
Floor	S.F.	1.90	1.00	2.90
Base	"	1.90	1.30	3.20
Synthetic latex mastic				
Floor	S.F.	3.30	1.60	4.90
Base	"	3.30	2.15	5.45

ACOUSTICAL TREATMENT

09510.10 CEILINGS AND WALLS

	UNIT	MAT.	INST.	TOTAL
Acoustical panels, suspension system not included				
Fiberglass panels				

ACOUSTICAL TREATMENT	UNIT	MAT.	INST.	TOTAL
09510.10 CEILINGS AND WALLS				
5/8" thick				
2' x 2'	S.F.	0.59	0.29	0.88
2' x 4'	"	0.59	0.26	0.85
3/4" thick				
2' x 2'	S.F.	0.86	0.29	1.15
2' x 4'	"	0.86	0.26	1.12
Glass cloth faced fiberglass panels				
3/4" thick	S.F.	1.85	0.32	2.17
1" thick	"	2.20	0.32	2.52
Mineral fiber panels				
5/8" thick				
2' x 2'	S.F.	0.65	0.29	0.94
2' x 4'	"	0.65	0.26	0.91
3/4" thick				
2' x 2'	S.F.	0.80	0.29	1.09
2' x 4'	"	0.80	0.26	1.06
For aluminum faced panels, add to material, 80%				
For vinyl faced panels, add to total, 125%				
For fire rated panels, add to material, 75%				
Wood fiber panels				
2" thick				
2' x 2'	S.F.	1.50	0.29	1.79
2' x 4'	"	1.50	0.26	1.76
For flameproofing, add to material, 10%				
For sculptured finish, add to material, 15%				
Air distributing panels				
3/4" thick	S.F.	1.50	0.64	2.14
5/8" thick	"	1.25	0.51	1.76
Acoustical tiles, suspension system not included				
Fiberglass tile, 12" x 12"				
5/8" thick	S.F.	0.33	0.47	0.80
3/4" thick	"	0.35	0.57	0.92
Glass cloth faced fiberglass tile				
3/4" thick	S.F.	1.45	0.57	2.02
3" thick	"	1.45	0.64	2.09
Mineral fiber tile, 12" x 12"				
5/8" thick				
Standard	S.F.	0.90	0.51	1.41
Vinyl faced	"	1.05	0.51	1.56
3/4" thick				
Standard	S.F.	1.15	0.51	1.66
Vinyl faced	"	1.30	0.51	1.81
Fire rated	"	2.30	0.51	2.81
Aluminum or mylar faced	"	2.65	0.51	3.16
Wood fiber tile, 12" x 12"				
1/2" thick	S.F.	0.75	0.51	1.26
3/4" thick	"	1.00	0.51	1.51
For flameproofing, add to material, 10%				
For sculptured 3 dimensional, add to material, 50%				
Metal pan units, 24 ga. steel				
12" x 12"	S.F.	5.15	0.57	5.72
12" x 24"	"	3.30	0.51	3.81

09 FINISHES

ACOUSTICAL TREATMENT

09510.10 CEILINGS AND WALLS

CEILINGS AND WALLS	UNIT	MAT.	INST.	TOTAL
Aluminum, .025" thick				
12" x 12"	S.F.	6.40	0.57	6.97
12" x 24"	"	3.45	0.51	3.96
Anodized aluminum, 0.25" thick				
12" x 12"	S.F.	7.45	0.57	8.02
12" x 24"	"	4.60	0.51	5.11
Stainless steel, 24 ga.				
12" x 12"	S.F.	10.20	0.57	10.77
12" x 24"	"	11.00	0.51	11.51
For flameproof sound absorbing pads, add to material	"			1.10
Metal ceiling systems				
.020" thick panels				
10', 12', and 16' lengths	S.F.	2.85	0.73	3.58
Custom lengths, 3' to 20'	"	2.85	0.73	3.58
.025" thick panels				
32 sf, 38 sf, and 52 sf pieces	S.F.	3.25	0.86	4.11
Custom lengths, 10 sf to 65 sf	"	3.35	0.86	4.21
Carriers, black, add	"			1.65
Recess filler strip, add	"			0.55
Custom lengths, add	"			0.83
Sound absorption walls, with fabric cover				
2-6" x 9' x 3/4"	S.F.	5.25	0.86	6.11
2' x 9' x 1"	"	6.40	0.86	7.26
Starter spline	L.F.	0.91	0.64	1.55
Internal spline	"	0.80	0.64	1.44
Acoustical treatment				
Barriers for plenums				
Leaded vinyl				
0.48 lb per sf	S.F.	1.65	1.20	2.85
0.87 lb per sf	"	1.65	1.30	2.95
Aluminum foil, fiberglass reinforcement				
Minimum	S.F.	0.65	0.86	1.51
Maximum	"	0.65	1.30	1.95
Aluminum mesh, paper backed	"	0.58	0.86	1.44
Fibered cement sheet, 3/16" thick	"	1.25	0.92	2.17
Sheet lead, 1/64" thick	"	1.55	0.64	2.19
Sound attenuation blanket				
1" thick	S.F.	0.35	2.55	2.90
1-1/2" thick	"	0.46	2.55	3.01
2" thick	"	0.63	2.55	3.18
3" thick	"	0.75	2.85	3.60
Ceiling suspension systems				
T bar system				
2' x 4'	S.F.	1.35	0.26	1.61
2' x 2'	"	1.55	0.29	1.84
Concealed Z bar suspension system, 12" module	"	1.25	0.43	1.68
For 1-1/2" carrier channels, 4' o.c., add	"			0.28
Carrier channel for recessed light fixtures	"			0.44

FLOORING

FLOORING	UNIT	MAT.	INST.	TOTAL
09550.10 WOOD FLOORING				
Wood strip flooring, unfinished				
Fir floor				
C and better				
Vertical grain	S.F.	2.20	0.86	3.06
Flat grain	"	2.10	0.86	2.96
Oak floor				
Minimum	S.F.	2.65	1.25	3.90
Average	"	2.90	1.20	4.10
Maximum	"	3.80	1.25	5.05
Maple floor				
25/32" x 2-1/4"				
Minimum	S.F.	2.55	1.25	3.80
Maximum	"	3.20	1.25	4.45
33/32" x 3-1/4"				
Minimum	S.F.	2.50	1.20	3.70
Maximum	"	3.05	1.20	4.25
Added costs				
For factory finish, add to material, 10%				
For random width floor, add to total, 20%				
For simulated pegs, add to total, 10%				
Wood block industrial flooring				
Creosoted				
2" thick	S.F.	2.40	0.68	3.08
2-1/2" thick	"	2.55	0.80	3.35
3" thick	"	2.65	0.86	3.51
Parquet, 5/16", white oak				
Finished	S.F.	6.10	1.30	7.40
Unfinished	"	2.05	1.30	3.35
Gym floor, 2 ply felt, 25/32" maple, finished, in mastic	"	4.60	1.45	6.05
Over wood sleepers	"	5.10	1.60	6.70
Finishing, sand, fill, finish, and wax	"	0.44	0.65	1.09
Refinish sand, seal, and 2 coats of polyurethane	"	0.78	0.86	1.64
Clean and wax floors	"	0.13	0.13	0.26
09630.10 UNIT MASONRY FLOORING				
Clay brick				
9 x 4-1/2 x 3" thick				
Glazed	S.F.	5.25	2.15	7.40
Unglazed	"	5.15	2.15	7.30
8 x 4 x 3/4" thick				
Glazed	S.F.	4.75	2.35	7.10
Unglazed	"	5.10	2.25	7.35
For herringbone pattern, add to labor, 15%				
09660.10 RESILIENT TILE FLOORING				
Solid vinyl tile, 1/8" thick, 12" x 12"				
Marble patterns	S.F.	2.00	0.65	2.65
Solid colors	"	2.10	0.65	2.75
Travertine patterns	"	2.50	0.65	3.15
Conductive resilient flooring, vinyl tile				
1/8" thick, 12" x 12"	S.F.	3.45	0.73	4.18

FLOORING	UNIT	MAT.	INST.	TOTAL
09665.10 RESILIENT SHEET FLOORING				
Vinyl sheet flooring				
Minimum	S.F.	2.25	0.26	2.51
Average	"	2.85	0.32	3.17
Maximum	"	5.05	0.43	5.48
Cove, to 6"	L.F.	0.64	0.52	1.16
Fluid applied resilient flooring				
Polyurethane, poured in place, 3/8" thick	S.F.	5.65	2.15	7.80
Vinyl sheet goods, backed				
0.070" thick	S.F.	2.30	0.32	2.62
0.093" thick	"	3.45	0.32	3.77
0.125" thick	"	4.05	0.32	4.37
0.250" thick	"	4.60	0.32	4.92
09678.10 RESILIENT BASE AND ACCESSORIES				
Wall base, vinyl				
Group 1				
4" high	L.F.	0.58	0.86	1.44
6" high	"	0.81	0.86	1.67
Group 2				
4" high	L.F.	0.51	0.86	1.37
6" high	"	0.61	0.86	1.47
Group 3				
4" high	L.F.	0.55	0.86	1.41
6" high	"	0.66	0.86	1.52
Stair accessories				
Treads, 1/4" x 12", rubber diamond surface				
Marbled	L.F.	4.15	2.15	6.30
Plain	"	4.05	2.15	6.20
Grit strip safety tread, 12" wide, colors				
3/16" thick	L.F.	5.80	2.15	7.95
5/16" thick	"	7.50	2.15	9.65
Risers, 7" high, 1/8" thick, colors				
Flat	L.F.	1.45	1.30	2.75
Coved	"	1.70	1.30	3.00
Nosing, rubber				
3/16" thick, 3" wide				
Black	L.F.	1.75	1.30	3.05
Colors	"	1.85	1.30	3.15
6" wide				
Black	L.F.	2.75	2.15	4.90
Colors	"	3.00	2.15	5.15

CARPET	UNIT	MAT.	INST.	TOTAL
09680.10 FLOOR LEVELING				
Repair and level floors to receive new flooring				
Minimum	S.Y.	0.47	0.86	1.33
Average	"	1.85	2.15	4.00
Maximum	"	2.30	2.55	4.85
09682.10 CARPET PADDING				
Carpet padding				
Foam rubber, waffle type, 0.3" thick	S.Y.	3.95	1.30	5.25
Jute padding				
Minimum	S.Y.	2.90	1.20	4.10
Average	"	3.50	1.30	4.80
Maximum	"	4.30	1.45	5.75
Sponge rubber cushion				
Minimum	S.Y.	3.60	1.20	4.80
Average	"	3.85	1.30	5.15
Maximum	"	4.10	1.45	5.55
Urethane cushion, 3/8" thick				
Minimum	S.Y.	2.30	1.20	3.50
Average	"	2.65	1.30	3.95
Maximum	"	2.90	1.45	4.35
09685.10 CARPET				
Carpet, acrylic				
24 oz., light traffic	S.Y.	18.90	1.45	20.35
28 oz., medium traffic	"	23.95	1.45	25.40
Residential				
Nylon				
15 oz., light traffic	S.Y.	13.60	1.45	15.05
28 oz., medium traffic	"	16.30	1.45	17.75
Commercial				
Nylon				
28 oz., medium traffic	S.Y.	17.40	1.45	18.85
35 oz., heavy traffic	"	20.80	1.45	22.25
Wool				
30 oz., medium traffic	S.Y.	28.40	1.45	29.85
36 oz., medium traffic	"	29.90	1.45	31.35
42 oz., heavy traffic	"	32.60	1.45	34.05
Carpet tile				
Foam backed		2.50	0.26	2.76
Minimum	S.F.	2.00	0.27	2.27
Average	"	2.30	0.30	2.60
Maximum	"	3.30	0.34	3.64
Tufted loop or shag		2.75	0.26	3.01
Minimum	S.F.	2.15	0.27	2.42
Average	"	2.55	0.30	2.85
Maximum	"	3.95	0.34	4.29
Clean and vacuum carpet				
Minimum	S.Y.	0.19	0.10	0.29
Average	"	0.31	0.17	0.48
Maximum	"	0.38	0.26	0.64

09 FINISHES

CARPET

09700.10 SPECIAL FLOORING	UNIT	MAT.	INST.	TOTAL
Epoxy flooring, marble chips				
Epoxy with colored quartz chips in 1/4" base	S.F.	6.95	1.45	8.40
Heavy duty epoxy topping, 3/16" thick	"	6.00	1.45	7.45
Epoxy terrazzo				
1/4" thick chemical resistant	S.F.	6.70	1.60	8.30

SPECIAL COATINGS

09800.10 FIREPROOFING	UNIT	MAT.	INST.	TOTAL
Sprayed fireproofing, fluted decking				
1 hr fire rating, 1/2" thick	S.F.	0.31	0.13	0.44
2 hr fire rating, 7/8" thick	"	0.55	0.18	0.73
Sprayed on roofs and beams				
1 hr fire rating, 1-3/8" thick	S.F.	0.87	0.21	1.08
2 hr fire rating, 1-1/2" thick	"	0.95	0.23	1.18
3 hr fire rating, 1-5/8" thick	"	1.00	0.23	1.23
Sprayed on beams				
1 hr fire rating, 7/16" thick	S.F.	0.30	0.13	0.43
2 hr fire rating, 1/2" thick	"	0.31	0.13	0.44
3 hr fire rating, 7/8" thick	"	0.56	0.16	0.72
4 hr fire rating, 1-1/4" thick	"	0.79	0.21	1.00
Sprayed on columns				
1 hr fire rating, 1-3/16" thick	S.F.	0.77	0.16	0.93
2 hr fire rating, 1-1/4" thick	"	0.78	0.21	0.99
3 hr fire rating, 1-7/8" thick	"	1.20	0.23	1.43
4 hr fire rating, 2-1/2" thick	"	1.60	0.29	1.89
Sprayed on flat surfaces				
2 hr fire rating, 5/8" thick	S.F.	0.41	0.12	0.53
3 hr fire rating, 7/8" thick	"	0.55	0.16	0.71
4 hr fire rating, 1-1/4" thick	"	0.77	0.21	0.98

PAINTING	UNIT	MAT.	INST.	TOTAL
09910.10 EXTERIOR PAINTING				
Exterior painting				
Wood surfaces, 1 coat primer, two coats paint				
Door and frame	EA.	3.30	43.90	47.20
Windows	S.F.	0.06	0.66	0.72
Wood trim	"	0.16	0.66	0.82
Wood siding	"	0.16	0.33	0.49
Hardboard surfaces				
One coat primer, two coats paint	S.F.	0.16	0.33	0.49
Asbestos cement surfaces				
One coat primer, two coats paint	S.F.	0.18	0.33	0.51
Galvanized surfaces, galvanized primer				
One coat primer, two coats paint	S.F.	0.17	0.31	0.48
Stucco surfaces, acrylic primer, acrylic latex paint				
One coat primer, two coats paint	S.F.	0.17	0.44	0.61
Concrete masonry unit surfaces, brush work				
One coat filler, one coat paint	S.F.	0.19	0.33	0.52
Two coats epoxy	"	0.28	0.44	0.72
Texture coating	"	0.25	0.26	0.51
Concrete surfaces				
One coat filler, one coat paint	S.F.	0.16	0.33	0.49
Two coats paint	"	0.23	0.44	0.67
Structural steel				
One field coat paint, brush work				
Light framing	S.F.	0.08	0.22	0.30
Heavy framing	"	0.08	0.13	0.21
One field coat paint, spray work				
Light framing	S.F.	0.06	0.07	0.13
Heavy framing	"	0.06	0.05	0.11
Miscellaneous steel items, spray work, one coat				
Exposed decking	S.F.	0.08	0.22	0.30
Joist	"	0.08	0.33	0.41
Columns	"	0.08	0.33	0.41
Pipes, one coat primer, one coat paint				
4" dia.	L.F.	0.06	0.33	0.39
8" dia.	"	0.16	0.44	0.60
12" dia.	"	0.29	0.66	0.95
Paint letters on pipe with brush	"	0.30	0.65	0.95
Paint pipe insulation cloth cover	S.F.	0.18	0.33	0.51
Sprinkler system piping	L.F.	0.19	0.22	0.41
Miscellaneous surfaces				
Stair pipe rails				
Two rails	L.F.	0.13	0.88	1.01
One rail	"	0.07	0.53	0.60
Stair to 4' wide, including rails, per riser	EA.	0.58	3.75	4.33
Gratings and frames	S.F.	0.13	0.88	1.01
Ladders	L.F.	0.16	0.75	0.91
Miscellaneous exposed metal	S.F.	0.12	0.26	0.38
09920.10 INTERIOR PAINTING				
Walls, concrete and masonry, brush, primer, acrylic				
One coat primer, one coat paint	S.F.	0.08	0.33	0.41

09 FINISHES

PAINTING

	UNIT	MAT.	INST.	TOTAL
09920.10 INTERIOR PAINTING				
Two coats paint	S.F.	0.13	0.44	0.57
Plywood, paint	"	0.05	0.15	0.20
Natural finish	"	0.06	0.16	0.22
Wood, paint	"	0.06	0.16	0.22
Natural finish	"	0.07	0.19	0.26
Metal				
One coat filler	S.F.	0.13	0.16	0.29
One coat primer, one coat paint	"	0.16	0.33	0.49
Two coats paint	"	0.20	0.44	0.64
Plaster or gypsum board, paint	"	0.07	0.15	0.22
Epoxy	"	0.10	0.16	0.26
Ceilings, one coat paint, wood	"	0.06	0.19	0.25
Concrete	"	0.10	0.16	0.26
Plaster	"	0.06	0.15	0.21
Miscellaneous metal, brushwork	"	0.12	0.33	0.45
09920.30 DOORS AND MILLWORK				
Painting, doors				
Minimum	S.F.	0.13	0.88	1.01
Average	"	0.19	1.30	1.49
Maximum	"	0.32	1.75	2.07
Cabinets, shelves, and millwork				
Minimum	S.F.	0.10	0.44	0.54
Average	"	0.13	0.75	0.88
Maximum	"	0.16	1.30	1.46
09920.60 WINDOWS				
Painting, windows				
Minimum	S.F.	0.06	0.53	0.59
Average	"	0.07	0.66	0.73
Maximum	"	0.10	1.05	1.15
09955.10 WALL COVERING				
Vinyl wall covering				
Medium duty	S.F.	0.92	0.38	1.30
Heavy duty	"	1.20	0.44	1.64
Over pipes and irregular shapes				
Lightweight, 13 oz.	S.F.	0.85	0.52	1.37
Medium weight, 25 oz.	"	0.96	0.58	1.54
Heavy weight, 34 oz.	"	1.10	0.65	1.75
Cork wall covering				
1' x 1' squares				
1/4" thick	S.F.	1.35	0.66	2.01
1/2" thick	"	1.50	0.66	2.16
3/4" thick	"	1.65	0.66	2.31
Wall fabrics				
Natural fabrics, grass cloths				
Minimum	S.F.	0.77	0.40	1.17
Average	"	0.96	0.44	1.40
Maximum	"	3.00	0.53	3.53
Flexible gypsum coated wall fabric, fire resistant	"	0.88	0.26	1.14

PAINTING

	UNIT	MAT.	INST.	TOTAL
09955.10 WALL COVERING				
Vinyl corner guards				
3/4" x 3/4" x 8'	EA.	4.15	3.30	7.45
2-3/4" x 2-3/4" x 4'	"	2.45	3.30	5.75
09980.10 PAINTING PREPARATION				
Cleaning, light				
Wood	S.F.	0.04	0.07	0.11
Plaster or gypsum wallboard	"	0.04	0.06	0.10
Prepare sprinkler piping for painting	L.F.	0.03	0.05	0.08
Prepare insulated pipe for painting	S.F.	0.03	0.07	0.10
Normal painting prep, masonry and concrete				
Unpainted	S.F.	0.04	0.04	0.08
Painted	"	0.04	0.07	0.11
Plaster or gypsum				
Unpainted	S.F.	0.04	0.04	0.08
Painted	"	0.04	0.07	0.11
Wood				
Unpainted	S.F.	0.04	0.04	0.08
Painted	"	0.04	0.07	0.11
Painted steel, light rusting	"	0.04	0.09	0.13
Sandblasting				
Brush off blast	S.F.	0.13	0.18	0.31
Commercial blast	"	0.25	0.44	0.69
Near white metal blast	"	0.36	0.75	1.11
White metal blast	"	0.48	0.88	1.36
09980.15 PAINT				
Paint, enamel				
600 sf per gal.	GAL			23.65
550 sf per gal.	"			18.25
500 sf per gal.	"			13.75
450 sf per gal.	"			12.65
350 sf per gal.	"			12.10
Filler, 60 sf per gal.	"			14.55
Latex, 400 sf per gal.	"			11.45
Aluminum				
400 sf per gal.	GAL			15.95
500 sf per gal.	"			30.80
Red lead, 350 sf per gal.	"			27.50
Primer				
400 sf per gal.	GAL			18.25
300 sf per gal.	"			18.15
Latex base, interior, white	"			15.20
Sealer and varnish				
400 sf per gal.	GAL			14.00
425 sf per gal.	"			20.50
600 sf per gal.	"			27.50

SPECIALTIES	UNIT	MAT.	INST.	TOTAL
10110.10 CHALKBOARDS				
Chalkboard, metal frame, 1/4" thick				
48"x60"	EA.	173.00	25.70	198.70
48"x96"	"	277.00	28.60	305.60
48"x144"	"	427.00	32.20	459.20
48"x192"	"	543.00	36.70	579.70
Liquid chalkboard				
48"x60"	EA.	196.00	25.70	221.70
48"x96"	"	312.00	28.60	340.60
48"x144"	"	485.00	32.20	517.20
48"x192"	"	612.00	36.70	648.70
Map rail, deluxe	L.F.	3.45	1.30	4.75
Average	PCT.			2.75
10165.10 TOILET PARTITIONS				
Toilet partition, plastic laminate				
Ceiling mounted	EA.	606.00	90.50	696.50
Floor mounted	"	549.00	68.00	617.00
Metal				
Ceiling mounted	EA.	404.00	90.50	494.50
Floor mounted	"	370.00	68.00	438.00
Wheel chair partition, plastic laminate				
Ceiling mounted	EA.	764.00	90.50	854.50
Floor mounted	"	715.00	68.00	783.00
Painted metal				
Ceiling mounted	EA.	572.00	90.50	662.50
Floor mounted	"	528.00	68.00	596.00
Urinal screen, plastic laminate				
Wall hung	EA.	277.00	33.90	310.90
Floor mounted	"	243.00	33.90	276.90
Porcelain enameled steel, floor mounted	"	308.00	33.90	341.90
Painted metal, floor mounted	"	208.00	33.90	241.90
Stainless steel, floor mounted	"	385.00	33.90	418.90
Metal toilet partitions				
Toilet partitions, front door and side divider, floor mounted				
Porcelain enameled steel	EA.	660.00	68.00	728.00
Painted steel	"	393.00	68.00	461.00
Stainless steel	"	759.00	68.00	827.00
10185.10 SHOWER STALLS				
Shower receptors				
Precast, terrazzo				
32" x 32"	EA.	196.00	25.00	221.00
32" x 48"	"	264.00	30.00	294.00
Concrete				
32" x 32"	EA.	88.00	25.00	113.00
48" x 48"	"	132.00	33.30	165.30
Shower door, trim and hardware				
Economy, 24" wide, chrome frame, tempered glass	EA.	139.00	30.00	169.00
Porcelain enameled steel, flush	"	266.00	30.00	296.00
Baked enameled steel, flush	"	156.00	30.00	186.00
Aluminum frame, tempered glass, 48" wide, sliding	"	341.00	37.50	378.50

SPECIALTIES	UNIT	MAT.	INST.	TOTAL
10185.10 SHOWER STALLS				
Folding	EA.	319.00	37.50	356.50
Aluminum frame and tempered glass, molded plastic				
Complete with receptor and door				
32" x 32"	EA.	455.00	75.00	530.00
36" x 36"	"	495.00	75.00	570.00
40" x 40"	"	572.00	85.50	657.50
Shower compartment, precast concrete receptor				
Single entry type				
Porcelain enameled steel	EA.	1,390	300.00	1,690
Baked enameled steel	"	808.00	300.00	1,108
Stainless steel	"	1,560	300.00	1,860
Double entry type				
Porcelain enameled steel	EA.	4,040	375.00	4,415
Baked enameled steel	"	2,540	375.00	2,915
Stainless steel	"	4,040	375.00	4,415
10190.10 CUBICLES				
Hospital track				
Ceiling hung	L.F.	2.90	2.85	5.75
Suspended	"	3.25	3.65	6.90
Hospital metal dividers, galvanized steel				
Baked enamel finish				71.50
54" high				
10" glass light	L.F.	76.00	12.85	88.85
14" glass light	"	76.00	12.85	88.85
24" glass light	"	72.50	12.85	85.35
60" high				
10" glass light	L.F.	80.50	14.30	94.80
14" glass light	"	80.50	14.30	94.80
24" glass light	"	76.00	14.30	90.30
Stainless steel				121.00
54" high				
10" glass light	L.F.	128.00	14.30	142.30
14" glass light	"	140.00	14.30	154.30
24" glass light	"	169.00	14.30	183.30
60" high				
14" glass light	L.F.	146.00	16.10	162.10
14" glass light	"	146.00	16.10	162.10
24" glass light	"	181.00	16.10	197.10
10210.10 VENTS AND WALL LOUVERS				
Block vent, 8"x16"x4" aluminum, w/screen, mill finish	EA.	86.50	9.50	96.00
Standard	"	46.20	8.95	55.15
Vents w/screen, 4" deep, 8" wide, 5" high				
Modular	EA.	59.00	8.95	67.95
Aluminum gable louvers	S.F.	10.70	4.75	15.45
Vent screen aluminum, 4" wide, continuous	L.F.	2.85	0.95	3.80
Louvers, aluminum, anodized, fixed blade				
Horizontal line	S.F.	31.80	7.15	38.95
Vertical line	"	31.80	7.15	38.95
Wall louver, aluminum mill finish				

SPECIALTIES	UNIT	MAT.	INST.	TOTAL
10210.10 VENTS AND WALL LOUVERS				
Under, 2 sf	S.F.	22.05	3.60	25.65
2 to 4 sf	"	18.90	3.20	22.10
5 to 10 sf	"	17.95	3.20	21.15
Galvanized steel				
Under 2 sf	S.F.	21.40	3.60	25.00
2 to 4 sf	"	15.10	3.20	18.30
5 to 10 sf	"	14.15	3.20	17.35
10225.10 DOOR LOUVERS				
Fixed, 1" thick, enameled steel				
8"x8"	EA.	34.60	3.25	37.85
12"x8"	"	38.30	3.25	41.55
12"x12"	"	43.30	3.70	47.00
16"x12"	"	58.00	3.95	61.95
18"x12"	"	58.00	6.45	64.45
20"x8"	"	70.50	3.70	74.20
20"x12"	"	72.50	7.35	79.85
20"x16"	"	75.00	8.55	83.55
20"x20"	"	77.00	10.30	87.30
24"x12"	"	64.50	8.55	73.05
24"x16"	"	68.00	9.20	77.20
24"x18"	"	79.50	9.90	89.40
24"x20"	"	82.50	10.70	93.20
24"x24"	"	82.50	11.70	94.20
26"x26"	"	98.50	16.10	114.60
10270.40 ACCESS & PEDESTAL FLOOR				
Panels, no covering, 2'x2'				
Plain	S.F.	5.20	0.34	5.54
Perforated	"	7.80	13.55	21.35
Pedestals				
For 6" to 12" clearance	EA.	4.35	2.70	7.05
Stringers				
2'	L.F.	0.69	1.30	1.99
6'	"	0.69	0.90	1.59
Accessories				
Ramp assembly	S.F.	26.00	1.10	27.10
Elevated floor assembly	"	42.20	1.00	43.20
Handrail	L.F.	31.50	13.55	45.05
Fascia plate	"	14.75	6.80	21.55
For carpet tiles, add	S.F.			3.30
For vinyl flooring, add	"			4.10
RF shielding components, floor liner				
Hot rolled steel sheet				
14 ga.	S.F.	1.15	0.68	1.83
11 ga.	"	1.80	2.10	3.90

SPECIALTIES	UNIT	MAT.	INST.	TOTAL
10290.10 PEST CONTROL				
Termite control				
Under slab spraying				
Minimum	S.F.	0.10	0.05	0.15
Average	"	0.12	0.11	0.23
Maximum	"	0.18	0.21	0.39
10350.10 FLAGPOLES				
Installed in concrete base				
Fiberglass				
25' high	EA.	421.00	171.00	592.00
50' high	"	3,700	429.00	4,129
Aluminum				
25' high	EA.	690.00	171.00	861.00
50' high	"	2,860	429.00	3,289
Bonderized steel				
25' high	EA.	1,400	198.00	1,598
50' high	"	3,130	514.00	3,644
Freestanding tapered, fiberglass				
30' high	EA.	774.00	184.00	958.00
40' high	"	1,630	234.00	1,864
50' high	"	2,970	257.00	3,227
60' high	"	3,760	303.00	4,063
Wall mounted, with collar, brushed aluminum finish				
15' long	EA.	438.00	129.00	567.00
18' long	"	449.00	129.00	578.00
20' long	"	527.00	135.00	662.00
24' long	"	507.00	151.00	658.00
Outrigger, wall, including base				
10' long	EA.	140.00	171.00	311.00
20' long	"	185.00	214.00	399.00
10400.10 IDENTIFYING DEVICES				
Directory and bulletin boards				
Open face boards				
Chrome plated steel frame	S.F.	16.75	12.85	29.60
Aluminum framed	"	14.45	12.85	27.30
Bronze framed	"	16.50	12.85	29.35
Stainless steel framed	"	32.30	12.85	45.15
Tack board, aluminum framed	"	9.80	12.85	22.65
Visual aid board, aluminum framed	"	9.80	12.85	22.65
Glass encased boards, hinged and keyed				
Aluminum framed	S.F.	38.10	32.20	70.30
Bronze framed	"	41.60	32.20	73.80
Stainless steel framed	"	98.00	32.20	130.20
Chrome plated steel framed	"	96.00	32.20	128.20
Metal plaque				
Cast bronze	S.F.	248.00	21.45	269.45
Aluminum	"	231.00	21.45	252.45
Metal engraved plaque				
Porcelain steel	S.F.	318.00	21.45	339.45

SPECIALTIES	UNIT	MAT.	INST.	TOTAL
10400.10 IDENTIFYING DEVICES				
Stainless steel	S.F.	289.00	21.45	310.45
Brass	"	318.00	21.45	339.45
Aluminum	"	150.00	21.45	171.45
Metal built-up plaque				
Bronze	S.F.	243.00	25.70	268.70
Copper and bronze	"	248.00	25.70	273.70
Copper and aluminum	"	277.00	25.70	302.70
Metal nameplate plaques				
Cast bronze	S.F.	248.00	16.10	264.10
Cast aluminum	"	225.00	16.10	241.10
Engraved, 1-1/2" x 6"				
Bronze	EA.	77.50	16.10	93.60
Aluminum	"	77.50	16.10	93.60
Letters, on masonry or concrete, aluminum, satin finish				
1/2" thick				
2" high	EA.	7.50	10.30	17.80
4" high	"	8.65	12.85	21.50
6" high	"	11.55	14.30	25.85
3/4" thick				
8" high	EA.	14.45	16.10	30.55
10" high	"	19.90	18.35	38.25
1" thick				
12" high	EA.	25.40	21.45	46.85
14" high	"	31.80	25.70	57.50
16" high	"	43.30	32.20	75.50
For polished aluminum add, 15%				
For clear anodized aluminum add, 15%				
For colored anodic aluminum add, 30%				
For profiled and color enameled letters add, 50%				
Cast bronze, satin finish letters				
3/8" thick				
2" high	EA.	14.30	10.30	24.60
4" high	"	15.30	12.85	28.15
1/2" thick, 6" high	"	20.50	14.30	34.80
5/8" thick, 8" high	"	26.90	16.10	43.00
1" thick				
10" high	EA.	46.20	19.35	65.55
12" high	"	52.50	21.45	73.95
14" high	"	60.50	25.70	86.20
16" high	"	80.50	32.20	112.70
Interior door signs, adhesive, flexible				
2" x 8"	EA.	10.70	5.10	15.80
4" x 4"	"	10.70	5.10	15.80
6" x 7"	"	14.15	5.10	19.25
6" x 9"	"	17.80	5.10	22.90
10" x 9"	"	24.55	5.10	29.65
10" x 12"	"	31.80	5.10	36.90
Hard plastic type, no frame				
3" x 8"	EA.	26.60	5.10	31.70
4" x 4"	"	21.35	5.10	26.45
4" x 12"	"	26.60	5.10	31.70
Hard plastic type, with frame				

10 SPECIALTIES

SPECIALTIES	UNIT	MAT.	INST.	TOTAL
10400.10 IDENTIFYING DEVICES				
3" x 8"	EA.	75.00	5.10	80.10
4" x 4"	"	61.00	5.10	66.10
4" x 12"	"	98.00	5.10	103.10
10450.10 CONTROL				
Access control, 7' high, indoor or outdoor impenetrability				
Remote or card control, type B	EA.	2,610	354.00	2,964
Free passage, type B	"	2,380	354.00	2,734
Remote or card control, type AA	"	3,340	354.00	3,694
Free passage, type AA	"	3,000	354.00	3,354
10500.10 LOCKERS				
Locker bench, floor mounted, laminated maple				
4'	EA.	63.00	22.60	85.60
6'	"	91.50	22.60	114.10
Wardrobe locker, single tier type				
12" x 15" x 72"	EA.	105.00	27.10	132.10
18" x 15" x 72"	"	132.00	28.50	160.50
12" x 18" x 72"	"	120.00	30.10	150.10
18" x 18" x 72"	"	143.00	31.90	174.90
Double tier type				
12" x 15" x 36"	EA.	74.50	12.85	87.35
18" x 15" x 36"	"	76.50	12.85	89.35
12" x 18" x 36"	"	68.50	12.85	81.35
18" x 18" x 36"	"	80.00	12.85	92.85
Two person unit				
18" x 15" x 72"	EA.	206.00	42.90	248.90
18" x 18" x 72"	"	120.00	51.50	171.50
Duplex unit				
15" x 15" x 72"	EA.	200.00	25.70	225.70
15" x 21" x 72"	"	212.00	25.70	237.70
Basket lockers, basket sets with baskets				
24 basket set	SET	429.00	129.00	558.00
30 basket set	"	515.00	161.00	676.00
36 basket set	"	572.00	214.00	786.00
42 basket set	"	652.00	257.00	909.00
10520.10 FIRE PROTECTION				
Portable fire extinguishers				
Water pump tank type				
2.5 gal.				
Red enameled galvanized	EA.	71.50	14.15	85.65
Red enameled copper	"	81.00	14.15	95.15
Polished copper	"	123.00	14.15	137.15
Carbon dioxide type, red enamel steel				
Squeeze grip with hose and horn				
2.5 lb	EA.	37.90	14.15	52.05
5 lb	"	79.00	16.30	95.30
10 lb	"	132.00	21.20	153.20
15 lb	"	154.00	26.50	180.50

SPECIALTIES	UNIT	MAT.	INST.	TOTAL
10520.10 FIRE PROTECTION				
20 lb	EA.	176.00	26.50	202.50
Wheeled type				
125 lb	EA.	1,040	42.40	1,082
250 lb	"	974.00	42.40	1,016
500 lb	"	1,920	42.40	1,962
Dry chemical, pressurized type				
Red enameled steel				
2.5 lb	EA.	19.80	14.15	33.95
5 lb	"	30.80	16.30	47.10
10 lb	"	44.00	21.20	65.20
20 lb	"	77.00	26.50	103.50
30 lb	"	214.00	26.50	240.50
Chrome plated steel, 2.5 lb	"	44.00	14.15	58.15
Other type extinguishers				
2.5 gal, stainless steel, pressurized water tanks	EA.	44.00	14.15	58.15
Soda and acid type	"	87.50	14.15	101.65
Cartridge operated, water type	"	146.00	14.15	160.15
Loaded stream, water type	"	69.50	14.15	83.65
Foam type	"	57.50	14.15	71.65
40 gal, wheeled foam type	"	3,210	42.40	3,252
Fire extinguisher cabinets				
Enameled steel				
8" x 12" x 27"	EA.	114.00	42.40	156.40
8" x 16" x 38"	"	146.00	42.40	188.40
Aluminum				
8" x 12" x 27"	EA.	145.00	42.40	187.40
8" x 16" x 38"	"	174.00	42.40	216.40
8" x 12" x 27"	"	251.00	42.40	293.40
Stainless steel				
8" x 16" x 38"	EA.	321.00	42.40	363.40
10550.10 POSTAL SPECIALTIES				
Mail chutes				
Single mail chute				
Finished aluminum	L.F.	408.00	64.50	472.50
Bronze	"	600.00	64.50	664.50
Single mail chute receiving box				
Finished aluminum	EA.	640.00	129.00	769.00
Bronze	"	1,100	129.00	1,229
Twin mail chute, double parallel				
Finished aluminum	FLR	765.00	129.00	894.00
Bronze	"	1,100	129.00	1,229
Receiving box, 36" x 20" x 12"				
Finished aluminum	EA.	1,270	214.00	1,484
Bronze	"	1,840	214.00	2,054
Locked receiving mail box				
Finished aluminum	EA.	640.00	129.00	769.00
Bronze	"	1,090	129.00	1,219
Commercial postal accessories for mail chutes				
Letter slot, brass	EA.	210.00	42.90	252.90
Bulk mail slot, brass	"	595.00	42.90	637.90
Mail boxes				

SPECIALTIES	UNIT	MAT.	INST.	TOTAL
10550.10 POSTAL SPECIALTIES				
Residential postal accessories				
Letter slot	EA.	67.50	12.95	80.45
Rural letter box	"	46.40	32.40	78.80
Apartment house, keyed, 3.5" x 4.5" x 16"	"	67.50	8.65	76.15
Ranch style	"	82.00	12.85	94.85
Commercial postal accessories				
Letter box, with combination lock	EA.	51.00	9.20	60.20
Key lock	"	47.00	9.20	56.20
Mail box, aluminum w/glass front, 4x5				
Horizontal rear load	EA.	65.00	7.35	72.35
Vertical front load	"	49.90	7.35	57.25
10600.10 MOVABLE PARTITIONS				
Partition, movable office type, 2-1/2" thick, vinyl-gypsum	S.F.	8.10	1.30	9.40
Enameled steel frame, with 1/4" thick clear glass	"	11.00	1.30	12.30
Door frame and hardware for movable partitions	EA.	335.00	85.50	420.50
Add for acoustic movable partition	S.F.			0.66
Accordion partition, 12' high				
Vinyl	S.F.	2.75	4.30	7.05
Acoustical	"	4.15	4.30	8.45
Standard office cubicles, 8' high, steel framed				
Baked enamel finish				
100% flush	L.F.	81.00	6.45	87.45
75% flush and 25% glass	"	98.00	7.15	105.15
50% flush and 50% glass	"	141.00	7.15	148.15
100% glass	"	173.00	8.55	181.55
Natural hardwood panels				
100% flush	L.F.	86.50	8.55	95.05
50% flush and 50% glass	"	116.00	9.20	125.20
Plastic laminated panels				
100% flush	L.F.	104.00	8.55	112.55
75% flush and 25% glass	"	127.00	9.20	136.20
50% flush and 50% glass	"	150.00	9.20	159.20
Vinyl covered panels				
100% flush	L.F.	92.50	8.85	101.35
75% flush and 25% glass	"	116.00	9.55	125.55
50% and 50% glass	"	139.00	9.55	148.55
Aluminum framed				
Enameled or anodized aluminum panels				
100% flush	L.F.	69.50	6.45	75.95
75% flush and 25% glass	"	67.00	7.15	74.15
50% flush and 50% glass	"	62.50	7.15	69.65
Vinyl covered panels				
100% flush	L.F.	76.00	8.85	84.85
75% flush and 25% glass	"	93.50	9.55	103.05
50% flush and 50% glass	"	111.00	9.55	120.55
60" high partitions, steel framed				
Enameled panels	L.F.	37.00	5.70	42.70
Natural hardwood panels, two sides	"	162.00	6.00	168.00
Plastic laminated panels	"	81.00	6.00	87.00
Vinyl covered panels	"	110.00	5.70	115.70

SPECIALTIES	UNIT	MAT.	INST.	TOTAL
10600.10 MOVABLE PARTITIONS				
Aluminum framed				
Anodized or baked enamel panels	L.F.	53.50	5.70	59.20
Natural hardwood panels	"	167.00	6.00	173.00
Plastic laminated panels	"	82.00	6.00	88.00
Vinyl covered panels	"	107.00	5.70	112.70
Wire mesh partitions				
Wall panels				
4' x 7'	EA.	81.50	16.10	97.60
4' x 8'	"	91.00	17.15	108.15
4' x 10'	"	114.00	19.80	133.80
Wall filler panels				
1' x 7'	EA.	45.50	16.10	61.60
1' x 8'	"	47.80	17.15	64.95
1' x 10'	"	64.00	18.35	82.35
2' x 7'	"	45.50	16.10	61.60
2' x 8'	"	47.80	17.15	64.95
2' x 10'	"	57.00	18.35	75.35
3' x 7'	"	60.50	16.10	76.60
3' x 8'	"	66.50	17.15	83.65
3' x 10'	"	81.50	18.35	99.85
Ceiling panels				
10' x 2'	EA.	55.00	36.70	91.70
10' x 4'	"	106.00	51.50	157.50
Wall panel with service window				
5' wide				
7' high	EA.	262.00	16.10	278.10
8' high	"	274.00	17.15	291.15
10' high	"	286.00	18.35	304.35
Doors				
Sliding				
3' x 7'	EA.	292.00	64.50	356.50
3' x 8'	"	297.00	73.50	370.50
3' x 10'	"	303.00	103.00	406.00
4' x 7'	"	309.00	73.50	382.50
4' x 8'	"	315.00	103.00	418.00
4' x 10'	"	338.00	129.00	467.00
5' x 7'	"	332.00	103.00	435.00
5' x 8'	"	402.00	129.00	531.00
5' x 10'	"	466.00	129.00	595.00
Swing door				
3' x 7'	EA.	233.00	64.50	297.50
4' x 7'	"	262.00	73.50	335.50
Swing door, with 1' transom				
3' x 7'	EA.	257.00	73.50	330.50
4' x 7'	"	286.00	103.00	389.00
Swing door, with 3' transom				
3' x 7'	EA.	222.00	103.00	325.00
4' x 7'	"	245.00	129.00	374.00
10670.10 SHELVING				
Shelving, enamel, closed side and back, 12" x 36"				
5 shelves	EA.	104.00	42.90	146.90

10 SPECIALTIES

SPECIALTIES	UNIT	MAT.	INST.	TOTAL
10670.10 SHELVING				
8 shelves	EA.	127.00	57.00	184.00
Open				
5 shelves	EA.	69.50	45.20	114.70
8 shelves	"	69.50	57.00	126.50
Metal storage shelving, baked enamel				
7 shelf unit, 72" or 84" high				
10" shelf	L.F.	27.50	27.10	54.60
12" shelf	"	25.40	27.10	52.50
15" shelf	"	31.20	28.60	59.80
18" shelf	"	33.00	31.90	64.90
24" shelf	"	35.80	32.20	68.00
30" shelf	"	39.30	34.30	73.60
36" shelf	"	48.50	36.70	85.20
4 shelf unit, 40" high				
10" shelf	L.F.	23.10	22.60	45.70
12" shelf	"	24.25	23.40	47.65
15" shelf	"	31.20	25.70	56.90
18" shelf	"	31.20	28.50	59.70
24" shelf	"	35.80	30.10	65.90
3 shelf unit, 32" high				
10" shelf	L.F.	19.35	13.55	32.90
12" shelf	"	19.35	13.55	32.90
15" shelf	"	20.60	14.30	34.90
18" shelf	"	22.80	15.95	38.75
24" shelf	"	26.30	16.10	42.40
Single shelf unit, attached to masonry				
10" shelf	L.F.	7.05	4.50	11.55
12" shelf	"	7.75	4.70	12.45
15" shelf	"	8.10	4.95	13.05
18" shelf	"	9.25	5.55	14.80
24" shelf	"	11.00	5.60	16.60
For stainless steel, add to material, 150%				
For attachment to gypsum board, add to labor, 50%				
Built-in wood shelves				
Posts and trimmed plywood	L.F.	1.50	3.65	5.15
Solid clear pine	"	2.55	3.95	6.50
Closet shelf, pine with rod	"	1.70	3.95	5.65
For lumber edge band, add to material	"			0.83
For prefinished shelves, add to material, 225%				
10750.10 TELEPHONE ENCLOSURES				
Telephone enclosure, wall mounted, shelf, 28" x 30" x 15"	EA.	660.00	68.00	728.00
Directory shelf, stainless steel, 3 binders	"	550.00	45.20	595.20
10800.10 BATH ACCESSORIES				
Ash receiver, wall mounted, aluminum	EA.	112.00	13.55	125.55
Grab bar, 1-1/2" dia., stainless steel, wall mounted				
24" long	EA.	31.40	13.55	44.95
36" long	"	44.90	14.25	59.15
42" long	"	51.50	15.05	66.55
48" long	"	61.50	15.95	77.45

10 SPECIALTIES

SPECIALTIES	UNIT	MAT.	INST.	TOTAL
10800.10 BATH ACCESSORIES				
52" long	EA.	67.50	16.95	84.45
1" dia., stainless steel				
12" long	EA.	24.70	11.80	36.50
18" long	"	26.90	12.35	39.25
24" long	"	29.20	13.55	42.75
30" long	"	32.50	14.25	46.75
36" long	"	37.00	15.05	52.05
48" long	"	38.10	15.95	54.05
Hand dryer, surface mounted, 110 volt	"	352.00	33.90	385.90
Medicine cabinet, 16 x 22, baked enamel, steel, lighted	"	56.00	10.85	66.85
With mirror, lighted	"	101.00	18.10	119.10
Mirror, 1/4" plate glass, up to 10 sf	S.F.	5.40	2.70	8.10
Mirror, stainless steel frame				
18"x24"	EA.	84.00	9.05	93.05
18"x32"	"	90.00	10.85	100.85
18"x36"	"	95.50	13.55	109.05
24"x30"	"	112.00	13.55	125.55
24"x36"	"	129.00	15.05	144.05
24"x48"	"	224.00	22.60	246.60
24"x60"	"	236.00	27.10	263.10
30"x30"	"	129.00	27.10	156.10
30"x72"	"	359.00	33.90	392.90
48"x72"	"	376.00	45.20	421.20
With shelf, 18"x24"	"	129.00	10.85	139.85
Sanitary napkin dispenser, stainless steel, wall mounted	"	309.00	18.10	327.10
Shower rod, 1" diameter				
Chrome finish over brass	EA.	50.50	13.55	64.05
Stainless steel	"	48.80	13.55	62.35
Soap dish, stainless steel, wall mounted	"	61.50	18.10	79.60
Toilet tissue dispenser, stainless, wall mounted				
Single roll	EA.	30.30	6.80	37.10
Double roll	"	47.10	7.75	54.85
Towel dispenser, stainless steel				
Flush mounted	EA.	88.00	15.05	103.05
Surface mounted	"	82.50	13.55	96.05
Combination towel dispenser and waste receptacle	"	275.00	18.10	293.10
Towel bar, stainless steel				
18" long	EA.	25.80	10.85	36.65
24" long	"	30.30	12.35	42.65
30" long	"	32.50	13.55	46.05
36" long	"	35.90	15.05	50.95
Toothbrush and tumbler holder	"	21.90	9.05	30.95
Waste receptacle, stainless steel, wall mounted	"	185.00	22.60	207.60
10900.10 WARDROBE SPECIALTIES				
Hospital wardrobe units, 24" x 24" x 76", with door				
Baked enameled steel	EA.	1,320	143.00	1,463
Hardwood	"	715.00	143.00	858.00
Stainless steel	"	2,200	143.00	2,343
Plastic laminated	"	792.00	143.00	935.00
Dormitory wardrobe units, 24" x 76", with door				

10 SPECIALTIES

SPECIALTIES	UNIT	MAT.	INST.	TOTAL
10900.10 WARDROBE SPECIALTIES				
Hardwood	EA.	715.00	143.00	858.00
Plastic laminated	"	627.00	143.00	770.00
Hat and coat rack				
Single tier				
Baked enameled steel	L.F.	86.00	6.45	92.45
Stainless steel	"	98.00	6.45	104.45
Aluminum	"	93.50	6.45	99.95
Double tier				
Baked enameled steel	L.F.	143.00	7.35	150.35
Stainless steel	"	198.00	7.35	205.35
Aluminum	"	176.00	7.35	183.35

11 EQUIPMENT

ARCHITECTURAL EQUIPMENT

ARCHITECTURAL EQUIPMENT	UNIT	MAT.	INST.	TOTAL
11010.10 MAINTENANCE EQUIPMENT				
Vacuum cleaning system				
3 valves				
1.5 hp	EA.	813.00	301.00	1,114
2.5 hp	"	1,120	387.00	1,507
5 valves	"	1,510	542.00	2,052
7 valves	"	1,740	678.00	2,418
11020.10 SECURITY EQUIPMENT				
Bulletproof teller window				
4' x 4'	EA.	7,710	429.00	8,139
5' x 4'	"	8,530	514.00	9,044
Bulletproof partitions				
Up to 12' high, 2.5" thick	S.F.	34.40	1.70	36.10
Counter for banks				
Minimum	L.F.	441.00	51.50	492.50
Maximum	"	1,770	85.50	1,856
Drive-up window				
Minimum	EA.	5,300	367.00	5,667
Maximum	"	31,780	857.00	32,637
Night depository				
Minimum	EA.	7,300	367.00	7,667
Maximum	"	14,710	857.00	15,567
Office safes, 30" x 20" x 20", 1 hr rating	"	1,290	64.50	1,355
30" x 16" x 15", 2 hr rating	"	1,240	51.50	1,292
30" x 28" x 20", H&G rating	"	6,360	32.20	6,392
Service windows, pass through painted steel				
24" x 36"	EA.	3,240	257.00	3,497
48" x 40"	"	5,240	322.00	5,562
72" x 40"	"	6,120	514.00	6,634
Special doors and windows				
3' x 7' bulletproof door with frame	EA.	3,880	367.00	4,247
12" x 12" vision panel	"	2,240	184.00	2,424
Surveillance system				
Minimum	EA.	3,770	64.50	3,835
Maximum	"	29,420	257.00	29,677
Vault door, 3' wide, 6'6" high				
3-1/2" thick	EA.	14,120	3,220	17,340
7" thick	"	15,770	4,290	20,060
10" thick	"	20,010	5,140	25,150
Insulated vault door				
2 hr rating				
32" wide	EA.	2,300	257.00	2,557
40" wide	"	2,410	271.00	2,681
4 hr rating				
32" wide	EA.	2,380	286.00	2,666
40" wide	"	2,530	322.00	2,852
6 hr rating				
32" wide	EA.	2,470	286.00	2,756
40" wide	"	2,850	322.00	3,172
Insulated file room door				
1 hr rating				
32" wide	EA.	2,530	257.00	2,787

ARCHITECTURAL EQUIPMENT	UNIT	MAT.	INST.	TOTAL
11020.10 SECURITY EQUIPMENT				
40" wide	EA.	2,740	286.00	3,026
11060.10 THEATER EQUIPMENT				
Roll out stage, steel frame, wood floor				
Manual	S.F.	22.35	1.70	24.05
Electric	"	24.10	2.70	26.80
Portable stages				
8" high	S.F.	7.15	1.35	8.50
18" high	"	8.25	1.50	9.75
36" high	"	8.80	1.60	10.40
48" high	"	9.35	1.70	11.05
Band risers				
Minimum	S.F.	19.25	1.35	20.60
Maximum	"	32.20	1.35	33.55
Chairs for risers				
Minimum	EA.	55.00	0.96	55.96
Maximum	"	77.00	0.96	77.96
11080.10 POLICE EQUIPMENT				
Firing range equipment, rifle				
3 position	EA.	8,580	904.00	9,484
4 position	"	11,000	1,360	12,360
5 position	"	13,750	1,510	15,260
6 position	"	16,500	1,600	18,100
11090.10 CHECKROOM EQUIPMENT				
Motorized checkroom equipment				
No shelf system, 6'4" height				
7'6" length	EA.	1,650	271.00	1,921
14'6" length	"	1,870	271.00	2,141
28' length	"	2,530	271.00	2,801
One shelf, 6'8" height				
7'6" length	EA.	1,980	271.00	2,251
14'6" length	"	2,420	271.00	2,691
28' length	"	3,630	271.00	3,901
Two shelves, 7'5" height				
7'6" length	EA.	2,420	271.00	2,691
14'6" length	"	2,970	271.00	3,241
28' length	"	4,180	271.00	4,451
Three shelves, 8' height				
7'6" length	EA.	2,530	542.00	3,072
14'6" length	"	3,080	542.00	3,622
28' length	"	4,400	542.00	4,942
Four shelves, 8'7" height				
7'6" length	EA.	2,530	542.00	3,072
14'6" length	"	3,080	542.00	3,622
28' length	"	4,400	542.00	4,942
200 lb				

11 EQUIPMENT

ARCHITECTURAL EQUIPMENT

ARCHITECTURAL EQUIPMENT	UNIT	MAT.	INST.	TOTAL
11110.10 LAUNDRY EQUIPMENT				
High capacity, heavy duty				
Washer extractors				
135 lb				
Standard	EA.	37,920	226.00	38,146
Pass through	"	43,790	226.00	44,016
200 lb				
Standard	EA.	42,580	226.00	42,806
Pass through	"	48,410	226.00	48,636
110 lb dryer	"	6,620	226.00	6,846
Hand operated presser	"	5,160	301.00	5,461
Mushroom press	"	3,310	301.00	3,611
Spreader feeders				
2 station	EA.	49,590	301.00	49,891
4 station	"	54,870	542.00	55,412
Delivery carts				
12 bushel	EA.	330.00	3.40	333.40
16 bushel	"	381.00	3.45	384.45
18 bushel	"	381.00	3.65	384.65
30 bushel	"	555.00	4.30	559.30
40 bushel	"	690.00	5.15	695.15
Low capacity				
Pressers				
Air operated	EA.	4,100	108.00	4,208
Hand operated	"	3,340	108.00	3,448
Extractor, low capacity	"	4,770	108.00	4,878
Ironer, 48"	"	2,220	54.00	2,274
Coin washers				
10 lb capacity	EA.	1,000	54.00	1,054
20 lb capacity	"	2,520	54.00	2,574
Coin dryer	"	421.00	33.90	454.90
Coin dry cleaner, 20 lb	"	19,800	108.00	19,908
11161.10 LOADING DOCK EQUIPMENT				
Dock leveler, 10 ton capacity				
6' x 8'	EA.	3,370	271.00	3,641
7' x 8'	"	3,700	271.00	3,971
Bumpers, laminated rubber				
4-1/2" thick				
6" x 14"	EA.	52.50	5.15	57.65
6" x 36"	"	54.00	5.70	59.70
10" x 14"	"	56.00	6.45	62.45
10" x 24"	"	64.00	7.35	71.35
10" x 36"	"	73.00	8.55	81.55
12" x 14"	"	61.50	6.75	68.25
12" x 24"	"	66.00	8.05	74.05
12" x 36"	"	83.00	9.55	92.55
6" thick				
10" x 14"	EA.	66.00	7.35	73.35
10" x 24"	"	87.50	8.85	96.35
10" x 36"	"	90.00	12.85	102.85
Extruded rubber bumpers				

ARCHITECTURAL EQUIPMENT	UNIT	MAT.	INST.	TOTAL
11161.10 LOADING DOCK EQUIPMENT				
T-section, 22" x 22" x 3"	EA.	84.00	5.15	89.15
Molded rubber bumpers				
24" x 12" x 3" thick	EA.	50.50	12.85	63.35
Door seal, 12" x 12", vinyl covered	L.F.	28.10	6.45	34.55
Dock boards, heavy duty, 5' x 5'				
5000 lb				
Minimum	EA.	830.00	214.00	1,044
Maximum	"	1,310	214.00	1,524
9000 lb				
Minimum	EA.	942.00	214.00	1,156
Maximum	"	1,570	234.00	1,804
15,000 lb	"	1,050	234.00	1,284
Truck shelters				
Minimum	EA.	999.00	198.00	1,197
Maximum	"	1,590	367.00	1,957
11170.10 WASTE HANDLING				
Incinerator, electric				
100 lb/hr				
Minimum	EA.	14,580	278.00	14,858
Maximum	"	20,900	278.00	21,178
400 lb/hr				
Minimum	EA.	26,240	555.00	26,795
Maximum	"	41,800	555.00	42,355
1000 lb/hr				
Minimum	EA.	44,890	841.00	45,731
Maximum	"	60,500	841.00	61,341
Incinerator, medical-waste				
25 lb/hr, 2-7 x 4-0	EA.	13,410	555.00	13,965
50 lb/hr, 2-11 x 4-11	"	14,690	555.00	15,245
75 lb/hr, 3-8 x 5-0	"	21,220	1,110	22,330
100 lb/hr, 3-8 x 6-0	"	23,320	1,110	24,430
Industrial compactor				
1 cy	EA.	7,520	308.00	7,828
3 cy	"	14,810	397.00	15,207
5 cy	"	17,130	555.00	17,685
Trash chutes steel, including sprinklers				
18" dia.	L.F.	49.00	129.00	178.00
24" dia.	"	64.00	135.00	199.00
30" dia.	"	81.50	143.00	224.50
36" dia.	"	107.00	151.00	258.00
Refuse bottom hopper	EA.	1,460	143.00	1,603
11400.10 FOOD SERVICE EQUIPMENT				
Unit kitchens				
30" compact kitchen				
Refrigerator, with range, sink	EA.	2,410	139.00	2,549
Sink only	"	2,300	92.50	2,393
Range only	"	2,020	69.50	2,090
Cabinet for upper wall section	"	280.00	39.70	319.70

ARCHITECTURAL EQUIPMENT

11400.10 FOOD SERVICE EQUIPMENT

Description	UNIT	MAT.	INST.	TOTAL
Stainless shield, for rear wall	EA.	95.50	11.10	106.60
Side wall	"	93.00	11.10	104.10
42" compact kitchen				
Refrigerator with range, sink	EA.	3,000	154.00	3,154
Sink only	"	2,580	139.00	2,719
Cabinet for upper wall section	"	404.00	46.30	450.30
Stainless shield, for rear wall	"	123.00	11.55	134.55
Side wall	"	50.50	11.55	62.05
54" compact kitchen				
Refrigerator, oven, range, sink	EA.	3,700	198.00	3,898
Cabinet for upper wall section	"	533.00	55.50	588.50
Stainless shield, for				
Rear wall	EA.	185.00	12.60	197.60
Side wall	"	50.50	12.60	63.10
60" compact kitchen				
Refrigerator, oven, range, sink	EA.	4,260	198.00	4,458
Cabinet for upper wall section	"	606.00	55.50	661.50
Stainless shield, for				
Rear wall	EA.	185.00	12.60	197.60
Side wall	"	50.50	12.60	63.10
72" compact kitchen				
Refrigerator, oven, range, sink	EA.	4,100	231.00	4,331
Cabinet for upper wall section	"	1,180	55.50	1,236
Stainless shield for				
Rear wall	EA.	219.00	13.90	232.90
Side wall	"	50.50	13.90	64.40
Bake oven				
Single deck				
Minimum	EA.	1,320	34.70	1,355
Maximum	"	3,080	69.50	3,150
Double deck				
Minimum	EA.	1,540	46.30	1,586
Maximum	"	5,500	69.50	5,570
Triple deck				
Minimum	EA.	2,200	46.30	2,246
Maximum	"	6,600	92.50	6,693
Convection type oven, electric, 40" x 45" x 57"				
Minimum	EA.	2,200	34.70	2,235
Maximum	"	3,850	69.50	3,920
Broiler, without oven, 69" x 26" x 39"				
Minimum	EA.	1,980	34.70	2,015
Maximum	"	3,850	46.30	3,896
Coffee urns, 10 gallons				
Minimum	EA.	4,400	92.50	4,493
Maximum	"	7,150	139.00	7,289
Fryer, with submerger				
Single				
Minimum	EA.	660.00	55.50	715.50
Maximum	"	1,980	92.50	2,073
Double				
Minimum	EA.	1,100	69.50	1,170
Maximum	"	3,740	92.50	3,833

ARCHITECTURAL EQUIPMENT

11400.10 FOOD SERVICE EQUIPMENT

ARCHITECTURAL EQUIPMENT	UNIT	MAT.	INST.	TOTAL
Griddle, counter				
3' long				
Minimum	EA.	1,100	46.30	1,146
Maximum	"	3,300	55.50	3,356
5' long				
Minimum	EA.	1,650	69.50	1,720
Maximum	"	4,290	92.50	4,383
Kettles, steam, jacketed				
20 gallons				
Minimum	EA.	1,760	69.50	1,830
Maximum	"	4,180	139.00	4,319
40 gallons				
Minimum	EA.	2,090	69.50	2,160
Maximum	"	4,620	139.00	4,759
60 gallons				
Minimum	EA.	2,750	69.50	2,820
Maximum	"	6,160	139.00	6,299
Range				
Heavy duty, single oven, open top				
Minimum	EA.	1,540	34.70	1,575
Maximum	"	3,960	92.50	4,053
Fry top				
Minimum	EA.	1,980	34.70	2,015
Maximum	"	4,290	92.50	4,383
Steamers, electric				
27 kw				
Minimum	EA.	5,950	69.50	6,020
Maximum	"	11,500	92.50	11,593
18 kw				
Minimum	EA.	4,380	69.50	4,450
Maximum	"	7,570	92.50	7,663
Dishwasher, rack type				
Single tank, 190 racks/hr	EA.	9,350	139.00	9,489
Double tank				
234 racks/hr	EA.	13,410	154.00	13,564
265 racks/hr	"	14,920	185.00	15,105
Dishwasher, automatic 100 meals/hr	"	44,210	92.50	44,303
Disposals				
100 gal/hr	EA.	982.00	92.50	1,075
120 gal/hr	"	1,420	95.50	1,516
250 gal/hr	"	3,700	99.00	3,799
Exhaust hood for dishwasher, gutter 4 sides, s-steel				
4'x4'x2'	EA.	1,320	103.00	1,423
4'x7'x2'	"	2,200	111.00	2,311
Food preparation machines				
Vertical cutter mixers				
25 quart	EA.	5,780	92.50	5,873
40 quart	"	6,960	92.50	7,053
80 quart	"	7,970	139.00	8,109
130 quart	"	8,860	231.00	9,091
Choppers				
5 lb	EA.	2,000	69.50	2,070

ARCHITECTURAL EQUIPMENT

ARCHITECTURAL EQUIPMENT	UNIT	MAT.	INST.	TOTAL
11400.10 FOOD SERVICE EQUIPMENT				
16 lb	EA.	2,750	92.50	2,843
40 lb	"	3,810	139.00	3,949
Mixers, floor models				
20 quart	EA.	2,200	34.70	2,235
60 quart	"	6,600	34.70	6,635
80 quart	"	8,800	39.70	8,840
140 quart	"	9,900	55.50	9,956
Ice cube maker				
50 lb per day				
Minimum	EA.	880.00	278.00	1,158
Maximum	"	1,760	278.00	2,038
500 lb per day				
Minimum	EA.	3,480	463.00	3,943
Maximum	"	6,400	463.00	6,863
Ice flakers				
300 lb per day	EA.	2,360	278.00	2,638
600 lb per day	"	2,800	463.00	3,263
1000 lb per day	"	3,310	617.00	3,927
2000 lb per day	"	7,520	694.00	8,214
Refrigerated cases				
Dairy products				490.00
Multi deck type	L.F.	499.00	18.50	517.50
For rear sliding doors, add	"			22.00
Delicatessen case, service deli				
Single deck	L.F.	462.00	139.00	601.00
Multi deck	"	506.00	174.00	680.00
Meat case				
Single deck	L.F.	330.00	163.00	493.00
Multi deck	"	363.00	174.00	537.00
Produce case				
Single deck	L.F.	348.00	163.00	511.00
Multi deck	"	396.00	174.00	570.00
Bottle coolers				
6' long				
Minimum	EA.	1,260	555.00	1,815
Maximum	"	2,660	555.00	3,215
10' long				
Minimum	EA.	1,800	925.00	2,725
Maximum	"	3,530	925.00	4,455
Frozen food cases				
Chest type	L.F.	359.00	163.00	522.00
Reach-in, glass door	"	748.00	174.00	922.00
Island case, single	"	365.00	163.00	528.00
Multi deck	"	808.00	174.00	982.00
Ice storage bins				
500 lb capacity	EA.	999.00	397.00	1,396
1000 lb capacity	"	2,000	793.00	2,793
11450.10 RESIDENTIAL EQUIPMENT				
Compactor, 4 to 1 compaction	EA.	990.00	69.50	1,060
Dishwasher, built-in				

11 EQUIPMENT

ARCHITECTURAL EQUIPMENT	UNIT	MAT.	INST.	TOTAL
11450.10 RESIDENTIAL EQUIPMENT				
2 cycles	EA.	825.00	139.00	964.00
4 or more cycles	"	1,210	139.00	1,349
Disposal				
Garbage disposer	EA.	330.00	92.50	422.50
Heaters, electric, built-in				
Ceiling type	EA.	165.00	92.50	257.50
Wall type		162.00	89.50	251.50
Minimum	EA.	110.00	69.50	179.50
Maximum	"	220.00	92.50	312.50
Hood for range, 2-speed, vented				
30" wide	EA.	118.00	92.50	210.50
42" wide	"	462.00	92.50	554.50
Ice maker, automatic				
30 lb per day	EA.	572.00	39.70	611.70
50 lb per day	"	1,210	139.00	1,349
Folding access stairs, disappearing metal stair				
8' long	EA.	1,380	38.30	1,418
11' long	"	1,340	37.90	1,378
12' long	"	1,460	38.30	1,498
Wood frame, wood stair				
22" x 54" x 8'9" long	EA.	66.00	26.80	92.80
25" x 54" x 10' long	"	69.50	26.80	96.30
Ranges electric				
Built-in, 30", 1 oven	EA.	800.00	92.50	892.50
2 oven	"	1,210	92.50	1,303
Counter top, 4 burner, standard	"	440.00	69.50	509.50
With grill	"	1,320	69.50	1,390
Free standing, 21", 1 oven	"	506.00	92.50	598.50
30", 1 oven	"	660.00	55.50	715.50
2 oven	"	1,980	55.50	2,036
Water softener				
30 grains per gallon	EA.	660.00	92.50	752.50
70 grains per gallon	"	1,210	139.00	1,349
11470.10 DARKROOM EQUIPMENT				
Dryers				
36" x 25" x 68"	EA.	6,600	150.00	6,750
48" x 25" x 68"	"	6,820	150.00	6,970
Processors, film				
Black and white	EA.	9,820	150.00	9,970
Color negatives	"	10,940	150.00	11,090
Prints	"	11,440	150.00	11,590
Transparencies	"	12,230	150.00	12,380
Sinks with cabinet and/or stand				
5" sink with stand				
24" x 48"	EA.	722.00	75.00	797.00
32" x 64"	"	895.00	100.00	995.00
38" x 52"	"	895.00	100.00	995.00
42" x 132"	"	1,130	150.00	1,280
48" x 52"	"	1,090	150.00	1,240
5" sink with cabinet				

11 EQUIPMENT

ARCHITECTURAL EQUIPMENT	UNIT	MAT.	INST.	TOTAL
11470.10 DARKROOM EQUIPMENT				
24" x 48"	EA.	1,440	75.00	1,515
32" x 64"	"	1,560	100.00	1,660
38" x 52"	"	1,530	100.00	1,630
42" x 132"	"	3,180	150.00	3,330
48" x 52"	"	1,960	150.00	2,110
10" sink with stand				
24" x 48"	EA.	884.00	75.00	959.00
32" x 64"	"	1,030	100.00	1,130
38" x 52"	"	1,020	100.00	1,120
10" sink with cabinet				
24" x 48"	EA.	1,440	75.00	1,515
38" x 52"	"	1,720	100.00	1,820
11480.10 ATHLETIC EQUIPMENT				
Basketball backboard				
Fixed	EA.	495.00	322.00	817.00
Swing-up	"	962.00	514.00	1,476
Portable, hydraulic	"	6,490	129.00	6,619
Suspended type, standard	"	2,010	514.00	2,524
For glass backboard, add	"			660.00
For electrically operated, add	"			880.00
Bleacher, telescoping, manual				
15 tier, minimum	SEAT	78.00	5.15	83.15
Maximum	"	89.00	5.15	94.15
20 tier, minimum	"	78.00	5.70	83.70
Maximum	"	89.00	5.70	94.70
30 tier, minimum	"	78.00	8.55	86.55
Maximum	"	89.00	8.55	97.55
Boxing ring elevated, complete, 22' x 22'	EA.	8,520	2,570	11,090
Gym divider curtain				
Minimum	S.F.	1.95	0.34	2.29
Maximum	"	2.65	0.34	2.99
Scoreboards, single face				
Minimum	EA.	2,590	257.00	2,847
Maximum	"	11,000	1,290	12,290
Parallel bars, wall mounted				
Minimum	EA.	1,320	257.00	1,577
Maximum	"	2,310	429.00	2,739
11500.10 INDUSTRIAL EQUIPMENT				
Vehicular paint spray booth, solid back, 14'4" x 9'6"				
24' deep	EA.	8,770	271.00	9,041
26'6" deep	"	8,940	271.00	9,211
28'6" deep	"	9,140	271.00	9,411
Drive through, 14'9" x 9'6"				
24' deep	EA.	9,200	271.00	9,471
26'6" deep	"	9,420	271.00	9,691
28'6" deep	"	9,540	271.00	9,811
Water wash, paint spray booth				
5' x 11'2" x 10'8"	EA.	4,600	271.00	4,871
6' x 11'2" x 10'8"	"	4,710	271.00	4,981
8' x 11'2" x 10'8"	"	5,950	271.00	6,221

11 EQUIPMENT

ARCHITECTURAL EQUIPMENT	UNIT	MAT.	INST.	TOTAL
11500.10 INDUSTRIAL EQUIPMENT				
10' x 11'2" x 11'2"	EA.	6,620	271.00	6,891
12' x 12'2" x 11'2"	"	8,080	271.00	8,351
14' x 12'2" x 11'2"	"	9,540	271.00	9,811
16' x 12'2" x 11'2"	"	10,770	271.00	11,041
20' x 12'2" x 11'2"	"	12,310	271.00	12,581
Dry type spray booth, with paint arrestors				
5'4" x 7'2" x 6'8"	EA.	1,930	271.00	2,201
6'4" x 7'2" x 6'8"	"	2,330	271.00	2,601
8'4" x 7'2" x 9'2"	"	2,740	271.00	3,011
10'4" x 7'2" x 9'2"	"	3,130	271.00	3,401
12'4" x 7'6" x 9'2"	"	3,700	271.00	3,971
14'4" x 7'6" x 9'8"	"	4,220	271.00	4,491
16'4" x 7'7" x 9'8"	"	4,820	271.00	5,091
20'4" x 7'7" x 10'8"	"	5,410	271.00	5,681
Air compressor, electric				
1 hp				
115 volt	EA.	2,410	181.00	2,591
5 hp				
115 volt	EA.	4,990	271.00	5,261
230 volt	"	4,210	271.00	4,481
Hydraulic lifts				
8,000 lb capacity	EA.	370.00	678.00	1,048
11,000 lb capacity	"	5,270	1,080	6,350
24,000 lb capacity	"	7,410	1,810	9,220
Power tools				
Band saws				
10"	EA.	495.00	22.60	517.60
14"	"	880.00	27.10	907.10
Motorized shaper	"	473.00	20.85	493.85
Motorized lathe	"	550.00	22.60	572.60
Bench saws				
9" saw	EA.	1,010	18.10	1,028
10" saw	"	1,520	19.35	1,539
12" saw	"	1,960	22.60	1,983
Electric grinders				
1/3 hp	EA.	187.00	10.85	197.85
1/2 hp	"	264.00	11.80	275.80
3/4 hp	"	319.00	11.80	330.80
11600.10 LABORATORY EQUIPMENT				
Cabinets, base				
Minimum	L.F.	88.00	22.60	110.60
Maximum	"	154.00	22.60	176.60
Full storage, 7' high				
Minimum	L.F.	132.00	22.60	154.60
Maximum	"	198.00	22.60	220.60
Wall				
Minimum	L.F.	58.50	27.10	85.60
Maximum	"	99.00	27.10	126.10
Counter tops				78.50
Minimum	S.F.	12.10	3.40	15.50
Average	"	28.60	3.85	32.45

ARCHITECTURAL EQUIPMENT

ARCHITECTURAL EQUIPMENT	UNIT	MAT.	INST.	TOTAL
11600.10 LABORATORY EQUIPMENT				
Maximum	S.F.	47.30	4.50	51.80
Tables				
Open underneath	S.F.	66.00	13.55	79.55
Doors underneath	"	121.00	16.95	137.95
Medical laboratory equipment				
Analyzer				
Chloride	EA.	385.00	13.90	398.90
Blood	"	16,500	23.15	16,523
Bath, water, utility, countertop unit	"	275.00	27.80	302.80
Hot plate, lab, countertop	"	88.00	25.20	113.20
Stirrer	"	154.00	25.20	179.20
Incubator, anaerobic, 23x23x36"	"	4,400	139.00	4,539
Dry heat bath	"	330.00	46.30	376.30
Incinerator, for sterilizing	"	132.00	2.80	134.80
Meter, serum protein	"	594.00	3.45	597.45
Ph analog, general purpose	"	440.00	3.95	443.95
Refrigerator, blood bank, undercounter type 153 litres	"	3,190	46.30	3,236
5.4 cf, undercounter type	"	2,750	46.30	2,796
Refrigerator/freezer, 4.4 cf, undercounter type	"	308.00	46.30	354.30
Sealer, impulse, free standing, 20x12x4"	"	165.00	9.25	174.25
Timer, electric, 1-60 minutes, bench or wall mounted	"	45.10	15.40	60.50
Glassware washer - dryer, undercounter	"	5,120	347.00	5,467
Balance, torsion suspension, tabletop, 4.5 lb capacity	"	352.00	15.40	367.40
Binocular microscope, with in base illuminator	"	1,380	10.70	1,391
Centrifuge, table model, 19x16x13"	"	836.00	11.10	847.10
Clinical model, with four place head	"	429.00	6.15	435.15
11700.10 MEDICAL EQUIPMENT				
Hospital equipment, lights				
Examination, portable	EA.	693.00	23.15	716.15
Meters				
Air flow meter	EA.	40.70	15.40	56.10
Oxygen flow meters	"	27.50	11.55	39.05
Racks				
40 chart, revolving open frame; mobile caddy	EA.	802.00	22.10	824.10
Scales.				
Clinical, metric with measure rod, 350 lb	EA.	237.00	24.10	261.10
Physical therapy				
Chair, hydrotherapy	EA.	198.00	4.50	202.50
Diathermy, shortwave, portable, on casters	"	1,760	10.85	1,771
Exercise bicycle, floor standing, 35" x 15"	"	506.00	9.05	515.05
Hydrocollator, 4 pack, portable, 129 x 90 x 160"	"	198.00	3.85	201.85
Lamp, infrared, mobile with variable heat control	"	264.00	20.85	284.85
Ultra violet, base mounted	"	242.00	20.85	262.85
Mirror, posture training, 27" wide and 72" high	"	254.00	6.45	260.45
Parallel bars, adjustable	"	1,130	32.20	1,162
Platform mat 10'x6', 1" thick	"	318.00	6.45	324.45
Pulley, duplex, wall mounted	"	225.00	85.50	310.50
Rack, crutch, wall mounted, 66 x 16 x 13"	"	491.00	25.70	516.70
Stimulator, galvanic-faradic, hand held	"	57.00	1.80	58.80
Ultrasound, muscle stimulator, portable, 13x13x8"	"	1,320	2.25	1,322
Sandbag set, velcro straps, saddle bag type	"	71.00	3.65	74.65

ARCHITECTURAL EQUIPMENT	UNIT	MAT.	INST.	TOTAL
11700.10 MEDICAL EQUIPMENT				
Whirlpool, 85 gallon	EA.	1,760	136.00	1,896
65 gallon capacity	"	1,500	136.00	1,636
Radiology				
Radiographic table, motor driven tilting table	EA.	19,580	2,710	22,290
Fluoroscope image/tv system	"	55,000	5,420	60,420
Processor for washing and drying radiographs				
Water filter unit, 30" x 48-1/2" x 37-1/2"	EA.	17.60	463.00	480.60
Cassette transfer cabinet	"	508.00	22.10	530.10
Base storage cabinets, sectional design				
With back splash, 24" deep and 35" high	L.F.	1,270	22.10	1,292
Wall storage cabinets	"	99.50	33.20	132.70
Steam sterilizers				
For heat and moisture stable materials	EA.	1,980	27.80	2,008
For fast drying after sterilization	"	1,320	34.70	1,355
Compact unit	"	990.00	34.70	1,025
Semi-automatic	"	3,960	139.00	4,099
Floor loading				
Single door	EA.	74,800	231.00	75,031
Double door	"	94,600	278.00	94,878
Utensil washer, sanitizer	"	4,180	214.00	4,394
Automatic washer/sterilizer	"	15,400	555.00	15,955
16 x 16 x 26", including generator & accessories	"	14,850	925.00	15,775
Steam generator, elec., 10 kw to 180 kw	"	14,240	555.00	14,795
Surgical scrub				
Minimum	EA.	1,540	92.50	1,633
Maximum	"	3,250	92.50	3,343
Gas sterilizers				
Automatic, free standing, 21x19x29"	EA.	2,800	278.00	3,078
Surgical tables				
Minimum	EA.	7,450	379.00	7,829
Maximum	"	30,950	530.00	31,480
Surgical lights, ceiling mounted				
Minimum	EA.	2,750	463.00	3,213
Maximum	"	4,510	555.00	5,065
Water stills				
4 liters/hr	EA.	1,420	92.50	1,513
8 liters/hr	"	2,090	92.50	2,183
19 liters/hr	"	5,780	231.00	6,011
X-ray equipment				
Mobile unit				
Minimum	EA.	4,950	139.00	5,089
Maximum	"	24,200	278.00	24,478
Film viewers				
Minimum	EA.	87.00	44.20	131.20
Maximum	"	346.00	88.50	434.50
Autopsy table				
Minimum	EA.	2,750	278.00	3,028
Maximum	"	5,500	278.00	5,778
Incubators				
15 cf	EA.	2,530	139.00	2,669
29 cf	"	3,080	231.00	3,311
Infant transport, portable	"	6,600	146.00	6,746

11 EQUIPMENT

ARCHITECTURAL EQUIPMENT	UNIT	MAT.	INST.	TOTAL
11700.10 MEDICAL EQUIPMENT				
Beds				
Stretcher, with pad, 30" x 78"	EA.	918.00	66.50	984.50
Transfer, for patient transport	"	1,240	66.50	1,307
Headwall				
Aluminum, with back frame and console	EA.	2,090	139.00	2,229
11700.20 DENTAL EQUIPMENT				
Dental care equipment				
Drill console with accessories	EA.	8,250	463.00	8,713
Amalgamator	"	206.00	13.90	219.90
Lathe	"	215.00	9.25	224.25
Finish polisher	"	162.00	18.50	180.50
Model trimer	"	254.00	12.60	266.60
Motor, wall mounted	"	283.00	12.60	295.60
Cleaner, ultrasonic	"	1,070	27.80	1,098
Curing unit, bench mounted	"	1,420	46.30	1,466
Oral evacuation system, dual pump	"	8,030	34.70	8,065
Sterilizer, table top, self contained	"	560.00	15.40	575.40
Dental lights				
Light, floor or ceiling mounted	EA.	1,320	139.00	1,459
X-ray unit				
Portable	EA.	1,980	69.50	2,050
Wall mounted with remote control	"	4,180	231.00	4,411
Illuminator, single panel	"	6,010	397.00	6,407
X-ray film processor	"	990.00	231.00	1,221
Shield, portable x-ray, lead lined	"	209.00	18.50	227.50

12 FURNISHINGS

INTERIOR	UNIT	MAT.	INST.	TOTAL
12302.10 CASEWORK				
Kitchen base cabinet, prefinished, 24" deep, 35" high				
12"wide	EA.	108.00	27.10	135.10
18" wide	"	132.00	27.10	159.10
24" wide	"	144.00	30.10	174.10
27" wide	"	174.00	30.10	204.10
36" wide	"	196.00	33.90	229.90
48" wide	"	224.00	33.90	257.90
Corner cabinet, 36" wide	"	151.00	33.90	184.90
Wall cabinet, 12" deep, 12" high				
30" wide	EA.	86.00	27.10	113.10
36" wide	"	91.00	27.10	118.10
15" high				
30" wide	EA.	91.00	30.10	121.10
36" wide	"	98.50	30.10	128.60
24" high				
30" wide	EA.	108.00	30.10	138.10
36" wide	"	132.00	30.10	162.10
30" high				
12" wide	EA.	82.50	33.90	116.40
18" wide	"	94.00	33.90	127.90
24" wide	"	101.00	33.90	134.90
27" wide	"	120.00	33.90	153.90
30" wide	"	135.00	38.70	173.70
36" wide	"	142.00	38.70	180.70
Corner cabinet, 30" high				
24" wide	EA.	132.00	45.20	177.20
30" wide	"	159.00	45.20	204.20
36" wide	"	176.00	45.20	221.20
Wardrobe	"	407.00	68.00	475.00
Vanity with top, laminated plastic				
24" wide	EA.	252.00	68.00	320.00
30" wide	"	309.00	68.00	377.00
36" wide	"	337.00	90.50	427.50
48" wide	"	372.00	108.00	480.00
12390.10 COUNTER TOPS				
Stainless steel, counter top, with backsplash	S.F.	66.00	6.80	72.80
Acid-proof, kemrock surface	"	17.60	4.50	22.10
12500.10 WINDOW TREATMENT				
Drapery tracks, wall or ceiling mounted				
Basic traverse rod				
50 to 90"	EA.	28.10	12.95	41.05
84 to 156"	"	43.10	14.40	57.50
136 to 250"	"	51.00	14.30	65.30
165 to 312"	"	74.50	16.10	90.60
Traverse rod with stationary curtain rod				
30 to 50"	EA.	30.80	12.85	43.65
50 to 90"	"	37.10	12.85	49.95
84 to 156"	"	55.50	14.30	69.80
136 to 250"	"	65.00	16.10	81.10
Double traverse rod				

12 FURNISHINGS

INTERIOR	UNIT	MAT.	INST.	TOTAL
12500.10 WINDOW TREATMENT				
30 to 50"	EA.	37.10	12.85	49.95
50 to 84"	"	49.70	12.85	62.55
84 to 156"	"	62.00	14.30	76.30
136 to 250"	"	76.00	16.10	92.10
12510.10 BLINDS				
Venetian blinds				
2" slats	S.F.	2.65	0.65	3.30
1" slats	"	5.15	0.65	5.80
12690.40 FLOOR MATS				
Recessed entrance mat, 3/8" thick, aluminum link	S.F.	12.10	12.85	24.95
Steel, flexible	"	6.40	12.85	19.25

CONSTRUCTION	UNIT	MAT.	INST.	TOTAL
13056.10 VAULTS				
Floor safes				
Class C				
1.0 cf	EA.	408.00	21.45	429.45
1.3 cf	"	467.00	32.20	499.20
1.9 cf	"	525.00	42.90	567.90
5.2 cf	"	1,160	42.90	1,203
13121.10 PRE-ENGINEERED BUILDINGS				
Pre-engineered metal building, 40'x100'				
14' eave height	S.F.	6.80	2.05	8.85
16' eave height	"	7.35	2.35	9.70
20' eave height	"	8.45	3.05	11.50
60'x100'				
14' eave height	S.F.	6.05	2.05	8.10
16' eave height	"	6.35	2.35	8.70
20' eave height	"	7.15	3.05	10.20
80'x100'				
14' eave height	S.F.	6.00	2.05	8.05
16' eave height	"	6.35	2.35	8.70
20' eave height	"	7.20	3.05	10.25
100'x100'				
14' eave height	S.F.	6.00	2.05	8.05
16' eave height	"	6.35	2.35	8.70
20' eave height	"	7.00	3.05	10.05
100'x150'				
14' eave height	S.F.	5.65	2.05	7.70
16' eave height	"	6.00	2.35	8.35
20' eave height	"	6.55	3.05	9.60
120'x150'				
14' eave height	S.F.	5.45	2.05	7.50
16' eave height	"	5.70	2.35	8.05
20' eave height	"	6.30	3.05	9.35
140'x150'				
14' eave height	S.F.	5.20	2.05	7.25
16' eave height	"	5.40	2.35	7.75
20' eave height	"	5.95	3.05	9.00
160'x200'				
14' eave height	S.F.	4.85	2.05	6.90
16' eave height	"	4.95	2.35	7.30
20' eave height	"	5.50	3.05	8.55
200'x200'				
14' eave height	S.F.	4.85	2.05	6.90
16' eave height	"	5.05	2.35	7.40
20' eave height	"	5.40	3.05	8.45
Hollow metal door and frame, 6' x 7'	EA.			633.00
Sectional steel overhead door, manually operated				
8' x 8'	EA.			523.00
12' x 12'	"			798.00
Roll-up steel door, manually operated				
10' x 10'	EA.			968.00
12' x 12'	"			1,160
For gravity ridge ventilator with birdscreen	"			341.00

CONSTRUCTION	UNIT	MAT.	INST.	TOTAL
13121.10 PRE-ENGINEERED BUILDINGS				
9" throat x 10'	EA.			385.00
12" throat x 10'	"			413.00
For 20" rotary vent with damper	"			160.00
For 4' x 3' fixed louver	"			169.00
For 4' x 3' aluminum sliding window	"			182.00
For 3' x 9' fiberglass panels	"			71.50
Liner panel, 26 ga, painted steel	S.F.	1.10	0.71	1.81
Wall panel insulated, 26 ga. steel, foam core	"	3.65	0.71	4.36
Roof panel, 26 ga. painted steel	"	0.92	0.41	1.33
Plastic (sky light)	"	2.55	0.41	2.96
Insulation, 3-1/2" thick blanket, R11	"	0.71	0.19	0.90
13152.10 SWIMMING POOL EQUIPMENT				
Diving boards				
14' long				
Aluminum	EA.	874.00	118.00	992.00
Fiberglass	"	539.00	118.00	657.00
Ladders, heavy duty				
2 steps				
Minimum	EA.	346.00	42.40	388.40
Maximum	"	556.00	42.40	598.40
4 steps				
Minimum	EA.	385.00	53.00	438.00
Maximum	"	737.00	53.00	790.00
Lifeguard chair				
Minimum	EA.	814.00	212.00	1,026
Maximum	"	1,280	212.00	1,492
Lights, underwater				
12 volt, with transformer	EA.	253.00	53.00	306.00
110 volt				
Minimum	EA.	231.00	53.00	284.00
Maximum	"	759.00	53.00	812.00
Ground fault interrupter for 110 volt, each light	"	95.00	17.65	112.65
Pool covers				
Reinforced polyethylene	S.F.	0.43	1.65	2.08
Vinyl water tube				
Minimum	S.F.	0.48	1.65	2.13
Maximum	"	0.74	1.65	2.39
Slides with water tube				
Minimum	EA.	253.00	177.00	430.00
Maximum	"	737.00	177.00	914.00

ELEVATORS	UNIT	MAT.	INST.	TOTAL
14210.10 ELEVATORS				
Passenger elevators, electric, geared				
Based on a shaft of 6 stops and 6 openings				
50 fpm, 2000 lb	EA.	58,300	1,150	59,450
100 fpm, 2000 lb	"	60,500	1,280	61,780
150 fpm				
2000 lb	EA.	66,000	1,440	67,440
3000 lb	"	84,150	1,640	85,790
4000 lb	"	88,640	1,910	90,550
200 fpm				
2500 lb	EA.	81,910	1,570	83,480
3000 lb	"	84,150	1,690	85,840
4000 lb	"	88,860	1,830	90,690
250 fpm				
2500 lb	EA.	84,150	1,570	85,720
3000 lb	"	88,640	1,690	90,330
4000 lb	"	90,880	1,830	92,710
300 fpm				
2500 lb	EA.	84,260	1,570	85,830
3000 lb	"	87,630	1,690	89,320
4000 lb	"	91,110	1,100	92,210
For each additional; 50 fpm, add per stop, $2000				
500 lb, add per stop, $2000				
Opening, add per stop, $4500				
Stop, add per stop, $4000				
Bonderized steel door, add per opening, $150				
Colored aluminum door, add per opening, $850				
Stainless steel door, add per opening, $600				
Cast bronze door, add per opening, $1100				
Two speed door, add per opening, $360				
Bi-parting door, add per opening, $850				
Custom cab interior add, $4800				
Based on a shaft of 8 stops and 8 openings				
300 fpm				
3000 lb	EA.	100,530	2,300	102,830
3500 lb	"	102,660	2,300	104,960
4000 lb	"	108,270	2,550	110,820
5000 lb	"	122,070	2,740	124,810
400 fpm				
3000 lb	EA.	106,250	2,200	108,450
3500 lb	"	110,850	2,200	113,050
4000 lb	"	11,880	2,550	14,430
5000 lb	"	129,030	2,620	131,650
600 fpm				
3000 lb	EA.	151,810	2,440	154,250
3500 lb	"	153,940	2,620	156,560
4000 lb	"	155,960	2,680	158,640
5000 lb	"	162,690	2,750	165,440
800 fpm				
3000 lb	EA.	179,520	2,440	181,960
3500 lb	"	181,760	2,620	184,380
4000 lb	"	184,010	2,680	186,690
5000 lb	"	189,620	2,750	192,370

14 CONVEYING

ELEVATORS

14210.10 ELEVATORS

ELEVATORS	UNIT	MAT.	INST.	TOTAL
For each additional; 100 fpm add per stop, $11,000				
500 lb, add per stop, $5500				
Opening add per stop, $11,000				
Stop add per stop, $4800				
Bypass floor, add per each, $2000				
Bonderized steel door, add per opening, $150				
Colored aluminum door, add per opening, $900				
Stainless steel door, add per opening, $600				
Cast bronze door, add per opening, $600				
Two speed bi-parting door, add per opening, $1000				
Custom cab interior, add $5000				
Hydraulic, based on a shaft of 3 stops, 3 openings				
50 fpm				
2000 lb	EA.	40,730	917.00	41,647
2500 lb	"	43,590	917.00	44,507
3000 lb	"	45,990	957.00	46,947
100 fpm				
2000 lb	EA.	44,730	917.00	45,647
2500 lb	"	47,130	957.00	48,087
3000 lb	"	50,110	1,000	51,110
150 fpm				
2000 lb	EA.	48,280	917.00	49,197
2500 lb	"	52,850	957.00	53,807
3000 lb	"	56,510	1,050	57,560
For each additional; 50 fpm add per stop, $3500				
500 lb, add per stop, $3500				
Opening, add, $4200				
Stop, add per stop, $5300				
Bonderized steel door, add per opening, $400				
Colored aluminum door, add per opening, $1500				
Stainless steel door, add per opening, $650				
Cast bronze door, add per opening, $1200				
Two speed door, add per opening, $400				
Bi-parting door, add per opening, $900				
Custom cab interior, add per cab, $5000				
Small elevators, 4 to 6 passenger capacity				
Electric, push				
2 stops	EA.	14,870	917.00	15,787
3 stops	"	18,300	1,000	19,300
4 stops	"	21,160	1,100	22,260
Freight elevators, electric				
Based on a shaft of 6 stops and 6 openings				
50 fpm				
3500 lb	EA.	117,810	1,280	119,090
4000 lb	"	118,260	1,280	119,540
5000 lb	"	119,380	1,440	120,820
100 fpm				
3500 lb	EA.	123,980	1,440	125,420
4000 lb	"	124,540	1,440	125,980
5000 lb	"	121,510	1,640	123,150
200 fpm				
3500 lb	EA.	122,070	1,570	123,640

14 CONVEYING

ELEVATORS

ELEVATORS	UNIT	MAT.	INST.	TOTAL
14210.10 ELEVATORS				
4000 lb	EA.	122,630	1,570	124,200
5000 lb	"	123,760	1,830	125,590
For elevator with manual door, deduct 15%				
For variable voltage control, add 20%				
Based on shaft of 8 stops and 8 openings				
100 fpm				
4000 lb	EA.	124,320	1,380	125,700
6000 lb	"	120,050	1,410	121,460
8000 lb	"	122,300	1,490	123,790
150 fpm				
4000 lb	EA.	120,050	1,570	121,620
6000 lb	"	122,300	1,620	123,920
8000 lb	"	123,420	1,720	125,140
200 fpm				
4000 lb	EA.	121,180	1,830	123,010
6000 lb	"	123,420	1,900	125,320
8000 lb	"	129,030	2,000	131,030
For each additional; 50 fpm, add per stop, $2000				
500 lb, add per stop, $600				
Opening, add per stop, $7000				
Stop, add per stop, $5500				
For variable voltage, add 20%				
Hydraulic, based on 3 stops and 3 openings				
50 fpm				
3000 lb	EA.	50,340	786.00	51,126
4000 lb	"	54,910	815.00	55,725
6000 lb	"	64,060	846.00	64,906
100 fpm				
3000 lb	EA.	56,740	786.00	57,526
4000 lb	"	60,630	815.00	61,445
6000 lb	"	70,810	846.00	71,656
150 fpm				
3000 lb	EA.	62,350	786.00	63,136
4000 lb	"	66,920	815.00	67,735
6000 lb	"	76,760	846.00	77,606
For each additional; 50 fpm, add per stop, $2000				
500 lb, add per stop, $600				
Opening, add per stop, $5500				
Stop, add per stop, $5500				
For elevator with manual door deduct from total, 15%				
14300.10 ESCALATORS				
Escalators				
32" wide, floor to floor				
12' high	EA.	76,650	1,910	78,560
15' high	"	83,510	2,300	85,810
18' high	"	90,380	2,870	93,250
22' high	"	97,240	3,830	101,070
25' high	"	108,680	4,600	113,280
48" wide				
12' high	EA.	85,800	1,900	87,700
15' high	"	93,240	2,290	95,530

14 CONVEYING

ELEVATORS

14300.10 ESCALATORS	UNIT	MAT.	INST.	TOTAL
18' high	EA.	99,530	2,890	102,420
22' high	"	110,970	3,930	114,900
25' high	"	118,980	4,400	123,380

LIFTS

14410.10 PERSONNEL LIFTS	UNIT	MAT.	INST.	TOTAL
Electrically operated, 1 or 2 person lift				
With attached foot platforms				
3 stops	EA.			13,280
5 stops	"			16,400
7 stops	"			19,520
For each additional stop, add $1250				
Residential stair climber, per story	EA.	3,930	231.00	4,161

14450.10 VEHICLE LIFTS	UNIT	MAT.	INST.	TOTAL
Automotive hoist, one post, semi-hydraulic, 8,000 lb	EA.	3,090	1,100	4,190
Full hydraulic, 8,000 lb	"	2,950	1,100	4,050
2 post, semi-hydraulic, 10,000 lb	"	3,220	1,570	4,790
Full hydraulic				
10,000 lb	EA.	3,440	1,570	5,010
13,000 lb	"	4,260	2,750	7,010
18,500 lb	"	5,580	2,750	8,330
24,000 lb	"	7,520	2,750	10,270
26,000 lb	"	8,800	2,750	11,550
Pneumatic hoist, fully hydraulic				
11,000 lb	EA.	4,150	3,670	7,820
24,000 lb	"	7,290	3,670	10,960

MATERIAL HANDLING

14560.10 CHUTES	UNIT	MAT.	INST.	TOTAL
Linen chutes, stainless steel, with supports				
18" dia.	L.F.	60.50	2.05	62.55

14 CONVEYING

MATERIAL HANDLING

14560.10 CHUTES	UNIT	MAT.	INST.	TOTAL
24" dia.	L.F.	86.50	2.20	88.70
30" dia.	"	92.50	2.40	94.90
Hopper	EA.	1,250	19.05	1,269
Skylight	"	759.00	28.60	787.60
Sprinkler unit at top	"	255.00	31.70	286.70
For galvanized metal, deduct from material cost, 35%				
For aluminum, deduct from material cost, 25%				

14580.10 PNEUMATIC SYSTEMS	UNIT	MAT.	INST.	TOTAL
Pneumatic message tube system				
Average, 20 station job				
3" round system	E.A.	203,940	252.00	204,192
4" round system	"	254,920	278.00	255,198
6" round system	"	385,220	308.00	385,528
4" x 7" oval system	"	402,210	555.00	402,765
Trash and linen tube system				
10 stations	EA.	130,290	5,740	136,030
15 stations	"	164,280	7,660	171,940
20 stations	"	220,390	8,840	229,230
30 stations	"	288,910	10,440	299,350

HOISTS AND CRANES

14600.10 INDUSTRIAL HOISTS	UNIT	MAT.	INST.	TOTAL
Industrial hoists, electric, light to medium duty				
500 lb	EA.	4,470	139.00	4,609
1000 lb	"	4,680	146.00	4,826
2000 lb	"	4,800	154.00	4,954
3000 lb	"	4,890	163.00	5,053
4000 lb	"	5,530	174.00	5,704
5000 lb	"	7,820	185.00	8,005
6000 lb	"	8,080	191.00	8,271
7500 lb	"	8,640	198.00	8,838
10,000 lb	"	14,590	206.00	14,796
15,000 lb	"	21,650	214.00	21,864
20,000 lb	"	25,580	231.00	25,811
25,000 lb	"	28,050	252.00	28,302
30,000 lb	"	27,500	278.00	27,778
Heavy duty				
500 lb	EA.	6,590	139.00	6,729
1000 lb	"	6,790	146.00	6,936
2000 lb	"	7,550	154.00	7,704

HOISTS AND CRANES	UNIT	MAT.	INST.	TOTAL
14600.10 INDUSTRIAL HOISTS				
3000 lb	EA.	7,970	163.00	8,133
4000 lb	"	8,420	174.00	8,594
5000 lb	"	8,750	185.00	8,935
6000 lb	"	8,980	191.00	9,171
7500 lb	"	12,340	198.00	12,538
10,000 lb	"	12,900	206.00	13,106
15,000 lb	"	14,250	214.00	14,464
20,000 lb	"	17,620	231.00	17,851
25,000 lb	"	25,580	252.00	25,832
30,000 lb	"	26,700	278.00	26,978
Air powered hoists				
500 lb	EA.	2,470	133.00	2,603
1000 lb	"	2,750	133.00	2,883
2000 lb	"	3,000	140.00	3,140
4000 lb	"	3,510	156.00	3,666
6000 lb	"	4,180	204.00	4,384
Overhead traveling bridge crane				
Single girder, 20' span				
3 ton	EA.	14,020	550.00	14,570
5 ton	"	15,430	550.00	15,980
7.5 ton	"	19,520	550.00	20,070
10 ton	"	19,520	688.00	20,208
15 ton	"	24,120	688.00	24,808
30' span				
3 ton	EA.	14,920	550.00	15,470
5 ton	"	16,490	550.00	17,040
10 ton	"	21,090	688.00	21,778
15 ton	"	25,240	688.00	25,928
Double girder, 40' span				
3 ton	EA.	28,950	1,220	30,170
5 ton	"	29,730	1,220	30,950
7.5 ton	"	30,740	1,220	31,960
10 ton	"	32,310	1,570	33,880
15 ton	"	36,350	1,570	37,920
25 ton	"	53,860	1,570	55,430
50' span				
3 ton	EA.	31,580	1,220	32,800
5 ton	"	32,710	1,220	33,930
7.5 ton	"	33,880	1,220	35,100
10 ton	"	36,800	1,570	38,370
15 ton	"	42,640	1,570	44,210
25 ton	"	58,340	1,570	59,910
14650.10 JIB CRANES				
Self supporting, swinging 8' boom, 200 deg rotation				
1000 lb	EA.	1,290	238.00	1,528
2000 lb	"	1,350	238.00	1,588
3000 lb	"	1,570	476.00	2,046
4000 lb	"	1,800	476.00	2,276
6000 lb	"	2,020	476.00	2,496
10,000 lb	"	2,660	476.00	3,136
Wall mounted, 180 deg rotation				

14 CONVEYING

HOISTS AND CRANES	UNIT	MAT.	INST.	TOTAL
14650.10 JIB CRANES				
2000 lb	EA.	718.00	238.00	956.00
3000 lb	"	853.00	238.00	1,091
4000 lb	"	954.00	476.00	1,430
6000 lb	"	1,080	476.00	1,556
10,000 lb	"	2,130	476.00	2,606

15 MECHANICAL

BASIC MATERIALS	UNIT	MAT.	INST.	TOTAL
15100.10 SPECIALTIES				
Wall penetration				
Concrete wall, 6" thick				
2" dia.	EA.	0.00	7.05	7.05
4" dia.	"	0.00	10.60	10.60
8" dia.	"	0.00	15.15	15.15
12" thick				
2" dia.	EA.	0.00	9.65	9.65
4" dia.	"	0.00	15.15	15.15
8" dia.	"	0.00	23.55	23.55
15120.10 BACKFLOW PREVENTERS				
Backflow preventer, flanged, cast iron, with valves				
3" pipe	EA.	1,810	150.00	1,960
4" pipe	"	2,420	167.00	2,587
6" pipe	"	3,640	250.00	3,890
8" pipe	"	6,530	300.00	6,830
15140.10 PIPE HANGERS, HEAVY				
Hangers				
1/2" pipe, clevis pipe hanger				
Black steel	EA.	1.75	10.00	11.75
Galvanized	"	2.00	10.00	12.00
3/4" pipe, clevis pipe hanger				
Black steel	EA.	1.75	10.00	11.75
Galvanized	"	2.05	10.00	12.05
1" pipe, clevis pipe hanger				
Black steel	EA.	1.85	10.00	11.85
Galvanized	"	2.15	10.00	12.15
U bolt	"	1.15	3.00	4.15
2" pipe, clevis pipe hanger				
Black steel	EA.	2.75	10.00	12.75
Galvanized	"	3.10	10.00	13.10
3" pipe, clevis pipe hanger				
Black steel	EA.	4.65	10.00	14.65
Galvanized	"	8.25	10.00	18.25
4" pipe, clevis pipe hanger				
Black steel	EA.	6.45	10.00	16.45
Galvanized	"	8.15	10.00	18.15
6" pipe, clevis pipe hanger				
Black steel	EA.	10.05	12.00	22.05
Galvanized	"	14.95	12.00	26.95
12" pipe, clevis pipe hanger				
Black steel	EA.	16.05	12.00	28.05
Galvanized	"	22.75	12.00	34.75
Threaded rod, galvanized				
3/8"	L.F.			0.26
1/2"	"			0.45
Hex nuts, galvanized				
3/8"	EA.			0.03
1/2"	"			0.07

BASIC MATERIALS

	UNIT	MAT.	INST.	TOTAL
15140.10 PIPE HANGERS, HEAVY				
1"	EA.			0.50
C-clamp, steel, with lock nut				
3/8"	EA.	1.20	3.75	4.95
1/2"	"	1.45	3.75	5.20
15175.60 EXPANSION TANKS				
Expansion tank, 125 psi, galvanized steel				
20 gallon	EA.	189.00	37.50	226.50
30 gallon	"	204.00	50.00	254.00
65 gallon	"	334.00	75.00	409.00
80 gallon	"	334.00	85.50	419.50

INSULATION

	UNIT	MAT.	INST.	TOTAL
15260.10 FIBERGLASS PIPE INSULATION				
Fiberglass insulation on 1/2" pipe				
1" thick	L.F.	2.10	1.00	3.10
1-1/2" thick	"	2.40	1.25	3.65
3/4" pipe				
1" thick	L.F.	2.15	1.00	3.15
1-1/2" thick	"	2.55	1.25	3.80
1" pipe				
1" thick	L.F.	2.20	1.00	3.20
1-1/2" thick	"	2.75	1.25	4.00
2" pipe				
1" thick	L.F.	3.10	1.25	4.35
1-1/2" thick	"	3.65	1.35	5.00
3" pipe				
1" thick	L.F.	3.50	1.45	4.95
1-1/2" thick	"	4.30	1.50	5.80
6" pipe				
1" thick	L.F.	5.50	1.60	7.10
2" thick	"	8.25	1.65	9.90
10" pipe				
2" thick	L.F.	12.10	1.60	13.70
3" thick	"	17.60	1.65	19.25
15260.60 EXTERIOR PIPE INSULATION				
Fiberglass insulation, aluminum jacket				
1/2" pipe				
1" thick	L.F.	2.75	2.30	5.05
1-1/2" thick	"	3.25	2.50	5.75

INSULATION

15260.60 EXTERIOR PIPE INSULATION

	UNIT	MAT.	INST.	TOTAL
3/4" pipe				
1" thick	L.F.	2.85	2.30	5.15
1-1/2" thick	"	3.35	2.50	5.85
1" pipe				
1" thick	L.F.	2.95	2.30	5.25
1-1/2" thick	"	3.65	2.50	6.15
2" pipe				
1" thick	L.F.	3.75	2.75	6.50
1-1/2" thick	"	4.50	2.85	7.35
3" pipe				
1" thick	L.F.	4.20	3.00	7.20
1-1/2" thick	"	4.95	3.15	8.10
6" pipe				
1" thick	L.F.	5.45	3.35	8.80
2" thick	"	9.00	3.55	12.55
10" pipe				
2" thick	L.F.	13.75	3.35	17.10
3" thick	"	19.80	3.55	23.35

15260.90 PIPE INSULATION FITTINGS

	UNIT	MAT.	INST.	TOTAL
Insulation protection saddle				
1" thick covering				
1/2" pipe	EA.	3.40	12.00	15.40
3/4" pipe	"	3.40	12.00	15.40
1" pipe	"	3.40	12.00	15.40
2" pipe	"	4.75	12.00	16.75
3" pipe	"	5.30	13.65	18.95
6" pipe	"	6.70	18.75	25.45
1-1/2" thick covering				
3/4" pipe	EA.	5.50	12.00	17.50
1" pipe	"	5.50	12.00	17.50
2" pipe	"	6.15	12.00	18.15
3" pipe	"	6.95	12.00	18.95
6" pipe	"	10.25	18.75	29.00
10" pipe	"	12.85	25.00	37.85

15280.10 EQUIPMENT INSULATION

	UNIT	MAT.	INST.	TOTAL
Equipment insulation, 2" thick, cellular glass	S.F.	1.45	1.85	3.30
Urethane, rigid, field applied jacket, plastered finish	"	1.60	3.75	5.35
Fiberglass, rigid, with vapor barrier	"	1.45	1.65	3.10

15290.10 DUCTWORK INSULATION

	UNIT	MAT.	INST.	TOTAL
Fiberglass duct insulation, plain blanket				
1-1/2" thick	S.F.	0.23	0.38	0.61
2" thick	"	0.28	0.50	0.78
With vapor barrier				
1-1/2" thick	S.F.	0.30	0.38	0.68
2" thick	"	0.35	0.50	0.85
Rigid with vapor barrier				
2" thick	S.F.	1.10	1.00	2.10

15 MECHANICAL

INSULATION	UNIT	MAT.	INST.	TOTAL
15290.10 DUCTWORK INSULATION				
3" thick	S.F.	1.80	1.20	3.00
4" thick	"	2.35	1.50	3.85
6" thick	"	3.20	2.00	5.20
Weatherproof, polystyrene, 3" thick, w/vapor barrier	"	2.40	3.00	5.40
Urethane board with vapor barrier	"	3.35	3.75	7.10

FIRE PROTECTION	UNIT	MAT.	INST.	TOTAL
15330.10 WET SPRINKLER SYSTEM				
Sprinkler head, 212 deg, brass, exposed piping	EA.	4.55	12.00	16.55
Chrome, concealed piping	"	4.45	16.65	21.10
Water motor alarm	"	61.50	50.00	111.50
Fire department inlet connection	"	94.50	60.00	154.50
Wall plate for fire dept connection	"	46.20	25.00	71.20
Swing check valve flanged iron body, 4"	"	102.00	100.00	202.00
Check valve, 6"	"	456.00	150.00	606.00
Wet pipe valve, flange to groove, 4"	"	402.00	33.30	435.30
Flange to flange				
6"	EA.	506.00	50.00	556.00
8"	"	666.00	100.00	766.00
Alarm valve, flange to flange, (wet valve)				
4"	EA.	330.00	33.30	363.30
8"	"	566.00	250.00	816.00
Inspector's test connection	"	28.10	25.00	53.10
Wall hydrant, polished brass, 2-1/2" x 2-1/2", single	"	264.00	21.45	285.45
2-way	"	726.00	21.45	747.45
3-way	"	1,070	21.45	1,091
Wet valve trim, includes retard chamber & guages, 4"-6"	"	368.00	25.00	393.00
Retard pressure switch for wet systems	"	671.00	60.00	731.00
Air maintenance device	"	189.00	25.00	214.00
Wall hydrant non-freeze, 8" thick wall, vacuum breaker	"	15.60	15.00	30.60
12" thick wall	"	16.85	15.00	31.85
15330.50 DRY SPRINKLER SYSTEM				
Dry pipe valve, flange to flange				
4"	EA.	764.00	60.00	824.00
6"	"	990.00	75.00	1,065
Trim, 4" and 6", includes gauges	"	358.00	25.00	383.00
Field testing and flushing	"	0.00	250.00	250.00
Disinfection	"	0.00	250.00	250.00
Pressure switch double circuit, open/close contacts	"	178.00	75.00	253.00
Low air				

15 MECHANICAL

FIRE PROTECTION	UNIT	MAT.	INST.	TOTAL
15330.50 DRY SPRINKLER SYSTEM				
Supervisory unit	EA.	605.00	50.00	655.00
Pressure switch	"	190.00	25.00	215.00
15330.90 HALON SYSTEM				
Halon fire protection system, per computer room				
Minimum	EA.			2,200
Average	"			3,140
Maximum	"			4,950

PLUMBING	UNIT	MAT.	INST.	TOTAL
15410.05 C.I. PIPE, ABOVE GROUND				
No hub pipe				
1-1/2" pipe	L.F.	3.85	2.15	6.00
2" pipe	"	4.05	2.50	6.55
3" pipe	"	6.05	3.00	9.05
4" pipe	"	8.25	5.00	13.25
6" pipe	"	11.00	6.00	17.00
8" pipe	"	13.75	10.00	23.75
10" pipe	"	24.20	12.00	36.20
No hub fittings, 1-1/2" pipe				
1/4 bend	EA.	4.95	10.00	14.95
1/8 bend	"	3.30	10.00	13.30
Sanitary tee	"	6.05	15.00	21.05
Wye	"	6.05	15.00	21.05
2" pipe				
1/4 bend	EA.	5.05	12.00	17.05
1/8 bend	"	3.40	12.00	15.40
Sanitary tee	"	6.25	20.00	26.25
Coupling	"			6.40
Wye	"	6.25	25.00	31.25
3" pipe				
1/4 bend	EA.	6.40	15.00	21.40
1/8 bend	"	5.70	15.00	20.70
Sanitary tee	"	7.50	18.75	26.25
Coupling	"			7.70
Wye	"	7.70	25.00	32.70
4" pipe				
1/4 bend	EA.	8.35	15.00	23.35
1/8 bend	"	7.50	15.00	22.50
Sanitary tee	"	9.35	25.00	34.35
Coupling	"			8.25
Wye	"	10.45	25.00	35.45
8" deep	"	13.85	15.00	28.85

PLUMBING	UNIT	MAT.	INST.	TOTAL
15410.05 C.I. PIPE, ABOVE GROUND				
6" pipe				
1/4 bend	EA.	15.95	25.00	40.95
1/8 bend	"	14.30	25.00	39.30
Sanitary tee	"	26.40	30.00	56.40
Coupling	"			17.60
Wye	"	25.30	30.00	55.30
8" pipe				
1/4 bend	EA.	37.40	25.00	62.40
1/8 bend	"	24.20	25.00	49.20
Sanitary tee	"	61.50	37.50	99.00
Coupling	"			16.50
Wye	"	50.50	30.00	80.50
10" pipe				
1/4 bend	EA.	75.00	25.00	100.00
1/8 bend	"	50.50	25.00	75.50
Coupling	"			25.30
Wye	"	103.00	50.00	153.00
10x6" wye	"	83.50	50.00	133.50
15410.06 C.I. PIPE, BELOW GROUND				
No hub pipe				
1-1/2" pipe	L.F.	3.85	1.50	5.35
2" pipe	"	4.05	1.65	5.70
3" pipe	"	6.05	1.85	7.90
4" pipe	"	8.25	2.50	10.75
6" pipe	"	11.00	2.75	13.75
8" pipe	"	13.75	3.35	17.10
10" pipe	"	24.20	3.75	27.95
Fittings, 1-1/2"				
1/4 bend	EA.	4.95	8.55	13.50
1/8 bend	"	3.30	8.55	11.85
Wye	"	6.05	12.00	18.05
2"				
1/4 bend	EA.	5.05	10.00	15.05
1/8 bend	"	3.40	10.00	13.40
3"				
1/4 bend	EA.	6.40	12.00	18.40
1/8 bend	"	5.70	12.00	17.70
Wye	"	7.70	18.75	26.45
4"				
1/4 bend	EA.	8.35	12.00	20.35
1/8 bend	"	7.50	12.00	19.50
Wye	"	10.45	18.75	29.20
6"				
1/4 bend	EA.	15.95	18.75	34.70
1/8 bend	"	14.30	18.75	33.05
8"				
1/4 bend	EA.	37.40	18.75	56.15
1/8 bend	"	24.20	18.75	42.95
Wye	"	50.50	25.00	75.50
10"				
1/4 bend	EA.	75.00	18.75	93.75

15 MECHANICAL

PLUMBING	UNIT	MAT.	INST.	TOTAL
15410.06 C.I. PIPE, BELOW GROUND				
1/8 bend	EA.	50.50	18.75	69.25
Wye	"	103.00	37.50	140.50
15410.09 SERVICE WEIGHT PIPE				
Service weight pipe, single hub				
3" x 5'	EA.	20.90	6.40	27.30
4" x 5'	"	25.30	6.65	31.95
6" x 5'	"	42.90	7.50	50.40
1/8 bend				
3"	EA.	5.50	12.00	17.50
4"	"	7.70	13.65	21.35
6"	"	12.10	15.00	27.10
1/4 bend				
3"	EA.	6.95	12.00	18.95
4"	"	9.80	13.65	23.45
6"	"	15.95	15.00	30.95
Sweep				
3"	EA.	9.35	12.00	21.35
4"	"	14.20	13.65	27.85
6"	"	25.30	15.00	40.30
Sanitary T				
3"	EA.	10.95	21.45	32.40
4"	"	13.85	25.00	38.85
6"	"	27.50	27.30	54.80
Wye				
3"	EA.	11.65	16.65	28.30
4"	"	14.85	17.65	32.50
6"	"	31.90	21.45	53.35
15410.10 COPPER PIPE				
Type "K" copper				
1/2"	L.F.	2.10	0.94	3.04
3/4"	"	4.30	1.00	5.30
1"	"	5.40	1.05	6.45
1-1/4"	"	6.50	1.15	7.65
1-1/2"	"	8.25	1.25	9.50
2"	"	11.55	1.35	12.90
2-1/2"	"	13.75	1.50	15.25
3"	"	16.30	1.60	17.90
4"	"	22.00	1.65	23.65
DWV, copper				
1-1/4"	L.F.	5.15	1.25	6.40
1-1/2"	"	6.50	1.35	7.85
2"	"	8.60	1.50	10.10
3"	"	11.90	1.65	13.55
4"	"	20.35	1.85	22.20
6"	"	55.00	2.15	57.15
Refrigeration tubing, copper, sealed				
1/8"	L.F.	0.61	1.20	1.81
3/16"	"	0.66	1.25	1.91
1/4"	"	0.77	1.30	2.07
5/16"	"	0.99	1.35	2.34

PLUMBING	UNIT	MAT.	INST.	TOTAL
15410.10 COPPER PIPE				
3/8"	L.F.	1.10	1.45	2.55
1/2"	"	1.45	1.50	2.95
7/8"	"	2.75	1.70	4.45
1-1/8"	"	4.20	2.00	6.20
1-3/8"	"	6.95	2.30	9.25
Type "L" copper				
1/4"	L.F.	0.66	0.88	1.54
3/8"	"	0.99	0.88	1.87
1/2"	"	1.75	0.94	2.69
3/4"	"	2.20	1.00	3.20
1"	"	3.85	1.05	4.90
1-1/4"	"	5.30	1.15	6.45
1-1/2"	"	6.50	1.25	7.75
2"	"	8.80	1.35	10.15
Type "M" copper				
1/2"	L.F.	1.10	0.94	2.04
3/4"	"	1.65	1.00	2.65
1"	"	2.20	1.05	3.25
1-1/4"	"	3.30	1.15	4.45
2"	"	7.70	1.35	9.05
15410.11 COPPER FITTINGS				
Copper pipe fittings				
1/2"				
90 deg ell	EA.	0.66	6.65	7.31
45 deg ell	"	0.83	6.65	7.48
Tee	"	1.10	8.55	9.65
Cap	"	0.44	3.35	3.79
Coupling	"	0.50	6.65	7.15
Union	"	3.30	7.50	10.80
3/4"				
90 deg ell	EA.	1.45	7.50	8.95
45 deg ell	"	1.75	7.50	9.25
Tee	"	2.40	10.00	12.40
Cap	"	0.88	3.55	4.43
Coupling	"	0.99	7.50	8.49
Union	"	4.95	8.55	13.50
1"				
90 deg ell	EA.	3.30	10.00	13.30
45 deg ell	"	4.40	10.00	14.40
Tee	"	5.50	12.00	17.50
Cap	"	1.65	5.00	6.65
Coupling	"	2.20	10.00	12.20
Union	"	6.60	10.00	16.60
1-1/4"				
90 deg ell	EA.	4.60	8.55	13.15
45 deg ell	"	5.70	8.55	14.25
Tee	"	7.70	15.00	22.70
Cap	"	1.30	5.00	6.30
Union	"	11.00	10.70	21.70
1-1/2"				
90 deg ell	EA.	6.05	10.70	16.75

PLUMBING	UNIT	MAT.	INST.	TOTAL
15410.11 COPPER FITTINGS				
45 deg ell	EA.	7.15	10.70	17.85
Tee	"	9.90	16.65	26.55
Cap	"	1.30	5.00	6.30
Coupling	"	4.40	10.00	14.40
Union	"	16.50	13.65	30.15
2"				
90 deg ell	EA.	11.55	12.00	23.55
45 deg ell	"	10.45	18.75	29.20
Tee	"	17.05	18.75	35.80
Cap	"	2.75	6.00	8.75
Coupling	"	7.15	12.00	19.15
Union	"	18.15	15.00	33.15
2-1/2"				
90 deg ell	EA.	23.10	15.00	38.10
45 deg ell	"	19.80	15.00	34.80
Tee	"	22.55	21.45	44.00
Cap	"	5.55	7.50	13.05
Coupling	"	11.00	15.00	26.00
Union	"	33.00	16.65	49.65
15410.18 GLASS PIPE				
Glass pipe				
1-1/2" dia.	L.F.	10.45	6.00	16.45
2" dia.	"	12.10	6.65	18.75
3" dia.	"	21.45	7.50	28.95
4" dia.	"	30.80	8.55	39.35
6" dia.	"	54.00	10.00	64.00
15410.30 PVC/CPVC PIPE				
PVC schedule 40				
1/2" pipe	L.F.	0.55	1.25	1.80
3/4" pipe	"	0.66	1.35	2.01
1" pipe	"	0.86	1.50	2.36
1-1/4" pipe	"	0.97	1.65	2.62
1-1/2" pipe	"	0.99	1.85	2.84
2" pipe	"	1.30	2.15	3.45
2-1/2" pipe	"	2.40	2.50	4.90
3" pipe	"	2.75	3.00	5.75
4" pipe	"	3.85	3.75	7.60
6" pipe	"	7.70	7.50	15.20
8" pipe	"	12.65	10.00	22.65
PVC schedule 80 pipe				
1-1/2" pipe	L.F.	0.85	1.85	2.70
2" pipe	"	1.10	2.15	3.25
3" pipe	"	1.75	3.00	4.75
4" pipe	"	2.75	3.75	6.50
Polypropylene, acid resistant, DWV pipe				
Schedule 40				
1-1/2" pipe	L.F.	3.30	2.15	5.45
2" pipe	"	4.45	2.50	6.95
3" pipe	"	8.25	3.00	11.25

15 MECHANICAL

PLUMBING	UNIT	MAT.	INST.	TOTAL
15410.30 PVC/CPVC PIPE				
4" pipe	L.F.	11.65	3.75	15.40
6" pipe	"	20.95	7.50	28.45
Polyethylene pipe and fittings				
SDR-21				
3" pipe	L.F.	0.90	3.75	4.65
4" pipe	"	1.75	5.00	6.75
6" pipe	"	3.50	7.50	11.00
8" pipe	"	5.30	8.55	13.85
10" pipe	"	8.25	10.00	18.25
15410.33 ABS DWV PIPE				
Schedule 40 ABS				
1-1/2" pipe	L.F.	0.77	1.50	2.27
2" pipe	"	0.99	1.65	2.64
3" pipe	"	2.10	2.15	4.25
4" pipe	"	3.10	3.00	6.10
6" pipe	"	5.05	3.75	8.80
15410.35 PLASTIC PIPE				
Fiberglass reinforced pipe				
2" pipe	L.F.	4.05	2.30	6.35
3" pipe	"	5.60	2.50	8.10
4" pipe	"	7.05	2.75	9.80
6" pipe	"	9.45	3.00	12.45
8" pipe	"	18.70	5.00	23.70
10" pipe	"	30.80	6.00	36.80
12" pipe	"	35.20	7.50	42.70
15410.70 STAINLESS STEEL PIPE				
Stainless steel, schedule 40, threaded				
1/2" pipe	L.F.	3.50	4.30	7.80
1" pipe	"	6.15	4.60	10.75
1-1/2" pipe	"	8.80	5.00	13.80
2" pipe	"	13.20	5.45	18.65
2-1/2" pipe	"	17.60	6.00	23.60
3" pipe	"	26.40	6.65	33.05
4" pipe	"	33.00	7.50	40.50
15410.80 STEEL PIPE				
Black steel, extra heavy pipe, threaded				
1/2" pipe	L.F.	1.10	1.20	2.30
3/4" pipe	"	1.30	1.20	2.50
1" pipe	"	1.75	1.50	3.25
1-1/2" pipe	"	2.85	1.65	4.50
2-1/2" pipe	"	6.05	3.75	9.80
3" pipe	"	8.25	5.00	13.25
4" pipe	"	13.10	6.00	19.10
5" pipe	"	17.60	7.50	25.10
6" pipe	"	22.00	7.50	29.50
8" pipe	"	24.75	10.00	34.75

PLUMBING	UNIT	MAT.	INST.	TOTAL
15410.80 STEEL PIPE				
10" pipe	L.F.	33.00	12.00	45.00
12" pipe	"	41.80	15.00	56.80
Fittings, malleable iron, threaded, 1/2" pipe				
90 deg ell	EA.	1.30	10.00	11.30
45 deg ell	"	2.10	10.00	12.10
Tee	"	1.75	15.00	16.75
3/4" pipe				
90 deg ell	EA.	1.75	10.00	11.75
45 deg ell	"	2.85	15.00	17.85
Tee	"	2.90	15.00	17.90
1" pipe				
90 deg ell	EA.	2.85	12.00	14.85
45 deg ell	"	3.20	12.00	15.20
Tee	"	4.30	16.65	20.95
1-1/2" pipe				
90 deg ell	EA.	5.45	15.00	20.45
45 deg ell	"	6.35	15.00	21.35
Tee	"	7.85	21.45	29.30
2-1/2" pipe				
90 deg ell	EA.	22.00	37.50	59.50
45 deg ell	"	27.50	37.50	65.00
Tee	"	29.70	50.00	79.70
3" pipe				
90 deg ell	EA.	31.90	50.00	81.90
45 deg ell	"	36.30	50.00	86.30
Tee	"	40.70	75.00	115.70
4" pipe				
90 deg ell	EA.	64.00	60.00	124.00
45 deg ell	"	65.00	60.00	125.00
Tee	"	93.50	100.00	193.50
6" pipe				
90 deg ell	EA.	182.00	60.00	242.00
45 deg ell	"	220.00	60.00	280.00
Tee	"	253.00	100.00	353.00
8" pipe				
90 deg ell	EA.	189.00	120.00	309.00
45 deg ell	"	231.00	120.00	351.00
Tee	"	275.00	188.00	463.00
10" pipe				
90 deg ell	EA.	211.00	150.00	361.00
45 deg ell	"	248.00	250.00	498.00
Tee	"	314.00	188.00	502.00
12" pipe				
90 deg ell	EA.	248.00	188.00	436.00
45 deg ell	"	286.00	188.00	474.00
Tee	"	363.00	250.00	613.00
15410.82 GALVANIZED STEEL PIPE				
Galvanized pipe				
1/2" pipe	L.F.	1.20	3.00	4.20
3/4" pipe	"	1.55	3.75	5.30
1" pipe	"	2.20	4.30	6.50

PLUMBING	UNIT	MAT.	INST.	TOTAL
15410.82 GALVANIZED STEEL PIPE				
1-1/4" pipe	L.F.	2.65	5.00	7.65
1-1/2" pipe	"	3.20	6.00	9.20
2" pipe	"	4.60	7.50	12.10
2-1/2" pipe	"	5.70	10.00	15.70
3" pipe	"	9.90	10.70	20.60
4" pipe	"	15.40	12.50	27.90
6" pipe	"	21.45	25.00	46.45
15430.23 CLEANOUTS				
Cleanout, wall				
2"	EA.	51.00	20.00	71.00
3"	"	57.00	20.00	77.00
4"	"	76.00	25.00	101.00
6"	"	156.00	30.00	186.00
8"	"	176.00	37.50	213.50
Floor				
2"	EA.	76.50	25.00	101.50
3"	"	81.50	25.00	106.50
4"	"	108.00	30.00	138.00
6"	"	185.00	37.50	222.50
8"	"	231.00	42.90	273.90
15430.24 GREASE TRAPS				
Grease traps, cast iron, 3" pipe				
35 gpm, 70 lb capacity	EA.	1,810	300.00	2,110
50 gpm, 100 lb capacity	"	2,190	375.00	2,565
15430.25 HOSE BIBBS				
Hose bibb				
1/2"	EA.	4.45	10.00	14.45
3/4"	"	4.70	10.00	14.70
15430.60 VALVES				
Gate valve, 125 lb, bronze, soldered				
1/2"	EA.	8.25	7.50	15.75
3/4"	"	10.00	7.50	17.50
1"	"	14.20	10.00	24.20
1-1/2"	"	24.60	12.00	36.60
2"	"	31.90	15.00	46.90
2-1/2"	"	65.00	18.75	83.75
Threaded				
1/4", 125 lb	EA.	5.50	12.00	17.50
1/2"				
125 lb	EA.	8.25	12.00	20.25
300 lb	"	27.50	12.00	39.50
3/4"				
125 lb	EA.	10.00	12.00	22.00
300 lb	"	34.10	12.00	46.10

PLUMBING

15430.60 VALVES	UNIT	MAT.	INST.	TOTAL
1"				
125 lb	EA.	14.20	12.00	26.20
300 lb	"	49.50	15.00	64.50
1-1/2"				
125 lb	EA.	24.20	15.00	39.20
300 lb	"	93.50	16.65	110.15
2"				
125 lb	EA.	31.90	21.45	53.35
300 lb	"	132.00	25.00	157.00
Cast iron, flanged				
2", 150 lb	EA.	183.00	25.00	208.00
2-1/2"				
125 lb	EA.	176.00	25.00	201.00
250 lb	"	352.00	25.00	377.00
3"				
125 lb	EA.	209.00	30.00	239.00
250 lb	"	440.00	30.00	470.00
4"				
125 lb	EA.	253.00	42.90	295.90
250 lb	"	594.00	42.90	636.90
6"				
125 lb	EA.	385.00	60.00	445.00
250 lb	"	990.00	60.00	1,050
8"				
125 lb	EA.	770.00	75.00	845.00
250 lb	"	1,870	75.00	1,945
OS&Y, flanged				
2"				
125 lb	EA.	132.00	25.00	157.00
250 lb	"	308.00	25.00	333.00
2-1/2"				
125 lb	EA.	143.00	25.00	168.00
250 lb	"	352.00	30.00	382.00
3"				
125 lb	EA.	143.00	30.00	173.00
250 lb	"	385.00	30.00	415.00
4"				
125 lb	EA.	204.00	50.00	254.00
250 lb	"	594.00	50.00	644.00
6"				
125 lb	EA.	324.00	60.00	384.00
250 lb	"	957.00	60.00	1,017
Ball valve, bronze, 250 lb, threaded				
1/2"	EA.	6.75	12.00	18.75
3/4"	"	10.45	12.00	22.45
1"	"	11.55	15.00	26.55
1-1/4"	"	19.80	16.65	36.45
1-1/2"	"	30.80	18.75	49.55
2"	"	35.20	21.45	56.65
Angle valve, bronze, 150 lb, threaded				
1/2"	EA.	22.00	10.70	32.70
3/4"	"	30.80	12.00	42.80

PLUMBING	UNIT	MAT.	INST.	TOTAL
15430.60 VALVES				
1"	EA.	44.00	12.00	56.00
1-1/4"	"	55.50	15.00	70.50
1-1/2"	"	74.50	16.65	91.15
Balancing valve, with meter connections, circuit setter				
1/2"	EA.	37.20	12.00	49.20
3/4"	"	39.60	13.65	53.25
1"	"	50.50	15.00	65.50
1-1/4"	"	70.00	16.65	86.65
1-1/2"	"	82.50	20.00	102.50
2"	"	121.00	25.00	146.00
2-1/2"	"	231.00	30.00	261.00
3"	"	348.00	37.50	385.50
4"	"	492.00	50.00	542.00
Pressure reducing valve, bronze, threaded, 250 lb				
1/2"	EA.	207.00	18.75	225.75
3/4"	"	207.00	18.75	225.75
1"	"	238.00	18.75	256.75
1-1/4"	"	276.00	21.45	297.45
1-1/2"	"	311.00	25.00	336.00
Pressure regulating valve, bronze, class 300				
1"	EA.	446.00	18.75	464.75
1-1/2"	"	583.00	23.10	606.10
2"	"	686.00	30.00	716.00
3"	"	755.00	42.90	797.90
4"	"	1,010	60.00	1,070
5"	"	1,410	75.00	1,485
6"	"	1,420	100.00	1,520
Solar water temperature regulating valve				
3/4"	EA.	265.00	25.00	290.00
1"	"	271.00	30.00	301.00
1-1/4"	"	297.00	33.30	330.30
1-1/2"	"	332.00	37.50	369.50
2-1/2"	"	824.00	75.00	899.00
Tempering valve, threaded				
3/4"	EA.	45.80	10.00	55.80
1"	"	114.00	12.00	126.00
1-1/4"	"	137.00	15.00	152.00
1-1/2"	"	183.00	15.00	198.00
2"	"	229.00	18.75	247.75
2-1/2"	"	286.00	25.00	311.00
3"	"	355.00	30.00	385.00
4"	"	812.00	42.90	854.90
Thermostatic mixing valve, threaded				
1/2"	EA.	54.50	10.70	65.20
3/4"	"	57.50	12.00	69.50
1"	"	202.00	13.05	215.05
1-1/2"	"	220.00	15.00	235.00
2"	"	287.00	18.75	305.75
Sweat connection				
1/2"	EA.	64.50	10.70	75.20
3/4"	"	74.00	12.00	86.00
Mixing valve, sweat connection				

PLUMBING

	UNIT	MAT.	INST.	TOTAL
15430.60 VALVES				
1/2"	EA.	25.20	10.70	35.90
3/4"	"	28.00	12.00	40.00
Liquid level gauge, aluminum body				
3/4"	EA.	205.00	12.00	217.00
125 psi, pvc body				
3/4"	EA.	205.00	12.00	217.00
150 psi, crs body				
3/4"	EA.	156.00	12.00	168.00
1"	"	177.00	12.00	189.00
175 psi, bronze body, 1/2"	"	327.00	10.70	337.70
15430.65 VACUUM BREAKERS				
Vacuum breaker, atmospheric, threaded connection				
3/4"	EA.	18.25	12.00	30.25
1"	"	28.10	12.00	40.10
Anti-siphon, brass				
3/4"	EA.	17.60	12.00	29.60
1"	"	27.00	12.00	39.00
1-1/4"	"	45.80	15.00	60.80
1-1/2"	"	54.00	16.65	70.65
2"	"	74.50	18.75	93.25
15430.68 STRAINERS				
Strainer, Y pattern, 125 psi, cast iron body, threaded				
3/4"	EA.	9.95	10.70	20.65
1"	"	12.45	12.00	24.45
1-1/4"	"	17.05	15.00	32.05
1-1/2"	"	21.35	15.00	36.35
2"	"	32.50	18.75	51.25
250 psi, brass body, threaded				
3/4"	EA.	24.30	12.00	36.30
1"	"	34.50	12.00	46.50
1-1/4"	"	43.70	15.00	58.70
1-1/2"	"	63.00	15.00	78.00
2"	"	104.00	18.75	122.75
Cast iron body, threaded				
3/4"	EA.	16.05	12.00	28.05
1"	"	20.45	12.00	32.45
1-1/4"	"	27.30	15.00	42.30
1-1/2"	"	36.20	15.00	51.20
2"	"	45.90	18.75	64.65
15430.70 DRAINS, ROOF & FLOOR				
Floor drain, cast iron, with cast iron top				
2"	EA.	44.00	25.00	69.00
3"	"	67.00	25.00	92.00
4"	"	88.00	25.00	113.00
6"	"	162.00	30.00	192.00
Roof drain, cast iron				
2"	EA.	121.00	25.00	146.00
3"	"	132.00	25.00	157.00

PLUMBING	UNIT	MAT.	INST.	TOTAL
15430.70 DRAINS, ROOF & FLOOR				
4"	EA.	143.00	25.00	168.00
5"	"	204.00	30.00	234.00
6"	"	209.00	30.00	239.00
15430.80 TRAPS				
Bucket trap, threaded				
3/4"	EA.	64.00	18.75	82.75
1"	"	275.00	20.00	295.00
1-1/4"	"	326.00	23.10	349.10
1-1/2"	"	487.00	27.30	514.30
Inverted bucket steam trap, threaded				
3/4"	EA.	55.00	18.75	73.75
1"	"	212.00	18.75	230.75
1-1/4"	"	286.00	16.65	302.65
1-1/2"	"	297.00	25.00	322.00
Float trap, 15 psi				
3/4"	EA.	58.50	18.75	77.25
1"	"	70.50	20.00	90.50
1-1/4"	"	87.50	21.45	108.95
1-1/2"	"	122.00	25.00	147.00
2"	"	237.00	30.00	267.00
Float and thermostatic trap, 15 psi				
3/4"	EA.	61.50	18.75	80.25
1"	"	42.90	20.00	62.90
1-1/4"	"	90.50	21.45	111.95
1-1/2"	"	125.00	25.00	150.00
2"	"	237.00	30.00	267.00
Steam trap, cast iron body, threaded, 125 psi				
3/4"	EA.	94.50	18.75	113.25
1"	"	104.00	20.00	124.00
1-1/4"	"	157.00	21.45	178.45
1-1/2"	"	269.00	25.00	294.00

PLUMBING FIXTURES	UNIT	MAT.	INST.	TOTAL
15440.10 BATHS				
Bath tub, 5' long				
Minimum	EA.	323.00	100.00	423.00
Average	"	681.00	150.00	831.00
Maximum	"	1,020	300.00	1,320
6' long				
Minimum	EA.	370.00	100.00	470.00

PLUMBING FIXTURES	UNIT	MAT.	INST.	TOTAL
15440.10 BATHS				
Average	EA.	716.00	150.00	866.00
Maximum	"	1,120	300.00	1,420
Square tub, whirlpool, 4'x4'				
Minimum	EA.	1,120	150.00	1,270
Average	"	1,390	300.00	1,690
Maximum	"	2,990	375.00	3,365
5'x5'				
Minimum	EA.	1,140	150.00	1,290
Average	"	1,620	300.00	1,920
Maximum	"	3,120	375.00	3,495
6'x6'				
Minimum	EA.	1,330	150.00	1,480
Average	"	1,820	300.00	2,120
Maximum	"	3,810	375.00	4,185
For trim and rough-in				
Minimum	EA.	98.00	100.00	198.00
Average	"	139.00	150.00	289.00
Maximum	"	202.00	300.00	502.00
Average	"	330.00	75.00	405.00
Maximum	"	275.00	100.00	375.00
15440.15 FAUCETS				
Kitchen				
Minimum	EA.	86.50	50.00	136.50
Average	"	139.00	60.00	199.00
Maximum	"	167.00	75.00	242.00
Bath				
Minimum	EA.	139.00	50.00	189.00
Average	"	203.00	60.00	263.00
Maximum	"	219.00	75.00	294.00
Lavatory, domestic				
Minimum	EA.	89.00	50.00	139.00
Average	"	185.00	60.00	245.00
Maximum	"	277.00	75.00	352.00
Hospital, patient rooms				
Minimum	EA.	219.00	75.00	294.00
Average	"	312.00	100.00	412.00
Maximum	"	393.00	150.00	543.00
Operating room				
Minimum	EA.	254.00	75.00	329.00
Average	"	335.00	100.00	435.00
Maximum	"	474.00	150.00	624.00
Washroom				
Minimum	EA.	127.00	50.00	177.00
Average	"	208.00	60.00	268.00
Maximum	"	300.00	75.00	375.00
Handicapped				
Minimum	EA.	162.00	60.00	222.00
Average	"	225.00	75.00	300.00
Maximum	"	335.00	100.00	435.00
Shower				
Minimum	EA.	104.00	50.00	154.00

PLUMBING FIXTURES	UNIT	MAT.	INST.	TOTAL
15440.15 FAUCETS				
Average	EA.	185.00	60.00	245.00
Maximum	"	289.00	75.00	364.00
For trim and rough-in				
Minimum	EA.	46.20	60.00	106.20
Average	"	69.50	75.00	144.50
Maximum	"	104.00	150.00	254.00
15440.18 HYDRANTS				
Wall hydrant				
8" thick	EA.	44.00	50.00	94.00
12" thick	"	55.00	60.00	115.00
18" thick	"	66.00	66.50	132.50
24" thick	"	77.00	75.00	152.00
Ground hydrant				
2' deep	EA.	132.00	37.50	169.50
4' deep	"	143.00	42.90	185.90
6' deep	"	176.00	50.00	226.00
8' deep	"	209.00	75.00	284.00
15440.20 LAVATORIES				
Lavatory, counter top, porcelain enamel on cast iron				
Minimum	EA.	113.00	60.00	173.00
Average	"	167.00	75.00	242.00
Maximum	"	243.00	100.00	343.00
Wall hung, china				
Minimum	EA.	127.00	60.00	187.00
Average	"	173.00	75.00	248.00
Maximum	"	462.00	100.00	562.00
Handicapped				
Minimum	EA.	254.00	75.00	329.00
Average	"	300.00	100.00	400.00
Maximum	"	450.00	150.00	600.00
For trim and rough-in				
Minimum	EA.	110.00	75.00	185.00
Average	"	133.00	100.00	233.00
Maximum	"	231.00	150.00	381.00
15440.30 SHOWERS				
Shower, fiberglass, 36"x34"x84"				
Minimum	EA.	450.00	214.00	664.00
Average	"	624.00	300.00	924.00
Maximum	"	895.00	300.00	1,195
Steel, 1 piece, 36"x36"				
Minimum	EA.	416.00	214.00	630.00
Average	"	612.00	300.00	912.00
Maximum	"	705.00	300.00	1,005
Receptor, molded stone, 36"x36"				
Minimum	EA.	139.00	100.00	239.00
Average	"	225.00	150.00	375.00
Maximum	"	346.00	250.00	596.00
For trim and rough-in				
Minimum	EA.	106.00	136.00	242.00

PLUMBING FIXTURES

PLUMBING FIXTURES	UNIT	MAT.	INST.	TOTAL
15440.30 SHOWERS				
Average	EA.	133.00	167.00	300.00
Maximum	"	162.00	300.00	462.00
15440.40 SINKS				
Service sink, 24"x29"				
Minimum	EA.	416.00	75.00	491.00
Average	"	520.00	100.00	620.00
Maximum	"	693.00	150.00	843.00
Kitchen sink, single, stainless steel, single bowl				
Minimum	EA.	121.00	60.00	181.00
Average	"	162.00	75.00	237.00
Maximum	"	229.00	100.00	329.00
Double bowl				
Minimum	EA.	162.00	75.00	237.00
Average	"	185.00	100.00	285.00
Maximum	"	248.00	150.00	398.00
Porcelain enamel, cast iron, single bowl				
Minimum	EA.	122.00	60.00	182.00
Average	"	146.00	75.00	221.00
Maximum	"	200.00	100.00	300.00
Double bowl				
Minimum	EA.	167.00	75.00	242.00
Average	"	191.00	100.00	291.00
Maximum	"	254.00	150.00	404.00
Mop sink, 24"x36"x10"				
Minimum	EA.	300.00	60.00	360.00
Average	"	393.00	75.00	468.00
Maximum	"	485.00	100.00	585.00
Washing machine box				
Minimum	EA.	104.00	75.00	179.00
Average	"	139.00	100.00	239.00
Maximum	"	185.00	150.00	335.00
For trim and rough-in				
Minimum	EA.	162.00	100.00	262.00
Average	"	231.00	150.00	381.00
Maximum	"	289.00	200.00	489.00
15440.50 URINALS				
Urinal, flush valve, floor mounted				
Minimum	EA.	433.00	75.00	508.00
Average	"	508.00	100.00	608.00
Maximum	"	693.00	150.00	843.00
Wall mounted				
Minimum	EA.	127.00	75.00	202.00
Average	"	312.00	100.00	412.00
Maximum	"	416.00	150.00	566.00
For trim and rough-in				
Minimum	EA.	81.00	75.00	156.00
Average	"	121.00	150.00	271.00
Maximum	"	144.00	200.00	344.00

15 MECHANICAL

PLUMBING FIXTURES	UNIT	MAT.	INST.	TOTAL
15440.60 WATER CLOSETS				
Water closet flush tank, floor mounted				
Minimum	EA.	165.00	75.00	240.00
Average	"	420.00	100.00	520.00
Maximum	"	825.00	150.00	975.00
Handicapped				
Minimum	EA.	245.00	100.00	345.00
Average	"	431.00	150.00	581.00
Maximum	"	660.00	300.00	960.00
Bowl, with flush valve, floor mounted				
Minimum	EA.	280.00	75.00	355.00
Average	"	315.00	100.00	415.00
Maximum	"	632.00	150.00	782.00
Wall mounted				
Minimum	EA.	292.00	75.00	367.00
Average	"	338.00	100.00	438.00
Maximum	"	660.00	150.00	810.00
For trim and rough-in				
Minimum	EA.	99.00	75.00	174.00
Average	"	117.00	100.00	217.00
Maximum	"	146.00	150.00	296.00
15440.70 WATER HEATERS				
Water heater, electric				
6 gal	EA.	187.00	50.00	237.00
10 gal	"	198.00	50.00	248.00
15 gal	"	209.00	50.00	259.00
20 gal	"	231.00	60.00	291.00
30 gal	"	242.00	60.00	302.00
40 gal	"	253.00	60.00	313.00
52 gal	"	286.00	75.00	361.00
66 gal	"	440.00	75.00	515.00
80 gal	"	484.00	75.00	559.00
100 gal	"	594.00	100.00	694.00
120 gal	"	759.00	100.00	859.00
Oil fired				
20 gal	EA.	715.00	150.00	865.00
50 gal	"	1,040	214.00	1,254
15440.90 MISCELLANEOUS FIXTURES				
Electric water cooler				
Floor mounted	EA.	451.00	100.00	551.00
Wall mounted	"	550.00	100.00	650.00
Wash fountain				
Wall mounted	EA.	759.00	150.00	909.00
Circular, floor supported	"	1,430	300.00	1,730
Deluge shower and eye wash	"	539.00	150.00	689.00

PLUMBING FIXTURES	UNIT	MAT.	INST.	TOTAL
15440.95 FIXTURE CARRIERS				
Water fountain, wall carrier				
Minimum	EA.	88.00	30.00	118.00
Average	"	110.00	37.50	147.50
Maximum	"	132.00	50.00	182.00
Lavatory, wall carrier				
Minimum	EA.	88.00	30.00	118.00
Average	"	110.00	37.50	147.50
Maximum	"	132.00	50.00	182.00
Sink, industrial, wall carrier				
Minimum	EA.	93.50	30.00	123.50
Average	"	116.00	37.50	153.50
Maximum	"	138.00	50.00	188.00
Toilets, water closets, wall carrier				
Minimum	EA.	116.00	30.00	146.00
Average	"	132.00	37.50	169.50
Maximum	"	176.00	50.00	226.00
Floor support				
Minimum	EA.	33.00	25.00	58.00
Average	"	44.00	30.00	74.00
Maximum	"	55.00	37.50	92.50
Urinals, wall carrier				
Minimum	EA.	71.50	30.00	101.50
Average	"	93.50	37.50	131.00
Maximum	"	116.00	50.00	166.00
Floor support				
Minimum	EA.	49.50	25.00	74.50
Average	"	66.00	30.00	96.00
Maximum	"	82.50	37.50	120.00
15450.30 PUMPS				
In-line pump, bronze, centrifugal				
5 gpm, 20' head	EA.	422.00	18.75	440.75
20 gpm, 40' head	"	781.00	18.75	799.75
50 gpm				
50' head	EA.	1,040	37.50	1,078
Cast iron, centrifugal				
50 gpm, 200' head	EA.	2,270	37.50	2,308
100 gpm				
100' head	EA.	2,100	50.00	2,150
Centrifugal, close coupled, c.i., single stage				
50 gpm, 100' head	EA.	921.00	37.50	958.50
100 gpm, 100' head	"	1,120	50.00	1,170
Base mounted				
50 gpm, 100' head	EA.	1,500	37.50	1,538
100 gpm, 50' head	"	1,660	50.00	1,710
200 gpm, 100' head	"	2,120	75.00	2,195
300 gpm, 175' head	"	2,750	75.00	2,825
Condensate pump, simplex				
1000 sf EDR, 2 gpm	EA.	723.00	250.00	973.00
2000 sf EDR, 3 gpm	"	735.00	250.00	985.00
4000 sf EDR, 6 gpm	"	740.00	273.00	1,013
6000 sf EDR, 9 gpm	"	746.00	273.00	1,019

PLUMBING FIXTURES

PLUMBING FIXTURES	UNIT	MAT.	INST.	TOTAL
15450.30 PUMPS				
Duplex, bronze				
8000 sf EDR, 12 gpm	EA.	1,000	273.00	1,273
10,000 sf EDR, 15 gpm	"	1,010	375.00	1,385
15,000 sf EDR, 23 gpm	"	1,270	429.00	1,699
20,000 sf EDR, 30 gpm	"	1,500	600.00	2,100
25,000 sf EDR, 38 gpm	"	1,520	600.00	2,120
15450.40 STORAGE TANKS				
Hot water storage tank, cement lined				
10 gallon	EA.	2,630	100.00	2,730
70 gallon	"	2,620	150.00	2,770
200 gallon	"	2,710	214.00	2,924
900 gallon	"	3,110	375.00	3,485
1100 gallon	"	3,580	375.00	3,955
2000 gallon	"	7,160	375.00	7,535
15480.10 SPECIAL SYSTEMS				
Air compressor, air cooled, two stage				
5.0 cfm, 175 psi	EA.	2,250	600.00	2,850
10 cfm, 175 psi	"	2,750	667.00	3,417
20 cfm, 175 psi	"	3,800	714.00	4,514
50 cfm, 125 psi	"	5,530	789.00	6,319
80 cfm, 125 psi	"	7,920	857.00	8,777
Single stage, 125 psi				
1.0 cfm	EA.	2,080	429.00	2,509
1.5 cfm	"	2,110	429.00	2,539
2.0 cfm	"	2,160	429.00	2,589
Automotive compressor, hose reel, air and water, 50' hose	"	968.00	250.00	1,218
Lube equipment, 3 reel, with pumps	"	5,060	1,200	6,260
Tire changer				
Truck	EA.	7,920	429.00	8,349
Passenger car	"	1,870	231.00	2,101
Air hose reel, includes, 50' hose	"	495.00	231.00	726.00
Hose reel, 5 reel, motor oil, gear oil, lube, air & water	"	4,620	1,200	5,820
Water hose reel, 50' hose	"	495.00	231.00	726.00
Pump, air operated, for motor or gear oil, fits 55 gal drum	"	649.00	30.00	679.00
For chassis lube	"	1,060	30.00	1,090
Fuel dispensing pump, lighted dial, one product				
One hose	EA.	2,310	250.00	2,560
Two hose	"	4,070	250.00	4,320
Two products, two hose	"	4,290	250.00	4,540

15 MECHANICAL

HEATING & VENTILATING	UNIT	MAT.	INST.	TOTAL
15555.10 BOILERS				
Cast iron, gas fired, hot water				
115 mbh	EA.	1,100	957.00	2,057
175 mbh	"	1,320	1,040	2,360
235 mbh	"	1,980	1,150	3,130
940 mbh	"	5,720	2,300	8,020
1600 mbh	"	8,360	2,870	11,230
3000 mbh	"	17,600	3,830	21,430
6000 mbh	"	38,500	5,740	44,240
Steam				
115 mbh	EA.	1,910	957.00	2,867
175 mbh	"	2,430	1,040	3,470
235 mbh	"	2,660	1,150	3,810
940 mbh	"	6,710	2,300	9,010
1600 mbh	"	12,100	2,870	14,970
3000 mbh	"	23,100	3,830	26,930
6000 mbh	"	42,900	5,740	48,640
Electric, hot water				
115 mbh	EA.	3,120	574.00	3,694
175 mbh	"	3,850	574.00	4,424
235 mbh	"	5,170	574.00	5,744
940 mbh	"	11,000	1,150	12,150
1600 mbh	"	16,500	2,300	18,800
3000 mbh	"	28,600	2,870	31,470
6000 mbh	"	42,900	3,830	46,730
Steam				
115 mbh	EA.	3,580	574.00	4,154
175 mbh	"	4,740	574.00	5,314
235 mbh	"	6,010	574.00	6,584
940 mbh	"	16,500	1,150	17,650
1600 mbh	"	27,500	2,300	29,800
3000 mbh	"	31,900	2,870	34,770
6000 mbh	"	49,500	3,830	53,330
Oil fired, hot water				
115 mbh	EA.	1,730	766.00	2,496
175 mbh	"	2,190	884.00	3,074
235 mbh	"	2,540	1,040	3,580
940 mbh	"	8,140	1,910	10,050
1600 mbh	"	13,400	2,300	15,700
3000 mbh	"	23,560	2,870	26,430
6000 mbh	"	44,000	5,740	49,740
Steam				
115 mbh	EA.	1,960	766.00	2,726
175 mbh	"	2,370	884.00	3,254
235 mbh	"	3,120	1,040	4,160
940 mbh	"	10,050	1,910	11,960
1600 mbh	"	16,400	2,300	18,700
3000 mbh	"	28,410	2,870	31,280
6000 mbh	"	46,750	5,740	52,490
15610.10 FURNACES				
Electric, hot air				
40 mbh	EA.	520.00	150.00	670.00

HEATING & VENTILATING

HEATING & VENTILATING	UNIT	MAT.	INST.	TOTAL
15610.10 FURNACES				
60 mbh	EA.	606.00	158.00	764.00
80 mbh	"	664.00	167.00	831.00
100 mbh	"	901.00	176.00	1,077
125 mbh	"	993.00	182.00	1,175
160 mbh	"	1,440	188.00	1,628
200 mbh	"	2,660	194.00	2,854
400 mbh	"	4,740	200.00	4,940
Gas fired hot air				
40 mbh	EA.	520.00	150.00	670.00
60 mbh	"	549.00	158.00	707.00
80 mbh	"	635.00	167.00	802.00
100 mbh	"	664.00	176.00	840.00
125 mbh	"	722.00	182.00	904.00
160 mbh	"	866.00	188.00	1,054
200 mbh	"	1,540	194.00	1,734
400 mbh	"	6,600	200.00	6,800
Oil fired hot air				
40 mbh	EA.	635.00	150.00	785.00
60 mbh	"	791.00	158.00	949.00
80 mbh	"	866.00	167.00	1,033
100 mbh	"	1,010	176.00	1,186
125 mbh	"	1,130	182.00	1,312
160 mbh	"	1,210	188.00	1,398
200 mbh	"	2,770	194.00	2,964
400 mbh	"	9,590	200.00	9,790

REFRIGERATION

REFRIGERATION	UNIT	MAT.	INST.	TOTAL
15670.10 CONDENSING UNITS				
Air cooled condenser, single circuit				
3 ton	EA.	1,320	50.00	1,370
5 ton	"	2,200	50.00	2,250
7.5 ton	"	3,520	143.00	3,663
20 ton	"	7,420	150.00	7,570
25 ton	"	8,660	150.00	8,810
30 ton	"	9,980	150.00	10,130
With low ambient dampers				
3 ton	EA.	1,540	75.00	1,615
5 ton	"	2,420	75.00	2,495
7.5 ton	"	3,740	150.00	3,890
20 ton	"	7,700	200.00	7,900
25 ton	"	8,910	200.00	9,110
30 ton	"	10,230	200.00	10,430

15 MECHANICAL

REFRIGERATION

	UNIT	MAT.	INST.	TOTAL
15670.10 CONDENSING UNITS				
Dual circuit				
10 ton	EA.	3,300	150.00	3,450
15 ton	"	4,840	214.00	5,054
20 ton	"	8,250	214.00	8,464
25 ton	"	9,460	214.00	9,674
30 ton	"	11,000	214.00	11,214
With low ambient dampers				
15 ton	EA.	5,390	214.00	5,604
20 ton	"	8,800	214.00	9,014
25 ton	"	10,010	214.00	10,224
30 ton	"	10,780	214.00	10,994
15680.10 CHILLERS				
Chiller, reciprocal				
Air cooled, remote condenser, starter				
20 ton	EA.	15,400	383.00	15,783
25 ton	"	17,380	383.00	17,763
30 ton	"	18,370	383.00	18,753
40 ton	"	27,170	574.00	27,744
Water cooled, with starter				
20 ton	EA.	13,200	383.00	13,583
25 ton	"	14,630	383.00	15,013
30 ton	"	17,600	574.00	18,174
40 ton	"	24,200	574.00	24,774
Packaged, air cooled, with starter				
20 ton	EA.	14,740	287.00	15,027
25 ton	"	15,950	287.00	16,237
30 ton	"	18,590	287.00	18,877
40 ton	"	21,340	287.00	21,627

HEAT TRANSFER

	UNIT	MAT.	INST.	TOTAL
15780.10 COMPUTER ROOM A/C				
Air cooled, alarm, high efficiency filter, elec. heat				
3 ton	EA.	7,970	231.00	8,201
5 ton	"	8,550	250.00	8,800
7.5 ton	"	14,550	300.00	14,850
10 ton	"	15,010	375.00	15,385
15 ton	"	16,630	429.00	17,059
Steam heat				
3 ton	EA.	9,240	231.00	9,471
5 ton	"	9,820	250.00	10,070
7.5 ton	"	15,590	300.00	15,890

HEAT TRANSFER	UNIT	MAT.	INST.	TOTAL
15780.10 COMPUTER ROOM A/C				
10 ton	EA.	16,050	375.00	16,425
15 ton	"	17,790	429.00	18,219
Hot water heat				
3 ton	EA.	9,240	231.00	9,471
5 ton	"	9,820	250.00	10,070
7.5 ton	"	15,590	300.00	15,890
10 ton	"	16,050	375.00	16,425
15 ton	"	17,790	429.00	18,219
Air cooled condenser, low ambient damper				
3 ton	EA.	2,080	60.00	2,140
5 ton	"	2,310	75.00	2,385
7.5 ton	"	3,350	150.00	3,500
10 ton	"	3,350	214.00	3,564
15 ton	"	3,700	176.00	3,876
Water cooled, high efficiency filter, alarm, elec. heat				
3 ton	EA.	9,120	214.00	9,334
5 ton	"	9,820	250.00	10,070
7.5 ton	"	15,590	375.00	15,965
10 ton	"	16,170	429.00	16,599
15 ton	"	18,940	500.00	19,440
Steam heat				
3 ton	EA.	10,390	214.00	10,604
5 ton	"	11,880	250.00	12,130
7.5 ton	"	16,750	375.00	17,125
10 ton	"	17,320	429.00	17,749
15 ton	"	20,100	500.00	20,600
Hot water heat				
3 ton	EA.	10,390	214.00	10,604
5 ton	"	11,090	250.00	11,340
7.5 ton	"	16,750	375.00	17,125
10 ton	"	17,320	429.00	17,749
15 ton	"	20,100	500.00	20,600
15780.20 ROOFTOP UNITS				
Packaged, single zone rooftop unit, with roof curb				
2 ton	EA.	2,090	300.00	2,390
3 ton	"	2,310	300.00	2,610
4 ton	"	3,190	375.00	3,565
5 ton	"	3,520	500.00	4,020
7.5 ton	"	4,950	600.00	5,550
15820.10 DEHUMIDIFIERS				
Dessicant dehumidifier, 1125 cfm	EA.			11,000
15830.20 FAN COIL UNITS				
Fan coil unit, 2 pipe, complete				
200 cfm ceiling hung	EA.	554.00	100.00	654.00
Floor mounted	"	520.00	75.00	595.00
300 cfm, ceiling hung	"	589.00	120.00	709.00
Floor mounted	"	560.00	100.00	660.00

HEAT TRANSFER	UNIT	MAT.	INST.	TOTAL
15830.20 FAN COIL UNITS				
400 cfm, ceiling hung	EA.	617.00	143.00	760.00
Floor mounted	"	594.00	100.00	694.00
500 cfm, ceiling hung	"	722.00	150.00	872.00
Floor mounted	"	689.00	115.00	804.00
600 cfm, ceiling hung	"	912.00	166.00	1,078
Floor mounted	"	843.00	136.00	979.00
15830.70 UNIT HEATERS				
Steam unit heater, horizontal				
12,500 btuh, 200 cfm	EA.	185.00	50.00	235.00
17,000 btuh, 300 cfm	"	214.00	50.00	264.00
40,000 btuh, 500 cfm	"	260.00	50.00	310.00
60,000 btuh, 700 cfm	"	300.00	50.00	350.00
70,000 btuh, 1000 cfm	"	404.00	75.00	479.00
Vertical				
12,500 btuh, 200 cfm	EA.	202.00	50.00	252.00
17,000 btuh, 300 cfm	"	329.00	50.00	379.00
40,000 btuh, 500 cfm	"	433.00	50.00	483.00
60,000 btuh, 700 cfm	"	491.00	50.00	541.00
70,000 btuh, 1000 cfm	"	537.00	50.00	587.00
Gas unit heater, horizontal				
27,400 btuh	EA.	404.00	120.00	524.00
38,000 btuh	"	462.00	120.00	582.00
56,000 btuh	"	520.00	120.00	640.00
82,200 btuh	"	578.00	120.00	698.00
103,900 btuh	"	606.00	188.00	794.00
125,700 btuh	"	635.00	188.00	823.00
133,200 btuh	"	578.00	188.00	766.00
149,000 btuh	"	751.00	188.00	939.00
172,000 btuh	"	797.00	188.00	985.00
190,000 btuh	"	866.00	188.00	1,054
225,000 btuh	"	953.00	188.00	1,141
Hot water unit heater, horizontal				
12,500 btuh, 200 cfm	EA.	208.00	50.00	258.00
17,000 btuh, 300 cfm	"	266.00	50.00	316.00
25,000 btuh, 500 cfm	"	289.00	50.00	339.00
30,000 btuh, 700 cfm	"	335.00	50.00	385.00
50,000 btuh, 1000 cfm	"	375.00	75.00	450.00
60,000 btuh, 1300 cfm	"	474.00	75.00	549.00
Vertical				
12,500 btuh, 200 cfm	EA.	243.00	50.00	293.00
17,000 btuh, 300 cfm	"	318.00	50.00	368.00
25,000 btuh, 500 cfm	"	346.00	50.00	396.00
30,000 btuh, 700 cfm	"	381.00	50.00	431.00
50,000 btuh, 1000 cfm	"	433.00	50.00	483.00
60,000 btuh, 1300 cfm	"	474.00	50.00	524.00
Cabinet unit heaters, ceiling, exposed, hot water				
200 cfm	EA.	653.00	100.00	753.00
300 cfm	"	681.00	120.00	801.00
400 cfm	"	743.00	143.00	886.00
600 cfm	"	762.00	158.00	920.00
800 cfm	"	1,070	188.00	1,258

15 MECHANICAL

HEAT TRANSFER	UNIT	MAT.	INST.	TOTAL
15830.70 UNIT HEATERS				
1000 cfm	EA.	1,160	214.00	1,374
1200 cfm	"	1,250	250.00	1,500
2000 cfm	"	2,250	333.00	2,583

AIR HANDLING	UNIT	MAT.	INST.	TOTAL
15855.10 AIR HANDLING UNITS				
Air handling unit, medium pressure, single zone				
1500 cfm	EA.	1,980	188.00	2,168
3000 cfm	"	2,640	333.00	2,973
4000 cfm	"	2,970	375.00	3,345
5000 cfm	"	3,520	400.00	3,920
6000 cfm	"	3,960	429.00	4,389
7000 cfm	"	4,400	462.00	4,862
8500 cfm	"	5,170	500.00	5,670
10,500 cfm	"	5,830	600.00	6,430
12,500 cfm	"	7,480	667.00	8,147
Rooftop air handling units				
4950 cfm	EA.	6,930	333.00	7,263
7370 cfm	"	7,920	429.00	8,349
9790 cfm	"	9,350	500.00	9,850
14,300 cfm	"	13,200	429.00	13,629
21,725 cfm	"	18,700	429.00	19,129
33,000 cfm	"	26,400	500.00	26,900
15870.20 EXHAUST FANS				
Belt drive roof exhaust fans				
640 cfm, 2618 fpm	EA.	660.00	37.50	697.50
940 cfm, 2604 fpm	"	852.00	37.50	889.50
1050 cfm, 3325 fpm	"	759.00	37.50	796.50
1170 cfm, 2373 fpm	"	1,100	37.50	1,138
2440 cfm, 4501 fpm	"	869.00	37.50	906.50
2760 cfm, 4950 fpm	"	957.00	37.50	994.50
3890 cfm, 6769 fpm	"	1,090	37.50	1,128
2380 cfm, 3382 fpm	"	1,210	37.50	1,248
2880 cfm, 3859 fpm	"	1,260	37.50	1,298
3200 cfm, 4173 fpm	"	1,280	50.00	1,330
3660 cfm, 3437 fpm	"	1,300	50.00	1,350
Direct drive fans				
60 to 390 cfm	EA.	539.00	37.50	576.50
145 to 590 cfm	"	649.00	37.50	686.50
295 to 860 cfm	"	786.00	37.50	823.50

AIR HANDLING

15870.20 EXHAUST FANS

AIR HANDLING	UNIT	MAT.	INST.	TOTAL
235 to 1300 cfm	EA.	847.00	37.50	884.50
415 to 1630 cfm	"	957.00	37.50	994.50
590 to 2045 cfm	"	1,100	37.50	1,138

AIR DISTRIBUTION

15890.10 METAL DUCTWORK

AIR DISTRIBUTION	UNIT	MAT.	INST.	TOTAL
Rectangular duct				
Galvanized steel				
Minimum	Lb.	1.75	2.75	4.50
Average	"	2.00	3.35	5.35
Maximum	"	2.10	5.00	7.10
Aluminum				
Minimum	Lb.	4.60	6.00	10.60
Average	"	5.05	7.50	12.55
Maximum	"	5.50	10.00	15.50
Fittings				
Minimum	EA.	2.65	10.00	12.65
Average	"	5.30	15.00	20.30
Maximum	"	13.20	30.00	43.20
For work				
10-20' high, add per pound, $.30				
30-50', add per pound, $.50				

15890.30 FLEXIBLE DUCTWORK

AIR DISTRIBUTION	UNIT	MAT.	INST.	TOTAL
Flexible duct, 1.25" fiberglass				
5" dia.	L.F.	1.30	1.50	2.80
6" dia.	"	1.45	1.65	3.10
7" dia.	"	1.65	1.75	3.40
8" dia.	"	2.00	1.85	3.85
10" dia.	"	2.20	2.15	4.35
12" dia.	"	2.75	2.30	5.05
14" dia.	"	3.30	2.50	5.80
16" dia.	"	3.85	2.75	6.60
Flexible duct connector, 3" wide fabric	"	1.35	5.00	6.35

15895.10 ROOF CURBS

AIR DISTRIBUTION	UNIT	MAT.	INST.	TOTAL
8" high, insulated, with liner and raised can				
15" x 15"	EA.	56.00	15.00	71.00
21" x 21"	"	70.50	15.00	85.50
36" x 36"	"	104.00	21.45	125.45
48" x 48"	"	286.00	23.10	309.10
60" x 60"	"	528.00	30.00	558.00

15 MECHANICAL

AIR DISTRIBUTION	UNIT	MAT.	INST.	TOTAL
15895.10 ROOF CURBS				
72" x 72"	EA.	847.00	37.50	884.50
15910.10 DAMPERS				
Horizontal parallel aluminum backdraft damper				
12" x 12"	EA.	28.90	7.50	36.40
16" x 16"	"	40.40	8.55	48.95
24" x 24"	"	63.50	15.00	78.50
36" x 36"	"	146.00	21.45	167.45
48" x 48"	"	260.00	30.00	290.00
"Up", parallel dampers				
12" x 12"	EA.	51.00	7.50	58.50
16" x 16"	"	71.50	8.55	80.05
24" x 24"	"	88.00	15.00	103.00
36" x 36"	"	150.00	21.45	171.45
48" x 48"	"	293.00	30.00	323.00
"Down", parallel dampers				
12" x 12"	EA.	51.00	7.50	58.50
16" x 16"	"	71.50	8.55	80.05
24" x 24"	"	88.00	15.00	103.00
36" x 36"	"	150.00	21.45	171.45
48" x 48"	"	293.00	30.00	323.00
Fire damper, 1.5 hr rating				
12" x 12"	EA.	17.35	15.00	32.35
16" x 16"	"	27.70	15.00	42.70
24" x 24"	"	34.70	15.00	49.70
36" x 36"	"	58.00	30.00	88.00
48" x 48"	"	104.00	42.90	146.90
15940.10 DIFFUSERS				
Ceiling diffusers, round, baked enamel finish				
6" dia.	EA.	28.90	10.00	38.90
12" dia.	"	52.00	12.50	64.50
16" dia.	"	75.00	13.65	88.65
20" dia.	"	110.00	15.00	125.00
Rectangular				
6x6"	EA.	28.60	10.00	38.60
12x12"	"	50.50	15.00	65.50
18x18"	"	81.00	15.00	96.00
24x24"	"	133.00	18.75	151.75
Lay in, flush mounted, perforated face, with grid				
6x6/24x24	EA.	46.20	12.00	58.20
12x12/24x24	"	50.50	12.00	62.50
18x18/24x24	"	53.00	12.00	65.00
Two-way slot diffuser with balancing damper, 4'	"	55.00	30.00	85.00
15940.20 RELIEF VENTILATORS				
Intake ventilator, aluminum, with screen, no curbs				
12" x 12"	EA.	104.00	25.00	129.00
16" x 16"	"	143.00	30.00	173.00

15 MECHANICAL

AIR DISTRIBUTION	UNIT	MAT.	INST.	TOTAL
15940.20 RELIEF VENTILATORS				
36" x 36"	EA.	578.00	50.00	628.00
48" x 48"	"	946.00	60.00	1,006
15940.40 REGISTERS AND GRILLES				
Lay in flush mounted, perforated face, return				
6x6/24x24	EA.	24.25	12.00	36.25
8x8/24x24	"	24.25	12.00	36.25
9x9/24x24	"	24.25	12.00	36.25
10x10/24x24	"	24.25	12.00	36.25
12x12/24x24	"	24.25	12.00	36.25
Rectangular, ceiling return, single deflection				
10x10	EA.	11.00	15.00	26.00
12x12	"	13.20	15.00	28.20
14x14	"	16.50	15.00	31.50
16x8	"	13.20	15.00	28.20
16x16	"	13.20	15.00	28.20
24x12	"	37.40	15.00	52.40
24x18	"	49.50	15.00	64.50
36x24	"	99.00	16.65	115.65
36x30	"	143.00	16.65	159.65
Wall, return air register				
12x12	EA.	21.95	7.50	29.45
16x16	"	32.30	7.50	39.80
18x18	"	38.10	7.50	45.60
20x20	"	45.00	7.50	52.50
24x24	"	62.50	7.50	70.00
Ceiling, return air grille				
6x6	EA.	8.10	10.00	18.10
8x8	"	9.25	12.00	21.25
10x10	"	10.40	12.00	22.40
Ceiling, exhaust grille, aluminum egg crate				
6x6	EA.	10.40	10.00	20.40
8x8	"	10.40	12.00	22.40
10x10	"	11.55	12.00	23.55
12x12	"	15.00	15.00	30.00
14x14	"	19.65	15.00	34.65
16x16	"	23.10	15.00	38.10
18x18	"	27.70	15.00	42.70
15940.80 PENTHOUSE LOUVERS				
Penthouse louvers				
12" high, extruded aluminum, 4" louver				
6' perimeter	EA.	264.00	75.00	339.00
12' perimeter	"	638.00	75.00	713.00
20' perimeter	"	1,320	200.00	1,520
16" high x 4' perimeter	"	220.00	75.00	295.00
6' perimeter	"	308.00	75.00	383.00
12' perimeter	"	715.00	75.00	790.00
20' perimeter	"	1,540	200.00	1,740
20" high x 4' perimeter	"	286.00	75.00	361.00
6' perimeter	"	330.00	75.00	405.00
12' perimeter	"	803.00	75.00	878.00

15 MECHANICAL

AIR DISTRIBUTION

15940.80 PENTHOUSE LOUVERS	UNIT	MAT.	INST.	TOTAL
20' perimeter	EA.	1,650	200.00	1,850
24" high x 4' perimeter	"	308.00	75.00	383.00
6' perimeter	"	368.00	75.00	443.00
12' perimeter	"	896.00	75.00	971.00
20' perimeter	"	1,860	200.00	2,060

CONTROLS

15950.10 HVAC CONTROLS	UNIT	MAT.	INST.	TOTAL
Pressure gauge, direct reading gage cock and syphon	EA.	47.30	18.75	66.05
Control valve, 1", modulating				
2-way	EA.	484.00	25.00	509.00
3-way	"	550.00	37.50	587.50
Self contained control valve with sensing element, 3/4"	"	88.00	18.75	106.75
Instrument air system 2-2 hp compressors, rcvr refrigerant dryer	"			4,070
Thermostat primary control device	"			88.00
Humidistat primary control device	"			66.00
Timers primary control device, indoor/outdoor, 24 hour	"			132.00
Thermometer, dir. reading, 3 dial	"			66.00
Control dampers, round				
6" dia.	EA.	48.40	12.00	60.40
8" dia	"	71.50	12.00	83.50
10" dia	"	94.50	12.00	106.50
12" dia	"	121.00	12.00	133.00
12" dia	"	176.00	15.00	191.00
18" dia	"	198.00	15.00	213.00
20" dia	"	253.00	15.00	268.00
Rectangular, parallel blade standard leakage				
12" x 12"	EA.	36.30	15.00	51.30
16" x 16"	"	54.00	15.00	69.00
20" x 20"	"	64.00	15.00	79.00
48" x 48"	"	187.00	42.90	229.90
48" x 60"	"	236.00	50.00	286.00
48" x 72"	"	286.00	50.00	336.00
Low leakage				
12" x 12"	EA.	71.50	15.00	86.50
16" x 16"	"	91.50	15.00	106.50
36" x 36"	"	231.00	25.00	256.00
48" x 48"	"	407.00	42.90	449.90
48" x 72"	"	638.00	50.00	688.00
Rectangular, opposed horizontal blade				
12" x 12"	EA.	47.30	15.00	62.30
16" x 16"	"	66.00	15.00	81.00

15 MECHANICAL

CONTROLS	UNIT	MAT.	INST.	TOTAL
15950.10 HVAC CONTROLS				
24" x 24"	EA.	88.00	15.00	103.00
36" x 36"	"	143.00	25.00	168.00
48" x 72"	"	407.00	50.00	457.00

BASIC MATERIALS

16110.20 CONDUIT SPECIALTIES

BASIC MATERIALS	UNIT	MAT.	INST.	TOTAL
Rod beam clamp, 1/2"	EA.	3.15	1.75	4.90
Hanger rod				
3/8"	L.F.	0.68	1.40	2.08
1/2"	"	1.10	1.75	2.85
Hanger channel, 1-1/2"				
No holes	EA.	2.55	1.05	3.60
Holes	"	2.85	1.05	3.90
Channel strap				
1/2"	EA.	0.70	1.75	2.45
3/4"	"	0.83	1.75	2.58
1"	"	0.89	1.75	2.64
1-1/4"	"	1.00	2.80	3.80
1-1/2"	"	1.10	2.80	3.90
2"	"	1.30	2.80	4.10
2-1/2"	"	1.45	4.25	5.70
3"	"	1.55	4.25	5.80
3-1/2"	"	1.90	4.25	6.15
4"	"	2.15	5.05	7.20
5"	"	2.95	5.05	8.00
6"	"	3.55	5.05	8.60
Conduit penetrations, roof and wall, 8" thick				
1/2"	EA.	0.00	21.35	21.35
3/4"	"	0.00	21.35	21.35
1"	"	0.00	27.80	27.80
1-1/4"	"	0.00	27.80	27.80
1-1/2"	"	0.00	27.80	27.80
2"	"	0.00	55.50	55.50
2-1/2"	"	0.00	55.50	55.50
3"	"	0.00	55.50	55.50
3-1/2"	"	0.00	69.50	69.50
4"	"	0.00	69.50	69.50
Fireproofing, for conduit penetrations				
1/2"	EA.	1.45	17.35	18.80
3/4"	"	1.45	17.35	18.80
1"	"	1.50	17.35	18.85
1-1/4"	"	2.10	27.30	29.40
1-1/2"	"	2.10	25.20	27.30
2"	"	2.10	25.20	27.30
2-1/2"	"	4.20	31.20	35.40
3"	"	4.20	33.70	37.90
3-1/2"	"	5.05	43.40	48.45
4"	"	6.05	52.50	58.55
Aluminum conduit				
1/2"	L.F.	0.79	1.05	1.84
3/4"	"	1.05	1.40	2.45
1"	"	1.05	1.75	2.80
1-1/4"	"	1.95	2.05	4.00
1-1/2"	"	2.45	2.80	5.25
2"	"	3.30	3.10	6.40
2-1/2"	"	5.20	3.45	8.65
3"	"	6.80	3.70	10.50
3-1/2"	"	8.15	4.25	12.40

BASIC MATERIALS

BASIC MATERIALS	UNIT	MAT.	INST.	TOTAL
16110.20 CONDUIT SPECIALTIES				
4"	L.F.	9.65	5.05	14.70
5"	"	13.85	6.30	20.15
6"	"	18.25	6.95	25.20
16110.22 EMT CONDUIT				
EMT conduit				
1/2"	L.F.	0.17	1.05	1.22
3/4"	"	0.23	1.40	1.63
1"	"	0.37	1.75	2.12
1-1/4"	"	0.53	2.05	2.58
1-1/2"	"	0.62	2.80	3.42
2"	"	0.78	3.10	3.88
2-1/2"	"	1.75	3.45	5.20
3"	"	2.20	4.25	6.45
3-1/2"	"	3.10	5.05	8.15
4"	"	3.60	6.30	9.90
16110.23 FLEXIBLE CONDUIT				
Flexible conduit, steel				
3/8"	L.F.	0.19	1.05	1.24
1/2	"	0.20	1.05	1.25
3/4"	"	0.28	1.40	1.68
1"	"	0.55	1.40	1.95
1-1/4"	"	0.70	1.75	2.45
1-1/2"	"	0.92	2.05	2.97
2"	"	1.20	2.80	4.00
2-1/2"	"	1.40	3.10	4.50
3"	"	1.75	3.70	5.45
16110.24 GALVANIZED CONDUIT				
Galvanized rigid steel conduit				
1/2"	L.F.	0.76	1.40	2.16
3/4"	"	0.95	1.75	2.70
1"	"	1.30	2.05	3.35
1-1/4"	"	1.70	2.80	4.50
1-1/2"	"	2.05	3.10	5.15
2"	"	2.65	3.45	6.10
2-1/2"	"	4.20	5.05	9.25
3"	"	5.60	6.30	11.90
3-1/2"	"	5.65	6.60	12.25
4"	"	8.30	7.30	15.60
5"	"	17.65	9.90	27.55
6"	"	25.40	13.20	38.60
16110.27 PLASTIC COATED CONDUIT				
Rigid steel conduit, plastic coated				
1/2"	L.F.	2.15	1.75	3.90
3/4"	"	2.50	2.05	4.55

16 ELECTRICAL

BASIC MATERIALS	UNIT	MAT.	INST.	TOTAL
16110.27 PLASTIC COATED CONDUIT				
1"	L.F.	3.25	2.80	6.05
1-1/4"	"	4.10	3.45	7.55
1-1/2"	"	4.95	4.25	9.20
2"	"	6.50	5.05	11.55
2-1/2"	"	9.90	6.60	16.50
3"	"	12.45	7.70	20.15
3-1/2"	"	15.15	8.70	23.85
4"	"	18.55	10.70	29.25
5"	"	31.80	13.20	45.00
16110.28 STEEL CONDUIT				
Intermediate metal conduit (IMC)				
1/2"	L.F.	0.71	1.05	1.76
3/4"	"	0.85	1.40	2.25
1"	"	1.15	1.75	2.90
1-1/4"	"	1.50	2.05	3.55
1-1/2"	"	1.75	2.80	4.55
2"	"	2.45	3.10	5.55
2-1/2"	"	3.80	4.15	7.95
3"	"	5.15	5.05	10.20
3-1/2"	"	6.80	6.30	13.10
4"	"	8.10	6.60	14.70
16110.35 WIREMOLD				
Wiremold raceway with fittings, surface mounted				
#200	L.F.	0.46	1.05	1.51
#500	"	0.55	1.05	1.60
#700	"	0.61	1.40	2.01
#800	"	0.99	1.40	2.39
Fittings, #200, 90 degree flat elbow	EA.	0.88	1.75	2.63
Internal elbow	"	1.75	1.75	3.50
Extension adapter	"	3.10	2.05	5.15
#200, #500, #700				
Single pole switch and box	EA.	5.90	13.90	19.80
Duplex receptacle with box	"	6.80	10.70	17.50
#500, #700				
90 deg. flat elbow	EA.	0.76	2.80	3.56
Internal elbow	"	0.99	2.80	3.79
Junction box	"	5.00	4.65	9.65
Fixture box	"	5.05	4.65	9.70
Shallow switch and receptacle box	"	4.10	2.80	6.90
#800				
90 deg. flat elbow	EA.	0.94	2.80	3.74
Internal elbow	"	0.94	2.80	3.74
Junction box	"	2.95	4.25	7.20

16 ELECTRICAL

BASIC MATERIALS

BASIC MATERIALS	UNIT	MAT.	INST.	TOTAL
16110.80 WIREWAYS				
Wireway, hinge cover type				
2-1/2" x 2-1/2"				
1' section	EA.	7.70	5.35	13.05
2'	"	10.95	6.60	17.55
3'	"	14.90	8.70	23.60
5'	"	24.60	13.20	37.80
10'	"	51.00	23.15	74.15
4" x 4"				
1'	EA.	7.45	8.70	16.15
2'	"	12.40	8.70	21.10
3'	"	18.35	10.70	29.05
4'	"	24.95	10.70	35.65
10'	"	56.00	27.80	83.80
16120.43 COPPER CONDUCTORS				
Copper conductors, type THW, solid				
#14	L.F.	0.04	0.14	0.18
#12	"	0.07	0.17	0.24
#10	"	0.09	0.21	0.30
Stranded				
#14	L.F.	0.07	0.14	0.21
#12	"	0.09	0.17	0.26
#10	"	0.11	0.21	0.32
#8	"	0.18	0.28	0.46
#6	"	0.23	0.31	0.54
#4	"	0.35	0.35	0.70
#3	"	0.50	0.35	0.85
#2	"	0.61	0.42	1.03
#1	"	0.77	0.49	1.26
1/0	"	0.94	0.56	1.50
2/0	"	1.10	0.69	1.79
3/0	"	1.35	0.87	2.22
4/0	"	1.70	0.97	2.67
250 MCM	"	2.10	1.05	3.15
300 MCM	"	2.50	1.15	3.65
350 MCM	"	3.05	1.40	4.45
400 MCM	"	3.40	1.55	4.95
500 MCM	"	4.10	1.80	5.90
600 MCM	"	5.80	2.05	7.85
750 MCM	"	7.45	2.30	9.75
1000 MCM	"	8.25	2.65	10.90
THHN-THWN, solid				
#14	L.F.	0.04	0.14	0.18
#12	"	0.07	0.17	0.24
#10	"	0.09	0.21	0.30
Stranded				
#14	L.F.	0.07	0.14	0.21
#12	"	0.09	0.17	0.26
#10	"	0.11	0.21	0.32
#8	"	0.18	0.28	0.46
#6	"	0.23	0.31	0.54

BASIC MATERIALS

BASIC MATERIALS	UNIT	MAT.	INST.	TOTAL
16120.43 COPPER CONDUCTORS				
#4	L.F.	0.35	0.35	0.70
#2	"	0.61	0.42	1.03
#1	"	0.77	0.49	1.26
1/0	"	0.94	0.56	1.50
2/0	"	1.10	0.69	1.79
3/0	"	1.35	0.87	2.22
4/0	"	1.70	0.97	2.67
250 MCM	"	2.10	1.05	3.15
350 MCM	"	3.05	1.40	4.45
16120.47 SHEATHED CABLE				
Non-metallic sheathed cable				
Type NM cable with ground				
#14/2	L.F.	0.12	0.52	0.64
#12/2	"	0.18	0.56	0.74
#10/2	"	0.29	0.62	0.91
#8/2	"	0.59	0.69	1.28
#6/2	"	0.88	0.87	1.75
#14/3	"	0.20	0.90	1.10
#12/3	"	0.29	0.93	1.22
#10/3	"	0.43	0.94	1.37
#8/3	"	0.91	0.96	1.87
#6/3	"	1.25	0.97	2.22
#4/3	"	1.85	1.10	2.95
#2/3	"	2.75	1.20	3.95
Type U.F. cable with ground				
#14/2	L.F.	0.15	0.56	0.71
#12/2	"	0.21	0.66	0.87
#10/2	"	0.31	0.69	1.00
#8/2	"	0.91	0.79	1.70
#6/2	"	1.10	0.94	2.04
#14/3	"	0.22	0.69	0.91
#12/3	"	0.31	0.76	1.07
#10/3	"	0.44	0.87	1.31
#8/3	"	1.20	0.97	2.17
#6/3	"	1.70	1.10	2.80
16130.60 PULL AND JUNCTION BOXES				
4"				
Octagon box	EA.	1.05	3.95	5.00
Box extension	"	1.55	2.05	3.60
Plaster ring	"	0.83	2.05	2.88
Cover blank	"	0.35	2.05	2.40
Square box	"	1.30	3.95	5.25
Box extension	"	1.40	2.05	3.45
Plaster ring	"	0.65	2.05	2.70
Cover blank	"	0.41	2.05	2.46
4-11/16"				
Square box	EA.	2.30	3.95	6.25
Box extension	"	3.10	2.05	5.15

16 ELECTRICAL

BASIC MATERIALS

BASIC MATERIALS	UNIT	MAT.	INST.	TOTAL
16130.60 PULL AND JUNCTION BOXES				
Plaster ring	EA.	1.80	2.05	3.85
Cover blank	"	0.70	2.05	2.75
Switch and device boxes				
2 gang	EA.	6.65	3.95	10.60
3 gang	"	8.20	3.95	12.15
4 gang	"	11.60	5.55	17.15
Device covers				
2 gang	EA.	3.25	2.05	5.30
3 gang	"	3.85	2.05	5.90
4 gang	"	5.60	2.05	7.65
Handy box	"	1.00	3.95	4.95
Extension	"	1.35	2.05	3.40
Switch cover	"	0.41	2.05	2.46
Switch box with knockout	"	1.10	5.05	6.15
Weatherproof cover, spring type	"	3.85	2.80	6.65
Cover plate, dryer receptacle 1 gang plastic	"	1.80	3.45	5.25
For 4" receptacle, 2 gang	"	3.10	3.45	6.55
Duplex receptacle cover plate, plastic	"	0.35	2.05	2.40
4", vertical bracket box, 1-1/2" with				
RMX clamps	EA.	1.60	5.05	6.65
BX clamps	"	2.45	5.05	7.50
4", octagon device cover				
1 switch	EA.	1.10	2.05	3.15
1 duplex recept	"	1.05	2.05	3.10
4", square face bracket boxes, 1-1/2"				
RMX	EA.	2.65	5.05	7.70
BX	"	2.85	5.05	7.90
16130.65 PULL BOXES AND CABINETS				
Galvanized pull boxes, screw cover				
4x4x4	EA.	3.60	6.60	10.20
4x6x4	"	4.50	6.60	11.10
6x6x4	"	5.40	6.60	12.00
6x8x4	"	7.80	6.60	14.40
8x8x4	"	9.10	8.70	17.80
8x10x4	"	10.55	8.40	18.95
8x12x4	"	11.80	8.70	20.50
Screw cover				
10x10x4	EA.	11.80	10.70	22.50
12x12x6	"	18.00	15.40	33.40
12x15x6	"	20.40	15.40	35.80
12x18x6	"	23.85	17.35	41.20
15x18x6	"	27.00	19.85	46.85
18x24x6	"	48.90	21.35	70.25
18x30x6	"	54.00	25.20	79.20
24x36x6	"	84.00	25.20	109.20
16130.80 RECEPTACLES				
125 volt, 20a, duplex, grounding type, standard grade	EA.	4.75	6.95	11.70
Ground fault interrupter type	"	37.10	10.30	47.40

BASIC MATERIALS

BASIC MATERIALS	UNIT	MAT.	INST.	TOTAL
16130.80 RECEPTACLES				
250 volt, 20a, 2 pole, single receptacle, ground type	EA.	5.55	6.95	12.50
120/208v, 4 pole, single receptacle, twist lock				
20a	EA.	11.95	12.05	24.00
50a	"	22.75	12.05	34.80
125/250v, 3 pole, flush receptacle				
30a	EA.	7.60	10.30	17.90
50a	"	8.75	10.30	19.05
60a	"	35.30	12.05	47.35
277 v, 20a, 2 pole, grounding type, twist lock	"	8.75	6.95	15.70
Dryer receptacle, 250v, 30a/50a, 3 wire	"	14.65	10.30	24.95
Clock receptacle, 2 pole, grounding type	"	6.10	6.95	13.05
125v, 20a single recept. grounding type				
Standard grade	EA.	4.45	6.95	11.40
Specification	"	6.90	6.95	13.85
Hospital	"	7.10	6.95	14.05
Isolated ground orange	"	12.40	8.70	21.10
Duplex				
Specification grade	EA.	7.15	6.95	14.10
Hospital	"	9.25	6.95	16.20
Isolated ground orange	"	13.95	8.70	22.65
GFI hospital grade recepts, 20a, 125v, duplex	"	42.90	10.30	53.20
16198.10 ELECTRIC MANHOLES				
Precast, handhole, 4' deep				
2'x2'	EA.	283.00	121.00	404.00
3'x3'	"	385.00	193.00	578.00
4'x4'	"	826.00	356.00	1,182
Power manhole, complete, precast, 8' deep				
4'x4'	EA.	1,250	487.00	1,737
6'x6'	"	1,640	694.00	2,334
8'x8'	"	1,900	731.00	2,631
6' deep, 9' x 12'	"	2,210	868.00	3,078
Cast in place, power manhole, 8' deep				
4'x4'	EA.	1,470	487.00	1,957
6'x6'	"	1,810	694.00	2,504
8'x8'	"	2,040	731.00	2,771
16199.10 UTILITY POLES & FITTINGS				
Wood pole, creosoted				
25'	EA.	124.00	81.50	205.50
40'	"	566.00	132.00	698.00
55'	"	780.00	262.00	1,042
Treated, wood preservative, 6"x6"				
8'	EA.	11.35	17.35	28.70
16'	"	56.50	55.50	112.00
20'	"	113.00	69.50	182.50
Aluminum, brushed, no base				
8'	EA.	243.00	69.50	312.50
20'	"	657.00	111.00	768.00
40'	"	2,150	217.00	2,367

16 ELECTRICAL

BASIC MATERIALS

16199.10 UTILITY POLES & FITTINGS

BASIC MATERIALS	UNIT	MAT.	INST.	TOTAL
Steel, no base				
10'	EA.	294.00	87.00	381.00
20'	"	430.00	132.00	562.00
35'	"	1,250	217.00	1,467
Concrete, no base				
13'	EA.	453.00	191.00	644.00
30'	"	1,290	421.00	1,711
50'	"	2,950	631.00	3,581
60'	"	3,740	694.00	4,434

POWER GENERATION

16210.10 GENERATORS

POWER GENERATION	UNIT	MAT.	INST.	TOTAL
Diesel generator, with auto transfer switch				
50kw	EA.	18,420	1,070	19,490
125kw	"	28,600	1,740	30,340
300kw	"	46,800	3,470	50,270
750kw	"	137,500	6,940	144,440

16320.10 TRANSFORMERS

POWER GENERATION	UNIT	MAT.	INST.	TOTAL
Floor mounted, single phase, int. dry, 480v-120/240v				
3 kva	EA.	217.00	63.00	280.00
5 kva	"	294.00	107.00	401.00
7.5 kva	"	441.00	121.00	562.00
10 kva	"	499.00	132.00	631.00
15 kva	"	648.00	149.00	797.00
100 kva	"	2,360	402.00	2,762
Three phase, 480v-120/280v				
15 kva	EA.	824.00	209.00	1,033
30 kva	"	1,060	327.00	1,387
45 kva	"	1,410	375.00	1,785
225 kva	"	4,650	534.00	5,184

16350.10 CIRCUIT BREAKERS

POWER GENERATION	UNIT	MAT.	INST.	TOTAL
Molded case, 240v, 15-60a, bolt-on				
1 pole	EA.	7.55	8.70	16.25
2 pole	"	16.80	12.05	28.85
70-100a, 2 pole	"	44.40	18.50	62.90
15-60a, 3 pole	"	54.50	13.90	68.40
70-100a, 3 pole	"	79.50	21.35	100.85
480v, 2 pole				
15-60	EA.	121.00	10.30	131.30

POWER GENERATION

POWER GENERATION	UNIT	MAT.	INST.	TOTAL
16350.10 CIRCUIT BREAKERS				
70-100a	EA.	153.00	13.90	166.90
3 pole				
15-60a	EA.	153.00	13.90	166.90
70-100a	"	183.00	15.40	198.40
70-225a	"	403.00	21.35	424.35
Load center circuit breakers, 240v				
1 pole, 10-60a	EA.	7.35	8.70	16.05
2 pole				
10-60a	EA.	16.80	13.90	30.70
70-100a	"	16.15	23.15	39.30
110-150a	"	102.00	25.20	127.20
3 pole				
10-60a	EA.	57.50	17.35	74.85
70-100a	"	86.50	25.20	111.70
Load center, G.F.I. breakers, 240v				
1 pole, 15-30a	EA.	57.50	10.30	67.80
2 pole, 15-30a	"	102.00	13.90	115.90
Key operated breakers, 240v, 1 pole, 10-30a	"	9.60	10.30	19.90
Tandem breakers, 240v				
1 pole, 15-30a	EA.	15.60	13.90	29.50
2 pole, 15-30a	"	30.00	18.50	48.50
Bolt-on, G.F.I. breakers, 240v, 1 pole, 15-30a	"	71.50	12.05	83.55
16360.10 SAFETY SWITCHES				
Fused, 3 phase, 30 amp, 600v, heavy duty				
NEMA 1	EA.	107.00	39.70	146.70
NEMA 3r	"	182.00	39.70	221.70
NEMA 4	"	459.00	55.50	514.50
NEMA 12	"	177.00	60.50	237.50
60a				
NEMA 1	EA.	129.00	39.70	168.70
NEMA 3r	"	212.00	39.70	251.70
NEMA 4	"	516.00	55.50	571.50
NEMA 12	"	180.00	60.50	240.50
100a				
NEMA 1	EA.	240.00	60.50	300.50
NEMA 3r	"	333.00	60.50	393.50
NEMA 4	"	1,010	69.50	1,080
NEMA 12	"	282.00	87.00	369.00
200a				
NEMA 1	EA.	323.00	87.00	410.00
NEMA 3r	"	459.00	87.00	546.00
NEMA 4	"	1,410	95.50	1,506
NEMA 12	"	450.00	121.00	571.00
Non-fused, 240-600v, heavy duty, 3 phase, 30 amp				
NEMA 1	EA.	55.50	39.70	95.20
NEMA 3r	"	100.00	39.70	139.70
NEMA 4	"	405.00	60.50	465.50
NEMA 12	"	123.00	60.50	183.50
60a				
NEMA1	EA.	100.00	39.70	139.70
NEMA 3r	"	177.00	39.70	216.70

16 ELECTRICAL

POWER GENERATION

	UNIT	MAT.	INST.	TOTAL
16360.10 SAFETY SWITCHES				
NEMA 4	EA.	482.00	60.50	542.50
NEMA 12	"	153.00	60.50	213.50
100a				
NEMA 1	EA.	156.00	60.50	216.50
NEMA 3r	"	252.00	60.50	312.50
NEMA 4	"	983.00	87.00	1,070
NEMA 12	"	219.00	87.00	306.00
200a, NEMA 1	"	268.00	87.00	355.00
600a, NEMA 12	"	1,300	427.00	1,727
16365.10 FUSES				
Fuse, one-time, 250v				
30a	EA.	0.88	1.75	2.63
60a	"	1.10	1.75	2.85
100a	"	4.75	1.75	6.50
200a	"	11.90	1.75	13.65
400a	"	21.40	1.75	23.15
600a	"	30.90	1.75	32.65
600v				
30a	EA.	3.55	1.75	5.30
60a	"	4.75	1.75	6.50
100a	"	10.70	1.75	12.45
200a	"	20.20	1.75	21.95
400a	"	39.20	1.75	40.95
16395.10 GROUNDING				
Ground rods, copper clad, 1/2" x				
6'	EA.	6.55	23.15	29.70
8'	"	8.50	25.20	33.70
10'	"	10.45	34.70	45.15
5/8" x				
5'	EA.	6.65	21.35	28.00
6'	"	7.90	25.20	33.10
8'	"	9.80	34.70	44.50
10'	"	12.75	43.40	56.15
3/4" x				
8'	EA.	15.85	25.20	41.05
10'	"	18.80	27.80	46.60
Ground rod clamp				
5/8"	EA.	2.10	4.25	6.35
3/4"	"	2.70	4.25	6.95

SERVICE AND DISTRIBUTION

16425.10 SWITCHBOARDS

	UNIT	MAT.	INST.	TOTAL
Switchboard, 90" high, no main disconnect, 208/120v				
400a	EA.	1,340	275.00	1,615
600a	"	2,070	278.00	2,348
1000a	"	2,610	278.00	2,888
1200a	"	2,790	347.00	3,137
1600a	"	3,060	414.00	3,474
2000a	"	3,290	487.00	3,777
2500a	"	3,630	555.00	4,185
240/480v				
600a	EA.	2,380	283.00	2,663
800a	"	2,610	283.00	2,893
1600a	"	3,290	414.00	3,704
2000a	"	3,510	487.00	3,997
2500a	"	3,740	555.00	4,295
3000a	"	4,310	957.00	5,267
4000a	"	5,210	1,030	6,240

16470.10 PANELBOARDS

	UNIT	MAT.	INST.	TOTAL
3 phase, 480/277v, main lugs only, 120a, 30 circuits	EA.	515.00	121.00	636.00
277/480v, 4 wire, flush surface				
225a, 30 circuits	EA.	549.00	139.00	688.00
400a, 30 circuits	"	629.00	174.00	803.00
600a, 42 circuits	"	783.00	209.00	992.00
208/120v, main circuit breaker, 3 phase, 4 wire				
100a				
12 circuits	EA.	451.00	177.00	628.00
20 circuits	"	554.00	219.00	773.00
30 circuits	"	817.00	244.00	1,061
400a				
30 circuits	EA.	1,720	514.00	2,234
42 circuits	"	2,060	555.00	2,615
600a, 42 circuits	"	2,290	631.00	2,921
120/208v, flush, 3 ph., 4 wire, main only				
100a				
12 circuits	EA.	315.00	177.00	492.00
20 circuits	"	434.00	219.00	653.00
30 circuits	"	629.00	244.00	873.00
400a				
30 circuits	EA.	1,260	514.00	1,774
42 circuits	"	1,830	555.00	2,385
600a, 42 circuits	"	2,860	631.00	3,491

16480.10 MOTOR CONTROLS

	UNIT	MAT.	INST.	TOTAL
Motor generator set, 3 phase, 480/277v, w/controls				
10kw	EA.	6,920	957.00	7,877
15kw	"	9,020	1,070	10,090
20kw	"	10,010	1,110	11,120
40kw	"	14,120	1,320	15,440
100kw	"	23,100	2,140	25,240
200kw	"	52,800	2,520	55,320
300kw	"	66,000	2,780	68,780

SERVICE AND DISTRIBUTION

16480.10 MOTOR CONTROLS

	UNIT	MAT.	INST.	TOTAL
2 pole, 230 volt starter, w/NEMA-1				
1 hp, 9 amp, size 00	EA.	100.00	34.70	134.70
2 hp, 18amp, size 0	"	128.00	34.70	162.70
3 hp, 27amp, size 1	"	139.00	34.70	173.70
5 hp, 45amp, size 1p	"	179.00	34.70	213.70
7-1/2 hp, 45a, size 2	"	302.00	34.70	336.70
15 hp, 90a, size 3	"	460.00	34.70	494.70

16490.10 SWITCHES

	UNIT	MAT.	INST.	TOTAL
Fused interrupter load, 35kv				
20A				
1 pole	EA.	13,200	555.00	13,755
2 pole	"	14,300	591.00	14,891
3 way	"	15,400	591.00	15,991
4 way	"	16,500	631.00	17,131
30a, 1 pole	"	13,200	555.00	13,755
3 way	"	15,400	591.00	15,991
4 way	"	16,500	631.00	17,131
Weatherproof switch, including box & cover, 20a				
1 pole	EA.	14,300	555.00	14,855
2 pole	"	15,400	591.00	15,991
3 way	"	16,500	631.00	17,131
4 way	"	17,600	631.00	18,231
Specification grade toggle switches, 20a, 120-277v				
Single pole	EA.	9.00	6.95	15.95
Double pole	"	10.35	10.30	20.65
3 way	"	9.55	8.70	18.25
Switch plates, plastic ivory				
1 gang	EA.	0.34	2.80	3.14
2 gang	"	0.69	3.45	4.14
3 gang	"	1.05	4.15	5.20
6 gang	"	4.60	6.30	10.90
Stainless steel				
1 gang	EA.	1.15	2.80	3.95
2 gang	"	2.25	3.45	5.70
4 gang	"	5.40	5.05	10.45
6 gang	"	9.25	5.05	14.30
Brass				
1 gang	EA.	2.50	2.80	5.30
2 gang	"	4.95	3.45	8.40
4 gang	"	12.10	5.05	17.15
6 gang	"	17.90	6.30	24.20

16490.20 TRANSFER SWITCHES

	UNIT	MAT.	INST.	TOTAL
Automatic transfer switch 600v, 3 pole				
30a	EA.	1,320	121.00	1,441
100a	"	2,750	165.00	2,915
400a	"	6,050	347.00	6,397
800a	"	10,120	631.00	10,751
1200a	"	16,500	793.00	17,293

16 ELECTRICAL

SERVICE AND DISTRIBUTION

	UNIT	MAT.	INST.	TOTAL
16490.20 TRANSFER SWITCHES				
2600a	EA.	38,500	1,460	39,960
16490.80 SAFETY SWITCHES				
Safety switch, 600v, 3 pole, heavy duty, NEMA-1				
30a	EA.	108.00	34.70	142.70
60a	"	129.00	39.70	168.70
100a	"	252.00	55.50	307.50
200a	"	323.00	87.00	410.00
400a	"	930.00	191.00	1,121
600a	"	1,520	278.00	1,798
800a	"	2,710	365.00	3,075
1200a	"	3,360	496.00	3,856

LIGHTING

	UNIT	MAT.	INST.	TOTAL
16510.05 INTERIOR LIGHTING				
Recessed fluorescent fixtures, 2'x2'				
2 lamp	EA.	45.30	25.20	70.50
4 lamp	"	61.00	25.20	86.20
2 lamp w/flange	"	57.00	34.70	91.70
4 lamp w/flange	"	72.00	34.70	106.70
1'x4'				
2 lamp	EA.	46.10	23.15	69.25
3 lamp	"	63.50	23.15	86.65
2 lamp w/flange	"	57.00	25.20	82.20
3 lamp w/flange	"	74.50	25.20	99.70
2'x4'				
2 lamp	EA.	57.00	25.20	82.20
3 lamp	"	66.50	25.20	91.70
4 lamp	"	65.00	25.20	90.20
2 lamp w/flange	"	67.50	34.70	102.20
3 lamp w/flange	"	77.00	34.70	111.70
4 lamp w/flange	"	76.00	34.70	110.70
4'x4'				
4 lamp	EA.	136.00	34.70	170.70
6 lamp	"	160.00	34.70	194.70
8 lamp	"	168.00	34.70	202.70
4 lamp w/flange	"	164.00	52.50	216.50
6 lamp w/flange	"	207.00	52.50	259.50
8 lamp, w/flange	"	195.00	52.50	247.50
Surface mounted incandescent fixtures				

16 ELECTRICAL

LIGHTING

LIGHTING	UNIT	MAT.	INST.	TOTAL
16510.05 INTERIOR LIGHTING				
40w	EA.	41.20	23.15	64.35
75w	"	44.80	23.15	67.95
100w	"	53.00	23.15	76.15
150w	"	62.50	23.15	85.65
Pendant				
40w	EA.	44.80	27.80	72.60
75w	"	49.50	27.80	77.30
100w	"	56.50	27.80	84.30
150w	"	64.50	27.80	92.30
Recessed incandescent fixtures				
40w	EA.	76.00	52.50	128.50
75w	"	79.50	52.50	132.00
100w	"	81.50	52.50	134.00
150w	"	93.50	52.50	146.00
Exit lights, 120v				
Recessed	EA.	59.00	43.40	102.40
Back mount	"	41.20	25.20	66.40
Universal mount	"	47.10	25.20	72.30
Emergency battery units, 6v-120v, 50 unit	"	88.00	52.50	140.50
With 1 head	"	102.00	52.50	154.50
With 2 heads	"	118.00	52.50	170.50
Light track single circuit				
2'	EA.	20.30	17.35	37.65
4'	"	37.90	17.35	55.25
8'	"	67.50	34.70	102.20
12'	"	107.00	52.50	159.50
Surface mounted fluorescent, wrap around lens				
1 lamp	EA.	41.60	27.80	69.40
2 lamps	"	45.20	30.80	76.00
4 lamps	"	71.50	34.70	106.20
Wall mounted fluorescent				
2-20w lamps	EA.	33.30	17.35	50.65
2-30w lamps	"	35.70	17.35	53.05
2-40w lamps	"	41.60	23.15	64.75
Indirect, with wood shielding, 2049w lamps				
4'	EA.	71.50	34.70	106.20
8'	"	101.00	55.50	156.50
Industrial fluorescent, 2 lamp				
4'	EA.	71.50	25.20	96.70
8'	"	113.00	46.30	159.30
Strip fluorescent				
4'				
1 lamp	EA.	23.80	23.15	46.95
2 lamps	"	29.70	23.15	52.85
8'				
1 lamp	EA.	47.60	25.20	72.80
2 lamps	"	53.50	30.80	84.30
Wire guard for strip fixture, 4' long	"	11.90	12.05	23.95
Strip fluorescent, 8' long, two 4' lamps	"	71.50	46.30	117.80
With four 4' lamps	"	95.00	55.50	150.50
Wet location fluorescent, plastic housing				
4' long				

LIGHTING

LIGHTING	UNIT	MAT.	INST.	TOTAL
16510.05 INTERIOR LIGHTING				
1 lamp	EA.	83.50	34.70	118.20
2 lamps	"	89.00	46.30	135.30
8' long				
2 lamps	EA.	147.00	55.50	202.50
4 lamps	"	201.00	60.50	261.50
Parabolic troffer, 2'x2'				
With 2 "U" lamps	EA.	101.00	34.70	135.70
With 3 "U" lamps	"	119.00	39.70	158.70
2'x4'				
With 2 40w lamps	EA.	135.00	39.70	174.70
With 3 40w lamps	"	154.00	46.30	200.30
With 4 40w lamps	"	166.00	46.30	212.30
1'x4'				
With 1 t12 lamp, 9 cell	EA.	101.00	25.20	126.20
With 2 t12 lamps	"	113.00	30.80	143.80
With 1 t12 lamp, 20 cell	"	113.00	25.20	138.20
With 2 t12 lamps	"	124.00	30.80	154.80
Steel sided surface fluorescent, 2'x4'				
3 lamps	EA.	89.00	46.30	135.30
4 lamps	"	95.00	46.30	141.30
Outdoor sign fluor., 1 lamp, remote ballast				
4' long	EA.	1,700	209.00	1,909
6' long	"	2,040	278.00	2,318
Recess mounted, commercial, 2'x2', 13" high				
100w	EA.	566.00	139.00	705.00
250w	"	623.00	156.00	779.00
High pressure sodium, hi-bay open				
400w	EA.	311.00	60.50	371.50
1000w	"	538.00	84.00	622.00
Enclosed				
400w	EA.	484.00	84.00	568.00
1000w	"	713.00	103.00	816.00
Metal halide hi-bay, open				
400w	EA.	283.00	60.50	343.50
1000w	"	481.00	84.00	565.00
Enclosed				
400w	EA.	453.00	84.00	537.00
1000w	"	680.00	103.00	783.00
High pressure sodium, low bay, surface mounted				
100w	EA.	198.00	34.70	232.70
150w	"	214.00	39.70	253.70
250w	"	243.00	46.30	289.30
400w	"	311.00	55.50	366.50
Metal halide, low bay, pendant mounted				
175w	EA.	214.00	46.30	260.30
250w	"	283.00	55.50	338.50
400w	"	351.00	77.00	428.00
Indirect luminare, square, metal halide, freestanding				
175w	EA.	566.00	34.70	600.70
250w	"	578.00	34.70	612.70
400w	"	623.00	34.70	657.70
High pressure sodium				

LIGHTING	UNIT	MAT.	INST.	TOTAL
16510.05 INTERIOR LIGHTING				
150w	EA.	578.00	34.70	612.70
250w	"	606.00	34.70	640.70
400w	"	657.00	34.70	691.70
Round, metal halide				
175w	EA.	554.00	34.70	588.70
250w	"	566.00	34.70	600.70
400w	"	600.00	34.70	634.70
High pressure sodium				
150w	EA.	543.00	34.70	577.70
250w	"	588.00	34.70	622.70
400w	"	612.00	34.70	646.70
Wall mounted, metal halide				
175w	EA.	441.00	87.00	528.00
250w	"	475.00	87.00	562.00
400w	"	781.00	111.00	892.00
High pressure sodium				
150w	EA.	509.00	87.00	596.00
250w	"	623.00	87.00	710.00
400w	"	815.00	111.00	926.00
Wall pack lithonia, high pressure sodium				
35w	EA.	135.00	30.80	165.80
55w	"	153.00	34.70	187.70
150w	"	169.00	55.50	224.50
250w	"	214.00	60.50	274.50
Low pressure sodium				
35w	EA.	192.00	60.50	252.50
55w	"	294.00	69.50	363.50
Wall pack hubbell, high pressure sodium				
35w	EA.	146.00	30.80	176.80
150w	"	186.00	55.50	241.50
250w	"	238.00	60.50	298.50
Compact fluorescent				
2-7w	EA.	90.00	34.70	124.70
2-13w	"	101.00	46.30	147.30
1-18w	"	124.00	46.30	170.30
Handball & racquet ball court, 2'x2', metal halide				
250w	EA.	317.00	87.00	404.00
400w	"	396.00	95.50	491.50
High pressure sodium				
250w	EA.	351.00	87.00	438.00
400w	"	385.00	95.50	480.50
Bollard light, 42" w/found., high pressure sodium				
70w	EA.	538.00	89.50	627.50
100w	"	550.00	89.50	639.50
150w	"	554.00	89.50	643.50
Light fixture lamps				
Lamp				
20w med. bipin base, cool white, 24"	EA.	4.10	5.05	9.15
30w cool white, rapid start, 36"	"	5.35	5.05	10.40
40w cool white "U", 3"	"	10.25	5.05	15.30
40w cool white, rapid start, 48"	"	2.45	5.05	7.50
70w high pressure sodium, mogul base	"	53.00	6.95	59.95

LIGHTING

LIGHTING	UNIT	MAT.	INST.	TOTAL
16510.05 INTERIOR LIGHTING				
75w slimline, 96"	EA.	6.85	6.95	13.80
100w				
Incandescent, 100a, inside frost	EA.	2.10	3.45	5.55
Mercury vapor, clear, mogul base	"	25.90	6.95	32.85
High pressure sodium, mogul base	"	56.50	6.95	63.45
150w				
Par 38 flood or spot, incandescent	EA.	6.35	3.45	9.80
High pressure sodium, 1/2 mogul base	"	59.00	6.95	65.95
175w				
Mercury vapor, clear, mogul base	EA.	44.80	6.95	51.75
Mercury halide, clear, mogul base	"	20.00	6.95	26.95
250w				
High pressure sodium, mogul base	EA.	62.50	6.95	69.45
Mercury vapor, clear, mogul base	"	35.30	6.95	42.25
Metal halide, clear, mogul base	"	56.50	6.95	63.45
High pressure sodium, mogul base	"	62.50	6.95	69.45
400w				
Mercury vapor, clear, mogul base	EA.	28.30	6.95	35.25
Metal halide, clear, mogul base	"	53.00	6.95	59.95
High pressure sodium, mogul base	"	67.00	6.95	73.95
1000w				
Mercury vapor, clear, mogul base	EA.	61.50	8.70	70.20
High pressure sodium, mogul base	"	157.00	8.70	165.70
16510.30 EXTERIOR LIGHTING				
Exterior light fixtures				
Rectangle, high pressure sodium				
70w	EA.	356.00	87.00	443.00
100w	"	367.00	89.50	456.50
150w	"	396.00	89.50	485.50
250w	"	532.00	95.50	627.50
400w	"	600.00	121.00	721.00
Flood, rectangular, high pressure sodium				
70w	EA.	367.00	87.00	454.00
100w	"	385.00	89.50	474.50
150w	"	425.00	89.50	514.50
400w	"	634.00	121.00	755.00
1000w	"	962.00	156.00	1,118
Round				
400w	EA.	657.00	121.00	778.00
1000w	"	1,030	156.00	1,186
Round, metal halide				
400w	EA.	725.00	121.00	846.00
1000w	"	1,080	156.00	1,236
Light fixture arms, cobra head, 6', high press. sodium				
100w	EA.	396.00	69.50	465.50
150w	"	623.00	87.00	710.00
250w	"	651.00	87.00	738.00
400w	"	673.00	103.00	776.00
Flood, metal halide				
400w	EA.	691.00	121.00	812.00

16 ELECTRICAL

LIGHTING	UNIT	MAT.	INST.	TOTAL
16510.30 EXTERIOR LIGHTING				
1000w	EA.	1,050	156.00	1,206
1500w	"	1,420	209.00	1,629
Mercury vapor				
250w	EA.	407.00	95.50	502.50
400w	"	704.00	121.00	825.00
Incandescent				
300w	EA.	73.50	60.50	134.00
500w	"	124.00	69.50	193.50
1000w	"	135.00	111.00	246.00
16610.30 UNINTERRUPTIBLE POWER				
Uninterruptible power systems, (U.P.S.)	EA.	10,940	278.00	11,218
5 kva	"	13,530	382.00	13,912
7.5 kva	"	16,500	555.00	17,055
10 kva	"	23,100	763.00	23,863
15 kva	"	26,330	925.00	27,255
20 kva	"	35,200	834.00	36,034
25 kva	"	43,890	868.00	44,758
30 kva	"	46,640	901.00	47,541
35 kva	"	49,430	938.00	50,368
40 kva	"	52,250	971.00	53,221
45 kva	"	54,980	1,010	55,990
50 kva	"	57,750	1,040	58,790
62.5 kva	"	66,440	1,110	67,550
75 kva	"	75,130	1,210	76,340
100 kva	"	85,470	1,250	86,720
150 kva	"	106,260	1,740	108,000
200 kva	"	110,000	1,910	111,910
300 kva	"	110,000	2,590	112,590
400 kva	"	110,000	3,120	113,120
500 kva	"	110,000	3,800	113,800

COMMUNICATIONS	UNIT	MAT.	INST.	TOTAL
16720.10 FIRE ALARM SYSTEMS				
Master fire alarm box, pedestal mounted	EA.	3,970	555.00	4,525
Master fire alarm box	"	2,040	209.00	2,249
Box light	"	113.00	17.35	130.35
Ground assembly for box	"	56.50	23.15	79.65
Bracket for pole type box	"	73.50	25.20	98.70
Pull station				
Waterproof	EA.	56.50	17.35	73.85
Manual	"	34.00	13.90	47.90

COMMUNICATIONS	UNIT	MAT.	INST.	TOTAL
16720.10 FIRE ALARM SYSTEMS				
Horn, waterproof	EA.	51.00	34.70	85.70
Interior alarm	"	34.00	25.20	59.20
Coded transmitter, automatic	"	600.00	69.50	669.50
Control panel, 8 zone	"	1,020	278.00	1,298
Battery charger and cabinet	"	363.00	69.50	432.50
Batteries, nickel cadmium or lead calcium	"	141.00	174.00	315.00
CO2 pressure switch connection	"	56.50	25.20	81.70
Annunciator panels				
Fire detection annunciator, remote type, 8-zone	EA.	204.00	63.00	267.00
12-zone	"	261.00	69.50	330.50
16-zone	"	328.00	87.00	415.00
Fire alarm systems				
Bell	EA.	62.50	21.35	83.85
Weatherproof bell	"	77.00	23.15	100.15
Horn	"	36.30	25.20	61.50
Siren	"	368.00	69.50	437.50
Chime	"	45.30	21.35	66.65
Audio/visual	"	65.50	25.20	90.70
Strobe light	"	60.50	25.20	85.70
Smoke detector	"	96.50	23.15	119.65
Heat detection	"	17.00	17.35	34.35
Thermal detector	"	15.85	17.35	33.20
Ionization detector	"	79.50	18.50	98.00
Duct detector	"	266.00	95.50	361.50
Test switch	"	45.30	17.35	62.65
Remote indicator	"	22.90	19.85	42.75
Door holder	"	98.50	25.20	123.70
Telephone jack	"	17.00	10.30	27.30
Fireman phone	"	243.00	34.70	277.70
Speaker	"	48.50	27.80	76.30
Remote fire alarm annunciator panel				
24 zone	EA.	1,250	231.00	1,481
48 zone	"	2,490	451.00	2,941
Control panel				
12 zone	EA.	849.00	103.00	952.00
16 zone	"	1,110	154.00	1,264
24 zone	"	1,700	231.00	1,931
48 zone	"	3,170	555.00	3,725
Power supply	"	198.00	52.50	250.50
Status command	"	5,380	174.00	5,554
Printer	"	2,100	52.50	2,153
Transponder	"	173.00	31.20	204.20
Transformer	"	124.00	23.15	147.15
Transceiver	"	169.00	25.20	194.20
Relays	"	68.00	17.35	85.35
Flow switch	"	220.00	69.50	289.50
Tamper switch	"	133.00	103.00	236.00
End of line resistor	"	9.60	12.05	21.65
Printed ckt. card	"	85.00	17.35	102.35
Central processing unit	"	11,100	214.00	11,314
UPS backup to c.p.u.	"	10,760	312.00	11,072
Smoke detector, fixed temp. & rate of rise comb.	"	169.00	55.50	224.50

COMMUNICATIONS

16720.50 SECURITY SYSTEMS

	UNIT	MAT.	INST.	TOTAL
Sensors				
Balanced magnetic door switch, surface mounted	EA.	81.00	17.35	98.35
With remote test	"	109.00	34.70	143.70
Flush mounted	"	77.50	64.50	142.00
Mounted bracket	"	5.95	12.05	18.00
Mounted bracket spacer	"	5.35	12.05	17.40
Photoelectric sensor, for fence				
6 beam	EA.	8,710	95.50	8,806
9 beam	"	10,640	148.00	10,788
Photoelectric sensor, 12 volt dc				
500' range	EA.	261.00	55.50	316.50
800' range	"	285.00	69.50	354.50
Capacitance wire grid kit				
Surface	EA.	71.50	34.70	106.20
Duct	"	53.50	55.50	109.00
Tube grid kit	"	89.00	17.35	106.35
Vibration sensor, 30 max per zone	"	107.00	17.35	124.35
Audio sensor, 30 max per zone	"	119.00	17.35	136.35
Inertia sensor				
Outdoor	EA.	83.00	25.20	108.20
Indoor	"	53.50	17.35	70.85
Ultrasonic transmitter, 20 max per zone				
Omni-directional	EA.	59.50	55.50	115.00
Directional	"	65.50	46.30	111.80
Transceiver				
Omni-directional	EA.	65.50	34.70	100.20
Directional	"	71.50	34.70	106.20
Passive infra-red sensor, 20 max per zone	"	463.00	55.50	518.50
Access/secure control unit, balanced magnetic switch	"	285.00	55.50	340.50
Photoelectric sensor	"	475.00	55.50	530.50
Photoelectric fence sensor	"	486.00	55.50	541.50
Capacitance sensor	"	564.00	60.50	624.50
Audio and vibration sensor	"	498.00	55.50	553.50
Inertia sensor	"	664.00	55.50	719.50
Ultrasonic sensor	"	760.00	60.50	820.50
Infra-red sensor	"	475.00	69.50	544.50
Monitor panel, with access/secure tone, standard	"	320.00	60.50	380.50
High security	"	475.00	69.50	544.50
Emergency power indicator	"	195.00	17.35	212.35
Monitor rack with 115v power supply				
1 zone	EA.	267.00	34.70	301.70
10 zone	"	1,370	87.00	1,457
Monitor cabinet, wall mounted				
1 zone	EA.	416.00	34.70	450.70
5 zone	"	1,490	55.50	1,546
10 zone	"	653.00	60.50	713.50
20 zone	"	2,070	69.50	2,140
Floor mounted, 50 zone	"	2,150	139.00	2,289
Security system accessories				
Tamper assembly for monitor cabinet	EA.	53.50	15.40	68.90
Monitor panel blank	"	7.15	12.05	19.20
Audible alarm	"	60.50	17.35	77.85

COMMUNICATIONS

16720.50 SECURITY SYSTEMS	UNIT	MAT.	INST.	TOTAL
Audible alarm control	EA.	249.00	12.05	261.05
Termination screw, terminal cabinet				
25 pair	EA.	178.00	55.50	233.50
50 pair	"	285.00	87.00	372.00
150 pair	"	469.00	174.00	643.00
Universal termination, cabinets & panel				
Remote test	EA.	41.60	60.50	102.10
No remote test	"	29.70	25.20	54.90
High security line supervision termination	"	212.00	34.70	246.70
Door cord for capacitance sensor, 12"	"	7.15	17.35	24.50
Insulation block kit for capacitance sensor	"	33.30	12.05	45.35
Termination block for capacitance sensor	"	7.25	12.05	19.30
Guard alert display	"	772.00	21.35	793.35
Uninterrupted power supply				
Plug-in 40kva transformer				
12 volt	EA.	29.70	12.05	41.75
18 volt	"	20.20	12.05	32.25
24 volt	"	14.25	12.05	26.30
Test relay	"	45.80	12.05	57.85
Coaxial cable, 50 ohm	L.F.	0.19	0.21	0.40
Door openers	EA.	53.50	17.35	70.85
Push buttons				
Standard	EA.	11.90	12.05	23.95
Weatherproof	"	17.85	15.40	33.25
Bells	"	45.20	25.20	70.40
Horns				
Standard	EA.	47.60	34.70	82.30
Weatherproof	"	89.00	43.40	132.40
Chimes	"	69.00	23.15	92.15
Flasher	"	50.00	21.35	71.35
Motion detectors	"	207.00	52.50	259.50
Intercom units	"	47.60	25.20	72.80
Remote annunciator	"	2,140	174.00	2,314

16740.10 TELEPHONE SYSTEMS	UNIT	MAT.	INST.	TOTAL
Communication cable				
25 pair	L.F.	0.35	0.90	1.25
100 pair	"	1.80	0.99	2.79
400 pair	"	5.95	1.55	7.50
Cable tap in manhole or junction box				
25 pair cable	EA.	3.05	132.00	135.05
100 pair cable	"	12.10	524.00	536.10
400 pair cable	"	48.50	2,140	2,189
Cable terminations, manhole or junction box				
25 pair cable	EA.	3.05	130.00	133.05
100 pair cable	"	12.10	524.00	536.10
400 pair cable	"	36.40	2,140	2,176

16780.50 TELEVISION SYSTEMS	UNIT	MAT.	INST.	TOTAL
TV outlet, self terminating, w/cover plate	EA.	6.55	10.70	17.25
Thru splitter	"	14.25	55.50	69.75

16 ELECTRICAL

COMMUNICATIONS	UNIT	MAT.	INST.	TOTAL
16780.50 TELEVISION SYSTEMS				
End of line	EA.	11.90	46.30	58.20
In line splitter multitap				
4 way	EA.	23.80	63.00	86.80
2 way	"	17.85	59.00	76.85
Equipment cabinet	"	59.50	55.50	115.00
Antenna				
Broad band uhf	EA.	119.00	121.00	240.00
Lightning arrester	"	32.10	25.20	57.30
TV cable	L.F.	0.26	0.17	0.43
16850.10 ELECTRIC HEATING				
Baseboard heater				
2', 375w	EA.	34.30	34.70	69.00
3', 500w	"	42.30	34.70	77.00
4', 750w	"	52.50	39.70	92.20
5', 935w	"	70.00	46.30	116.30
6', 1125w	"	78.00	55.50	133.50
7', 1310w	"	97.00	63.00	160.00
8', 1500w	"	109.00	69.50	178.50
9', 1680w	"	120.00	77.00	197.00
10', 1875w	"	125.00	79.50	204.50
Unit heater wall mounted				
1500w	EA.	160.00	55.50	215.50
2500w	"	172.00	63.00	235.00
4000w	"	229.00	79.50	308.50
Thermostat				
Integral	EA.	26.30	17.35	43.65
Line voltage	"	26.30	17.35	43.65
Electric heater connection	"	1.15	8.70	9.85

Man-Hour Tables

The man-hour productivities used to develop the labor costs are listed in the following section of this book. These productivities represent typical installation labor for thousands of construction items. The data takes into account all activities involved in normal construction under commonly experienced working conditions. As with the Costbook pages, these items are listed according to the CSI MASTERFORMAT. In order to best use the information in this book, please review this sample page and read the "Features in this Book" section.

CSI MASTERFORMAT Division

CSI Broadscope Category

CSI Mediumscope Category
(First 5 Digits)

Detailed Description
Complete descriptions of items may include information listed above a particular line. Review of the whole category is recommended for a complete description.

Unit of Measurement
Each item is defined in terms of the common estimating unit. Quantities listed are defined as man-hour per unit.

Man-Hours
Man-hour quantities represent typical installation times and take into account all activities involved in normal construction under commonly experienced working conditions.

08 DOORS AND WINDOWS

HARDWARE	UNIT	MAN/ HOURS
08710.20 LOCKSETS		
Latchset, passage or closet latch	EA.	0.667
Lockset		
Privacy (bath or bathroom)	EA.	0.667
Entry lock	"	0.667
Classroom lock (outside key, operated)	"	0.667
Storeroom lock	"	0.667
Bored locks and latches, satin chrome plated		
Latchset passage or closet latch	EA.	0.667
Lockset		
Privacy (bath or bedroom)	EA.	0.667
Entry lock	"	0.667
Classroom lock	"	0.667
Corridor lock	"	0.667
Miscellaneous locks		
Exit lock with alarm, single door	EA.	3.200
Electric strike		
Rim mounted wrought steel	EA.	2.000
Mortised, wrought steel with bronze plating	"	3.200
Dead bolt		
Bored, wrought brass, keyed both sides	EA.	1.333
Mortised, cast brass	"	1.333
Lockset, [illegible]er, mechanical	"	0.800
08710.30 CLOSERS		
Door closers		
Surface mounted, traditional type, parallel arm		
Standard	EA.	1.000
Heavy duty	"	1.000
Modern type, parallel arm, standard duty	"	1.000
Overhead, concealed, pivot hung, single acting		
Interior	EA.	1.000
Exterior	"	1.000
Floor concealed, single acting, offset, pivoted		
Interior	EA.	2.667
Exterior	"	2.667
Chrome plated or bright brass with brass base		
Offset		
Pivot hinges, top pivot, mortised	EA.	1.000
Intermediate pivot hinges, surface mounted	"	1.000
Pivot hinges, bottom pivot	"	3.200
08710.40 DOOR TRIM		
Door bumper, bronze, wall type	EA.	0.160
Wall type, 4" dia. with convex rubber pad, aluminum	"	0.160
Floor type		
Aluminum	EA.	0.160
Brass	"	0.160
Door holders		
Wall type, bronze	EA.	0.160

HARDWARE	UNIT	MAN/ HOURS
08710.40 DOOR TRIM		
Overhead	EA.	0.400
Floor type	"	0.400
Plunger type	"	0.400
Wall type, aluminum	"	0.400
Surface bolt	"	0.160
Panic device, unlabeled, rim type	"	2.000
Rim type with thumb piece	"	2.000
Mortise	"	2.000
Vertical rod	"	2.000
Labeled, rim type	"	2.000
Mortise	"	2.000
Vertical rod	"	2.000
Silencers, rubber type	"	0.016
Dust proof strike with plate, brass	"	0.267
Flush bolt, lever extension, brass, rated	"	0.160
Surface bolt with strike, brass, 6" long	"	0.160
Door coordinator, labeled, brass, satin chrome	"	0.571
Door plates		
Kick plate, .050" aluminum, 3 beveled edges		
10" x 28"	EA.	0.400
10" x 30"	"	0.400
10" x 34"	"	0.400
10" x 38"	"	0.400
Push plate, 4" x 16", .050"		
Aluminum	EA.	0.160
Bronze	"	0.160
Stainless steel	"	0.160
Armor plate, 40" x 34"	"	0.320
Pull handle, 4" x 16", .050"		
Aluminum	EA.	0.160
Bronze	"	0.160
Stainless steel	"	0.160
Hasp assembly		
3"	EA.	0.133
4-1/2"	"	0.178
6"	"	0.229
Electro-magnetic door holder		
Wall mounted	EA.	2.667
Floor mounted	"	2.667
Smoke detector door holder		
Photo electric type	EA.	2.667
Ionization type	"	2.667
Pneumatic operators, activated by rubber mats		
Swing		
Single	EA.	6.667
Double	"	10.000
Sliding		
Single	EA.	6.667
Double	"	10.000

02 SITEWORK

SOIL TESTS

02010.10 SOIL BORING	UNIT	MAN/HOURS
Borings, uncased, stable earth		
2-1/2" dia.	L.F.	0.300
4" dia.	"	0.343
Cased, including samples		
2-1/2" dia.	L.F.	0.400
4" dia.	"	0.686
Drilling in rock		
No sampling	L.F.	0.632
With casing and sampling	"	0.800
Test pits		
Light soil	EA.	4.000
Heavy soil	"	6.000

DEMOLITION

02060.10 BUILDING DEMOLITION	UNIT	MAN/HOURS
Building, complete with disposal		
Wood frame	C.F.	0.003
Concrete	"	0.004
Steel frame	"	0.005
Rubbish handling		
Load in dumpster or truck		
Minimum	C.F.	0.018
Maximum	"	0.027
For use of elevators, add		
Minimum	C.F.	0.004
Maximum	"	0.008
Rubbish hauling		
Hand loaded on trucks, 2 mile trip	C.Y.	0.320
Machine loaded on trucks, 2 mile trip	"	0.240

HIGHWAY DEMOLITION

02065.10 PAVEMENT DEMOLITION	UNIT	MAN/HOURS
Bituminous pavement, up to 3" thick		
On streets	S.Y.	0.096
On pipe trench	"	0.060
Concrete pavement, 6" thick		
No reinforcement	S.Y.	0.160
With wire mesh	"	0.240
With rebars	"	0.300
9" thick		
No reinforcement	S.Y.	0.200
With wire mesh	"	0.300
With rebars	"	0.400
12" thick		
No reinforcement	S.Y.	0.240
With wire mesh	"	0.343
With rebars	"	0.480
Sidewalk, 4" thick, with disposal	"	0.080
Removal of pavement markings by waterblasting	S.F.	0.004

02065.15 SAW CUTTING PAVEMENT	UNIT	MAN/HOURS
Pavement, bituminous		
2" thick	L.F.	0.016
3" thick	"	0.020
4" thick	"	0.025
5" thick	"	0.027
6" thick	"	0.029
Concrete pavement, with wire mesh		
4" thick	L.F.	0.031
5" thick	"	0.033
6" thick	"	0.036
8" thick	"	0.040
10" thick	"	0.044
Plain concrete, unreinforced		
4" thick	L.F.	0.027
5" thick	"	0.031
6" thick	"	0.033
8" thick	"	0.036
10" thick	"	0.040

02065.80 CURB & GUTTER	UNIT	MAN/HOURS
Removal, plain concrete curb	L.F.	0.080
Plain concrete curb and 2' gutter	"	0.107

02065.85 GUARDRAILS	UNIT	MAN/HOURS
Remove standard guardrail		
Steel	L.F.	0.120
Wood	"	0.096

HIGHWAY DEMOLITION

02075.80 CORE DRILLING	UNIT	MAN/HOURS
Concrete		
6" thick		
3" dia.	EA.	0.571
4" dia.	"	0.667
6" dia.	"	0.800
8" dia.	"	1.333
8" thick		
3" dia.	EA.	0.800
4" dia.	"	1.000
6" dia.	"	1.143
8" dia.	"	1.600
10" thick		
3" dia.	EA.	1.000
4" dia.	"	1.143
6" dia.	"	1.333
8" dia.	"	2.000
12" thick		
3" dia.	EA.	1.333
4" dia.	"	1.600
6" dia.	"	2.000
8" dia.	"	2.667

SITE DEMOLITION

02105.10 CATCH BASINS/MANHOLES	UNIT	MAN/HOURS
Abandon catch basin or manhole (fill with sand)	EA.	4.800
Remove and reset frame and cover	"	2.400
Remove catch basin, to 10' deep		
Masonry	EA.	6.000
Concrete	"	8.000

02105.20 FENCES	UNIT	MAN/HOURS
Remove fencing		
Chain link, 8' high		
For disposal	L.F.	0.040
For reuse	"	0.100
Wood		
4' high	EA.	0.027
6' high	"	0.032
8' high	EA.	0.040
Masonry		
8" thick		
4' high	EA.	0.080
6' high	"	0.100
8' high	"	0.114
12" thick		
4' high	EA.	0.133
6' high	"	0.160
8' high	"	0.200
12' high	"	0.267

02105.30 HYDRANTS	UNIT	MAN/HOURS
Remove fire hydrant	EA.	4.000
Remove and reset fire hydrant	"	12.000

02105.42 DRAINAGE PIPING	UNIT	MAN/HOURS
Remove drainage pipe, not including excavation		
12" dia.	L.F.	0.100
18" dia.	"	0.126
24" dia.	"	0.160
36" dia.	"	0.200

02105.43 GAS PIPING	UNIT	MAN/HOURS
Remove welded steel pipe, not including excavation		
4" dia.	L.F.	0.150
5" dia.	"	0.240
6" dia.	"	0.300
8" dia.	"	0.480
10" dia.	EA.	0.600

02105.45 SANITARY PIPING	UNIT	MAN/HOURS
Remove sewer pipe, not including excavation		
4" dia.	L.F.	0.096
6" dia.	"	0.109
8" dia.	"	0.120
10" dia.	"	0.126
12" dia.	"	0.133
15" dia.	"	0.141
18" dia.	"	0.160
24" dia.	"	0.200
30" dia.	"	0.240
36" dia.	"	0.300

02 SITEWORK

SITE DEMOLITION

02105.48 WATER PIPING

	UNIT	MAN/HOURS
Remove water pipe, not including excavation		
4" dia.	L.F.	0.109
6" dia.	"	0.114
8" dia.	"	0.126
10" dia.	"	0.133
12" dia.	"	0.141
14" dia.	"	0.150
16" dia.	"	0.160
18" dia.	"	0.171
20" dia.	"	0.185
Remove valves		
6"	EA.	1.200
10"	"	1.333
14"	"	1.500
18"	"	2.000

02105.60 UNDERGROUND TANKS

	UNIT	MAN/HOURS
Remove underground storage tank, and backfill		
50 to 250 gals	EA.	8.000
600 gals	"	8.000
1000 gals	"	12.000
4000 gals	"	19.200
5000 gals	"	19.200
10,000 gals	"	32.000
12,000 gals	"	40.000
15,000 gals	"	48.000
20,000 gals	"	60.000

02105.66 SEPTIC TANKS

	UNIT	MAN/HOURS
Remove septic tank		
1000 gals	EA.	2.000
2000 gals	"	2.400
5000 gals	"	3.000
15,000 gals	"	24.000
25,000 gals	"	32.000
40,000 gals	"	48.000

02105.80 WALLS, EXTERIOR

	UNIT	MAN/HOURS
Concrete wall		
Light reinforcing		
6" thick	S.F.	0.160
8" thick	"	0.171
10" thick	"	0.200
12" thick	"	0.240
Masonry		
No reinforcing		
8" thick	S.F.	0.069

SITE DEMOLITION

02105.80 WALLS, EXTERIOR

	UNIT	MAN/HOURS
12" thick	S.F.	0.080
16" thick	"	0.096
Horizontal reinforcing		
8" thick	S.F.	0.077
12" thick	"	0.086
16" thick	"	0.109
Remove concrete headwall		
15" pipe	EA.	1.200
18" pipe	"	1.333
24" pipe	"	1.600
30" pipe	"	1.714
36" pipe	"	1.846
48" pipe	"	2.000
60" pipe	"	2.400

02110.10 CLEARING AND GRUBBING

	UNIT	MAN/HOURS
Clear wooded area		
Light density	ACRE	60.000
Medium density	"	80.000
Heavy density	"	96.000

02110.50 TREE CUTTING & CLEARING

	UNIT	MAN/HOURS
Cut trees and clear out stumps		
9" to 12" dia.	EA.	4.800
To 24" dia.	"	6.000
24" dia. and up	"	8.000
Loading and trucking		
For machine load, per load, round trip		
1 mile	EA.	0.960
3 mile	"	1.091
5 mile	"	1.200
10 mile	"	1.600
20 mile	"	2.400
Hand loaded, round trip		
1 mile	EA.	2.000
3 mile	"	2.286
5 mile	"	2.667
10 mile	"	3.200
20 mile	"	4.000
Tree trimming for pole line construction		
Light cutting	L.F.	0.012
Medium cutting	"	0.016
Heavy cutting	"	0.024

02 SITEWORK

DEWATERING

02144.10 WELLPOINT SYSTEMS

Description	UNIT	MAN/HOURS
Pumping, gas driven, 50' hose		
3" header pipe	DAY	8.000
6" header pipe	"	8.000
Wellpoint system per job		
6" header pipe	L.F.	0.032
8" header pipe	"	0.040
10" header pipe	"	0.053
Jetting wellpoint system		
14' long	EA.	0.533
18' long	"	0.667
Sand filter for wellpoints	L.F.	0.013
Replacement of wellpoint components	EA.	0.160

SHORING AND UNDERPINNING

02162.10 TRENCH SHEETING

Description	UNIT	MAN/HOURS
Closed timber, including pull and salvage, excavation		
8' deep	S.F.	0.064
10' deep	"	0.067
12' deep	"	0.071
14' deep	"	0.075
16' deep	"	0.080
18' deep	"	0.098
20' deep	"	0.091

02170.10 COFFERDAMS

Description	UNIT	MAN/HOURS
Cofferdam, steel, pull and salvage, driven from shore		
15' deep	S.F.	0.137
20' deep	"	0.128
25' deep	"	0.120
30' deep	"	0.113
40' deep	"	0.107
Driven from barge		
20' deep	S.F.	0.148
30' deep	"	0.137
40' deep	"	0.128
50' deep	"	0.120

EARTHWORK

02210.10 HAULING MATERIAL

Description	UNIT	MAN/HOURS
Haul material by 10 cy dump truck, round trip distance		
1 mile	C.Y.	0.044
2 mile	"	0.053
5 mile	"	0.073
10 mile	"	0.080
20 mile	"	0.089
30 mile	"	0.107
Site grading, cut & fill, sandy clay, 200' haul, 75 hp dozer	"	0.032
Spread topsoil by equipment on site	"	0.036
Site grading (cut and fill to 6") less than 1 acre		
75 hp dozer	C.Y.	0.053
1.5 cy backhoe/loader	"	0.080

02210.30 BULK EXCAVATION

Description	UNIT	MAN/HOURS
Excavation, by small dozer		
Large areas	C.Y.	0.016
Small areas	"	0.027
Trim banks	"	0.040
Drag line		
1-1/2 cy bucket		
Sand or gravel	C.Y.	0.040
Light clay	"	0.053
Heavy clay	"	0.060
Unclassified	"	0.064
2 cy bucket		
Sand or gravel	C.Y.	0.037
Light clay	"	0.048
Heavy clay	"	0.053
Unclassified	"	0.056
2-1/2 cy bucket		
Sand or gravel	C.Y.	0.034
Light clay	"	0.044
Heavy clay	"	0.048
Unclassified	"	0.051
3 cy bucket		
Sand or gravel	C.Y.	0.030
Light clay	"	0.040
Heavy clay	"	0.044
Unclassified	"	0.046
Hydraulic excavator		
1 cy capacity		
Light material	C.Y.	0.040
Medium material	"	0.048
Wet material	"	0.060
Blasted rock	"	0.069
1-1/2 cy capacity		
Light material	C.Y.	0.010
Medium material	"	0.013
Wet material	"	0.016

02 SITEWORK

EARTHWORK

02210.30 BULK EXCAVATION

EARTHWORK	UNIT	MAN/HOURS
Blasted rock	C.Y.	0.020
2 cy capacity		
Light material	C.Y.	0.009
Medium material	"	0.011
Wet material	"	0.013
Blasted rock	"	0.016
Wheel mounted front-end loader		
7/8 cy capacity		
Light material	C.Y.	0.020
Medium material	"	0.023
Wet material	"	0.027
Blasted rock	"	0.032
1-1/2 cy capacity		
Light material	C.Y.	0.011
Medium material	"	0.012
Wet material	"	0.013
Blasted rock	"	0.015
2-1/2 cy capacity		
Light material	C.Y.	0.009
Medium material	"	0.010
Wet material	"	0.011
Blasted rock	"	0.011
3-1/2 cy capacity		
Light material	C.Y.	0.009
Medium material	"	0.009
Wet material	"	0.010
Blasted rock	"	0.011
6 cy capacity		
Light material	C.Y.	0.005
Medium material	"	0.006
Wet material	"	0.006
Blasted rock	"	0.007
Track mounted front-end loader		
1-1/2 cy capacity		
Light material	C.Y.	0.013
Medium material	"	0.015
Wet material	"	0.016
Blasted rock	"	0.018
2-3/4 cy capacity		
Light material	C.Y.	0.008
Medium material	"	0.009
Wet material	"	0.010
Blasted rock	"	0.011

02220.10 BORROW

EARTHWORK	UNIT	MAN/HOURS
Borrow fill, F.O.B. at pit		
Sand, haul to site, round trip		
10 mile	C.Y.	0.080
20 mile	C.Y.	0.133
30 mile	"	0.200
Place borrow fill and compact		
Less than 1 in 4 slope	C.Y.	0.040
Greater than 1 in 4 slope	"	0.053

02220.40 BUILDING EXCAVATION

EARTHWORK	UNIT	MAN/HOURS
Structural excavation, unclassified earth		
3/8 cy backhoe	C.Y.	0.107
3/4 cy backhoe	"	0.080
1 cy backhoe	"	0.067
Foundation backfill and compaction by machine	"	0.160

02220.50 UTILITY EXCAVATION

EARTHWORK	UNIT	MAN/HOURS
Trencher, sandy clay, 8" wide trench		
18" deep	L.F.	0.018
24" deep	"	0.020
36" deep	"	0.023
Trench backfill, 95% compaction		
Tamp by hand	C.Y.	0.500
Vibratory compaction	"	0.400
Trench backfilling, with borrow sand, place & compact	"	0.400

02220.60 TRENCHING

EARTHWORK	UNIT	MAN/HOURS
Trenching and continuous footing excavation		
By gradall		
1 cy capacity		
Light soil	C.Y.	0.023
Medium soil	"	0.025
Heavy/wet soil	"	0.027
Loose rock	"	0.029
Blasted rock	"	0.031
By hydraulic excavator		
1/2 cy capacity		
Light soil	C.Y.	0.027
Medium soil	"	0.029
Heavy/wet soil	"	0.032
Loose rock	"	0.036
Blasted rock	"	0.040
1 cy capacity		
Light soil	C.Y.	0.019
Medium soil	"	0.020
Heavy/wet soil	"	0.021
Loose rock	"	0.023
Blasted rock	"	0.025
1-1/2 cy capacity		
Light soil	C.Y.	0.017

02 SITEWORK

EARTHWORK	UNIT	MAN/HOURS
02220.60 TRENCHING		
Medium soil	C.Y.	0.018
Heavy/wet soil	"	0.019
Loose rock	"	0.020
Blasted rock	"	0.021
2 cy capacity		
Light soil	C.Y.	0.016
Medium soil	"	0.017
Heavy/wet soil	"	0.018
Loose rock	"	0.019
Blasted rock	"	0.020
2-1/2 cy capacity		
Light soil	C.Y.	0.015
Medium soil	"	0.015
Heavy/wet soil	"	0.016
Loose rock	"	0.017
Blasted rock	"	0.018
Trencher, chain, 1' wide to 4' deep		
Light soil	C.Y.	0.020
Medium soil	"	0.023
Heavy soil	"	0.027
Hand excavation		
Bulk, wheeled 100'		
Normal soil	C.Y.	0.889
Sand or gravel	"	0.800
Medium clay	"	1.143
Heavy clay	"	1.600
Loose rock	"	2.000
Trenches, up to 2' deep		
Normal soil	C.Y.	1.000
Sand or gravel	"	0.889
Medium clay	"	1.333
Heavy clay	"	2.000
Loose rock	"	2.667
Trenches, to 6' deep		
Normal soil	C.Y.	1.143
Sand or gravel	"	1.000
Medium clay	"	1.600
Heavy clay	"	2.667
Loose rock	"	4.000
Backfill trenches		
With compaction		
By hand	C.Y.	0.667
By 60 hp tracked dozer	"	0.020
By 200 hp tracked dozer	"	0.009
By small front-end loader	"	0.023
Spread dumped fill or gravel, no compaction		
6" layers	S.Y.	0.013
12" layers	"	0.016
Compaction in 6" layers		

EARTHWORK	UNIT	MAN/HOURS
02220.60 TRENCHING		
By hand with air tamper	S.Y.	0.016
Backfill trenches, sand bedding, no compaction		
By hand	C.Y.	0.667
By small front-end loader	"	0.023
02220.70 ROADWAY EXCAVATION		
Roadway excavation		
1/4 mile haul	C.Y.	0.016
2 mile haul	"	0.027
5 mile haul	"	0.040
Excavation of open ditches	"	0.011
Trim banks, swales or ditches	S.Y.	0.013
Bulk swale excavation by dragline		
Small jobs	C.Y.	0.060
Large jobs	"	0.034
Spread base course	"	0.020
Roll and compact	"	0.027
02220.71 BASE COURSE		
Base course, crushed stone		
3" thick	S.Y.	0.004
4" thick	"	0.004
6" thick	"	0.005
8" thick	"	0.005
10" thick	"	0.006
12" thick	"	0.007
Base course, bank run gravel		
4" deep	S.Y.	0.004
6" deep	"	0.005
8" deep	"	0.005
10" deep	"	0.005
12" deep	"	0.006
Prepare and roll sub base		
Minimum	S.Y.	0.004
Maximum	"	0.007
02220.90 HAND EXCAVATION		
Excavation		
To 2' deep		
Normal soil	C.Y.	0.889
Sand and gravel	"	0.800
Medium clay	"	1.000
Heavy clay	"	1.143
Loose rock	"	1.333
To 6' deep		
Normal soil	C.Y.	1.143
Sand and gravel	"	1.000
Medium clay	"	1.333

02 SITEWORK

EARTHWORK	UNIT	MAN/HOURS
02220.90 HAND EXCAVATION		
Heavy clay	C.Y.	1.600
Loose rock	"	2.000
Backfilling foundation without compaction, 6" lifts	"	0.500
Compaction of backfill around structures or in trench		
By hand with air tamper	C.Y.	0.571
By hand with vibrating plate tamper	"	0.533
1 ton roller	"	0.400
Miscellaneous hand labor		
Trim slopes, sides of excavation	S.F.	0.001
Trim bottom of excavation	"	0.002
Excavation around obstructions and services	C.Y.	2.667
02240.05 SOIL STABILIZATION		
Straw bale secured with rebar	L.F.	0.027
Filter barrier, 18" high filter fabric	"	0.080
Sediment fence, 36" fabric with 6" mesh	"	0.100
Soil stabilization with tar paper, burlap, straw and stakes	S.F.	0.001
02240.30 GEOTEXTILE		
Filter cloth, light reinforcement		
Woven		
12'-6" wide x 50' long	S.F.	0.001
Various lengths	"	0.001
Non-woven		
14'-8" wide x 430' long	S.F.	0.001
Various lengths	"	0.001
02270.10 SLOPE PROTECTION		
Gabions, stone filled		
6" deep	S.Y.	0.200
9" deep	"	0.229
12" deep	"	0.267
18" deep	"	0.320
36" deep	"	0.533
02270.40 RIPRAP		
Riprap		
Crushed stone blanket, max size 2-1/2"	TON	0.533
Stone, quarry run, 300 lb. stones	"	0.492
400 lb. stones	"	0.457
500 lb. stones	"	0.427
750 lb. stones	"	0.400
Dry concrete riprap in bags 3" thick, 80 lb. per bag	BAG	0.027

EARTHWORK	UNIT	MAN/HOURS
02280.20 SOIL TREATMENT		
Soil treatment, termite control pretreatment, under slabs		0.004
Under slabs	S.F.	0.004
By walls	"	0.005
02290.30 WEED CONTROL		
Weed control, bromicil, 15 lb./acre, wettable powder	ACRE	4.000
Vegetation control, by application of plant killer	S.Y.	0.003
Weed killer, lawns and fields	"	0.002

TUNNELING	UNIT	MAN/HOURS
02300.10 PIPE JACKING		
Pipe casing, horizontal jacking		
18" dia.	L.F.	0.711
21" dia.	"	0.762
24" dia.	"	0.800
27" dia.	"	0.800
30" dia.	"	0.842
36" dia.	"	0.914
42" dia.	"	1.000
48" dia.	"	1.067

PILES AND CAISSONS	UNIT	MAN/HOURS
02360.50 PRESTRESSED PILING		
Prestressed concrete piling, less than 60' long		
10" sq.	L.F.	0.040
12" sq.	"	0.042
14" sq.	"	0.043
16" sq.	"	0.044

PILES AND CAISSONS	UNIT	MAN/ HOURS
02360.50 PRESTRESSED PILING		
18" sq.	L.F.	0.047
20" sq.	"	0.048
24" sq.	"	0.049
More than 60' long		
12" sq.	L.F.	0.034
14" sq.	"	0.035
16" sq.	"	0.036
18" sq.	"	0.036
20" sq.	"	0.037
24" sq.	"	0.038
Straight cylinder, less than 60' long		
12" dia.	L.F.	0.044
14" dia.	"	0.045
16" dia.	"	0.046
18" dia.	"	0.047
20" dia.	"	0.048
24" dia.	"	0.049
More than 60' long		
12" dia.	L.F.	0.035
14" dia.	"	0.036
16" dia.	"	0.036
18" dia.	"	0.037
20" dia.	"	0.038
24" dia.	"	0.038
Concrete sheet piling		
12" thick x 20' long	S.F.	0.096
25' long	"	0.087
30' long	"	0.080
35' long	"	0.074
40' long	"	0.069
16" thick x 40' long	"	0.053
45' long	"	0.051
50' long	"	0.048
55' long	"	0.046
60' long	"	0.044
02360.60 STEEL PILES		
H-section piles		
8x8		
36 lb/ft		
30' long	L.F.	0.080
40' long	"	0.064
50' long	"	0.053
10x10		
42 lb/ft		
30' long	L.F.	0.080
40' long	"	0.064
50' long	"	0.053

PILES AND CAISSONS	UNIT	MAN/ HOURS
02360.60 STEEL PILES		
57 lb/ft		
30' long	L.F.	0.080
40' long	"	0.064
50' long	"	0.053
12x12		
53 lb/ft		
30' long	L.F.	0.087
40' long	"	0.069
50' long	"	0.053
74 lb/ft		
30' long	L.F.	0.087
40' long	"	0.069
50' long	"	0.053
14x14		
73 lb/ft		
40' long	L.F.	0.087
50' long	"	0.069
60' long	"	0.053
89 lb/ft		
40' long	L.F.	0.087
50' long	"	0.069
60' long	"	0.053
102 lb/ft		
40' long	L.F.	0.087
50' long	"	0.069
60' long	"	0.053
117 lb/ft		
40' long	L.F.	0.091
50' long	"	0.071
60' long	"	0.055
Splice		
8"	EA.	1.333
10"	"	1.600
12"	"	1.600
14"	"	2.000
Tapered friction piles, with fluted steel casing, up to 50'		
With 4000 psi concrete no reinforcing		
12" dia.	L.F.	0.048
14" dia.	"	0.049
16" dia.	"	0.051
18" dia.	"	0.056
02360.65 STEEL PIPE PILES		
Concrete filled, 3000# concrete, up to 40'		
8" dia.	L.F.	0.069
10" dia.	"	0.071
12" dia.	"	0.074
14" dia.	"	0.077

PILES AND CAISSONS

02360.65 STEEL PIPE PILES	UNIT	MAN/HOURS
16" dia.	L.F.	0.080
18" dia.	"	0.083
Pipe piles, non-filled		
8" dia.	L.F.	0.053
10" dia.	"	0.055
12" dia.	"	0.056
14" dia.	"	0.060
16" dia.	"	0.062
18" dia.	"	0.064
Splice		
8" dia.	EA.	1.600
10" dia.	"	1.600
12" dia.	"	2.000
14" dia.	"	2.000
16" dia.	"	2.667
18" dia.	"	2.667
Standard point		
8" dia.	EA.	1.600
10" dia.	"	1.600
12" dia.	"	2.000
14" dia.	"	2.000
16" dia.	"	2.667
18" dia.	"	2.667
Heavy duty point		
8" dia.	EA.	2.000
10" dia.	"	2.000
12" dia.	"	2.667
14" dia.	"	2.667
16" dia.	"	3.200
18" dia.	"	3.200

02360.70 STEEL SHEET PILING	UNIT	MAN/HOURS
Steel sheet piling, driven and pulled, 12" wide		
20' long	S.F.	0.096
35' long	"	0.069
50' long	"	0.048
Over 50' long	"	0.044

02360.80 WOOD AND TIMBER PILES	UNIT	MAN/HOURS
Treated wood piles, 12" butt, 8" tip		
25' long	L.F.	0.096
30' long	"	0.080
35' long	"	0.069
40' long	"	0.060
12" butt, 7" tip		
40' long	L.F.	0.060
45' long	"	0.053
50' long	"	0.048
55' long	L.F.	0.044
60' long	"	0.040

02380.10 CAISSONS	UNIT	MAN/HOURS
Caisson, including 3000# concrete, in stable ground		
18" dia.	L.F.	0.192
24" dia.	"	0.200
30" dia.	"	0.240
36" dia.	"	0.274
48" dia.	"	0.320
60" dia.	"	0.436
72" dia.	"	0.533
84" dia.	"	0.686
Wet ground, casing required but pulled		
18" dia.	L.F.	0.240
24" dia.	"	0.267
30" dia.	"	0.300
36" dia.	"	0.320
48" dia.	"	0.400
60" dia.	"	0.533
72" dia.	"	0.800
84" dia.	"	1.200
Soft rock		
18" dia.	L.F.	0.686
24" dia.	"	1.200
30" dia.	"	1.600
36" dia.	"	2.400
48" dia.	"	3.200
60" dia.	"	4.800
72" dia.	"	5.333
84" dia.	"	6.000

RAILROAD WORK

02450.10 RAILROAD WORK	UNIT	MAN/HOURS
Rail		
90 lb	L.F.	0.010
100 lb	"	0.010
115 lb	"	0.010
132 lb	"	0.010

RAILROAD WORK	UNIT	MAN/HOURS
02450.10 RAILROAD WORK		
Rail relay		
90 lb	L.F.	0.010
100 lb	"	0.010
115 lb	"	0.010
132 lb	"	0.010
New angle bars, per pair		
90 lb	EA.	0.012
100 lb	"	0.012
115 lb	"	0.012
132 lb	"	0.012
Angle bar relay		
90 lb	EA.	0.012
100 lb	"	0.012
115 lb	"	0.012
132 lb	"	0.012
New tie plates		
90 lb	EA.	0.009
100 lb	"	0.009
115 lb	"	0.009
132 lb	"	0.009
Tie plate relay		
100 lb	EA.	0.009
115 lb	"	0.009
132 lb	"	0.009
Track accessories		
Wooden cross ties, 8'	EA.	0.060
Concrete cross ties, 8'	"	0.120
Tie plugs, 5"	"	0.006
Track bolts and nuts, 1"	"	0.006
Lockwashers, 1"	"	0.004
Track spikes, 6"	"	0.024
Wooden switch ties	B.F.	0.006
Rail anchors	EA.	0.022
Ballast	TON	0.120
Gauge rods	EA.	0.096
Compromise splice bars	"	0.160
Turnout		
90 lb	EA.	24.000
100 lb	"	24.000
110 lb	"	24.000
115 lb	"	24.000
132 lb	"	24.000
Turnout relay		
90 lb	EA.	24.000
100 lb	"	24.000
110 lb	"	24.000
115 lb	"	24.000
132 lb	"	24.000
Railroad track in place, complete		

RAILROAD WORK	UNIT	MAN/HOURS
02450.10 RAILROAD WORK		
New rail		
90 lb	L.F.	0.240
100 lb	"	0.240
110 lb	"	0.240
115 lb	"	0.240
132 lb	"	0.240
Rail relay		
90 lb	L.F.	0.240
100 lb	"	0.240
110 lb	"	0.240
115 lb	"	0.240
132 lb	"	0.240
No. 8 turnout		
90 lb	EA.	32.000
100 lb	"	32.000
110 lb	"	32.000
115 lb	"	32.000
132 lb	"	32.000
No. 8 turnout relay		
90 lb	EA.	32.000
100 lb	"	32.000
110 lb	"	32.000
115 lb	"	32.000
132 lb	"	32.000
Railroad crossings, asphalt, based on 8" thick x 20'		
Including track and approach		
12' roadway	EA.	6.000
15' roadway	"	6.857
18' roadway	"	8.000
21' roadway	"	9.600
24' roadway	"	12.000
Precast concrete inserts		
12' roadway	EA.	2.400
15' roadway	"	3.000
18' roadway	"	4.000
21' roadway	"	4.800
24' roadway	"	5.333
Molded rubber, with headers		
12' roadway	EA.	2.400
15' roadway	"	3.000
18' roadway	"	4.000
21' roadway	"	4.800
24' roadway	"	5.333

PAVING AND SURFACING

02510.20 ASPHALT SURFACES

	UNIT	MAN/ HOURS
Asphalt wearing surface, for flexible pavement		
1" thick	S.Y.	0.016
1-1/2" thick	"	0.019
2" thick	"	0.024
3" thick	"	0.032
Bituminous sidewalk, no base		
2" thick	S.Y.	0.028
3" thick	"	0.030

02520.10 CONCRETE PAVING

	UNIT	MAN/ HOURS
Concrete paving, reinforced, 5000 psi concrete		
6" thick	S.Y.	0.150
7" thick	"	0.160
8" thick	"	0.171
9" thick	"	0.185
10" thick	"	0.200
11" thick	"	0.218
12" thick	"	0.240
15" thick	"	0.300
Concrete paving, for pipe trench, reinforced		
7" thick	S.Y.	0.240
8" thick	"	0.267
9" thick	"	0.300
10" thick	"	0.343
Fibrous concrete		
5" thick	S.Y.	0.185
8" thick	"	0.200
Roller compacted concrete, (RCC), place and compact		
8" thick	S.Y.	0.240
12" thick	"	0.300
Steel edge forms up to		
12" deep	L.F.	0.027
15" deep	"	0.032
Paving finishes		
Belt dragged	S.Y.	0.040
Curing	"	0.008

02545.10 ASPHALT REPAIR

	UNIT	MAN/ HOURS
Coal tar emulsion seal coat, rubber additive, fuel resistant	S.Y.	0.011
Bituminous surface treatment, single	"	0.008
Double	"	0.001
Bituminous prime coat	"	0.001
Tack coat	"	0.001
Crack sealing, concrete paving	L.F.	0.005
Bituminous paving for pipe trench, 4" thick	S.Y.	0.160
Polypropylene, nonwoven paving fabric	"	0.004
Rubberized asphalt	"	0.073
Asphalt slurry seal	"	0.047

PAVING AND SURFACING

02580.10 PAVEMENT MARKINGS

	UNIT	MAN/ HOURS
Pavement line marking, paint		
4" wide	L.F.	0.002
6" wide	"	0.004
8" wide	"	0.007
Reflective paint, 4" wide	"	0.007
Airfield markings, retro-reflective		
White	L.F.	0.007
Yellow	"	0.007
Preformed tape, 4" wide		
Inlaid reflective	L.F.	0.001
Reflective paint	"	0.002
Thermoplastic		
White	L.F.	0.004
Yellow	"	0.004
12" wide, thermoplastic, white	"	0.011
Directional arrows, reflective preformed tape	EA.	0.800
Messages, reflective preform tape (per letter)	"	0.400
Handicap symbol, preformed tape	"	0.800
Parking stall painting	"	0.160

UTILITIES

02605.30 MANHOLES

	UNIT	MAN/ HOURS
Precast sections, 48" dia.		2.000
Base section	EA.	2.000
1'0" riser	"	1.600
1'4" riser	"	1.714
2'8" riser	"	1.846
4'0" riser	"	2.000
2'8" cone top	"	2.400
Precast manholes, 48" dia.		
4' deep	EA.	4.800
6' deep	"	6.000
7' deep	"	6.857
8' deep	"	8.000
10' deep	"	9.600
Cast-in-place, 48" dia., with frame and cover		
5' deep	EA.	12.000
6' deep	"	13.714
8' deep	"	16.000

UTILITIES	UNIT	MAN/HOURS
02605.30 MANHOLES		
10' deep	EA.	19.200
Brick manholes, 48" dia. with cover, 8" thick		
4' deep	EA.	8.000
6' deep	"	8.889
8' deep	"	10.000
10' deep	"	11.429
12' deep	"	13.333
14' deep	"	16.000
Inverts for manholes		
Single channel	EA.	3.200
Triple channel	"	4.000
Frames and covers, 24" diameter		
300 lb	EA.	0.800
400 lb	"	0.889
500 lb	"	1.143
Watertight, 350 lb	"	2.667
For heavy equipment, 1200 lb	"	4.000
Steps for manholes		
7" x 9"	EA.	0.160
8" x 9"	"	0.178
Curb inlet, 4' throat, cast in place		
12"-30" pipe	EA.	12.000
36"-48" pipe	"	13.714
Raise exist frame and cover, repaving	"	4.800
02610.10 CAST IRON FLANGED PIPE		
Cast iron flanged sections		
4" pipe, with one bolt set		
3' section	EA.	0.218
4' section	"	0.240
5' section	"	0.267
6' section	"	0.300
8' section	"	0.343
10' section	"	0.480
12' section	"	0.800
15' section	"	1.200
18' section	"	1.600
6" pipe, with one bolt set		
3' section	EA.	0.240
4' section	"	0.282
5' section	"	0.320
6' section	"	0.369
8' section	"	0.533
10' section	"	0.600
12' section	"	0.800
15' section	"	1.200
18' section	"	1.714
8" pipe, with one bolt set		
3' section	EA.	0.300

UTILITIES	UNIT	MAN/HOURS
02610.10 CAST IRON FLANGED PIPE		
4' section	EA.	0.343
5' section	"	0.400
6' section	"	0.480
8' section	"	0.686
10' section	"	0.800
12' section	"	1.200
15' section	"	1.600
18' section	"	2.000
10" pipe, with one bolt set		
3' section	EA.	0.308
4' section	"	0.353
5' section	"	0.414
6' section	"	0.500
8' section	"	0.727
10' section	"	0.857
12' section	"	1.333
15' section	"	1.714
18' section	"	2.400
12" pipe, with one bolt set		
3' section	EA.	0.333
4' section	"	0.387
5' section	"	0.462
6' section	"	0.545
8' section	"	0.800
10' section	"	0.923
12' section	"	1.500
15' section	"	2.000
18' section	"	2.667
02610.11 CAST IRON FITTINGS		
Mechanical joint, with 2 bolt kits		
90 deg bend		
4"	EA.	0.533
6"	"	0.615
8"	"	0.800
10"	"	1.143
12"	"	1.600
14"	"	2.000
16"	"	2.667
45 deg bend		
4"	EA.	0.533
6"	"	0.615
8"	"	0.800
10"	"	1.143
12"	"	1.600
14"	"	2.000
16"	"	2.667
Tee, with 3 bolt kits		
4" x 4"	EA.	0.800

UTILITIES	UNIT	MAN/ HOURS
02610.11 CAST IRON FITTINGS		
6" x 6"	EA.	1.000
8" x 8"	"	1.333
10" x 10"	"	2.000
12" x 12"	"	2.667
Wye, with 3 bolt kits		
6" x 6"	EA.	1.000
8" x 8"	"	1.333
10" x 10"	"	2.000
12" x 12"	"	2.667
Reducer, with 2 bolt kits		
6" x 4"	EA.	1.000
8" x 6"	"	1.333
10" x 8"	"	2.000
12" x 10"	"	2.667
Flanged, 90 deg bend, 125 lb.		
4"	EA.	0.667
6"	"	0.800
8"	"	1.000
10"	"	1.333
12"	"	2.000
14"	"	2.667
16"	"	2.667
Tee		
4"	EA.	1.000
6"	"	1.143
8"	"	1.333
10"	"	1.600
12"	"	2.000
14"	"	2.667
16"	"	4.000
02610.13 GATE VALVES		
Gate valve, (AWWA) mechanical joint, with adjustable box		
4" valve	EA.	0.800
6" valve	"	0.960
8" valve	"	1.200
10" valve	"	1.412
12" valve	"	1.714
14" valve	"	2.000
16" valve	"	2.182
18" valve	"	2.400
Flanged, with box, post indicator (AWWA)		
4" valve	EA.	0.960
6" valve	"	1.091
8" valve	"	1.333
10" valve	"	1.600
12" valve	"	2.000
14" valve	"	2.400
16" valve	"	3.000

UTILITIES	UNIT	MAN/ HOURS
02610.15 WATER METERS		
Water meter, displacement type		
1"	EA.	0.800
1-1/2"	"	0.889
2"	"	1.000
02610.17 CORPORATION STOPS		
Stop for flared copper service pipe		
3/4"	EA.	0.400
1"	"	0.444
1-1/4"	"	0.533
1-1/2"	"	0.667
2"	"	0.800
02610.40 DUCTILE IRON PIPE		
Ductile iron pipe, cement lined, slip-on joints		
4"	L.F.	0.067
6"	"	0.071
8"	"	0.075
10"	"	0.080
12"	"	0.096
14"	"	0.120
16"	"	0.133
18"	"	0.150
20"	"	0.171
Mechanical joint pipe		
4"	L.F.	0.092
6"	"	0.100
8"	"	0.109
10"	"	0.120
12"	"	0.160
14"	"	0.185
16"	"	0.218
18"	"	0.240
20"	"	0.267
Fittings, mechanical joint		
90 degree elbow		
4"	EA.	0.533
6"	"	0.615
8"	"	0.800
10"	"	1.143
12"	"	1.600
14"	"	2.000
16"	"	2.667
18"	"	3.200
20"	"	4.000
45 degree elbow		
4"	EA.	0.533
6"	"	0.615
8"	"	0.800

UTILITIES	UNIT	MAN/ HOURS
02610.40 DUCTILE IRON PIPE		
10"	EA.	1.143
12"	"	1.600
14"	"	2.000
16"	"	2.667
18"	"	4.000
20"	"	4.000
Tee		
4"x3"	EA.	1.000
4"x4"	"	1.000
6"x3"	"	1.143
6"x4"	"	1.143
6"x6"	"	1.143
8"x4"	"	1.333
8"x6"	"	1.333
8"x8"	"	1.333
10"x4"	"	1.600
10"x6"	"	1.600
10"x8"	"	1.600
10"x10"	"	1.600
12"x4"	"	2.000
12"x6"	"	2.000
12"x8"	"	2.000
12"x10"	"	2.000
12"x12"	"	2.133
14"x4"	"	2.286
14"x6"	"	2.286
14"x8"	"	2.286
14"x10"	"	2.286
14"x12"	"	2.462
14"x14"	"	2.462
16"x4"	"	2.667
16"x6"	"	2.667
16"x8"	"	2.667
16"x10"	"	2.667
16"x12"	"	2.667
16"x14"	"	2.667
16"x16"	"	2.667
18"x6"	"	2.909
18"x8"	"	2.909
18"x10"	"	2.909
18"x12"	"	2.909
18"x14"	"	2.909
18"x16"	"	2.909
18"x18"	"	2.909
20"x6"	"	3.200
20"x8"	"	3.200
20"x10"	"	3.200
20"x12"	"	3.200
20"x14"	"	3.200

UTILITIES	UNIT	MAN/ HOURS
02610.40 DUCTILE IRON PIPE		
20"x16"	EA.	3.200
20"x18"	"	3.200
20"x20"	"	3.200
Cross		
4"x3"	EA.	1.333
4"x4"	"	1.333
6"x3"	"	1.600
6"x4"	"	1.600
6"x6"	"	1.600
8"x4"	"	1.778
8"x6"	"	1.778
8"x8"	"	1.778
10"x4"	"	2.000
10"x6"	"	2.000
10"x8"	"	2.000
10"x10"	"	2.000
12"x4"	"	2.286
12"x6"	"	2.286
12"x8"	"	2.286
12"x10"	"	2.462
12"x12"	"	2.462
14"x4"	"	2.667
14"x6"	"	2.667
14"x8"	"	2.667
14"x10"	"	2.667
14"x12"	"	2.909
14"x14"	"	2.909
16"x4"	"	3.200
16"x6"	"	3.200
16"x8"	"	3.200
16"x10"	"	3.200
16"x12"	"	3.200
16"x14"	"	3.200
16"x16"	"	3.200
18"x6"	"	3.556
18"x8"	"	3.556
18"x10"	"	3.556
18"x12"	"	3.556
18"x14"	"	3.556
18"x16"	"	3.556
18"x18"	"	3.556
20"x6"	"	3.810
20"x8"	"	3.810
20"x10"	"	3.810
20"x12"	"	3.810
20"x14"	"	3.810
20"x16"	"	3.810
20"x18"	"	4.000
20"x20"	"	4.000

UTILITIES	UNIT	MAN/ HOURS
02610.60 PLASTIC PIPE		
PVC, class 150 pipe		
4" dia.	L.F.	0.060
6" dia.	"	0.065
8" dia.	"	0.069
10" dia.	"	0.075
12" dia.	"	0.080
Schedule 40 pipe		
1-1/2" dia.	L.F.	0.047
2" dia.	"	0.050
2-1/2" dia.	"	0.053
3" dia.	"	0.057
4" dia.	"	0.067
6" dia.	"	0.080
90 degree elbows		
1"	EA.	0.133
1-1/2"	"	0.133
2"	"	0.145
2-1/2"	"	0.160
3"	"	0.178
4"	"	0.200
6"	"	0.267
45 degree elbows		
1"	EA.	0.133
1-1/2"	"	0.133
2"	"	0.145
2-1/2"	"	0.160
3"	"	0.178
4"	"	0.200
6"	"	0.267
Tees		
1"	EA.	0.160
1-1/2"	"	0.160
2"	"	0.178
2-1/2"	"	0.200
3"	"	0.229
4"	"	0.267
6"	"	0.320
Couplings		
1"	EA.	0.133
1-1/2"	"	0.133
2"	"	0.145
2-1/2"	"	0.160
3"	"	0.178
4"	"	0.200
6"	"	0.267
Drainage pipe		
PVC schedule 80		
1" dia.	L.F.	0.047
1-1/2" dia.	"	0.047

UTILITIES	UNIT	MAN/ HOURS
02610.60 PLASTIC PIPE		
ABS, 2" dia.	L.F.	0.050
2-1/2" dia.	"	0.053
3" dia.	"	0.057
4" dia.	"	0.067
6" dia.	"	0.080
8" dia.	"	0.063
10" dia.	"	0.075
12" dia.	"	0.080
90 degree elbows		
1"	EA.	0.133
1-1/2"	"	0.133
2"	"	0.145
2-1/2"	"	0.160
3"	"	0.178
4"	"	0.200
6"	"	0.267
45 degree elbows		
1"	EA.	0.133
1-1/2"	"	0.133
2"	"	0.145
2-1/2"	"	0.160
3"	"	0.178
4"	"	0.200
6"	"	0.267
Tees		
1"	EA.	0.160
1-1/2"	"	0.160
2"	"	0.178
2-1/2"	"	0.200
3"	"	0.229
4"	"	0.267
6"	"	0.320
Couplings		
1"	EA.	0.133
1-1/2"	"	0.133
2"	"	0.145
2-1/2"	"	0.160
3"	"	0.178
4"	"	0.200
6"	"	0.267
Pressure pipe		
PVC, class 200 pipe		
3/4"	L.F.	0.040
1"	"	0.042
1-1/4"	"	0.044
1-1/2"	"	0.047
2"	"	0.050
2-1/2"	"	0.053
3"	"	0.057

02 SITEWORK

UTILITIES	UNIT	MAN/HOURS
02610.60 PLASTIC PIPE		
4"	L.F.	0.067
6"	"	0.080
8"	"	0.069
90 degree elbows		
3/4"	EA.	0.133
1"	"	0.133
1-1/4"	"	0.133
1-1/2"	"	0.133
2"	"	0.145
2-1/2"	"	0.160
3"	"	0.178
4"	"	0.200
6"	"	0.267
8"	"	0.400
45 degree elbows		
3/4"	EA.	0.133
1"	"	0.133
1-1/4"	"	0.133
1-1/2"	"	0.133
2"	"	0.145
2-1/2"	"	0.160
3"	"	0.178
4"	"	0.200
6"	"	0.267
8"	"	0.400
Tees		
3/4"	EA.	0.160
1"	"	0.160
1-1/4"	"	0.160
1-1/2"	"	0.160
2"	"	0.178
2-1/2"	"	0.200
3"	"	0.229
4"	"	0.267
6"	"	0.320
8"	"	0.444
Couplings		
3/4"	EA.	0.133
1"	"	0.133
1-1/4"	"	0.133
1-1/2"	"	0.133
2"	"	0.145
2-1/2"	"	0.160
3"	"	0.178
4"	"	0.178
6"	"	0.200
8"	"	0.267

UTILITIES	UNIT	MAN/HOURS
02610.90 VITRIFIED CLAY PIPE		
Vitrified clay pipe, extra strength		
6" dia.	L.F.	0.109
8" dia.	"	0.114
10" dia.	"	0.120
12" dia.	"	0.160
15" dia.	"	0.240
18" dia.	"	0.267
24" dia.	"	0.343
30" dia.	"	0.480
36" dia.	"	0.686
02630.10 TAPPING SADDLES & SLEEVES		
Tapping saddle, tap size to 2"		
4" saddle	EA.	0.400
6" saddle	"	0.500
8" saddle	"	0.667
10" saddle	"	0.800
12" saddle	"	1.143
14" saddle	"	1.600
Tapping sleeve		
4x4	EA.	0.533
6x4	"	0.615
6x6	"	0.615
8x4	"	0.800
8x6	"	0.800
10x4	"	0.960
10x6	"	0.960
10x8	"	0.960
10x10	"	1.000
12x4	"	1.000
12x6	"	1.091
12x8	"	1.200
12x10	"	1.333
12x12	"	1.500
Tapping valve, mechanical joint		
4" valve	EA.	3.000
6" valve	"	4.000
8" valve	"	6.000
10" valve	"	8.000
12" valve	"	12.000
Tap hole in pipe		
4" hole	EA.	1.000
6" hole	"	1.600
8" hole	"	2.667
10" hole	"	3.200
12" hole	"	4.000

UTILITIES	UNIT	MAN/ HOURS
02640.15 VALVE BOXES		
Valve box, adjustable, for valves up to 20"		
3' deep	EA.	0.267
4' deep	"	0.320
5' deep	"	0.400
02640.19 THRUST BLOCKS		
Thrust block, 3000# concrete		
1/4 c.y.	EA.	1.333
1/2 c.y.	"	1.600
3/4 c.y.	"	2.667
1 c.y.	"	5.333
02645.10 FIRE HYDRANTS		
Standard, 3 way post, 6" mechanical joint		
2' deep	EA.	8.000
4' deep	"	9.600
6' deep	"	12.000
8' deep	"	13.714
02665.10 CHILLED WATER SYSTEMS		
Chilled water pipe, 2" thick insulation, w/casing		
Align and tack weld on sleepers		
1-1/2" dia.	L.F.	0.022
3" dia.	"	0.034
4" dia.	"	0.048
6" dia.	"	0.060
8" dia.	"	0.069
10" dia.	"	0.080
12" dia.	"	0.096
14" dia.	"	0.104
16" dia.	"	0.120
Align and tack weld on trench bottom		
18" dia.	L.F.	0.133
20" dia.	"	0.150
Preinsulated fittings		
Align and tack weld on sleepers		
Elbows		
1-1/2"	EA.	0.500
3"	"	0.800
4"	"	1.000
6"	"	1.333
8"	"	1.600
Tees		
1-1/2"	EA.	0.533
3"	"	0.889
4"	"	1.143
6"	"	1.600

UTILITIES	UNIT	MAN/ HOURS
02665.10 CHILLED WATER SYSTEMS		
8"	EA.	2.000
Reducers		
3"	EA.	0.667
4"	"	0.800
6"	"	1.000
8"	"	1.333
Anchors, not including concrete		
4"	EA.	1.000
6"	"	1.000
Align and tack weld on trench bottom		
Elbows		
10"	EA.	1.500
12"	"	1.714
14"	"	1.846
16"	"	2.000
18"	"	2.182
20"	"	2.400
Tees		
10"	EA.	1.500
12"	"	1.714
14"	"	1.846
16"	"	2.000
18"	"	2.182
20"	"	2.400
Reducers		
10"	EA.	1.000
12"	"	1.091
14"	"	1.200
16"	"	1.333
18"	"	1.500
20"	"	1.714
Anchors, not including concrete		
10"	EA.	1.000
12"	"	1.091
14"	"	1.200
16"	"	1.333
18"	"	1.500
20"	"	1.714
02670.10 WELLS		
Domestic water, drilled and cased		
4" dia.	L.F.	1.600
6" dia.	"	1.846
8" dia.	"	2.400
02685.10 GAS DISTRIBUTION		
Gas distribution lines		
Polyethylene, 60 psi coils		
1-1/4" dia.	L.F.	0.053

UTILITIES	UNIT	MAN/ HOURS
02685.10 GAS DISTRIBUTION		
1-1/2" dia.	L.F.	0.057
2" dia.	"	0.067
3" dia.	"	0.080
30' pipe lengths		
3" dia.	L.F.	0.089
4" dia.	"	0.100
6" dia.	"	0.133
8" dia.	"	0.160
Steel, schedule 40, plain end		
1" dia.	L.F.	0.067
2" dia.	"	0.073
3" dia.	"	0.080
4" dia.	"	0.160
5" dia.	"	0.171
6" dia.	"	0.200
8" dia.	"	0.218
Natural gas meters, direct digital reading, threaded		
250 cfh @ 5 lbs	EA.	1.600
425 cfh @ 10 lbs	"	1.600
800 cfh @ 20 lbs	"	2.000
1000 cfh @ 25 lbs	"	2.000
1,400 cfh @ 100 lbs	"	2.667
2,300 cfh @ 100 lbs	"	4.000
5,000 cfh @ 100 lbs	"	8.000
Gas pressure regulators		
Threaded		
3/4"	EA.	1.000
1"	"	1.333
1-1/4"	"	1.333
1-1/2"	"	1.333
2"	"	1.600
Flanged		
3"	EA.	2.000
4"	"	2.667
02690.10 STORAGE TANKS		
Oil storage tank, underground		
Steel		
500 gals	EA.	3.000
1,000 gals	"	4.000
4,000 gals	"	8.000
5,000 gals	"	12.000
10,000 gals	"	24.000
Fiberglass, double wall		
550 gals	EA.	4.000
1,000 gals	"	4.000
2,000 gals	"	6.000
4,000 gals	"	12.000
6,000 gals	"	16.000

UTILITIES	UNIT	MAN/ HOURS
02690.10 STORAGE TANKS		
8,000 gals	EA.	24.000
10,000 gals	"	30.000
12,000 gals	"	40.000
15,000 gals	"	53.333
20,000 gals	"	60.000
Above ground		
Steel		
275 gals	EA.	2.400
500 gals	"	4.000
1,000 gals	"	4.800
1,500 gals	"	6.000
2,000 gals	"	8.000
5,000 gals	"	12.000
Fill cap	"	0.800
Vent cap	"	0.800
Level indicator	"	0.800
02695.80 STEAM METERS		
In-line turbine, direct reading, 300 lb, flanged		
2"	EA.	1.000
3"	"	1.333
4"	"	1.600
Threaded, 2"		
5" line	EA.	8.000
6" line	"	8.000
8" line	"	8.000
10" line	"	8.000
12" line	"	8.000
14" line	"	8.000
16" line	"	8.000

SEWERAGE AND DRAINAGE	UNIT	MAN/ HOURS
02720.10 CATCH BASINS		
Standard concrete catch basin		
Cast in place, 3'8" x 3'8", 6" thick wall		
2' deep	EA.	6.000
3' deep	"	6.000
4' deep	"	8.000
5' deep	"	8.000
6' deep	"	9.600

SEWERAGE AND DRAINAGE

02720.10 CATCH BASINS

	UNIT	MAN/HOURS
4'x4', 8" thick wall, cast in place		
2' deep	EA.	6.000
3' deep	"	6.000
4' deep	"	8.000
5' deep	"	8.000
6' deep	"	9.600
Frames and covers, cast iron		
Round		
24" dia.	EA.	2.000
26" dia.	"	2.000
28" dia.	"	2.000
Rectangular		
23"x23"	EA.	2.000
27"x20"	"	2.000
24"x24"	"	2.000
26"x26"	"	2.000
Curb inlet frames and covers		
27"x27"	EA.	2.000
24"x36"	"	2.000
24"x25"	"	2.000
24"x22"	"	2.000
20"x22"	"	2.000
Airfield catch basin frame and grating, galvanized		
2'x4'	EA.	2.000
2'x2'	"	2.000

02720.40 STORM DRAINAGE

	UNIT	MAN/HOURS
Concrete pipe		
Plain, bell and spigot joint, class II		
6" pipe	L.F.	0.109
8" pipe	"	0.120
10" pipe	"	0.126
12" pipe	"	0.133
15" pipe	"	0.141
18" pipe	"	0.150
21" pipe	"	0.160
24" pipe	"	0.171
Reinforced, class III, tongue and groove joint		
12" pipe	L.F.	0.133
15" pipe	"	0.141
18" pipe	"	0.150
21" pipe	"	0.160
24" pipe	"	0.171
27" pipe	"	0.185
30" pipe	"	0.200
36" pipe	"	0.218
42" pipe	"	0.240
48" pipe	"	0.267

02720.40 STORM DRAINAGE

	UNIT	MAN/HOURS
54" pipe	L.F.	0.300
60" pipe	"	0.343
66" pipe	"	0.400
72" pipe	"	0.480
Flared end-section, concrete		
12" pipe	L.F.	0.133
15" pipe	"	0.141
18" pipe	"	0.150
24" pipe	"	0.171
30" pipe	"	0.200
36" pipe	"	0.218
42" pipe	"	0.240
48" pipe	"	0.267
54" pipe	"	0.300
Corrugated metal pipe, coated, paved invert		
16 ga.		
8" pipe	L.F.	0.080
10" pipe	"	0.083
12" pipe	"	0.086
15" pipe	"	0.092
18" pipe	"	0.100
21" pipe	"	0.109
24" pipe	"	0.120
30" pipe	"	0.133
36" pipe	"	0.150
12 ga., 48" pipe	"	0.171
10 ga.		
60" pipe	L.F.	0.200
72" pipe	"	0.240
Galvanized or aluminum, plain		
16 ga.		
8" pipe	L.F.	0.080
10" pipe	"	0.083
12" pipe	"	0.086
15" pipe	"	0.092
18" pipe	"	0.100
24" pipe	"	0.120
30" pipe	"	0.133
36" pipe	"	0.150
12 ga., 48" pipe	"	0.171
10 ga., 60" pipe	"	0.200
Galvanized or aluminum, coated oval arch		
16 ga.		
17" x 13"	L.F.	0.109
21" x 15"	"	0.120
14 ga.		
28" x 20"	L.F.	0.133
35" x 24"	"	0.171
12 ga.		

SEWERAGE AND DRAINAGE	UNIT	MAN/ HOURS
02720.40 STORM DRAINAGE		
42" x 29"	L.F.	0.200
57" x 38"	"	0.240
64" x 43"	"	0.253
Oval arch culverts, plain		
16 ga.		
17" x 13"	L.F.	0.109
21" x 15"	"	0.120
14 ga.		
28" x 20"	L.F.	0.133
35" x 24"	"	0.171
12 ga.		
57" x 38"	L.F.	0.200
64" x 43"	"	0.240
71" x 47"	"	0.253
Nestable corrugated metal pipe		
16 ga.		
10" pipe	L.F.	0.083
12" pipe	"	0.086
15" pipe	"	0.092
18" pipe	"	0.100
24" pipe	"	0.120
30" pipe	"	0.133
14 ga., 36" pipe	"	0.150
Headwalls, cast in place, 30 deg wingwall		
12" pipe	EA.	2.000
15" pipe	"	2.000
18" pipe	"	2.286
24" pipe	"	2.286
30" pipe	"	2.667
36" pipe	"	4.000
42" pipe	"	4.000
48" pipe	"	5.333
54" pipe	"	6.667
60" pipe	"	8.000
4" cleanout for storm drain		
4" pipe	EA.	1.000
6" pipe	"	1.000
8" pipe	"	1.000
Connect new drain line		
To existing manhole	EA.	2.667
To new manhole	"	1.600
02720.70 UNDERDRAIN		
Drain tile, clay		
6" pipe	L.F.	0.053
8" pipe	"	0.056
12" pipe	"	0.060
Porous concrete, standard strength		

SEWERAGE AND DRAINAGE	UNIT	MAN/ HOURS
02720.70 UNDERDRAIN		
6" pipe	L.F.	0.053
8" pipe	"	0.056
12" pipe	"	0.060
15" pipe	"	0.067
18" pipe	"	0.080
Corrugated metal pipe, perforated type		
6" pipe	L.F.	0.060
8" pipe	"	0.063
10" pipe	"	0.067
12" pipe	"	0.071
18" pipe	"	0.075
Perforated clay pipe		
6" pipe	L.F.	0.069
8" pipe	"	0.071
12" pipe	"	0.073
Drain tile, concrete		
6" pipe	L.F.	0.053
8" pipe	"	0.056
12" pipe	"	0.060
Perforated rigid PVC underdrain pipe		
4" pipe	L.F.	0.040
6" pipe	"	0.048
8" pipe	"	0.053
10" pipe	"	0.060
12" pipe	"	0.069
Underslab drainage, crushed stone		
3" thick	S.F.	0.008
4" thick	"	0.009
6" thick	"	0.010
8" thick	"	0.010
Plastic filter fabric for drain lines	"	0.008
Gravel fill in trench, crushed or bank run, 1/2" to 3/4"	C.Y.	0.600
02730.10 SANITARY SEWERS		
Clay		
6" pipe	L.F.	0.080
8" pipe	"	0.086
10" pipe	"	0.092
12" pipe	"	0.100
PVC		
4" pipe	L.F.	0.060
6" pipe	"	0.063
8" pipe	"	0.067
10" pipe	"	0.071
12" pipe	"	0.075
Cleanout		
4" pipe	EA.	1.000
6" pipe	"	1.000
8" pipe	"	1.000

02 SITEWORK

SEWERAGE AND DRAINAGE

SEWERAGE AND DRAINAGE	UNIT	MAN/ HOURS
02730.10 SANITARY SEWERS		
Connect new sewer line		
To existing manhole	EA.	2.667
To new manhole	"	1.600
02740.10 DRAINAGE FIELDS		
Perforated PVC pipe, for drain field		
4" pipe	L.F.	0.053
6" pipe	"	0.057
02740.50 SEPTIC TANKS		
Septic tank, precast concrete		
1000 gals	EA.	4.000
2000 gals	"	6.000
5000 gals	"	12.000
25,000 gals	"	48.000
40,000 gals	"	80.000
Leaching pit, precast concrete, 72" diameter		
3' deep	EA.	3.000
6' deep	"	3.429
8' deep	"	4.000
02760.10 PIPELINE RESTORATION		
Relining existing water main		
6" dia.	L.F.	0.240
8" dia.	"	0.253
10" dia.	"	0.267
12" dia.	"	0.282
14" dia.	"	0.300
16" dia.	"	0.320
18" dia.	"	0.343
20" dia.	"	0.369
24" dia.	"	0.400
36" dia.	"	0.480
48" dia.	"	0.533
72" dia.	"	0.600
Replacing in line gate valves		
6" valve	EA.	3.200
8" valve	"	4.000
10" valve	"	4.800
12" valve	"	6.000
16" valve	"	6.857
18" valve	"	8.000
20" valve	"	9.600
24" valve	"	12.000
36" valve	"	16.000

SITE IMPROVEMENTS

SITE IMPROVEMENTS	UNIT	MAN/ HOURS
02830.10 CHAIN LINK FENCE		
Chain link fence, 9 ga., galvanized, with posts 10' o.c.		
4' high	L.F.	0.057
5' high	"	0.073
6' high	"	0.100
7' high	"	0.123
8' high	"	0.160
For barbed wire with hangers, add		
3 strand	L.F.	0.040
6 strand	"	0.067
Corner or gate post, 3" post		
4' high	EA.	0.267
5' high	"	0.296
6' high	"	0.348
7' high	"	0.400
8' high	"	0.444
4" post		
4' high	EA.	0.296
5' high	"	0.348
6' high	"	0.400
7' high	"	0.444
8' high	"	0.500
Gate with gate posts, galvanized, 3' wide		
4' high	EA.	2.000
5' high	"	2.667
6' high	"	2.667
7' high	"	4.000
8' high	"	4.000
Fabric, galvanized chain link, 2" mesh, 9 ga.		
4' high	L.F.	0.027
5' high	"	0.032
6' high	"	0.040
8' high	"	0.053
Line post, no rail fitting, galvanized, 2-1/2" dia.		
4' high	EA.	0.229
5' high	"	0.250
6' high	"	0.267
7' high	"	0.320
8' high	"	0.400
1-7/8" H beam		
4' high	EA.	0.229
5' high	"	0.250
6' high	"	0.267
7' high	"	0.320
8' high	"	0.400
2-1/4" H beam		
4' high	EA.	0.229
5' high	"	0.250
6' high	"	0.267
7' high	"	0.320

SITE IMPROVEMENTS	UNIT	MAN/ HOURS
02830.10 CHAIN LINK FENCE		
8' high	EA.	0.400
Vinyl coated, 9 ga., with posts 10' o.c.		
4' high	L.F.	0.057
5' high	"	0.073
6' high	"	0.100
7' high	"	0.123
8' high	"	0.160
For barbed wire w/hangers, add		
3 strand	L.F.	0.040
6 Strand	"	0.067
Corner, or gate post, 4' high		
3" dia.	EA.	0.267
4" dia.	"	0.267
6" dia.	"	0.320
Gate, with posts, 3' wide		
4' high	EA.	2.000
5' high	"	2.667
6' high	"	2.667
7' high	"	4.000
8' high	"	4.000
Line post, no rail fitting, 2-1/2" dia.		
4' high	EA.	0.229
5' high	"	0.250
6' high	"	0.267
7' high	"	0.320
8' high	"	0.400
Corner post, no top rail fitting, 4" dia.		
4' high	EA.	0.267
5' high	"	0.296
6' high	"	0.348
7' high	"	0.400
8' high	"	0.444
Fabric, vinyl, chain link, 2" mesh, 9 ga.		
4' high	L.F.	0.027
5' high	"	0.032
6' high	"	0.040
8' high	"	0.053
Swing gates, galvanized, 4' high		
Single gate		
3' wide	EA.	2.000
4' wide	"	2.000
Double gate		
10' wide	EA.	3.200
12' wide	"	3.200
14' wide	"	3.200
16' wide	"	3.200
18' wide	"	4.571
20' wide	"	4.571
22' wide	"	4.571

SITE IMPROVEMENTS	UNIT	MAN/ HOURS
02830.10 CHAIN LINK FENCE		
24' wide	EA.	5.333
26' wide	"	5.333
28' wide	"	6.400
30' wide	"	6.400
5' high		
Single gate		
3' wide	EA.	2.667
4' wide	"	2.667
Double gate		
10' wide	EA.	4.000
12' wide	"	4.000
14' wide	"	4.000
16' wide	"	4.000
18' wide	"	4.571
20' wide	"	4.571
22' wide	"	4.571
24' wide	"	5.333
26' wide	"	5.333
28' wide	"	6.400
30' wide	"	6.400
6' high		
Single gate		
3' wide	EA.	2.667
4' wide	"	2.667
Double gate		
10' wide	EA.	4.000
12' wide	"	4.000
14' wide	"	4.000
16' wide	"	4.000
18' wide	"	4.571
20' wide	"	4.571
22' wide	"	4.571
24' wide	"	5.333
26' wide	"	5.333
28' wide	"	6.400
30' wide	"	6.400
7' high		
Single gate		
3' wide	EA.	4.000
4' wide	"	4.000
Double gate		
10' wide	EA.	5.333
12' wide	"	5.333
14' wide	"	5.333
16' wide	"	5.333
18' wide	"	6.400
20' wide	"	6.400
22' wide	"	6.400
24' wide	"	8.000

SITE IMPROVEMENTS	UNIT	MAN/ HOURS
02830.10 CHAIN LINK FENCE		
26' wide	EA.	8.000
28' wide	"	10.000
30' wide	"	10.000
8' high		
Single gate		
3' wide	EA.	4.000
4' wide	"	4.000
Double gate		
10' wide	EA.	5.333
12' wide	"	5.333
14' wide	"	5.333
16' wide	"	5.333
18' wide	"	6.400
20' wide	"	6.400
22' wide	"	6.400
24' wide	"	8.000
26' wide	"	8.000
28' wide	"	10.000
30' wide	"	10.000
Vinyl coated swing gates, 4' high		
Single gate		
3' wide	EA.	2.000
4' wide	"	2.000
Double gate		
10' wide	EA.	3.200
12' wide	"	3.200
14' wide	"	3.200
16' wide	"	3.200
18' wide	"	4.571
20' wide	"	4.571
22' wide	"	4.571
24' wide	"	5.333
26' wide	"	5.333
28' wide	"	6.400
30' wide	"	6.400
5' high		
Single gate		
3' wide	EA.	2.667
4' wide	"	2.667
Double gate		
10' wide	EA.	4.000
12' wide	"	4.000
14' wide	"	4.000
16' wide	"	4.000
18' wide	"	4.571
20' wide	"	4.571
22' wide	"	4.571
24' wide	"	5.333
26' wide	"	5.333

SITE IMPROVEMENTS	UNIT	MAN/ HOURS
02830.10 CHAIN LINK FENCE		
28' wide	EA.	6.400
30' wide	"	6.400
6' high		
Single gate		
3' wide	EA.	2.667
4' wide	"	2.667
Double gate		
10' wide	EA.	4.000
12' wide	"	4.000
14' wide	"	4.000
16' wide	"	4.000
18' wide	"	4.571
20' wide	"	4.571
22' wide	"	4.571
24' wide	"	5.333
26' wide	"	5.333
28' wide	"	6.400
30' wide	"	6.400
7' high		
Single gate		
3' wide	EA.	4.000
4' wide	"	4.000
Double gate		
10' wide	EA.	5.333
12' wide	"	5.333
14' wide	"	5.333
16' wide	"	5.333
18' wide	"	6.400
20' wide	"	6.400
22' wide	"	6.400
24' wide	"	8.000
26' wide	"	8.000
28' wide	"	10.000
30' wide	"	10.000
8' high		
Single gate		
3' wide	EA.	4.000
4' wide	"	4.000
Double gate		
10' wide	EA.	5.333
12' wide	"	5.333
14' wide	"	5.333
16' wide	"	5.333
18' wide	"	6.400
20' wide	"	6.400
22' wide	"	6.400
24' wide	"	8.000
28' wide	"	8.000
30' wide	"	10.000

SITE IMPROVEMENTS

02830.10 CHAIN LINK FENCE

SITE IMPROVEMENTS	UNIT	MAN/ HOURS
Motor operator for gates, no wiring	EA.	
Drilling fence post holes		
In soil		
By hand	EA.	0.400
By machine auger	"	0.200
In rock		
By jackhammer	EA.	2.667
By rock drill	"	0.800
Aluminum privacy slats, installed vertically	S.F.	0.020
Post hole, dig by hand	EA.	0.533
Set fence post in concrete	"	0.400

02830.70 RECREATIONAL COURTS

SITE IMPROVEMENTS	UNIT	MAN/ HOURS
Walls, galvanized steel		
8' high	L.F.	0.160
10' high	"	0.178
12' high	"	0.211
Vinyl coated		
8' high	L.F.	0.160
10' high	"	0.178
12' high	"	0.211
Gates, galvanized steel		
Single, 3' transom		
3'x7'	EA.	4.000
4'x7'	"	4.571
5'x7'	"	5.333
6'x7'	"	6.400
Double, 3' transom		
10'x7'	EA.	16.000
12'x7'	"	17.778
14'x7'	"	20.000
Double, no transom		
10'x10'	EA.	13.333
12'x10'	"	16.000
14'x10'	"	17.778
Vinyl coated		
Single, 3' transom		
3'x7'	EA.	4.000
4'x7'	"	4.571
5'x7'	"	5.333
6'x7'	"	6.400
Double, 3'		
10'x7'	EA.	16.000
12'x7'	"	17.778
14'x7'	"	20.000
Double, no transom		
10'x10'	EA.	13.333
12'x10'	"	16.000

SITE IMPROVEMENTS

02830.70 RECREATIONAL COURTS

SITE IMPROVEMENTS	UNIT	MAN/ HOURS
14'x10'	EA.	17.778
Wire and miscellaneous metal fences		
Chicken wire, post 4' o.c.		
2" mesh		
4' high	L.F.	0.040
6' high	"	0.053
Galvanized steel		
12 gauge, 2" by 4" mesh, posts 5' o.c.		
3' high	L.F.	0.040
5' high	"	0.050
14 gauge, 1" by 2" mesh, posts 5' o.c.		
3' high	L.F.	0.040
5' high	"	0.050
Vinyl coated		
12'high	L.F.	0.211

02840.30 GUARDRAILS

SITE IMPROVEMENTS	UNIT	MAN/ HOURS
Pipe bollard, steel pipe, concrete filled, painted		
6" dia.	EA.	0.667
8" dia.	"	1.000
12" dia.	"	2.667
Corrugated steel, guardrail, galvanized	L.F.	0.040
End section, wrap around or flared	EA.	0.800
Timber guardrail, 4" x 8"	L.F.	0.030
Guard rail, 3 cables, 3/4" dia.		
Steel posts	L.F.	0.300
Wood posts	"	0.267
Steel box beam		
6" x 6"	L.F.	0.133
6" x 8"	"	0.050
Concrete posts	EA.	0.400
Barrel type impact barrier	"	0.800
Light shield, 6' high	L.F.	0.160

02840.40 PARKING BARRIERS

SITE IMPROVEMENTS	UNIT	MAN/ HOURS
Timber, treated, 8' long		
4" x 4"	EA.	0.667
6" x 6"	"	0.800
Precast concrete, 4' long, with dowels		
12" x 6"	EA.	0.400
12" x 8"	"	0.444

02840.60 SIGNAGE

SITE IMPROVEMENTS	UNIT	MAN/ HOURS
Traffic signs		
Reflectorized signs per OSHA standards, including post		
Stop, 24"x24"	EA.	0.533
Yield, 30" triangle	"	0.533
Speed limit, 12"x18"	"	0.533

SITE IMPROVEMENTS	UNIT	MAN/ HOURS
02840.60 SIGNAGE		
Directional, 12"x18"	EA.	0.533
Exit, 12"x18"	"	0.533
Entry, 12"x18"	"	0.533
Warning, 24"x24"	"	0.533
Informational, 12"x18"	"	0.533
Handicap parking, 12"x18"	"	0.533
02860.40 RECREATIONAL FACILITIES		
Bleachers, outdoor, portable, per seat		
10 tiers		
Minimum	EA.	0.150
Maximum	"	0.200
20 tiers		
Minimum	EA.	0.141
Maximum	"	0.185
Grandstands, fixed, wood seat, steel frame, per seat		
15 tiers		
Minimum	EA.	0.240
Maximum	"	0.400
30 tiers		
Minimum	EA.	0.218
Maximum	"	0.343
Seats		
Seat backs only		
Fiberglass	EA.	0.080
Steel and wood seat	"	0.080
Seat restoration, fiberglass on wood		
Seats	EA.	0.160
Plain bench, no backs	"	0.067
Benches		
Park, precast concrete with backs		
4' long	EA.	2.667
8' long	"	4.000
Fiberglass, with backs		
4' long	EA.	2.000
8' long	"	2.667
Wood, with backs and fiberglass supports		
4' long	EA.	2.000
8' long	"	2.667
Steel frame, 6' long		
All steel	EA.	2.000
Hardwood boards	"	2.000
Players bench (no back), steel frame, fir seat, 10' long	"	2.667

SITE IMPROVEMENTS	UNIT	MAN/ HOURS
02860.40 RECREATIONAL FACILITIES		
Gravel and cinders over stone base	S.Y.	0.060
Rubber-cork base resilient pavement	"	0.480
For colored surfaces, add	"	0.048
Colored rubberized asphalt	"	0.600
Artificial resilient mat over asphalt	"	1.200
Tennis courts		
Bituminous pavement, 2-1/2" thick	S.Y.	0.150
Colored sealer, acrylic emulsion		
3 coats	S.Y.	0.053
For 2 color seal coating, add	"	0.008
For preparing old courts, add	"	0.005
Net, nylon, 42' long	EA.	1.000
Paint markings on asphalt, 2 coats	"	8.000
Playground equipment		
Basketball backboard		
Minimum	EA.	2.000
Maximum	"	2.286
Bike rack, 10' long	"	1.600
Golf shelter, fiberglass	"	2.000
Ground socket for movable posts		
Minimum	EA.	0.500
Maximum	"	0.500
Horizontal monkey ladder, 14' long	"	1.333
Posts, tether ball	"	0.400
Multiple purpose, 10' long	"	0.800
See-saw, steel		
Minimum	EA.	3.200
Average	"	4.000
Maximum	"	5.333
Slide		
Minimum	EA.	6.400
Maximum	"	7.273
Swings, plain seats		
8' high		
Minimum	EA.	5.333
Maximum	"	6.154
12' high		
Minimum	EA.	6.154
Maximum	"	8.889
02870.10 PREFABRICATED PLANTERS		
Concrete precast, circular		
24" dia., 18" high	EA.	0.800
42" dia., 30" high	"	1.000
Fiberglass, circular		
36" dia., 27" high	EA.	0.400
60" dia., 39" high	"	0.444
Tapered, circular		

02 SITEWORK

SITE IMPROVEMENTS

02870.10 PREFABRICATED PLANTERS	UNIT	MAN/HOURS
24" dia., 36" high	EA.	0.364
40" dia., 36" high	"	0.400
Square		
2' by 2', 17" high	EA.	0.364
4' by 4', 39" high	"	0.444
Rectangular		
4' by 1', 18" high	EA.	0.400

LANDSCAPING

02910.10 SHRUB & TREE MAINTENANCE	UNIT	MAN/HOURS
Moving shrubs on site		
12" ball	EA.	1.000
24" ball	"	1.333
3' high	"	0.800
4' high	"	0.889
5' high	"	1.000
18" spread	"	1.143
30" spread	"	1.333
Moving trees on site		
24" ball	EA.	1.200
48" ball	"	1.600
Trees		
3' high	EA.	0.480
6' high	"	0.533
8' high	"	0.600
10' high	"	0.800
Palm trees		
7' high	EA.	0.600
10' high	"	0.800
20' high	"	2.400
40' high	"	4.800
Guying trees		
4" dia.	EA.	0.400
8" dia.	"	0.500

02920.10 TOPSOIL	UNIT	MAN/HOURS
Spread topsoil, with equipment		
Minimum	C.Y.	0.080
Maximum	"	0.100

LANDSCAPING

02920.10 TOPSOIL	UNIT	MAN/HOURS
By hand		
Minimum	C.Y.	0.800
Maximum	"	1.000
Area preparation for seeding (grade, rake and clean)		
Square yard	S.Y.	0.006
By acre	ACRE	32.000
Remove topsoil and stockpile on site		
4" deep	C.Y.	0.067
6" deep	"	0.062
Spreading topsoil from stock pile		
By loader	C.Y.	0.073
By hand	"	0.800
Top dress by hand	S.Y.	0.008
Place imported top soil		
By loader		
4" deep	S.Y.	0.008
6" deep	"	0.009
By hand		
4" deep	S.Y.	0.089
6" deep	"	0.100
Plant bed preparation, 18" deep		
With backhoe/loader	S.Y.	0.020
By hand	"	0.080

02930.30 SEEDING	UNIT	MAN/HOURS
Mechanical seeding, 175 lb/acre		
By square yard	S.Y.	0.002
By acre	ACRE	8.000
450 lb/acre		
By square yard	S.Y.	0.002
By acre	ACRE	10.000
Seeding by hand, 10 lb per 100 s.y.		
By square yard	S.Y.	0.003
By acre	ACRE	13.333
Reseed disturbed areas	S.F.	0.004

02950.10 PLANTS	UNIT	MAN/HOURS
Euonymus coloratus, 18" (purple wintercreeper)	EA.	0.133
Hedera Helix, 2-1/4" pot (English ivy)	"	0.133
Liriope muscari, 2" clumps	"	0.080
Santolina, 12"	"	0.080
Vinca major or minor, 3" pot	"	0.080
Cortaderia argentia, 2 gallon (pampas grass)	"	0.080
Ophiopogan japonicus, 1 quart (4" pot)	"	0.080
Ajuga reptans, 2-3/4" pot (carpet bugle)	"	0.080
Pachysandra terminalis, 2-3/4" pot (Japanese spurge)	"	0.080

02 SITEWORK

LANDSCAPING	UNIT	MAN/ HOURS
02950.30 SHRUBS		
Juniperus conferia litoralis, 18"-24" (Shore Juniper)	"	0.320
Horizontalis plumosa, 18"-24" (Andorra Juniper)	"	0.320
Sabina tamar-iscfolia-tamarix juniper, 18"-24"	"	0.320
Chin San Hose, 18"-24" (San Hose Juniper)	"	0.320
Sargenti, 18"-24" (Sargent's Juniper)	"	0.320
Nandina domestica, 18"-24" (Heavenly Bamboo)	"	0.320
Raphiolepis Indica Springtime, 18"-24" Indian Hawthorn	"	0.320
Osmanthus Heterophyllus Gulftide, 18"-24" (Osmanthus)	"	0.320
Ilex Cornuta Burfordi Nana, 18"-24" (Dwarf Burford Holly)	"	0.320
Glabra, 18"-24" (Inkberry Holly)	"	0.320
Azalea, Indica types, 18"-24"	"	0.320
Kurume types, 18"-24"	"	0.320
Berberis Julianae, 18"-24" (Wintergreen Barberry)	"	0.320
Pieris Japonica Japanese, 18"-24" (Japanese Pieris)	"	0.320
Ilex Cornuta Rotunda, 18"-24" (Dwarf Chinese Holly)	"	0.320
Juniperus Horizontalis Plumosa, 24"-30" (Andorra Juniper)	"	0.400
Rhodopendrow Hybrids, 24"-30"	"	0.400
Aucuba Japonica Varigata, 24"-30" (Gold Dust Aucuba)	"	0.400
Ilex Crenata Willow Leaf, 24"-30" (Japanese Holly)	"	0.400
Cleyera Japonica, 30"-36" (Japanese Cleyera)	"	0.500
Pittosporum Tobira, 30"-36"	"	0.500
Prumus Laurocerasus, 30"-36"	"	0.500
Ilex Cornuta Burfordi, 30"-36" (Burford Holly)	"	0.500
Abelia Grandiflora, 24"-36" (Yew Podocarpus)	"	0.400
Podocarpos Macrophylla, 24"-36" (Yew Podocarpus)	"	0.400
Pyracantha Coccinea Lalandi, 3'-4' (Firethorn)	"	0.500
Photinia Frazieri, 3'-4' (Red Photinia)	"	0.500
Forsythia Suspensa, 3'-4' (Weeping Forsythia)	"	0.500
Camellia Japonica, 3'-4' (Common Camellia)	"	0.500
Juniperus Chin Torulosa, 3'-4' (Hollywood Juniper)	"	0.500
Cupressocyparis Leylandi, 3'-4'	"	0.500
Ilex Opaca Fosteri, 5'-6' (Foster's Holly)	"	0.667
Opaca, 5'-6' (American Holly)	"	0.667
Nyrica Cerifera, 4'-5' (Southern Wax Myrtles)	"	0.571
Ligustrum Japonicum, 4'-5' (Japanese Privet)	"	0.571
02950.60 TREES		
Cornus Florida, 5'-6' (White flowering Dogwood)	EA.	0.667
Prunus Serrulata Kwanzan, 6'-8' (Kwanzan Cherry)	"	0.800
Caroliniana, 6'-8' (Carolina Cherry Laurel)	"	0.800
Cercis Canadensis, 6'-8' (Eastern Redbud)	"	0.800
Koelreuteria Paniculata, 8'-10' (Goldenrain tree)	"	1.000
Acer Platanoides, 1-3/4"-2" (11'-13') (Norway Maple)	"	1.333
Rubrum, 1-3/4"-2" (11'-13') (Red Maple)	"	1.333
Saccharum, 1-3/4"-2" (Sugar Maple)	"	1.333
Fraxinus Pennsylvanica, 1-3/4"-2" Laneolata-Green Ash	"	1.333
Celtis Occidentalis, 1-3/4"-2" (American Hackberry)	"	1.333
Glenditsia Triacantos Inermis, 2"	"	1.333

LANDSCAPING	UNIT	MAN/ HOURS
02950.60 TREES		
Prunus Cerasifera 'Thundercloud', 6'-8'	EA.	0.800
Yeodensis, 6'-8' (Yoshino Cherry)	"	0.800
Lagerstroemia Indica, 8'-10' (Crapemyrtle)	"	1.000
Crataegus Phaenopyrum, 8'-10' Washington Hawthorn	"	1.000
Quercus Borealis, 1-3/4"-2" (Northern Red Oak)	"	1.333
Quercus Acutissima, 1-3/4"-2" (8'-10') (Sawtooth Oak)	"	1.333
Saliz Babylonica, 1-3/4"-2" (Weeping Willow)	"	1.333
Tilia Cordata Greenspire, 1-3/4"-2" (10'-12')	"	1.333
Malus, 2"-2-1/2" (8'-10') (Flowering Crabapple)	"	1.333
Platanus Occidentalis, (12'-14')	"	1.600
Pyrus Calleryana Bradford, 2"-2-1/2" (Bradford Pear)	"	1.333
Quercus Palustris, 2"-2-1/2" (12'-14') (Pin Oak)	"	1.333
Phellos, 2-1/2"-3" (Willow Oak)	"	1.600
Nigra, 2"-2-1/2" (Water Oak)	"	1.333
Magnolia Soulangeana, 4'-5' (Saucer Magnolia)	"	0.667
Grandiflora, 6'-8' (Southern Magnolia)	"	0.800
Cedrus Deodara, 10'-12' (Deodare Cedar)	"	1.333
Ginkgo Biloba, 10'-12' (2"-2-1/2") (Maidenhair Tree)	"	1.333
Pinus Thunbergi, 5'-6' (Japanese Black Pine)	"	0.667
Strobus, 6'-8' (White Pine)	"	0.800
Taeda, 6'-8' (Loblolly Pine)	"	0.800
Quercus Virginiana, 2"-2-1/2" (live oak)	"	1.600
02970.10 FERTILIZING		
Fertilizing (23#/1000 sf)		
By square yard	S.Y.	0.002
By acre	ACRE	10.000
Liming (70#/1000 sf)		
By square yard	S.Y.	0.003
By acre	ACRE	13.333
02980.10 LANDSCAPE ACCESSORIES		
Steel edging, 3/16" x 4"	L.F.	0.010
Landscaping stepping stones, 15"x15", white	EA.	0.040
Wood chip mulch	C.Y.	0.533
2" thick	S.Y.	0.016
4" thick	"	0.023
6" thick	"	0.029
Gravel mulch, 3/4" stone	C.Y.	0.800
White marble chips, 1" deep	S.F.	0.008
Peat moss		
2" thick	S.Y.	0.018
4" thick	"	0.027
6" thick	"	0.033
Landscaping timbers, treated lumber		
4" x 4"	L.F.	0.027
6" x 6"	"	0.029
8" x 8"	"	0.033

FORMWORK	UNIT	MAN/ HOURS
03110.05 BEAM FORMWORK		
Beam forms, job built		
Beam bottoms		
1 use	S.F.	0.133
2 uses	"	0.127
3 uses	"	0.123
4 uses	"	0.118
5 uses	"	0.114
Beam sides		
1 use	S.F.	0.089
2 uses	"	0.084
3 uses	"	0.080
4 uses	"	0.076
5 uses	"	0.073
03110.10 BOX CULVERT FORMWORK		
Box culverts, job built		
6' x 6'		
1 use	S.F.	0.080
2 uses	"	0.076
3 uses	"	0.073
4 uses	"	0.070
5 uses	"	0.067
8' x 12'		
1 use	S.F.	0.067
2 uses	"	0.064
3 uses	"	0.062
4 uses	"	0.059
5 uses	"	0.057
03110.15 COLUMN FORMWORK		
Column, square forms, job built		
8" x 8" columns		
1 use	S.F.	0.160
2 uses	"	0.154
3 uses	"	0.148
4 uses	"	0.143
5 uses	"	0.138
12" x 12" columns		
1 use	S.F.	0.145
2 uses	"	0.140
3 uses	"	0.136
4 uses	"	0.131
5 uses	"	0.127
16" x 16" columns		
1 use	S.F.	0.133
2 uses	"	0.129
3 uses	"	0.125

FORMWORK	UNIT	MAN/ HOURS
03110.15 COLUMN FORMWORK		
4 uses	S.F.	0.121
5 uses	"	0.118
24" x 24" columns		
1 use	S.F.	0.123
2 uses	"	0.119
3 uses	"	0.116
4 uses	"	0.113
5 uses	"	0.110
36" x 36" columns		
1 use	S.F.	0.114
2 uses	"	0.111
3 uses	"	0.108
4 uses	"	0.105
5 uses	"	0.103
Round fiber forms, 1 use		
10" dia.	L.F.	0.160
12" dia.	"	0.163
14" dia.	"	0.170
16" dia.	"	0.178
18" dia.	"	0.190
24" dia.	"	0.205
30" dia.	"	0.222
36" dia.	"	0.242
42" dia.	"	0.267
03110.18 CURB FORMWORK		
Curb forms		
Straight, 6" high		
1 use	L.F.	0.080
2 uses	"	0.076
3 uses	"	0.073
4 uses	"	0.070
5 uses	"	0.067
Curved, 6" high		
1 use	L.F.	0.100
2 uses	"	0.094
3 uses	"	0.089
4 uses	"	0.085
5 uses	"	0.082
03110.20 ELEVATED SLAB FORMWORK		
Elevated slab formwork		
Slab, with drop panels		
1 use	S.F.	0.064
2 uses	"	0.062
3 uses	"	0.059
4 uses	"	0.057
5 uses	"	0.055

03 CONCRETE

FORMWORK	UNIT	MAN/HOURS
03110.20 ELEVATED SLAB FORMWORK		
Floor slab, hung from steel beams		
1 use	S.F.	0.062
2 uses	"	0.059
3 uses	"	0.057
4 uses	"	0.055
5 uses	"	0.053
Floor slab, with pans or domes		
1 use	S.F.	0.073
2 uses	"	0.070
3 uses	"	0.067
4 uses	"	0.064
5 uses	"	0.062
Equipment curbs, 12" high		
1 use	L.F.	0.080
2 uses	"	0.076
3 uses	"	0.073
4 uses	"	0.070
5 uses	"	0.067
03110.25 EQUIPMENT PAD FORMWORK		
Equipment pad, job built		
1 use	S.F.	0.100
2 uses	"	0.094
3 uses	"	0.089
4 uses	"	0.084
5 uses	"	0.080
03110.35 FOOTING FORMWORK		
Wall footings, job built, continuous		
1 use	S.F.	0.080
2 uses	"	0.076
3 uses	"	0.073
4 uses	"	0.070
5 uses	"	0.067
Column footings, spread		
1 use	S.F.	0.100
2 uses	"	0.094
3 uses	"	0.089
4 uses	"	0.084
5 uses	"	0.080
03110.50 GRADE BEAM FORMWORK		
Grade beams, job built		
1 use	S.F.	0.080
2 uses	"	0.076
3 uses	"	0.073
4 uses	"	0.070
5 uses	"	0.067

FORMWORK	UNIT	MAN/HOURS
03110.53 PILE CAP FORMWORK		
Pile cap forms, job built		
Square		
1 use	S.F.	0.100
2 uses	"	0.094
3 uses	"	0.089
4 uses	"	0.084
5 uses	"	0.080
Triangular		
1 use	S.F.	0.114
2 uses	"	0.107
3 uses	"	0.100
4 uses	"	0.094
5 uses	"	0.089
03110.55 SLAB/MAT FORMWORK		
Mat foundations, job built		
1 use	S.F.	0.100
2 uses	"	0.094
3 uses	"	0.089
4 uses	"	0.084
5 uses	"	0.080
Edge forms		
6" high		
1 use	L.F.	0.073
2 uses	"	0.070
3 uses	"	0.067
4 uses	"	0.064
5 uses	"	0.062
12" high		
1 use	L.F.	0.080
2 uses	"	0.076
3 uses	"	0.073
4 uses	"	0.070
5 uses	"	0.067
Formwork for openings		
1 use	S.F.	0.160
2 uses	"	0.145
3 uses	"	0.133
4 uses	"	0.123
5 uses	"	0.114
03110.60 STAIR FORMWORK		
Stairway forms, job built		
1 use	S.F.	0.160
2 uses	"	0.145
3 uses	"	0.133
4 uses	"	0.123

03 CONCRETE

FORMWORK	UNIT	MAN/ HOURS
03110.60 STAIR FORMWORK		
5 uses	S.F.	0.114
Stairs, elevated		
1 use	S.F.	0.160
2 uses	"	0.133
3 uses	"	0.114
4 uses	"	0.107
5 uses	"	0.100
03110.65 WALL FORMWORK		
Wall forms, exterior, job built		
Up to 8' high wall		
1 use	S.F.	0.080
2 uses	"	0.076
3 uses	"	0.073
4 uses	"	0.070
5 uses	"	0.067
Over 8' high wall		
1 use	S.F.	0.100
2 uses	"	0.094
3 uses	"	0.089
4 uses	"	0.084
5 uses	"	0.080
Over 16' high wall		
1 use	S.F.	0.114
2 uses	"	0.107
3 uses	"	0.100
4 uses	"	0.094
5 uses	"	0.089
Radial wall forms		
1 use	S.F.	0.123
2 uses	"	0.114
3 uses	"	0.107
4 uses	"	0.100
5 uses	"	0.094
Retaining wall forms		
1 use	S.F.	0.089
2 uses	"	0.084
3 uses	"	0.080
4 uses	"	0.076
5 uses	"	0.073
Radial retaining wall forms		
1 use	S.F.	0.133
2 uses	"	0.123
3 uses	"	0.114
4 uses	"	0.107
5 uses	"	0.100
Column pier and pilaster		
1 use	S.F.	0.160

FORMWORK	UNIT	MAN/ HOURS
03110.65 WALL FORMWORK		
2 uses	S.F.	0.145
3 uses	"	0.133
4 uses	"	0.123
5 uses	"	0.114
Interior wall forms		
Up to 8' high		
1 use	S.F.	0.073
2 uses	"	0.070
3 uses	"	0.067
4 uses	"	0.064
5 uses	"	0.062
Over 8' high		
1 use	S.F.	0.089
2 uses	"	0.084
3 uses	"	0.080
4 uses	"	0.076
5 uses	"	0.073
Over 16' high		
1 use	S.F.	0.100
2 uses	"	0.094
3 uses	"	0.089
4 uses	"	0.084
5 uses	"	0.080
Radial wall forms		
1 use	S.F.	0.107
2 uses	"	0.100
3 uses	"	0.094
4 uses	"	0.089
5 uses	"	0.084
Curved wall forms, 24" sections		
1 use	S.F.	0.160
2 uses	"	0.145
3 uses	"	0.133
4 uses	"	0.123
5 uses	"	0.114
PVC form liner, per side, smooth finish		
1 use	S.F.	0.067
2 uses	"	0.064
3 uses	"	0.062
4 uses	"	0.057
5 uses	"	0.053
03110.90 MISCELLANEOUS FORMWORK		
Keyway forms (5 uses)		
2 x 4	L.F.	0.040
2 x 6	"	0.044
Bulkheads		
Walls, with keyways		

FORMWORK	UNIT	MAN/ HOURS
03110.90 MISCELLANEOUS FORMWORK		
2 piece	L.F.	0.073
3 piece	"	0.080
Elevated slab, with keyway		
2 piece	L.F.	0.067
3 piece	"	0.073
Ground slab, with keyway		
2 piece	L.F.	0.057
3 piece	"	0.062
Chamfer strips		
Wood		
1/2" wide	L.F.	0.018
3/4" wide	"	0.018
1" wide	"	0.018
PVC		
1/2" wide	L.F.	0.018
3/4" wide	"	0.018
1" wide	"	0.018
Radius		
1"	L.F.	0.019
1-1/2"	"	0.019
Reglets		
Galvanized steel, 24 ga.	L.F.	0.032
Metal formwork		
Straight edge forms		
4" high	L.F.	0.050
6" high	"	0.053
8" high	"	0.057
12" high	"	0.062
16" high	"	0.067
Curb form, S-shape		
12" x		
1'-6"	L.F.	0.114
2'	"	0.107
2'-6"	"	0.100
3'	"	0.089

REINFORCEMENT	UNIT	MAN/ HOURS
03210.05 BEAM REINFORCING		
Beam-girders		
#3 - #4	TON	20.000
#5 - #6	"	16.000
#7 - #8	"	13.333
#9 - #10	"	11.429
#11 - #12	"	10.667
#13 - #14	"	10.000
Galvanized		
#3 - #4	TON	20.000
#5 - #6	"	16.000
#7 - #8	"	13.333
#9 - #10	"	11.429
#11 - #12	"	10.667
#13 - #14	"	10.000
Bond beams		
#3 - #4	TON	26.667
#5 - #6	"	20.000
#7 - #8	"	17.778
Galvanized		
#3 - #4	TON	26.667
#5 - #6	"	20.000
#7 - #8	"	17.778
03210.10 BOX CULVERT REINFORCING		
Box culverts		
#3 - #4	TON	10.000
#5 - #6	"	8.889
#7 - #8	"	8.000
#9 - #10	"	7.273
#11 - #12	"	6.667
Galvanized		
#3 - #4	TON	10.000
#5 - #6	"	8.889
#7 - #8	"	8.000
#9 - #10	"	7.273
#11 - #12	"	6.667
03210.15 COLUMN REINFORCING		
Columns		
#3 - #4	TON	22.857
#5 - #6	"	17.778
#7 - #8	"	16.000
#9 - #10	"	14.545
#11 - #12	"	13.333
#13 - #14	"	12.308
#15 - #16	"	11.429
Galvanized		
#3 - #4	TON	22.857
#5 - #6	"	17.778

REINFORCEMENT	UNIT	MAN/ HOURS
03210.15 COLUMN REINFORCING		
#7 - #8	TON	16.000
#9 - #10	"	14.545
#11 - #12	"	13.333
#13 - #14	"	12.308
#15 - #16	"	11.429
Spirals		
8" to 24" dia.	TON	20.000
24" to 48" dia.	"	17.778
48" to 84" dia.	"	16.000
03210.20 ELEVATED SLAB REINFORCING		
Elevated slab		
#3 - #4	TON	10.000
#5 - #6	"	8.889
#7 - #8	"	8.000
#9 - #10	"	7.273
#11 - #12	"	6.667
Galvanized		
#3 - #4	TON	10.000
#5 - #6	"	8.889
#7 - #8	"	8.000
#9 - #10	"	7.273
#11 - #12	"	6.667
03210.25 EQUIP. PAD REINFORCING		
Equipment pad		
#3 - #4	TON	16.000
#5 - #6	"	14.545
#7 - #8	"	13.333
#9 - #10	"	12.308
#11 - #12	"	11.429
03210.35 FOOTING REINFORCING		
Footings		
Grade 50		
#3 - #4	TON	13.333
#5 - #6	"	11.429
#7 - #8	"	10.000
#9 - #10	"	8.889
Grade 60		
#3 - #4	TON	13.333
#5 - #6	"	11.429
#7 - #8	"	10.000
#9 - #10	"	8.889
Grade 70		
#3 - #4	TON	13.333
#5 - #6	"	11.429
#7 - #8	"	10.000

REINFORCEMENT	UNIT	MAN/ HOURS
03210.35 FOOTING REINFORCING		
#9 - #10	TON	8.889
#11- #12	"	8.000
Straight dowels, 24" long		
1" dia.	EA.	0.080
3/4" dia.	"	0.080
5/8" dia.	"	0.067
1/2" dia.	"	0.057
03210.45 FOUNDATION REINFORCING		
Foundations		
#3 - #4	TON	13.333
#5 - #6	"	11.429
#7 - #8	"	10.000
#9 - #10	"	8.889
#11 - #12	"	8.000
Galvanized		
#3 - #4	TON	13.333
#5 - #6	"	11.429
#7 - #8	"	10.000
#9 - #10	"	8.889
#11 - #12	"	8.000
03210.50 GRADE BEAM REINFORCING		
Grade beams		
#3 - #4	TON	12.308
#5 - #6	"	10.667
#7 - #8	"	9.412
#9 - #10	"	8.421
#11 - #12	"	7.619
Galvanized		
#3 - #4	TON	12.308
#5 - #6	"	10.667
#7 - #8	"	9.412
#9 - #10	"	8.421
#11 - #12	"	7.619
03210.53 PILE CAP REINFORCING		
Pile caps		
#3 - #4	TON	20.000
#5 - #6	"	17.778
#7 - #8	"	16.000
#9 - #10	"	14.545
#11 - #12	"	13.333
Galvanized		
#3 - #4	TON	20.000
#5 - #6	"	17.778
#7 - #8	"	16.000
#9 - #10	"	14.545

03 CONCRETE

REINFORCEMENT	UNIT	MAN/HOURS
03210.53 PILE CAP REINFORCING		
#11 - #12	TON	13.333
03210.55 SLAB/MAT REINFORCING		
Bars, slabs		
#3 - #4	TON	13.333
#5 - #6	"	11.429
#7 - #8	"	10.000
#9 - #10	"	8.889
#11 - #12	"	8.000
Galvanized		
#3 - #4	TON	13.333
#5 - #6	"	11.429
#7 - #8	"	10.000
#9 - #10	"	8.889
#11 - #12	"	8.000
Wire mesh, slabs		
Galvanized		
4x4		
W1.4xW1.4	S.F.	0.005
W2.0xW2.0	"	0.006
W2.9xW2.9	"	0.006
W4.0xW4.0	"	0.007
6x6		
W1.4xW1.4	S.F.	0.004
W2.0xW2.0	"	0.004
W2.9xW2.9	"	0.005
W4.0xW4.0	"	0.005
Standard		
2x2		
W.9xW.9	S.F.	0.005
4x4		
W1.4xW1.4	S.F.	0.005
W2.0xW2.0	"	0.006
W2.9xW2.9	"	0.006
W4.0xW4.0	"	0.007
6x6		
W1.4xW1.4	S.F.	0.004
W2.0xW2.0	"	0.004
W2.9xW2.9	"	0.005
W4.0xW4.0	"	0.005
03210.60 STAIR REINFORCING		
Stairs		
#3 - #4	TON	16.000
#5 - #6	"	13.333
#7 - #8	"	11.429

REINFORCEMENT	UNIT	MAN/HOURS
03210.60 STAIR REINFORCING		
#9 - #10	TON	10.000
Galvanized		
#3 - #4	TON	16.000
#5 - #6	"	13.333
#7 - #8	"	11.429
#9 - #10	"	10.000
03210.65 WALL REINFORCING		
Walls		
#3 - #4	TON	11.429
#5 - #6	"	10.000
#7 - #8	"	8.889
#9 - #10	"	8.000
Galvanized		
#3 - #4	TON	11.429
#5 - #6	"	10.000
#7 - #8	"	8.889
#9 - #10	"	8.000
Masonry wall (horizontal)		
#3 - #4	TON	32.000
#5 - #6	"	26.667
Galvanized		
#3 - #4	TON	32.000
#5 - #6	"	26.667
Masonry wall (vertical)		
#3 - #4	TON	40.000
#5 - #6	"	32.000
Galvanized		
#3 - #4	TON	40.000
#5 - #6	"	32.000

ACCESSORIES	UNIT	MAN/HOURS
03250.40 CONCRETE ACCESSORIES		
Expansion joint, poured		
Asphalt		
1/2" x 1"	L.F.	0.016
1" x 2"	"	0.017
Liquid neoprene, cold applied		
1/2" x 1"	L.F.	0.016
1" x 2"	"	0.018

ACCESSORIES	UNIT	MAN/HOURS
03250.40 CONCRETE ACCESSORIES		
Polyurethane, 2 parts		
1/2" x 1"	L.F.	0.027
1" x 2"	"	0.029
Rubberized asphalt, cold		
1/2" x 1"	L.F.	0.016
1" x 2"	"	0.017
Hot, fuel resistant		
1/2" x 1"	L.F.	0.016
1" x 2"	"	0.017
Expansion joint, premolded, in slabs		
Asphalt		
1/2" x 6"	L.F.	0.020
1" x 12"	"	0.027
Cork		
1/2" x 6"	L.F.	0.020
1" x 12"	"	0.027
Neoprene sponge		
1/2" x 6"	L.F.	0.020
1" x 12"	"	0.027
Polyethylene foam		
1/2" x 6"	L.F.	0.020
1" x 12"	"	0.027
Polyurethane foam		
1/2" x 6"	L.F.	0.020
1" x 12"	"	0.027
Polyvinyl chloride foam		
1/2" x 6"	L.F.	0.020
1" x 12"	"	0.027
Rubber, gray sponge		
1/2" x 6"	L.F.	0.020
1" x 12"	"	0.027
Asphalt felt control joints or bond breaker, screed joints		
4" slab	L.F.	0.016
6" slab	"	0.018
8" slab	"	0.020
10" slab	"	0.023
Keyed cold expansion and control joints, 24 ga.		
4" slab	L.F.	0.050
5" slab	"	0.050
6" slab	"	0.053
8" slab	"	0.057
10" slab	"	0.062
Waterstops		
Polyvinyl chloride		
Ribbed		
3/16" thick x		
4" wide	L.F.	0.040
6" wide	"	0.044
1/2" thick x		

ACCESSORIES	UNIT	MAN/HOURS
03250.40 CONCRETE ACCESSORIES		
9" wide	L.F.	0.050
Ribbed with center bulb		
3/16" thick x 9" wide	L.F.	0.050
3/8" thick x 9" wide	"	0.050
Dumbbell type, 3/8" thick x 6" wide	"	0.044
Plain, 3/8" thick x 9" wide	"	0.050
Center bulb, 3/8" thick x 9" wide	"	0.050
Rubber		
Flat dumbbell		
3/8" thick x		
6" wide	L.F.	0.044
9" wide	"	0.050
Center bulb		
3/8" thick x		
6" wide	L.F.	0.044
9" wide	"	0.050
Vapor barrier		
4 mil polyethylene	S.F.	0.003
6 mil polyethylene	"	0.003
Gravel porous fill, under floor slabs, 3/4" stone	C.Y.	1.333
Reinforcing accessories		
Beam bolsters		
1-1/2" high, plain	L.F.	0.008
Galvanized	"	0.008
3" high		
Plain	L.F.	0.010
Galvanized	"	0.010
Slab bolsters		
1" high		
Plain	L.F.	0.004
Galvanized	"	0.004
2" high		
Plain	L.F.	0.004
Galvanized	"	0.004
Chairs, high chairs		
3" high		
Plain	EA.	0.020
Galvanized	"	0.020
5" high		
Plain	EA.	0.021
Galvanized	"	0.021
8" high		
Plain	EA.	0.023
Galvanized	"	0.023
12" high		
Plain	EA.	0.027
Galvanized	"	0.027
Continuous, high chair		
3" high		

ACCESSORIES	UNIT	MAN/HOURS
03250.40 CONCRETE ACCESSORIES		
Plain	L.F.	0.005
Galvanized	"	0.005
5" high		
Plain	L.F.	0.006
Galvanized	"	0.006
8" high		
Plain	L.F.	0.006
Galvanized	"	0.006
12" high		
Plain	L.F.	0.007
Galvanized	"	0.007

CAST-IN-PLACE CONCRETE	UNIT	MAN/HOURS
03350.10 CONCRETE FINISHES		
Floor finishes		
Broom	S.F.	0.011
Screed	"	0.010
Darby	"	0.010
Steel float	"	0.013
Granolithic topping		
1/2" thick	S.F.	0.036
1" thick	"	0.040
2" thick	"	0.044
Wall finishes		
Burlap rub, with cement paste	S.F.	0.013
Float finish	"	0.020
Etch with acid	"	0.013
Sandblast		
Minimum	S.F.	0.016
Maximum	"	0.016
Bush hammer		
Green concrete	S.F.	0.040
Cured concrete	"	0.062
Break ties and patch holes	"	0.016
Carborundum		
Dry rub	S.F.	0.027
Wet rub	"	0.040
Floor hardeners		
Metallic		

CAST-IN-PLACE CONCRETE	UNIT	MAN/HOURS
03350.10 CONCRETE FINISHES		
Light service	S.F.	0.010
Heavy service	"	0.013
Non-metallic		
Light service	S.F.	0.010
Heavy service	"	0.013
Rusticated concrete finish		
Beveled edge	L.F.	0.044
Square edge	"	0.057
Solid board concrete finish		
Standard	S.F.	0.067
Rustic	"	0.080
03360.10 PNEUMATIC CONCRETE		
Pneumatic applied concrete (gunite)		
2" thick	S.F.	0.030
3" thick	"	0.040
4" thick	"	0.048
Finish surface		
Minimum	S.F.	0.040
Maximum	"	0.080
03370.10 CURING CONCRETE		
Sprayed membrane		
Slabs	S.F.	0.002
Walls	"	0.002
Curing paper		
Slabs	S.F.	0.002
Walls	"	0.002
Burlap		
7.5 oz.	S.F.	0.003
12 oz.	"	0.003

PLACING CONCRETE	UNIT	MAN/HOURS
03380.05 BEAM CONCRETE		
Beams and girders		
2500# or 3000# concrete		
By crane	C.Y.	0.960

03 CONCRETE

PLACING CONCRETE

03380.05 BEAM CONCRETE

PLACING CONCRETE	UNIT	MAN/HOURS
By pump	C.Y.	0.873
By hand buggy	"	0.800
3500# or 4000# concrete		
By crane	C.Y.	0.960
By pump	"	0.873
By hand buggy	"	0.800
5000# concrete		
By crane	C.Y.	0.960
By pump	"	0.873
By hand buggy	"	0.800
Bond beam, 3000# concrete		
By pump		
8" high		
4" wide	L.F.	0.019
6" wide	"	0.022
8" wide	"	0.024
10" wide	"	0.027
12" wide	"	0.030
16" high		
8" wide	L.F.	0.030
10" wide	"	0.034
12" wide	"	0.040
By crane		
8" high		
4" wide	L.F.	0.021
6" wide	"	0.023
8" wide	"	0.024
10" wide	"	0.027
12" wide	"	0.030
16" high		
8" wide	L.F.	0.030
10" wide	"	0.032
12" wide	"	0.037

03380.15 COLUMN CONCRETE

PLACING CONCRETE	UNIT	MAN/HOURS
Columns		
2500# or 3000# concrete		
By crane	C.Y.	0.873
By pump	"	0.800
3500# or 4000# concrete		
By crane	C.Y.	0.873
By pump	"	0.800
5000# concrete		
By crane	C.Y.	0.873
By pump	"	0.800

PLACING CONCRETE

03380.20 ELEVATED SLAB CONCRETE

PLACING CONCRETE	UNIT	MAN/HOURS
Elevated slab		
2500# or 3000# concrete		
By crane	C.Y.	0.480
By pump	"	0.369
By hand buggy	"	0.800
3500# or 4000# concrete		
By crane	C.Y.	0.480
By pump	"	0.369
By hand buggy	"	0.800
5000# concrete		
By crane	C.Y.	0.480
By pump	"	0.369
By hand buggy	"	0.800
Topping		
2500# or 3000# concrete		
By crane	C.Y.	0.480
By pump	"	0.369
By hand buggy	"	0.800
3500# or 4000# concrete		
By crane	C.Y.	0.480
By pump	"	0.369
By hand buggy	"	0.800
5000# concrete		
By crane	C.Y.	0.480
By pump	"	0.369
By hand buggy	"	0.800

03380.25 EQUIPMENT PAD CONCRETE

PLACING CONCRETE	UNIT	MAN/HOURS
Equipment pad		
2500# or 3000# concrete		
By chute	C.Y.	0.267
By pump	"	0.686
By crane	"	0.800
3500# or 4000# concrete		
By chute	C.Y.	0.267
By pump	"	0.686
By crane	"	0.800
5000# concrete		
By chute	C.Y.	0.267
By pump	"	0.686
By crane	"	0.800

03380.35 FOOTING CONCRETE

PLACING CONCRETE	UNIT	MAN/HOURS
Continuous footing		
2500# or 3000# concrete		
By chute	C.Y.	0.267
By pump	"	0.600
By crane	"	0.686

03 CONCRETE

PLACING CONCRETE

03380.35 FOOTING CONCRETE	UNIT	MAN/HOURS
3500# or 4000# concrete		
By chute	C.Y.	0.267
By pump	"	0.600
By crane	"	0.686
5000# concrete		
By chute	C.Y.	0.267
By pump	"	0.600
By crane	"	0.686
Spread footing		
2500# or 3000# concrete		
Under 5 cy		
By chute	C.Y.	0.267
By pump	"	0.640
By crane	"	0.738
Over 5 cy		
By chute	C.Y.	0.200
By pump	"	0.565
By crane	"	0.640
3500# or 4000# concrete		
Under 5 c.y.		
By chute	C.Y.	0.267
By pump	"	0.640
By crane	"	0.738
Over 5 c.y.		
By chute	C.Y.	0.200
By pump	"	0.565
By crane	"	0.640
5000# concrete		
Under 5 c.y.		
By chute	C.Y.	0.267
By pump	"	0.640
By crane	"	0.738
Over 5 c.y.		
By chute	C.Y.	0.200
By pump	"	0.565
By crane	"	0.640

03380.50 GRADE BEAM CONCRETE	UNIT	MAN/HOURS
Grade beam		
2500# or 3000# concrete		
By chute	C.Y.	0.267
By crane	"	0.686
By pump	"	0.600
By hand buggy	"	0.800
3500# or 4000# concrete		
By chute	C.Y.	0.267
By crane	"	0.686
By pump	"	0.600
By hand buggy	"	0.800

PLACING CONCRETE

03380.50 GRADE BEAM CONCRETE	UNIT	MAN/HOURS
5000# concrete		
By chute	C.Y.	0.267
By crane	"	0.686
By pump	"	0.600
By hand buggy	"	0.800

03380.53 PILE CAP CONCRETE	UNIT	MAN/HOURS
Pile cap		
2500# or 3000 concrete		
By chute	C.Y.	0.267
By crane	"	0.800
By pump	"	0.686
By hand buggy	"	0.800
3500# or 4000# concrete		
By chute	C.Y.	0.267
By crane	"	0.800
By pump	"	0.686
By hand buggy	"	0.800
5000# concrete		
By chute	C.Y.	0.267
By crane	"	0.800
By pump	"	0.686
By hand buggy	"	0.800

03380.55 SLAB/MAT CONCRETE	UNIT	MAN/HOURS
Slab on grade		
2500# or 3000# concrete		
By chute	C.Y.	0.200
By crane	"	0.400
By pump	"	0.343
By hand buggy	"	0.533
3500# or 4000# concrete		
By chute	C.Y.	0.200
By crane	"	0.400
By pump	"	0.343
By hand buggy	"	0.533
5000# concrete		
By chute	C.Y.	0.200
By crane	"	0.400
By pump	"	0.343
By hand buggy	"	0.533
Foundation mat		
2500# or 3000# concrete, over 20 cy		
By chute	C.Y.	0.160
By crane	"	0.343
By pump	"	0.300
By hand buggy	"	0.400

PLACING CONCRETE	UNIT	MAN/HOURS
03380.58 SIDEWALKS		
Walks, cast in place with wire mesh, base not incl.		
4" thick	S.F.	0.027
5" thick	"	0.032
6" thick	"	0.040
03380.60 STAIR CONCRETE		
Stairs		
2500# or 3000# concrete		
By chute	C.Y.	0.267
By crane	"	0.800
By pump	"	0.686
By hand buggy	"	0.800
3500# or 4000# concrete		
By chute	C.Y.	0.267
By crane	"	0.800
By pump	"	0.686
By hand buggy	"	0.800
5000# concrete		
By chute	C.Y.	0.267
By crane	"	0.800
By pump	"	0.686
By hand buggy	"	0.800
03380.65 WALL CONCRETE		
Walls		
2500# or 3000# concrete		
To 4'		
By chute	C.Y.	0.229
By crane	"	0.800
By pump	"	0.738
To 8'		
By crane	C.Y.	0.873
By pump	"	0.800
To 16'		
By crane	C.Y.	0.960
By pump	"	0.873
Over 16'		
By crane	C.Y.	1.067
By pump	"	0.960
3500# or 4000# concrete		
To 4'		
By chute	C.Y.	0.229
By crane	"	0.800
By pump	"	0.738
To 8'		
By crane	C.Y.	0.873
By pump	"	0.800
To 16'		
By crane	C.Y.	0.960

PLACING CONCRETE	UNIT	MAN/HOURS
03380.65 WALL CONCRETE		
By pump	C.Y.	0.873
Over 16'		
By crane	C.Y.	1.067
By pump	"	0.960
5000# concrete		
To 4'		
By chute	C.Y.	0.229
By crane	"	0.800
By pump	"	0.738
To 8'		
By crane	C.Y.	0.873
By pump	"	0.800
To 16'		
By crane	C.Y.	0.960
By pump	"	0.873
Filled block (CMU)		
3000# concrete, by pump		
4" wide	S.F.	0.034
6" wide	"	0.040
8" wide	"	0.048
10" wide	"	0.056
12" wide	"	0.069
Pilasters, 3000# concrete	C.F.	0.960
Wall cavity, 2" thick, 3000# concrete	S.F.	0.032

PRECAST CONCRETE	UNIT	MAN/HOURS
03400.10 PRECAST BEAMS		
Prestressed, double tee, 24" deep, 8' wide		
35' span		
115 psf	S.F.	0.008
140 psf	"	0.008
40' span		
80 psf	S.F.	0.009
143 psf	"	0.009
45' span		
50 psf	S.F.	0.007
70 psf	"	0.007
100 psf	"	0.007
130 psf	"	0.007
50' span		

PRECAST CONCRETE	UNIT	MAN/HOURS
03400.10 PRECAST BEAMS		
75 psf	S.F.	0.007
100 psf	"	0.007
Precast beams, girders and joists		
1000 lb/lf live load		
10' span	L.F.	0.160
20' span	"	0.096
30' span	"	0.080
3000 lb/lf live load		
10' span	L.F.	0.160
20' span	"	0.096
30' span	"	0.080
5000 lb/lf live load		
10' span	L.F.	0.160
20' span	"	0.096
30' span	"	0.080
03400.20 PRECAST COLUMNS		
Prestressed concrete columns		
10" x 10"		
10' long	EA.	0.960
15' long	"	1.000
20' long	"	1.067
25' long	"	1.143
30' long	"	1.200
12" x 12"		
20' long	EA.	1.200
25' long	"	1.297
30' long	"	1.371
16" x 16"		
20' long	EA.	1.200
25' long	"	1.297
30' long	"	1.371
20" x 20"		
20' long	EA.	1.263
25' long	"	1.333
30' long	"	1.412
24" x 24"		
20' long	EA.	1.333
25' long	"	1.412
30' long	"	1.500
28" x 28"		
20' long	EA.	1.500
25' long	"	1.600
30' long	"	1.714
32" x 32"		
20' long	EA.	1.600
25' long	"	1.714
30' long	"	1.846

PRECAST CONCRETE	UNIT	MAN/HOURS
03400.20 PRECAST COLUMNS		
36" x 36"		
20' long	EA.	1.714
25' long	"	1.846
30' long	"	2.000
03400.30 PRECAST SLABS		
Prestressed flat slab		
6" thick, 4' wide		
20' span		
80 psf	S.F.	0.020
110 psf	"	0.020
25' span		
80 psf	S.F.	0.019
Cored slab		
6" thick, 4' wide		
20' span		
80 psf	S.F.	0.020
100 psf	"	0.020
130 psf	"	0.020
8" thick, 4' wide		
25' span		
70 psf	S.F.	0.019
125 psf	"	0.019
170 psf	"	0.019
30' span		
70 psf	S.F.	0.016
90 psf	"	0.016
35' span		
70 psf	S.F.	0.015
10" thick, 4' wide		
30' span		
75 psf	S.F.	0.016
100 psf	"	0.016
130 psf	"	0.016
35' span		
60 psf	S.F.	0.015
80 psf	"	0.015
120 psf	"	0.015
40' span		
65 psf	S.F.	0.012
Slabs, roof and floor members, 4' wide		
6" thick, 25' span	S.F.	0.019
8" thick, 30' span	"	0.015
10" thick, 40' span	"	0.013
Tee members		
Multiple tee, roof and floor		
Minimum	S.F.	0.012
Maximum	"	0.024
Double tee wall member		

03 CONCRETE

PRECAST CONCRETE	UNIT	MAN/HOURS
03400.30 PRECAST SLABS		
Minimum	S.F.	0.014
Maximum	"	0.027
Single tee		
Short span, roof members		
Minimum	S.F.	0.015
Maximum	"	0.030
Long span, roof members		
Minimum	S.F.	0.012
Maximum	"	0.024
03400.40 PRECAST WALLS		
Wall panel, 8' x 20'		
Gray cement		
Liner finish		
4" wall	S.F.	0.014
5" wall	"	0.014
6" wall	"	0.015
8" wall	"	0.015
Sandblast finish		
4" wall	S.F.	0.014
5" wall	"	0.014
6" wall	"	0.015
8" wall	"	0.015
White cement		
Liner finish		
4" wall	S.F.	0.014
5" wall	"	0.014
6" wall	"	0.015
8" wall	"	0.015
Sandblast finish		
4" wall	S.F.	0.014
5" wall	"	0.014
6" wall	"	0.015
8" wall	"	0.015
Double tee wall panel, 24" deep		
Gray cement		
Liner finish	S.F.	0.016
Sandblast finish	"	0.016
White cement		
Form liner finish	S.F.	0.016
Sandblast finish	"	0.016
Partition panels		
4" wall	S.F.	0.016
5" wall	"	0.016
6" wall	"	0.016
8" wall	"	0.016
Cladding panels		
4" wall	S.F.	0.017

PRECAST CONCRETE	UNIT	MAN/HOURS
03400.40 PRECAST WALLS		
5" wall	S.F.	0.017
6" wall	"	0.017
8" wall	"	0.017
Sandwich panel, 2.5" cladding panel, 2" insulation		
5" wall	S.F.	0.017
6" wall	"	0.017
8" wall	"	0.017
03400.90 PRECAST SPECIALTIES		
Precast concrete, coping, 4' to 8' long		
12" wide	L.F.	0.060
10" wide	"	0.069
Splash block, 30"x12"x4"	EA.	0.400
Stair unit, per riser	"	0.400
Sun screen and trellis, 8' long, 12" high		
4" thick blades	EA.	0.300
5" thick blades	"	0.300
6" thick blades	"	0.320
8" thick blades	"	0.320
Bearing pads for precast members, 2" wide strips		
1/8" thick	L.F.	0.003
1/4" thick	"	0.003
1/2" thick	"	0.003
3/4" thick	"	0.004
1" thick	"	0.004
1-1/2" thick	"	0.004

CEMENTITOUS TOPPINGS	UNIT	MAN/HOURS
03550.10 CONCRETE TOPPINGS		
Gypsum fill		
2" thick	S.F.	0.005
2-1/2" thick	"	0.005
3" thick	"	0.005
3-1/2" thick	"	0.005
4" thick	"	0.006
Formboard		
Mineral fiber board		
1" thick	S.F.	0.020

03 CONCRETE

CEMENTITOUS TOPPINGS

03550.10 CONCRETE TOPPINGS

CEMENTITOUS TOPPINGS	UNIT	MAN/HOURS
1-1/2" thick	S.F.	0.023
Cement fiber board		
1" thick	S.F.	0.027
1-1/2" thick	"	0.031
Glass fiber board		
1" thick	S.F.	0.020
1-1/2" thick	"	0.023
Poured deck		
Vermiculite or perlite		
1 to 4 mix	C.Y.	0.800
1 to 6 mix	"	0.738
Vermiculite or perlite		
2" thick		
1 to 4 mix	S.F.	0.005
1 to 6 mix	"	0.005
3" thick		
1 to 4 mix	S.F.	0.007
1 to 6 mix	"	0.007
Concrete plank, lightweight		
2" thick	S.F.	0.024
2-1/2" thick	"	0.024
3-1/2" thick	"	0.027
4" thick	"	0.027
Channel slab, lightweight, straight		
2-3/4" thick	S.F.	0.024
3-1/2" thick	"	0.024
3-3/4" thick	"	0.024
4-3/4" thick	"	0.027
Gypsum plank		
2" thick	S.F.	0.024
3" thick	"	0.024
Cement fiber, T and G planks		
1" thick	S.F.	0.022
1-1/2" thick	"	0.022
2" thick	"	0.024
2-1/2" thick	"	0.024
3" thick	"	0.024
3-1/2" thick	"	0.027
4" thick	"	0.027
Bulb tees, sub-purlins and grout		
6' span	S.F.	0.027
8' span	"	0.024

GROUT

03600.10 GROUTING

GROUT	UNIT	MAN/HOURS
Grouting for bases		
Nonshrink		
Metallic grout		
1" deep	S.F.	0.160
2" deep	"	0.178
Non-metallic grout		
1" deep	S.F.	0.160
2" deep	"	0.178
Fluid type		
Non-metallic		
1" deep	S.F.	0.160
2" deep	"	0.178
Grouting for joints		
Portland cement grout (1 cement to 3 sand, by volume)		
1/2" joint thickness		
6" wide joints	L.F.	0.027
8" wide joints	"	0.032
1" joint thickness		
4" wide joints	L.F.	0.025
6" wide joints	"	0.028
8" wide joints	"	0.033
Nonshrink, nonmetallic grout		
1/2" joint thickness		
4" wide joint	L.F.	0.023
6" wide joint	"	0.027
8" wide joint	"	0.032
1" joint thickness		
4" wide joint	L.F.	0.025
6" wide joint	"	0.028
8" wide joint	"	0.033

CONCRETE RESTORATION

03730.10 CONCRETE REPAIR

CONCRETE RESTORATION	UNIT	MAN/HOURS
Epoxy grout floor patch, 1/4" thick	S.F.	0.080
Epoxy gel grout	"	0.800
Injection valve, 1 way, threaded plastic	EA.	0.160
Grout crack seal, 2 component	C.F.	0.800
Grout, non shrink	"	0.800
Concrete, epoxy modified		
Sand mix	C.F.	0.320

CONCRETE RESTORATION	UNIT	MAN/ HOURS
03730.10 CONCRETE REPAIR		
Gravel mix	C.F.	0.296
Concrete repair		
Soffit repair		
16" wide	L.F.	0.160
18" wide	"	0.167
24" wide	"	0.178
30" wide	"	0.190
32" wide	"	0.200
Edge repair		
2" spall	L.F.	0.200
3" spall	"	0.211
4" spall	"	0.216
6" spall	"	0.222
8" spall	"	0.235
9" spall	"	0.267
Crack repair, 1/8" crack	"	0.080
Reinforcing steel repair		
1 bar, 4 ft		
#4 bar	L.F.	0.100
#5 bar	"	0.100
#6 bar	"	0.107
#8 bar	"	0.107
#9 bar	"	0.114
#11 bar	"	0.114
Pile repairs		
Polyethylene wrap		
30 mil thick		
60" wide	S.F.	0.267
72" wide	"	0.320
60 mil thick		
60" wide	S.F.	0.267
80" wide	"	0.364
Pile spall, average repair 3'		
18" x 18"	EA.	0.667
20" x 20"	"	0.800

04 MASONRY

MORTAR AND GROUT	UNIT	MAN/ HOURS
04100.10 MASONRY GROUT		
Grout, non shrink, non-metallic, trowelable	C.F.	0.016
Grout door frame, hollow metal		
Single	EA.	0.600
Double	"	0.632
Grout-filled concrete block (CMU)		
4" wide	S.F.	0.020
6" wide	"	0.022
8" wide	"	0.024
12" wide	"	0.025
Grout-filled individual cmu cells		
4" wide	L.F.	0.012
6" wide	"	0.012
8" wide	"	0.012
10" wide	"	0.014
12" wide	"	0.014
Bond beams or lintels, 8" deep		
6" thick	L.F.	0.022
8" thick	"	0.024
10" thick	"	0.027
12" thick	"	0.030
Cavity walls		
2" thick	S.F.	0.032
3" thick	"	0.032
4" thick	"	0.034
6" thick	"	0.040
04150.10 MASONRY ACCESSORIES		
Foundation vents	EA.	0.320
Bar reinforcing		
Horizontal		
#3 - #4	Lb.	0.032
#5 - #6	"	0.027
Vertical		
#3 - #4	Lb.	0.040
#5 - #6	"	0.032
Horizontal joint reinforcing		
Truss type		
4" wide, 6" wall	L.F.	0.003
6" wide, 8" wall	"	0.003
8" wide, 10" wall	"	0.003
10" wide, 12" wall	"	0.004
12" wide, 14" wall	"	0.004
Ladder type		
4" wide, 6" wall	L.F.	0.003
6" wide, 8" wall	"	0.003
8" wide, 10" wall	"	0.003
10" wide, 12" wall	"	0.003
Rectangular wall ties		

MORTAR AND GROUT	UNIT	MAN/ HOURS
04150.10 MASONRY ACCESSORIES		
3/16" dia., galvanized		
2" x 6"	EA.	0.013
2" x 8"	"	0.013
2" x 10"	"	0.013
2" x 12"	"	0.013
4" x 6"	"	0.016
4" x 8"	"	0.016
4" x 10"	"	0.016
4" x 12"	"	0.016
1/4" dia., galvanized		
2" x 6"	EA.	0.013
2" x 8"	"	0.013
2" x 10"	"	0.013
2" x 12"	"	0.013
4" x 6"	"	0.016
4" x 8"	"	0.016
4" x 10"	"	0.016
4" x 12"	"	0.016
"Z" type wall ties, galvanized		
6" long		
1/8" dia.	EA.	0.013
3/16" dia.	"	0.013
1/4" dia.	"	0.013
8" long		
1/8" dia.	EA.	0.013
3/16" dia.	"	0.013
1/4" dia.	"	0.013
10" long		
1/8" dia.	EA.	0.013
3/16" dia.	"	0.013
1/4" dia.	"	0.013
Dovetail anchor slots		
Galvanized steel, filled		
24 ga.	L.F.	0.020
20 ga.	"	0.020
16 oz. copper, foam filled	"	0.020
Dovetail anchors		
16 ga.		
3-1/2" long	EA.	0.013
5-1/2" long	"	0.013
12 ga.		
3-1/2" long	EA.	0.013
5-1/2" long	"	0.013
Dovetail, triangular galvanized ties, 12 ga.		
3" x 3"	EA.	0.013
5" x 5"	"	0.013
7" x 7"	"	0.013
7" x 9"	"	0.013
Brick anchors		

MORTAR AND GROUT

04150.10 MASONRY ACCESSORIES	UNIT	MAN/ HOURS
Corrugated, 3-1/2" long		
16 ga.	EA.	0.013
12 ga.	"	0.013
Non-corrugated, 3-1/2" long		
16 ga.	EA.	0.013
12 ga.	"	0.013
Cavity wall anchors, corrugated, galvanized		
5" long		
16 ga.	EA.	0.013
12 ga.	"	0.013
7" long		
28 ga.	EA.	0.013
24 ga.	"	0.013
22 ga.	"	0.013
16 ga.	"	0.013
Mesh ties, 16 ga., 3" wide		
8" long	EA.	0.013
12" long	"	0.013
20" long	"	0.013
24" long	"	0.013

04150.20 MASONRY CONTROL JOINTS	UNIT	MAN/ HOURS
Control joint, cross shaped PVC	L.F.	0.020
Closed cell joint filler		
1/2"	L.F.	0.020
3/4"	"	0.020
Rubber, for		
4" wall	L.F.	0.020
6" wall	"	0.021
8" wall	"	0.022
PVC, for		
4" wall	L.F.	0.020
6" wall	"	0.021
8" wall	"	0.022

04150.50 MASONRY FLASHING	UNIT	MAN/ HOURS
Through-wall flashing		
5 oz. coated copper	S.F.	0.067
0.030" elastomeric	"	0.053

UNIT MASONRY

04210.10 BRICK MASONRY	UNIT	MAN/ HOURS
Standard size brick, running bond		
Face brick, red (6.4/sf)		
Veneer	S.F.	0.133
Cavity wall	"	0.114
9" solid wall	"	0.229
Common brick (6.4/sf)		
Select common for veneers	S.F.	0.133
Back-up		
4" thick	S.F.	0.100
8" thick	"	0.160
Firewall		
12" thick	S.F.	0.267
16" thick	"	0.364
Glazed brick (7.4/sf)		
Veneer	S.F.	0.145
Buff or gray face brick (6.4/sf)		
Veneer	S.F.	0.133
Cavity wall	"	0.114
Jumbo or oversize brick (3/sf)		
4" veneer	S.F.	0.080
4" back-up	"	0.067
8" back-up	"	0.114
12" firewall	"	0.200
16" firewall	"	0.267
Norman brick, red face, (4.5/sf)		
4" veneer	S.F.	0.100
Cavity wall	"	0.089
Chimney, standard brick, including flue		
16" x 16"	L.F.	0.800
16" x 20"	"	0.800
16" x 24"	"	0.800
20" x 20"	"	1.000
20" x 24"	"	1.000
20" x 32"	"	1.143
Window sill, face brick on edge	"	0.200

04210.20 STRUCTURAL TILE	UNIT	MAN/ HOURS
Structural glazed tile		
6T series, 5-1/2" x 12"		
Glazed on one side		
2" thick	S.F.	0.080
4" thick	"	0.080
6" thick	"	0.089
8" thick	"	0.100
Glazed on two sides		
4" thick	S.F.	0.100
6" thick	"	0.114
Special shapes		

UNIT MASONRY	UNIT	MAN/ HOURS
04210.20 STRUCTURAL TILE		
Group 1	S.F.	0.160
Group 2	"	0.160
Group 3	"	0.160
Group 4	"	0.160
Group 5	"	0.160
Fire rated		
4" thick, 1 hr rating	S.F.	0.080
6" thick, 2 hr rating	"	0.089
8W series, 8" x 16"		
Glazed on one side		
2" thick	S.F.	0.053
4" thick	"	0.053
6" thick	"	0.062
8" thick	"	0.062
Glazed on two sides		
4" thick	S.F.	0.067
6" thick	"	0.080
8" thick	"	0.080
Special shapes		
Group 1	S.F.	0.114
Group 2	"	0.114
Group 3	"	0.114
Group 4	"	0.114
Group 5	"	0.114
Fire rated		
4" thick, 1 hr rating	S.F.	0.114
6" thick, 2 hr rating	"	0.114
04210.60 PAVERS, MASONRY		
Brick walk laid on sand, sand joints		
Laid flat, (4.5 per sf)	S.F.	0.089
Laid on edge, (7.2 per sf)	"	0.133
Precast concrete patio blocks		
2" thick		
Natural	S.F.	0.027
Colors	"	0.027
Exposed aggregates, local aggregate		
Natural	S.F.	0.027
Colors	"	0.027
Granite or limestone aggregate	"	0.027
White tumblestone aggregate	"	0.027
Stone pavers, set in mortar		
Bluestone		
1" thick		
Irregular	S.F.	0.200
Snapped rectangular	"	0.160
1-1/2" thick, random rectangular	"	0.200
2" thick, random rectangular	"	0.229

UNIT MASONRY	UNIT	MAN/ HOURS
04210.60 PAVERS, MASONRY		
Slate		
Natural cleft		
Irregular, 3/4" thick	S.F.	0.229
Random rectangular		
1-1/4" thick	S.F.	0.200
1-1/2" thick	"	0.222
Granite blocks		
3" thick, 3" to 6" wide		
4" to 12" long	S.F.	0.267
6" to 15" long	"	0.229
Crushed stone, white marble, 3" thick	"	0.016
04220.10 CONCRETE MASONRY UNITS		
Hollow, load bearing		
4"	S.F.	0.059
6"	"	0.062
8"	"	0.067
10"	"	0.073
12"	"	0.080
Solid, load bearing		
4"	S.F.	0.059
6"	"	0.062
8"	"	0.067
10"	"	0.073
12"	"	0.080
Back-up block, 8" x 16"		
2"	S.F.	0.046
4"	"	0.047
6"	"	0.050
8"	"	0.053
10"	"	0.057
12"	"	0.062
Foundation wall, 8" x 16"		
6"	S.F.	0.057
8"	"	0.062
10"	"	0.067
12"	"	0.073
Solid		
6"	S.F.	0.062
8"	"	0.067
10"	"	0.073
12"	"	0.080
Exterior, styrofoam inserts, standard weight, 8" x 16"		
6"	S.F.	0.062
8"	"	0.067
10"	"	0.073
12"	"	0.080
Lightweight		

04 MASONRY

UNIT MASONRY

04220.10 CONCRETE MASONRY UNITS

	UNIT	MAN/ HOURS
6"	S.F.	0.062
8"	"	0.067
10"	"	0.073
12"	"	0.080
Acoustical slotted block		
4"	S.F.	0.073
6"	"	0.073
8"	"	0.080
Filled cavities		
4"	S.F.	0.089
6"	"	0.094
8"	"	0.100
Hollow, split face		
4"	S.F.	0.059
6"	"	0.062
8"	"	0.067
10"	"	0.073
12"	"	0.080
Split rib profile		
4"	S.F.	0.073
6"	"	0.073
8"	"	0.080
10"	"	0.080
12"	"	0.080
High strength block, 3500 psi		
2"	S.F.	0.059
4"	"	0.062
6"	"	0.062
8"	"	0.067
10"	"	0.073
12"	"	0.080
Solar screen concrete block		
4" thick		
6" x 6"	S.F.	0.178
8" x 8"	"	0.160
12" x 12"	"	0.123
8" thick		
8" x 16"	S.F.	0.114
Glazed block, cove base, glazed 1 side		
2"	L.F.	0.089
4"	"	0.089
6"	"	0.100
8"	"	0.100
Single face		
2"	S.F.	0.067
4"	"	0.067
6"	"	0.073
8"	"	0.080
10"	"	0.089
12"	S.F.	0.094
Double face		
4"	S.F.	0.084
6"	"	0.089
8"	"	0.100
Corner or bullnose		
2"	EA.	0.100
4"	"	0.114
6"	"	0.114
8"	"	0.133
10"	"	0.145
12"	"	0.160
Gypsum unit masonry		
Partition blocks (12"x30")		
Solid		
2"	S.F.	0.032
Hollow		
3"	S.F.	0.032
4"	"	0.033
6"	"	0.036
Vertical reinforcing		
4' o.c., add 5% to labor		
2'8" o.c., add 15% to labor		
Interior partitions, add 10% to labor		

04220.90 BOND BEAMS & LINTELS

	UNIT	MAN/ HOURS
Bond beam, no grout or reinforcement		
8" x 16" x		
4" thick	L.F.	0.062
6" thick	"	0.064
8" thick	"	0.067
10" thick	"	0.070
12" thick	"	0.073
Beam lintel, no grout or reinforcement		
8" x 16" x		
10" thick	L.F.	0.080
12" thick	"	0.089
Precast masonry lintel		
6 lf, 8" high x		
4" thick	L.F.	0.133
6" thick	"	0.133
8" thick	"	0.145
10" thick	"	0.145
10 lf, 8" high x		
4" thick	L.F.	0.080
6" thick	"	0.080
8" thick	"	0.089
10" thick	"	0.089
Steel angles and plates		

UNIT MASONRY

04220.90 BOND BEAMS & LINTELS	UNIT	MAN/HOURS
Minimum	Lb.	0.011
Maximum	"	0.020
Various size angle lintels		
1/4" stock		
3" x 3"	L.F.	0.050
3" x 3-1/2"	"	0.050
3/8" stock		
3" x 4"	L.F.	0.050
3-1/2" x 4"	"	0.050
4" x 4"	"	0.050
5" x 3-1/2"	"	0.050
6" x 3-1/2"	"	0.050
1/2" stock		
6" x 4"	L.F.	0.050

04240.10 CLAY TILE	UNIT	MAN/HOURS
Hollow clay tile, for back-up, 12" x 12"		
Scored face		
Load bearing		
4" thick	S.F.	0.057
6" thick	"	0.059
8" thick	"	0.062
10" thick	"	0.064
12" thick	"	0.067
Non-load bearing		
3" thick	S.F.	0.055
4" thick	"	0.057
6" thick	"	0.059
8" thick	"	0.062
12" thick	"	0.067
Partition, 12" x 12"		
In walls		
3" thick	S.F.	0.067
4" thick	"	0.067
6" thick	"	0.070
8" thick	"	0.073
10" thick	"	0.076
12" thick	"	0.080
Clay tile floors		
4" thick	S.F.	0.044
6" thick	"	0.047
8" thick	"	0.050
10" thick	"	0.053
12" thick	"	0.057
Terra cotta		
Coping, 10" or 12" wide, 3" thick	L.F.	0.160

UNIT MASONRY

04270.10 GLASS BLOCK	UNIT	MAN/HOURS
Glass block, 4" thick		
6" x 6"	S.F.	0.267
8" x 8"	"	0.200
12" x 12"	"	0.160
Replacement glass blocks, 4" x 8" x 8"		
Minimum	S.F.	0.800
Maximum	"	1.600

04295.10 PARGING/MASONRY PLASTER	UNIT	MAN/HOURS
Parging		
1/2" thick	S.F.	0.053
3/4" thick	"	0.067
1" thick	"	0.080

STONE

04400.10 STONE	UNIT	MAN/HOURS
Rubble stone		
Walls set in mortar		
8" thick	S.F.	0.200
12" thick	"	0.320
18" thick	"	0.400
24" thick	"	0.533
Dry set wall		
8" thick	S.F.	0.133
12" thick	"	0.200
18" thick	"	0.267
24" thick	"	0.320
Cut stone		
Imported marble		
Facing panels		
3/4" thick	S.F.	0.320
1-1/2" thick	"	0.364
2-1/4" thick	"	0.444
Base		
1" thick		
4" high	L.F.	0.400
6" high	"	0.400
Columns, solid		
Plain faced	C.F.	16.000

STONE	UNIT	MAN/HOURS
04400.10 STONE		
Fluted	C.F.	16.000
Flooring, travertine, minimum	S.F.	0.123
Average	"	0.160
Maximum	"	0.178
Domestic marble		
Facing panels		
7/8" thick	S.F.	0.320
1-1/2" thick	"	0.364
2-1/4" thick	"	0.444
Stairs		
12" treads	L.F.	0.400
6" risers	"	0.267
Thresholds, 7/8" thick, 3' long, 4" to 6" wide		
Plain	EA.	0.667
Beveled	"	0.667
Window sill		
6" wide, 2" thick	L.F.	0.320
Stools		
5" wide, 7/8" thick	L.F.	0.320
Limestone panels up to 12' x 5', smooth finish		
2" thick	S.F.	0.096
3" thick	"	0.096
4" thick	"	0.096
Miscellaneous limestone items		
Steps, 14" wide, 6" deep	L.F.	0.533
Coping, smooth finish	C.F.	0.267
Sills, lintels, jambs, smooth finish	"	0.320
Granite veneer facing panels, polished		
7/8" thick		
Black	S.F.	0.320
Gray	"	0.320
Base		
4" high	L.F.	0.160
6" high	"	0.178
Curbing, straight, 6" x 16"	"	0.400
Radius curbs, radius over 5'	"	0.533
Ashlar veneer		
4" thick, random	S.F.	0.320
Pavers, 4" x 4" split		
Gray	S.F.	0.160
Pink	"	0.160
Black	"	0.160
Slate, panels		
1" thick	S.F.	0.320
2" thick	"	0.364
Sills or stools		
1" thick		
6" wide	L.F.	0.320
10" wide	"	0.348

STONE	UNIT	MAN/HOURS
04400.10 STONE		
2" thick		
6" wide	L.F.	0.364
10" wide	"	0.400

MASONRY RESTORATION	UNIT	MAN/HOURS
04520.10 RESTORATION AND CLEANING		
Masonry cleaning		
Washing brick		
Smooth surface	S.F.	0.013
Rough surface	"	0.018
Steam clean masonry		
Smooth face		
Minimum	S.F.	0.010
Maximum	"	0.015
Rough face		
Minimum	S.F.	0.013
Maximum	"	0.020
Sandblast masonry		
Minimum	S.F.	0.016
Maximum	"	0.027
Pointing masonry		
Brick	S.F.	0.032
Concrete block	"	0.023
Cut and repoint		
Brick		
Minimum	S.F.	0.040
Maximum	"	0.080
Stone work	L.F.	0.062
Cut and recaulk		
Oil base caulks	L.F.	0.053
Butyl caulks	"	0.053
Polysulfides and acrylics	"	0.053
Silicones	"	0.053
Cement and sand grout on walls, to 1/8" thick		
Minimum	S.F.	0.032
Maximum	"	0.040
Brick removal and replacement		
Minimum	EA.	0.100
Average	"	0.133

04 MASONRY

MASONRY RESTORATION	UNIT	MAN/ HOURS
04520.10 RESTORATION AND CLEANING		
Maximum	EA.	0.400
04550.10 REFRACTORIES		
Flue liners		
Rectangular		
8" x 12"	L.F.	0.133
12" x 12"	"	0.145
12" x 18"	"	0.160
16" x 16"	"	0.178
18" x 18"	"	0.190
20" x 20"	"	0.200
24" x 24"	"	0.229
Round		
18" dia.	L.F.	0.190
24" dia.	"	0.229

05 METALS

METAL FASTENING

05050.10 STRUCTURAL WELDING

Item	UNIT	MAN/HOURS
Welding		
Single pass		
1/8"	L.F.	0.040
3/16"	"	0.053
1/4"	"	0.067
Miscellaneous steel shapes		
Plain	Lb.	0.002
Galvanized	"	0.003
Plates		
Plain	Lb.	0.002
Galvanized	"	0.003

05050.95 METAL LINTELS

Item	UNIT	MAN/HOURS
Lintels, steel		
Plain	Lb.	0.020
Galvanized	"	0.020

05120.10 STRUCTURAL STEEL

Item	UNIT	MAN/HOURS
Beams and girders, A-36		
Welded	TON	4.800
Bolted	"	4.364
Columns		
Pipe		
6" dia.	Lb.	0.005
12" dia.	"	0.004
Purlins and girts		
Welded	TON	8.000
Bolted	"	6.857
Column base plates		
Up to 150 lb each	Lb.	0.005
Over 150 lb each	"	0.007
Structural pipe		
3" to 5" o.d.	TON	9.600
6" to 12" o.d.	"	6.857
Structural tube		
6" square		
Light sections	TON	9.600
Heavy sections	"	6.857
6" wide rectangular		
Light sections	TON	8.000
Heavy sections	"	6.000
Greater than 6" wide rectangular		
Light sections	TON	8.000
Heavy sections	"	6.000
Miscellaneous structural shapes		
Steel angle	TON	12.000
Steel plate	"	8.000

METAL FASTENING

05120.10 STRUCTURAL STEEL

Item	UNIT	MAN/HOURS
Trusses, field welded		
60 lb/lf	TON	6.000
100 lb/lf	"	4.800
150 lb/lf	"	4.000
Bolted		
60 lb/lf	TON	5.333
100 lb/lf	"	4.364
150 lb/lf	"	3.692

05200.10 METAL JOISTS

Item	UNIT	MAN/HOURS
Joist		
DLH series	TON	3.200
K series	"	3.200
LH series	"	3.200

05300.10 METAL DECKING

Item	UNIT	MAN/HOURS
Roof, 1-1/2" deep, non-composite		
16 ga.		
Primed	S.F.	0.008
Galvanized	"	0.008
18 ga.		
Primed	S.F.	0.008
Galvanized	"	0.008
20 ga.		
Primed	S.F.	0.008
Galvanized	"	0.008
22 ga.		
Primed	S.F.	0.008
Galvanized	"	0.008
Open type decking, galvanized		
1-1/2" deep		
18 ga.	S.F.	0.008
20 ga.	"	0.008
22 ga.	"	0.008
3" deep		
16 ga.	S.F.	0.009
18 ga.	"	0.009
20 ga.	"	0.009
22 ga.	"	0.009
4-1/2" deep		
16 ga.	S.F.	0.010
18 ga.	"	0.010
6" deep		
16 ga.	S.F.	0.011
18 ga.	"	0.011
7-1/2" deep		
16 ga.	S.F.	0.011
18 ga.	"	0.011
Cellular type		

05 METALS

METAL FASTENING

05300.10 METAL DECKING	UNIT	MAN/ HOURS
1-1/2" deep, galvanized		
18-18 ga.	S.F.	0.010
22-18 ga.	"	0.010
3" deep, galvanized		
16-16 ga.	S.F.	0.011
18-16 ga.	"	0.011
18-18 ga.	"	0.011
20-18 ga.	"	0.011
4-1/2" deep, galvanized		
16-16 ga.	S.F.	0.011
18-16 ga.	"	0.011
18-18 ga.	"	0.011
20-18 ga.	"	0.011
Composite deck, non-cellular, galvanized		
1-1/2" deep		
18 ga.	S.F.	0.009
20 ga.	"	0.009
22 ga.	"	0.009
3" deep		
18 ga.	S.F.	0.009
20 ga.	"	0.009
22 ga.	"	0.009
Slab form floor deck		
9/16" deep		
28 ga.	S.F.	0.009
1-5/16" deep		
24 ga.	S.F.	0.009
22 ga.	"	0.009

COLD FORMED FRAMING

05410.10 METAL FRAMING	UNIT	MAN/ HOURS
Furring channel, galvanized		
Beams and columns, 3/4"		
12" o.c.	S.F.	0.080
16" o.c.	"	0.073
Walls, 3/4"		
12" o.c.	S.F.	0.040
16" o.c.	"	0.033
24" o.c.	"	0.027
1-1/2"		

COLD FORMED FRAMING

05410.10 METAL FRAMING	UNIT	MAN/ HOURS
12" o.c.	S.F.	0.040
16" o.c.	"	0.033
24" o.c.	"	0.027
Stud, load bearing		
16" o.c.		
16 ga.		
2-1/2"	S.F.	0.036
3-5/8"	"	0.036
4"	"	0.036
6"	"	0.040
18 ga.		
2-1/2"	S.F.	0.036
3-5/8"	"	0.036
4"	"	0.036
6"	"	0.040
8"	"	0.040
20 ga.		
2-1/2"	S.F.	0.036
3-5/8"	"	0.036
4"	"	0.036
6"	"	0.040
8"	"	0.040
24" o.c.		
16 ga.		
2-1/2"	S.F.	0.031
3-5/8"	"	0.031
4"	"	0.031
6"	"	0.033
8"	"	0.033
18 ga.		
2-1/2"	S.F.	0.031
3-5/8"	"	0.031
4"	"	0.031
6"	"	0.033
8"	"	0.033
20 ga.		
2-1/2"	S.F.	0.031
3-5/8"	"	0.031
4"	"	0.031
6"	"	0.033
8"	"	0.033

05 METALS

METAL FABRICATIONS	UNIT	MAN/ HOURS
05510.10 STAIRS		
Stock unit, steel, complete, per riser		
Tread		
3'-6" wide	EA.	1.000
4' wide	"	1.143
5' wide	"	1.333
Metal pan stair, cement filled, per riser		
3'-6" wide	EA.	0.800
4' wide	"	0.889
5' wide	"	1.000
Landing, steel pan	S.F.	0.200
Cast iron tread, steel stringers, stock units, per riser		
Tread		
3'-6" wide	EA.	1.000
4' wide	"	1.143
5' wide	"	1.333
Stair treads, abrasive, 12" x 3'-6"		
Cast iron		
3/8"	EA.	0.400
1/2"	"	0.400
Cast aluminum		
5/16"	EA.	0.400
3/8"	"	0.400
1/2"	"	0.400
05515.10 LADDERS		
Ladder, 18" wide		0.533
With cage	L.F.	0.533
Without cage	"	0.400
05520.10 RAILINGS		
Railing, pipe		
1-1/4" diameter, welded steel		
2-rail		
Primed	L.F.	0.160
Galvanized	"	0.160
3-rail		
Primed	L.F.	0.200
Galvanized	"	0.200
Wall mounted, single rail, welded steel		
Primed	L.F.	0.123
Galvanized	"	0.123
1-1/2" diameter, welded steel		
2-rail		
Primed	L.F.	0.160
Galvanized	"	0.160
3-rail		
Primed	L.F.	0.200

METAL FABRICATIONS	UNIT	MAN/ HOURS
05520.10 RAILINGS		
Galvanized	L.F.	0.200
Wall mounted, single rail, welded steel		
Primed	L.F.	0.123
Galvanized	"	0.123
2" diameter, welded steel		
2-rail		
Primed	L.F.	0.178
Galvanized	"	0.178
3-rail		
Primed	L.F.	0.229
Galvanized	"	0.229
Wall mounted, single rail, welded steel		
Primed	L.F.	0.133
Galvanized	"	0.133
05530.10 METAL GRATING		
Floor plate, checkered, steel		
1/4"		
Primed	S.F.	0.011
Galvanized	"	0.011
3/8"		
Primed	S.F.	0.012
Galvanized	"	0.012
Aluminum grating, pressure-locked bearing bars		
3/4" x 1/8"	S.F.	0.020
1" x 1/8"	"	0.020
1-1/4" x 1/8"	"	0.020
1-1/4" x 3/16"	"	0.020
1-1/2" x 1/8"	"	0.020
1-3/4" x 3/16"	"	0.020
Miscellaneous expenses		
Cutting		
Minimum	L.F.	0.053
Maximum	"	0.080
Banding		
Minimum	L.F.	0.133
Maximum	"	0.160
Toe plates		
Minimum	L.F.	0.160
Maximum	"	0.200
Steel grating, primed		
3/4" x 1/8"	S.F.	0.027
1" x 1/8"	"	0.027
1-1/4" x 1/8"	"	0.027
1-1/4" x 3/16"	"	0.027
1-1/2" x 1/8"	"	0.027
1-3/4" x 3/16"	"	0.027
Galvanized		
3/4" x 1/8"	S.F.	0.027

05 METALS

METAL FABRICATIONS	UNIT	MAN/HOURS
05530.10 METAL GRATING		
1" x 1/8"	S.F.	0.027
1-1/4" x 1/8"	"	0.027
1-1/4" x 3/16"	"	0.027
1-1/2" x 1/8"	"	0.027
1-3/4" x 3/16"	"	0.027
Miscellaneous expenses		
Cutting		
Minimum	L.F.	0.057
Maximum	"	0.089
Banding		
Minimum	L.F.	0.145
Maximum	"	0.178
Toe plates		
Minimum	L.F.	0.178
Maximum	"	0.229
05540.10 CASTINGS		
Miscellaneous castings		
Light sections	Lb.	0.016
Heavy sections	"	0.011
Manhole covers and frames		
Regular, city type		
18" dia.		
100 lb	EA.	1.600
24" dia.		
200 lb	EA.	1.600
300 lb	"	1.778
400 lb	"	1.778
26" dia., 475 lb	"	2.000
30" dia., 600 lb	"	2.286
8" square, 75 lb	"	0.320
24" square		
126 lb	EA.	1.600
500 lb	"	2.000
Watertight type		
20" dia., 200 lb	EA.	2.000
24" dia., 350 lb	"	2.667
Steps, cast iron		
7" x 9"	EA.	0.133
8" x 9"	"	0.133
Manhole covers and frames, aluminum		
12" x 12"	EA.	0.320
18" x 18"	"	0.320
24" x 24"	"	0.400
Corner protection		
Steel angle guard with anchors		
2" x 2" x 3/16"	L.F.	0.114
2" x 3" x 1/4"	"	0.114

METAL FABRICATIONS	UNIT	MAN/HOURS
05540.10 CASTINGS		
3" x 3" x 5/16"	L.F.	0.114
3" x 4" x 5/16"	"	0.123
4" x 4" x 5/16"	"	0.123

MISC. FABRICATIONS	UNIT	MAN/HOURS
05580.10 METAL SPECIALTIES		
Kick plate		
4" high x 1/4" thick		
Primed	L.F.	0.160
Galvanized	"	0.160
6" high x 1/4" thick		
Primed	L.F.	0.178
Galvanized	"	0.178
05700.10 ORNAMENTAL METAL		
Railings, vertical square bars, 6" o.c., with shaped top rails		
Steel	L.F.	0.400
Aluminum	"	0.400
Bronze	"	0.533
Stainless steel	"	0.533
Laminated metal or wood handrails with metal supports		
2-1/2" round or oval shape	L.F.	0.400
Grilles and louvers		
Fixed type louvers		
4 through 10 sf	S.F.	0.133
Over 10 sf	"	0.100
Movable type louvers		
4 through 10 sf	S.F.	0.133
Over 10 sf	"	0.100
Aluminum louvers		
Residential use, fixed type, with screen		
8" x 8"	EA.	0.400
12" x 12"	"	0.400
12" x 18"	"	0.400
14" x 24"	"	0.400
18" x 24"	"	0.400
30" x 24"	"	0.444

MISC. FABRICATIONS	UNIT	MAN/ HOURS
05800.10 EXPANSION CONTROL		
Expansion joints with covers, floor assembly type		
With 1" space		
Aluminum	L.F.	0.133
Bronze	"	0.133
Stainless steel	"	0.133
With 2" space		
Aluminum	L.F.	0.133
Bronze	"	0.133
Stainless steel	"	0.133
Ceiling and wall assembly type		
With 1" space		
Aluminum	L.F.	0.160
Bronze	"	0.160
Stainless steel	"	0.160
With 2" space		
Aluminum	L.F.	0.160
Bronze	"	0.160
Stainless steel	"	0.160
Exterior roof and wall, aluminum		
Roof to roof		
With 1" space	L.F.	0.133
With 2" space	"	0.133
Roof to wall		
With 1" space	L.F.	0.145
With 2" space	"	0.145
Flat wall to wall		
With 1" space	L.F.	0.133
With 2" space	"	0.133
Corner to flat wall		
With 1" space	L.F.	0.160
With 2" in space	"	0.160

FASTENERS AND ADHESIVES

06050.10 ACCESSORIES

	UNIT	MAN/HOURS
Column/post base, cast aluminum		
4" x 4"	EA.	0.200
6" x 6"	"	0.200
Bridging, metal, per pair		
12" o.c.	EA.	0.080
16" o.c.	"	0.073
Anchors		
Bolts, threaded two ends, with nuts and washers		
1/2" dia.		
4" long	EA.	0.050
7-1/2" long	"	0.050
3/4" dia.		
7-1/2" long	EA.	0.050
15" long	"	0.050
Framing anchors		
10 gauge	EA.	0.067
Bolts, carriage		
1/4 x 4	EA.	0.080
5/16 x 6	"	0.084
3/8 x 6	"	0.084
1/2 x 6	"	0.084
Joist and beam hangers		
18 ga.		
2 x 4	EA.	0.080
2 x 6	"	0.080
2 x 8	"	0.080
2 x 10	"	0.089
2 x 12	"	0.100
16 ga.		
3 x 6	EA.	0.089
3 x 8	"	0.089
3 x 10	"	0.094
3 x 12	"	0.107
3 x 14	"	0.114
4 x 6	"	0.089
4 x 8	"	0.089
4 x 10	"	0.094
4 x 12	"	0.107
4 x 14	"	0.114
Rafter anchors, 18 ga., 1-1/2" wide		
5-1/4" long	EA.	0.067
10-3/4" long	"	0.067
Shear plates		
2-5/8" dia.	EA.	0.062
4" dia.	"	0.067
Sill anchors		
Embedded in concrete	EA.	0.080
Split rings		
2-1/2" dia.	EA.	0.089
4" dia.	EA.	0.100
Strap ties, 14 ga., 1-3/8" wide		
12" long	EA.	0.067
18" long	"	0.073
24" long	"	0.080
36" long	"	0.089
Toothed rings		
2-5/8" dia.	EA.	0.133
4" dia.	"	0.160

ROUGH CARPENTRY

06110.10 BLOCKING

	UNIT	MAN/HOURS
Steel construction		
Walls		
2x4	L.F.	0.053
2x6	"	0.062
2x8	"	0.067
2x10	"	0.073
2x12	"	0.080
Ceilings		
2x4	L.F.	0.062
2x6	"	0.073
2x8	"	0.080
2x10	"	0.089
2x12	"	0.100
Wood construction		
Walls		
2x4	L.F.	0.044
2x6	"	0.050
2x8	"	0.053
2x10	"	0.057
2x12	"	0.062
Ceilings		
2x4	L.F.	0.050
2x6	"	0.057
2x8	"	0.062
2x10	"	0.067
2x12	"	0.073

ROUGH CARPENTRY

06110.20 CEILING FRAMING

ROUGH CARPENTRY	UNIT	MAN/ HOURS
Ceiling joists		
12" o.c.		
2x4	S.F.	0.019
2x6	"	0.020
2x8	"	0.021
2x10	"	0.022
2x12	"	0.024
16" o.c.		
2x4	S.F.	0.015
2x6	"	0.016
2x8	"	0.017
2x10	"	0.017
2x12	"	0.018
24" o.c.		
2x4	S.F.	0.013
2x6	"	0.013
2x8	"	0.014
2x10	"	0.015
2x12	"	0.016
Headers and nailers		
2x4	L.F.	0.026
2x6	"	0.027
2x8	"	0.029
2x10	"	0.031
2x12	"	0.033
Sister joists for ceilings		
2x4	L.F.	0.057
2x6	"	0.067
2x8	"	0.080
2x10	"	0.100
2x12	"	0.133

06110.30 FLOOR FRAMING

ROUGH CARPENTRY	UNIT	MAN/ HOURS
Floor joists		
12" o.c.		
2x6	S.F.	0.016
2x8	"	0.016
2x10	"	0.017
2x12	"	0.017
2x14	"	0.017
3x6	"	0.017
3x8	"	0.017
3x10	"	0.018
3x12	"	0.019
3x14	"	0.020
4x6	"	0.017
4x8	"	0.017
4x10	"	0.018
4x12	S.F.	0.019
4x14	"	0.020
16" o.c.		
2x6	S.F.	0.013
2x8	"	0.014
2x10	"	0.014
2x12	"	0.014
2x14	"	0.015
3x6	"	0.014
3x8	"	0.014
3x10	"	0.015
3x12	"	0.015
3x14	"	0.016
4x6	"	0.014
4x8	"	0.014
4x10	"	0.015
4x12	"	0.015
4x14	"	0.016
Sister joists for floors		
2x4	L.F.	0.050
2x6	"	0.057
2x8	"	0.067
2x10	"	0.080
2x12	"	0.100
3x6	"	0.080
3x8	"	0.089
3x10	"	0.100
3x12	"	0.114
4x6	"	0.080
4x8	"	0.089
4x10	"	0.100
4x12	"	0.114

06110.40 FURRING

ROUGH CARPENTRY	UNIT	MAN/ HOURS
Furring, wood strips		
Walls		
On masonry or concrete walls		
1x2 furring		
12" o.c.	S.F.	0.025
16" o.c.	"	0.023
24" o.c.	"	0.021
1x3 furring		
12" o.c.	S.F.	0.025
16" o.c.	"	0.023
24" o.c.	"	0.021
On wood walls		
1x2 furring		
12" o.c.	S.F.	0.018
16" o.c.	"	0.016

06 WOOD AND PLASTICS

ROUGH CARPENTRY	UNIT	MAN/ HOURS
06110.40 FURRING		
24" o.c.	S.F.	0.015
1x3 furring		
12" o.c.	S.F.	0.018
16" o.c.	"	0.016
24" o.c.	"	0.015
Ceilings		
On masonry or concrete ceilings		
1x2 furring		
12" o.c.	S.F.	0.044
16" o.c.	"	0.040
24" o.c.	"	0.036
1x3 furring		
12" o.c.	S.F.	0.044
16" o.c.	"	0.040
24" o.c.	"	0.036
On wood ceilings		
1x2 furring		
12" o.c.	S.F.	0.030
16" o.c.	"	0.027
24" o.c.	"	0.024
1x3		
12" o.c.	S.F.	0.030
16" o.c.	"	0.027
24" o.c.	"	0.024
06110.50 ROOF FRAMING		
Roof framing		
Rafters, gable end		
0-2 pitch (flat to 2-in-12)		
12" o.c.		
2x4	S.F.	0.017
2x6	"	0.017
2x8	"	0.018
2x10	"	0.019
2x12	"	0.020
16" o.c.		
2x6	S.F.	0.014
2x8	"	0.015
2x10	"	0.015
2x12	"	0.016
24" o.c.		
2x6	S.F.	0.012
2x8	"	0.013
2x10	"	0.013
2x12	"	0.013
4-6 pitch (4-in-12 to 6-in-12)		
12" o.c.		
2x4	S.F.	0.017
2x6	"	0.018

ROUGH CARPENTRY	UNIT	MAN/ HOURS
06110.50 ROOF FRAMING		
2x8	S.F.	0.019
2x10	"	0.020
2x12	"	0.021
16" o.c.		
2x6	S.F.	0.015
2x8	"	0.015
2x10	"	0.016
2x12	"	0.017
24" o.c.		
2x6	S.F.	0.013
2x8	"	0.013
2x10	"	0.014
2x12	"	0.015
8-12 pitch (8-in-12 to 12-in-12)		
12" o.c.		
2x4	S.F.	0.018
2x6	"	0.019
2x8	"	0.020
2x10	"	0.021
2x12	"	0.022
16" o.c.		
2x6	S.F.	0.015
2x8	"	0.016
2x10	"	0.017
2x12	"	0.017
24" o.c.		
2x6	S.F.	0.013
2x8	"	0.013
2x10	"	0.014
2x12	"	0.014
Ridge boards		
2x6	L.F.	0.040
2x8	"	0.044
2x10	"	0.050
2x12	"	0.057
Hip rafters		
2x6	L.F.	0.029
2x8	"	0.030
2x10	"	0.031
2x12	"	0.032
Jack rafters		
4-6 pitch (4-in-12 to 6-in-12)		
16" o.c.		
2x6	S.F.	0.024
2x8	"	0.024
2x10	"	0.026
2x12	"	0.027
24" o.c.		
2x6	S.F.	0.018

ROUGH CARPENTRY	UNIT	MAN/HOURS
06110.50 ROOF FRAMING		
2x8	S.F.	0.019
2x10	"	0.020
2x12	"	0.020
8-12 pitch (8-in-12 to 12-in-12)		
16" o.c.		
2x6	S.F.	0.025
2x8	"	0.026
2x10	"	0.027
2x12	"	0.028
24" o.c.		
2x6	S.F.	0.019
2x8	"	0.020
2x10	"	0.020
2x12	"	0.021
Sister rafters		
2x4	L.F.	0.057
2x6	"	0.067
2x8	"	0.080
2x10	"	0.100
2x12	"	0.133
Fascia boards		
2x4	L.F.	0.040
2x6	"	0.040
2x8	"	0.044
2x10	"	0.044
2x12	"	0.050
Cant strips		
Fiber		
3x3	L.F.	0.023
4x4	"	0.024
Wood		
3x3	L.F.	0.024
06110.60 SLEEPERS		
Sleepers, over concrete		
12" o.c.		
1x2	S.F.	0.018
1x3	"	0.019
2x4	"	0.022
2x6	"	0.024
16" o.c.		
1x2	S.F.	0.016
1x3	"	0.016
2x4	"	0.019
2x6	"	0.020

ROUGH CARPENTRY	UNIT	MAN/HOURS
06110.65 SOFFITS		
Soffit framing		
2x3	L.F.	0.057
2x4	"	0.062
2x6	"	0.067
2x8	"	0.073
06110.70 WALL FRAMING		
Framing wall, studs		
12" o.c.		
2x3	S.F.	0.015
2x4	"	0.015
2x6	"	0.016
2x8	"	0.017
16" o.c.		
2x3	S.F.	0.013
2x4	"	0.013
2x6	"	0.013
2x8	"	0.014
24" o.c.		
2x3	S.F.	0.011
2x4	"	0.011
2x6	"	0.011
2x8	"	0.012
Plates, top or bottom		
2x3	L.F.	0.024
2x4	"	0.025
2x6	"	0.027
2x8	"	0.029
Headers, door or window		
2x6		
Single		
3' long	EA.	0.400
6' long	"	0.500
Double		
3' long	EA.	0.444
6' long	"	0.571
2x8		
Single		
4' long	EA.	0.500
8' long	"	0.615
Double		
4' long	EA.	0.571
8' long	"	0.727
2x10		
Single		
5' long	EA.	0.615
10' long	"	0.800
Double		

06 WOOD AND PLASTICS

ROUGH CARPENTRY	UNIT	MAN/ HOURS
06110.70 WALL FRAMING		
5' long	EA.	0.667
10' long	"	0.800
2x12		
Single		
6' long	EA.	0.615
12' long	"	0.800
Double		
6' long	EA.	0.727
12' long	"	0.889
06115.10 FLOOR SHEATHING		
Sub-flooring, plywood, CDX		
1/2" thick	S.F.	0.010
5/8" thick	"	0.011
3/4" thick	"	0.013
Structural plywood		
1/2" thick	S.F.	0.010
5/8" thick	"	0.011
3/4" thick	"	0.012
Board type subflooring		
1x6		
Minimum	S.F.	0.018
Maximum	"	0.020
1x8		
Minimum	S.F.	0.017
Maximum	"	0.019
1x10		
Minimum	S.F.	0.016
Maximum	"	0.018
Underlayment		
Hardboard, 1/4" tempered	S.F.	0.010
Plywood		
3/8" thick	S.F.	0.010
1/2" thick	"	0.011
5/8" thick	"	0.011
3/4" thick	"	0.012
06115.20 ROOF SHEATHING		
Sheathing		
3/8" thick	S.F.	0.010
1/2" thick	"	0.011
5/8" thick	"	0.011
3/4" thick	"	0.012
Structural plywood		
3/8" thick	S.F.	0.010
1/2" thick	"	0.011
5/8" thick	"	0.011

ROUGH CARPENTRY	UNIT	MAN/ HOURS
06115.20 ROOF SHEATHING		
3/4" thick	S.F.	0.012
06115.30 WALL SHEATHING		
Sheathing		
Plywood, CDX		
3/8" thick	S.F.	0.012
1/2" thick	"	0.012
5/8" thick	"	0.013
3/4" thick	"	0.015
Waferboard		
3/8" thick	S.F.	0.012
1/2" thick	"	0.012
5/8" thick	"	0.013
3/4" thick	"	0.015
Structural plywood		
3/8" thick	S.F.	0.012
1/2" thick	"	0.012
5/8" thick	"	0.013
3/4" thick	"	0.015
Gypsum, 1/2" thick	"	0.012
Asphalt impregnated fiberboard, 1/2" thick	"	0.012
06125.10 WOOD DECKING		
Decking, T&G solid		
Cedar		
3" thick	S.F.	0.020
4" thick	"	0.021
Fir		
3" thick	S.F.	0.020
4" thick	"	0.021
Southern yellow pine		
3" thick	S.F.	0.023
4" thick	"	0.025
White pine		
3" thick	S.F.	0.020
4" thick	"	0.021
06130.10 HEAVY TIMBER		
Mill framed structures		
Beams to 20' long		
Douglas fir		
6x8	L.F.	0.080
6x10	"	0.083
6x12	"	0.089
6x14	"	0.092
6x16	"	0.096

ROUGH CARPENTRY

06130.10 HEAVY TIMBER

	UNIT	MAN/HOURS
8x10	L.F.	0.083
8x12	"	0.089
8x14	"	0.092
8x16	"	0.096
Southern yellow pine		
6x8	L.F.	0.080
6x10	"	0.083
6x12	"	0.089
6x14	"	0.092
6x16	"	0.096
8x10	"	0.083
8x12	"	0.089
8x14	"	0.092
8x16	"	0.096
Columns to 12' high		
Douglas fir		
6x6	L.F.	0.120
8x8	"	0.120
10x10	"	0.133
12x12	"	0.133
Southern yellow pine		
6x6	L.F.	0.120
8x8	"	0.120
10x10	"	0.133
12x12	"	0.133
Posts, treated		
4x4	L.F.	0.032
6x6	"	0.040

06190.20 WOOD TRUSSES

	UNIT	MAN/HOURS
Truss, fink, 2x4 members		
3-in-12 slope		
24' span	EA.	0.686
26' span	"	0.686
28' span	"	0.727
30' span	"	0.727
34' span	"	0.774
38' span	"	0.774
5-in-12 slope		
24' span	EA.	0.706
28' span	"	0.727
30' span	"	0.750
32' span	"	0.750
40' span	"	0.800
Gable, 2x4 members		
5-in-12 slope		
24' span	EA.	0.706
26' span	"	0.706
28' span	"	0.727
30' span	EA.	0.750
32' span	"	0.750
36' span	"	0.774
40' span	"	0.800
King post type, 2x4 members		
4-in-12 slope		
16' span	EA.	0.649
18' span	"	0.667
24' span	"	0.706
26' span	"	0.706
30' span	"	0.750
34' span	"	0.750
38' span	"	0.774
42' span	"	0.828

FINISH CARPENTRY

06200.10 FINISH CARPENTRY

	UNIT	MAN/HOURS
Mouldings and trim		
Apron, flat		
9/16 x 2	L.F.	0.040
9/16 x 3-1/2	"	0.042
Base		
Colonial		
7/16 x 2-1/4	L.F.	0.040
7/16 x 3	"	0.040
7/16 x 3-1/4	"	0.040
9/16 x 3	"	0.042
9/16 x 3-1/4	"	0.042
11/16 x 2-1/4	"	0.044
Ranch		
7/16 x 2-1/4	L.F.	0.040
7/16 x 3-1/4	"	0.040
9/16 x 2-1/4	"	0.042
9/16 x 3	"	0.042
9/16 x 3-1/4	"	0.042
Casing		
11/16 x 2-1/2	L.F.	0.036
11/16 x 3-1/2	"	0.038
Chair rail		
9/16 x 2-1/2	L.F.	0.040

06 WOOD AND PLASTICS

FINISH CARPENTRY	UNIT	MAN/HOURS
06200.10 FINISH CARPENTRY		
9/16 x 3-1/2	L.F.	0.040
Closet pole		
1-1/8" dia.	L.F.	0.053
1-5/8" dia.	"	0.053
Cove		
9/16 x 1-3/4	L.F.	0.040
11/16 x 2-3/4	"	0.040
Crown		
9/16 x 1-5/8	L.F.	0.053
9/16 x 2-5/8	"	0.062
11/16 x 3-5/8	"	0.067
11/16 x 4-1/4	"	0.073
11/16 x 5-1/4	"	0.080
Drip cap		
1-1/16 x 1-5/8	L.F.	0.040
Glass bead		
3/8 x 3/8	L.F.	0.050
1/2 x 9/16	"	0.050
5/8 x 5/8	"	0.050
3/4 x 3/4	"	0.050
Half round		
1/2	L.F.	0.032
5/8	"	0.032
3/4	"	0.032
Lattice		
1/4 x 7/8	L.F.	0.032
1/4 x 1-1/8	"	0.032
1/4 x 1-3/8	"	0.032
1/4 x 1-3/4	"	0.032
1/4 x 2	"	0.032
Ogee molding		
5/8 x 3/4	L.F.	0.040
11/16 x 1-1/8	"	0.040
11/16 x 1-3/8	"	0.040
Parting bead		
3/8 x 7/8	L.F.	0.050
Quarter round		
1/4 x 1/4	L.F.	0.032
3/8 x 3/8	"	0.032
1/2 x 1/2	"	0.032
11/16 x 11/16	"	0.035
3/4 x 3/4	"	0.035
1-1/16 x 1-1/16	"	0.036
Railings, balusters		
1-1/8 x 1-1/8	L.F.	0.080
1-1/2 x 1-1/2	"	0.073
Screen moldings		
1/4 x 3/4	L.F.	0.067
5/8 x 5/16	"	0.067

FINISH CARPENTRY	UNIT	MAN/HOURS
06200.10 FINISH CARPENTRY		
Shoe		
7/16 x 11/16	L.F.	0.032
Sash beads		
1/2 x 3/4	L.F.	0.067
1/2 x 7/8	"	0.067
1/2 x 1-1/8	"	0.073
5/8 x 7/8	"	0.073
Stop		
5/8 x 1-5/8		
Colonial	L.F.	0.050
Ranch	"	0.050
Stools		
11/16 x 2-1/4	L.F.	0.089
11/16 x 2-1/2	"	0.089
11/16 x 5-1/4	"	0.100
Exterior trim, casing, select pine, 1x3	"	0.040
Douglas fir		
1x3	L.F.	0.040
1x4	"	0.040
1x6	"	0.044
1x8	"	0.050
Cornices, white pine, #2 or better		
1x2	L.F.	0.040
1x4	"	0.040
1x6	"	0.044
1x8	"	0.047
1x10	"	0.050
1x12	"	0.053
Shelving, pine		
1x8	L.F.	0.062
1x10	"	0.064
1x12	"	0.067
Plywood shelf, 3/4", with edge band, 12" wide	"	0.080
Adjustable shelf, and rod, 12" wide		
3' to 4' long	EA.	0.200
5' to 8' long	"	0.267
Prefinished wood shelves with brackets and supports		
8" wide		
3' long	EA.	0.200
4' long	"	0.200
6' long	"	0.200
10" wide		
3' long	EA.	0.200
4' long	"	0.200
6' long	"	0.200

06 WOOD AND PLASTICS

FINISH CARPENTRY

06220.10 MILLWORK

	UNIT	MAN/ HOURS
Countertop, laminated plastic		
25" x 7/8" thick		
Minimum	L.F.	0.200
Average	"	0.267
Maximum	"	0.320
25" x 1-1/4" thick		
Minimum	L.F.	0.267
Average	"	0.320
Maximum	"	0.400
Add for cutouts	EA.	0.500
Backsplash, 4" high, 7/8" thick	L.F.	0.160
Plywood, sanded, A-C		
1/4" thick	S.F.	0.027
3/8" thick	"	0.029
1/2" thick	"	0.031
A-D		
1/4" thick	S.F.	0.027
3/8" thick	"	0.029
1/2" thick	"	0.031
Base cabinets, 34-1/2" high, 24" deep, hardwood, no tops		
Minimum	L.F.	0.320
Average	"	0.400
Maximum	"	0.533
Wall cabinets		
Minimum	L.F.	0.267
Average	"	0.320
Maximum	"	0.400
Oil borne		
Water borne		

ARCHITECTURAL WOODWORK

06420.10 PANEL WORK

	UNIT	MAN/ HOURS
Hardboard, tempered, 1/4" thick		
Natural faced	S.F.	0.020
Plastic faced	"	0.023
Pegboard, natural	"	0.020
Plastic faced	"	0.023
Untempered, 1/4" thick		
Natural faced	S.F.	0.020
Plastic faced	S.F.	0.023
Pegboard, natural	"	0.020
Plastic faced	"	0.023
Plywood unfinished, 1/4" thick		
Birch		
Natural	S.F.	0.027
Select	"	0.027
Knotty pine	"	0.027
Cedar (closet lining)		
Standard boards T&G	S.F.	0.027
Particle board	"	0.027
Plywood, prefinished, 1/4" thick, premium grade		
Birch veneer	S.F.	0.032
Cherry veneer	"	0.032
Chestnut veneer	"	0.032
Lauan veneer	"	0.032
Mahogany veneer	"	0.032
Oak veneer (red)	"	0.032
Pecan veneer	"	0.032
Rosewood veneer	"	0.032
Teak veneer	"	0.032
Walnut veneer	"	0.032

06430.10 STAIRWORK

	UNIT	MAN/ HOURS
Risers, 1x8, 42" wide		
White oak	EA.	0.400
Pine	"	0.400
Treads, 1-1/16" x 9-1/2" x 42"		
White oak	EA.	0.500

06440.10 COLUMNS

	UNIT	MAN/ HOURS
Column, hollow, round		
12" diameter		
10' high	EA.	0.800
12' high	"	0.857
14' high	"	0.960
16' high	"	1.200
24" diameter		
16' high	EA.	1.200
18' high	"	1.263
20' high	"	1.263
22' high	"	1.333
24' high	"	1.333

07 THERMAL AND MOISTURE

MOISTURE PROTECTION

07100.10 WATERPROOFING

MOISTURE PROTECTION	UNIT	MAN/HOURS
Membrane waterproofing, elastomeric		
Butyl		
1/32" thick	S.F.	0.032
1/16" thick	"	0.033
Butyl with nylon		
1/32" thick	S.F.	0.032
1/16" thick	"	0.033
Neoprene		
1/32" thick	S.F.	0.032
1/16" thick	"	0.033
Neoprene with nylon		
1/32" thick	S.F.	0.032
1/16" thick	"	0.033
Bituminous membrane waterproofing, asphalt felt, 15 lb.		
One ply	S.F.	0.020
Two ply	"	0.024
Three ply	"	0.029
Four ply	"	0.033
Five ply	"	0.042
Modified asphalt membrane waterproofing, fibrous asphalt		
One ply	S.F.	0.033
Two ply	"	0.040
Three ply	"	0.044
Four ply	"	0.053
Five ply	"	0.064
Asphalt coated protective board		
1/8" thick	S.F.	0.020
1/4" thick	"	0.020
3/8" thick	"	0.020
1/2" thick	"	0.021
Cement protective board		
3/8" thick	S.F.	0.027
1/2" thick	"	0.027
Fluid applied, neoprene		0.020
50 mil	S.F.	0.027
90 mil	"	0.027
Tab extended polyurethane		
.050" thick	S.F.	0.020
Fluid applied rubber based polyurethane		
6 mil	S.F.	0.025
15 mil	"	0.020
Bentonite waterproofing, panels		
3/16" thick	S.F.	0.020
1/4" thick	"	0.020
5/8" thick	"	0.021
Granular admixtures, trowel on, 3/8" thick	"	0.020
Metallic oxide waterproofing, iron compound, troweled		
5/8" thick	S.F.	0.020
3/4" thick	"	0.023

07150.10 DAMPROOFING

MOISTURE PROTECTION	UNIT	MAN/HOURS
Silicone dampproofing, sprayed on		
Concrete surface		
1 coat	S.F.	0.004
2 coats	"	0.006
Concrete block		
1 coat	S.F.	0.005
2 coats	"	0.007
Brick		
1 coat	S.F.	0.006
2 coats	"	0.008

07160.10 BITUMINOUS DAMPPROOFING

MOISTURE PROTECTION	UNIT	MAN/HOURS
Building paper, asphalt felt		
15 lb	S.F.	0.032
30 lb	"	0.033
Asphalt dampproofing, troweled, cold, primer plus		
1 coat	S.F.	0.027
2 coats	"	0.040
3 coats	"	0.050
Fibrous asphalt dampproofing, hot troweled, primer plus		
1 coat	S.F.	0.032
2 coats	"	0.044
3 coats	"	0.057
Asphaltic paint dampproofing, per coat		
Brush on	S.F.	0.011
Spray on	"	0.009

07190.10 VAPOR BARRIERS

MOISTURE PROTECTION	UNIT	MAN/HOURS
Vapor barrier, polyethylene		
2 mil	S.F.	0.004
6 mil	"	0.004
8 mil	"	0.004
10 mil	"	0.004

INSULATION

07210.10 BATT INSULATION

INSULATION	UNIT	MAN/HOURS
Ceiling, fiberglass, unfaced		
3-1/2" thick, R11	S.F.	0.009
6" thick, R19	"	0.011

INSULATION	UNIT	MAN/ HOURS
07210.10 BATT INSULATION		
9" thick, R30	S.F.	0.012
Suspended ceiling, unfaced		
3-1/2" thick, R11	S.F.	0.009
6" thick, R19	"	0.010
9" thick, R30	"	0.011
Crawl space, unfaced		
3-1/2" thick, R11	S.F.	0.012
6" thick, R19	"	0.013
9" thick, R30	"	0.015
Wall, fiberglass		
Paper backed		
2" thick, R7	S.F.	0.008
3" thick, R8	"	0.009
4" thick, R11	"	0.009
6" thick, R19	"	0.010
Foil backed, 1 side		
2" thick, R7	S.F.	0.008
3" thick, R11	"	0.009
4" thick, R14	"	0.009
6" thick, R21	"	0.010
Foil backed, 2 sides		
2" thick, R7	S.F.	0.009
3" thick, R11	"	0.010
4" thick, R14	"	0.011
6" thick, R21	"	0.011
Unfaced		
2" thick, R7	S.F.	0.008
3" thick, R9	"	0.009
4" thick, R11	"	0.009
6" thick, R19	"	0.010
Mineral wool batts		
Paper backed		
2" thick, R6	S.F.	0.008
4" thick, R12	"	0.009
6" thick, R19	"	0.010
Fasteners, self adhering, attached to ceiling deck		
2-1/2" long	EA.	0.013
4-1/2" long	"	0.015
Capped, self-locking washers for fastening insulation	"	0.008
07210.20 BOARD INSULATION		
Insulation, rigid		
Fiberglass, roof		
0.75" thick, R2.78	S.F.	0.007
1.06" thick, R4.17	"	0.008
1.31" thick, R5.26	"	0.008
1.63" thick, R6.67	"	0.008
2.25" thick, R8.33	"	0.009

INSULATION	UNIT	MAN/ HOURS
07210.20 BOARD INSULATION		
Composite board, roof		
1-1/2" thick, R6.67	S.F.	0.008
1-5/8" thick, R7.69	"	0.008
2" thick, R10.0	"	0.009
2-1/4" thick, R12.50	"	0.009
2-1/2" thick, R14.29	"	0.010
2-3/4" thick, R16.67	"	0.011
3-1/4" thick, R20.00	"	0.011
Perlite board, roof		
1.00" thick, R2.78	S.F.	0.007
1.50" thick, R4.17	"	0.007
2.00" thick, R5.92	"	0.007
2.50" thick, R6.67	"	0.008
3.00" thick, R8.33	"	0.008
4.00" thick, R10.00	"	0.008
5.25" thick, R14.29	"	0.009
Rigid urethane		
Roof		
1" thick, R6.67	S.F.	0.007
1.20" thick, R8.33	"	0.007
1.50" thick, R11.11	"	0.007
2" thick, R14.29	"	0.007
2.25" thick, R16.67	"	0.008
Wall		
1" thick, R6.67	S.F.	0.008
1.5" thick, R11.11	"	0.009
2" thick, R14.29	"	0.009
Polystyrene		
Roof		
1.0" thick, R4.17	S.F.	0.007
1.5" thick, R6.26	"	0.007
2.0" thick, R8.33	"	0.007
Wall		
1.0" thick, R4.17	S.F.	0.008
1.5" thick, R6.26	"	0.009
2.0" thick, R8.33	"	0.009
Rigid board insulation, deck		
Mineral fiberboard		
1" thick, R3.0	S.F.	0.007
2" thick, R5.26	"	0.007
Fiberglass		
1" thick, R4.3	S.F.	0.007
2" thick, R8.5	"	0.007
Polystyrene		
1" thick, R5.4	S.F.	0.007
2" thick, R10.8	"	0.007
Urethane		
.75" thick, R5.4	S.F.	0.007
1" thick, R6.4	"	0.007

07 THERMAL AND MOISTURE

INSULATION	UNIT	MAN/HOURS
07210.20 BOARD INSULATION		
1.5" thick, R10.7	S.F.	0.007
2" thick, R14.3	"	0.007
Foamglass		
1" thick, R1.8	S.F.	0.007
2" thick, R5.26	"	0.007
Wood fiber		
1" thick, R3.85	S.F.	0.007
2" thick, R7.7	"	0.007
Particle board		
3/4" thick, R2.08	S.F.	0.007
1" thick, R2.77	"	0.007
2" thick, R5.50	"	0.007
07210.60 LOOSE FILL INSULATION		
Blown-in type		
Fiberglass		
5" thick, R11	S.F.	0.007
6" thick, R13	"	0.008
9" thick, R19	"	0.011
Rockwool, attic application		
6" thick, R13	S.F.	0.008
8" thick, R19	"	0.010
10" thick, R22	"	0.012
12" thick, R26	"	0.013
15" thick, R30	"	0.016
Poured type		
Fiberglass		
1" thick, R4	S.F.	0.005
2" thick, R8	"	0.006
3" thick, R12	"	0.007
4" thick, R16	"	0.008
Mineral wool		
1" thick, R3	S.F.	0.005
2" thick, R6	"	0.006
3" thick, R9	"	0.007
4" thick, R12	"	0.008
Vermiculite or perlite		
2" thick, R4.8	S.F.	0.006
3" thick, R7.2	"	0.007
4" thick, R9.6	"	0.008
Masonry, poured vermiculite or perlite		
4" block	S.F.	0.004
6" block	"	0.005
8" block	"	0.006
10" block	"	0.006
12" block	"	0.007

INSULATION	UNIT	MAN/HOURS
07210.70 SPRAYED INSULATION		
Foam, sprayed on		
Polystyrene		
1" thick, R4	S.F.	0.008
2" thick, R8	"	0.011
Urethane		
1" thick, R7.7	S.F.	0.008
2" thick, R15.4	"	0.011
07250.10 FIREPROOFING		
Sprayed on		
1" thick		
On beams	S.F.	0.018
On columns	"	0.016
On decks		
Flat surface	S.F.	0.008
Fluted surface	"	0.010
1-1/2" thick		
On beams	S.F.	0.023
On columns	"	0.020
On decks		
Flat surface	S.F.	0.010
Fluted surface	"	0.013

SHINGLES AND TILES	UNIT	MAN/HOURS
07310.10 ASPHALT SHINGLES		
Standard asphalt shingles, strip shingles		
210 lb/square	SQ.	0.800
235 lb/square	"	0.889
240 lb/square	"	1.000
260 lb/square	"	1.143
300 lb/square	"	1.333
385 lb/square	"	1.600
Roll roofing, mineral surface		
90 lb	SQ.	0.571
110 lb	"	0.667
140 lb	"	0.800

SHINGLES AND TILES

07310.30 METAL SHINGLES

SHINGLES AND TILES	UNIT	MAN/ HOURS
Aluminum, .020" thick		
Plain	SQ.	1.600
Colors	"	1.600
Steel, galvanized		
26 ga.		
Plain	SQ.	1.600
Colors	"	1.600
24 ga.		
Plain	SQ.	1.600
Colors	"	1.600
Porcelain enamel, 22 ga.		
Minimum	SQ.	2.000
Average	"	2.000
Maximum	"	2.000

07310.60 SLATE SHINGLES

SHINGLES AND TILES	UNIT	MAN/ HOURS
Slate shingles		
Pennsylvania		
Ribbon	SQ.	4.000
Clear	"	4.000
Vermont		
Black	SQ.	4.000
Grey	"	4.000
Green	"	4.000
Red	"	4.000
Replacement shingles		
Small jobs	EA.	0.267
Large jobs	S.F.	0.133

07310.70 WOOD SHINGLES

SHINGLES AND TILES	UNIT	MAN/ HOURS
Wood shingles, on roofs		
White cedar, #1 shingles		
4" exposure	SQ.	2.667
5" exposure	"	2.000
#2 shingles		
4" exposure	SQ.	2.667
5" exposure	"	2.000
Resquared and rebutted		
4" exposure	SQ.	2.667
5" exposure	"	2.000
On walls		
White cedar, #1 shingles		
4" exposure	SQ.	4.000
5" exposure	"	3.200
6" exposure	"	2.667
#2 shingles		
4" exposure	SQ.	4.000
5" exposure	"	3.200
6" exposure	"	2.667

07310.80 WOOD SHAKES

SHINGLES AND TILES	UNIT	MAN/ HOURS
Shakes, hand split, 24" red cedar, on roofs		
5" exposure	SQ.	4.000
7" exposure	"	3.200
9" exposure	"	2.667
On walls		
6" exposure	SQ.	4.000
8" exposure	"	3.200
10" exposure	"	2.667

ROOFING AND SIDING

07410.10 MANUFACTURED ROOFS

ROOFING AND SIDING	UNIT	MAN/ HOURS
Aluminum roof panels, for structural steel framing		
Corrugated		
Unpainted finish		
.024"	S.F.	0.020
.030"	"	0.020
Painted finish		
.024"	S.F.	0.020
.030"	"	0.020
V-beam		
Unpainted finish		
.032"	S.F.	0.020
.040"	"	0.020
.050"	"	0.020
Painted finish		
.032"	S.F.	0.020
.040"	"	0.020
.050"	"	0.020
Steel roof panels, for structural steel framing		
Corrugated, painted		
18 ga.	S.F.	0.020
20 ga.	"	0.020
22 ga.	"	0.020
Box rib, painted		
18 ga.	S.F.	0.021
20 ga.	"	0.021
22 ga.	"	0.021
4" rib, painted		
18 ga.	S.F.	0.022
20 ga.	"	0.022

ROOFING AND SIDING	UNIT	MAN/ HOURS
07410.10 MANUFACTURED ROOFS		
22 ga.	S.F.	0.022
Standing seam roof		
2" high seam, painted		
22 ga.	S.F.	0.032
24 ga.	"	0.032
26 ga.	"	0.032
07410.30 MANUFACTURED WALLS		
Sandwich panels with 1-1/2" fiberglass insulation		
Galvanized 18 ga. steel interior panels		
Exterior panels		
16 ga. aluminum	S.F.	0.107
18 ga. galvanized steel	"	0.107
20 ga. painted steel	"	0.107
20 ga. stainless steel	"	0.107
Metal liner panels, 1-3/8" thick, 24" wide		
Galvanized		
22 ga.	S.F.	0.027
20 ga.	"	0.027
18 ga.	"	0.027
Primed		
22 ga.	S.F.	0.027
20 ga.	"	0.027
18 ga.	"	0.027
07440.10 AGGREGATE COATED PANELS		
Dryvit type system		
1" thick	S.F.	0.027
1-1/2" thick	"	0.029
2" thick	"	0.033
07460.10 METAL SIDING PANELS		
Aluminum siding panels		
Corrugated		
Plain finish		
.024"	S.F.	0.032
.032"	"	0.032
Painted finish		
.024"	S.F.	0.032
.032"	"	0.032
V. beam		
Plain finish		
.032"	S.F.	0.032
.040"	"	0.032
.050"	"	0.032
Painted finish		
.032"	S.F.	0.032
.040"	"	0.032
.050"	"	0.032

ROOFING AND SIDING	UNIT	MAN/ HOURS
07460.10 METAL SIDING PANELS		
4" rib		
Plain finish		
.032"	S.F.	0.036
.040"	"	0.036
.050"	"	0.036
Painted finish		
.032"	S.F.	0.036
.040"	"	0.036
.050"	"	0.036
Steel siding panels		
Corrugated		
22 ga.	S.F.	0.053
24 ga.	"	0.053
26 ga.	"	0.053
Box rib		
20 ga.	S.F.	0.053
22 ga.	"	0.053
24 ga.	"	0.053
26 ga.	"	0.053
07460.50 PLASTIC SIDING		
Horizontal vinyl siding, solid		
8" wide		
Standard	S.F.	0.031
Insulated	"	0.031
10" wide		
Standard	S.F.	0.029
Insulated	"	0.029
Vinyl moldings for doors and windows	L.F.	0.032
07460.60 PLYWOOD SIDING		
Rough sawn cedar, 3/8" thick	S.F.	0.027
Fir, 3/8" thick	"	0.027
Texture 1-11, 5/8" thick		
Cedar	S.F.	0.029
Fir	"	0.029
Redwood	"	0.029
Southern Yellow Pine	"	0.029
07460.70 STEEL SIDING		
Ribbed, sheets, galvanized		
22 ga.	S.F.	0.032
24 ga.	"	0.032
26 ga.	"	0.032
28 ga.	"	0.032
Primed		
24 ga.	S.F.	0.032

07 THERMAL AND MOISTURE

ROOFING AND SIDING

ROOFING AND SIDING	UNIT	MAN/HOURS
07460.70 STEEL SIDING		
26 ga.	S.F.	0.032
28 ga.	"	0.032
07460.80 WOOD SIDING		
Beveled siding, cedar		
A grade		
1/2 x 6	S.F.	0.040
1/2 x 8	"	0.032
3/4 x 10	"	0.027
Clear		
1/2 x 6	S.F.	0.040
1/2 x 8	"	0.032
3/4 x 10	"	0.027
B grade		
1/2 x 6	S.F.	0.040
1/2 x 8	"	0.320
3/4 x 10	"	0.027
Board and batten		
Cedar		
1x6	S.F.	0.040
1x8	"	0.032
1x10	"	0.029
1x12	"	0.026
Pine		
1x6	S.F.	0.040
1x8	"	0.032
1x10	"	0.029
1x12	"	0.026
Redwood		
1x6	S.F.	0.040
1x8	"	0.032
1x10	"	0.029
1x12	"	0.026
Tongue and groove		
Cedar		
1x4	S.F.	0.044
1x6	"	0.042
1x8	"	0.040
1x10	"	0.038
Pine		
1x4	S.F.	0.044
1x6	"	0.042
1x8	"	0.040
1x10	"	0.038
Redwood		
1x4	S.F.	0.044
1x6	"	0.042
1x8	"	0.040
1x10	"	0.038

MEMBRANE ROOFING

MEMBRANE ROOFING	UNIT	MAN/HOURS
07510.10 BUILT-UP ASPHALT ROOFING		
Built-up roofing, asphalt felt, including gravel		
2 ply	SQ.	2.000
3 ply	"	2.667
4 ply	"	3.200
Walkway, for built-up roofs		
3' x 3' x		
1/2" thick	S.F.	0.027
3/4" thick	"	0.027
1" thick	"	0.027
Cant strip, 4" x 4"		
Treated wood	L.F.	0.023
Foamglass	"	0.020
Mineral fiber	"	0.020
New gravel for built-up roofing, 400 lb/sq	SQ.	1.600
Roof gravel (ballast)	C.Y.	4.000
Aluminum coating, top surfacing, for built-up roofing	SQ.	1.333
Remove & replace gravel, includes flood coat	"	2.667
07530.10 SINGLE-PLY ROOFING		
Elastic sheet roofing		
Neoprene, 1/16" thick	S.F.	0.010
EPDM rubber		
45 mil	S.F.	0.010
60 mil	"	0.010
PVC		
45 mil	S.F.	0.010
60 mil	"	0.010
Flashing		
Pipe flashing, 90 mil thick		
1" pipe	EA.	0.200
2" pipe	"	0.200
3" pipe	"	0.211
4" pipe	"	0.211
5" pipe	"	0.222
6" pipe	"	0.222
8" pipe	"	0.235
10" pipe	"	0.267
12" pipe	"	0.267
Neoprene flashing, 60 mil thick strip		
6" wide	L.F.	0.067
12" wide	"	0.100
18" wide	"	0.133
24" wide	"	0.200
Adhesives		
Mastic sealer, applied at joints only		
1/4" bead	L.F.	0.004
Fluid applied roofing		
Urethane, 2 components, elastomeric top membrane		

07 THERMAL AND MOISTURE

MEMBRANE ROOFING

07530.10 SINGLE-PLY ROOFING

	UNIT	MAN/HOURS
1" thick	S.F.	0.013
Vinyl liquid roofing, 2 coats, 2 mils per coat	"	0.011
Silicone roofing, 2 coats sprayed, 16 mil per coat	"	0.013
Inverted roof system		
Insulated membrane with coarse gravel ballast		
3 ply with 2" polystyrene	S.F.	0.013
Ballast, 3/4" through 1-1/2" dia. river gravel, 100lb/sf	"	0.800
Walkway for membrane roofs, 1/2" thick	"	0.027

FLASHING AND SHEET METAL

07610.10 METAL ROOFING

	UNIT	MAN/HOURS
Sheet metal roofing, copper, 16 oz, batten seam	SQ.	5.333
Standing seam	"	5.000
Aluminum roofing, natural finish		
Corrugated, on steel frame		
.0175" thick	SQ.	2.286
.0215" thick	"	2.286
.024" thick	"	2.286
.032" thick	"	2.286
V-beam, on steel frame		
.032" thick	SQ.	2.286
.040" thick	"	2.286
.050" thick	"	2.286
Ridge cap		
.019" thick	L.F.	0.027
Corrugated galvanized steel roofing, on steel frame		
28 ga.	SQ.	2.286
26 ga.	"	2.286
24 ga.	"	2.286
22 ga.	"	2.286
26 ga., factory insulated with 1" polystyrene	"	3.200
Ridge roll		
10" wide	L.F.	0.027
20" wide	"	0.032

07620.10 FLASHING AND TRIM

	UNIT	MAN/HOURS
Counter flashing		
Aluminum, .032"	S.F.	0.080
Stainless steel, .015"	"	0.080

FLASHING AND SHEET METAL

07620.10 FLASHING AND TRIM

	UNIT	MAN/HOURS
16 oz.	S.F.	0.080
20 oz.	"	0.080
24 oz.	"	0.080
32 oz.	"	0.080
Valley flashing		
Aluminum, .032"	S.F.	0.050
Stainless steel, .015	"	0.050
Copper		
16 oz.	S.F.	0.050
20 oz.	"	0.067
24 oz.	"	0.050
32 oz.	"	0.050
Base flashing		
Aluminum, .040"	S.F.	0.067
Stainless steel, .018"	"	0.067
16 oz.	"	0.067
20 oz.	"	0.050
24 oz.	"	0.067
32 oz.	"	0.067
Waterstop, "T" section, 22 ga.		
1-1/2" x 3"	L.F.	0.040
2" x 2"	"	0.040
4" x 3"	"	0.040
6" x 4"	"	0.040
8" x 4"	"	0.040
Scupper outlets		
10" x 10" x 4"	EA.	0.200
22" x 4" x 4"	"	0.200
8" x 8" x 5"	"	0.200
Flashing and trim, aluminum		
.019" thick	S.F.	0.057
.032" thick	"	0.057
.040" thick	"	0.062
Neoprene sheet flashing, .060" thick	"	0.050
Copper, paper backed		
2 oz.	S.F.	0.080
Drainage boots, roof, cast iron		
2 x 3	L.F.	0.100
3 x 4	"	0.100
4 x 5	"	0.107
4 x 6	"	0.107
5 x 7	"	0.114
6 x 6	EA.	0.200
8 x 8	"	0.200
8 x 10	"	0.200
8 x 12	"	0.200
Reglets, copper 10 oz.	L.F.	0.053
Stainless steel, .020"	"	0.053
Gravel stop		

FLASHING AND SHEET METAL	UNIT	MAN/HOURS
07620.10 FLASHING AND TRIM		
Aluminum, .032"		
4"	L.F.	0.027
6"	"	0.027
8"	"	0.031
10"	"	0.031
Copper, 16 oz.		
4"	L.F.	0.027
6"	"	0.027
8"	"	0.031
10"	"	0.031
07620.20 GUTTERS AND DOWNSPOUTS		
Copper gutter and downspout		
Downspouts, 16 oz. copper		
Round		
3" dia.	L.F.	0.053
4" dia.	"	0.053
Rectangular, corrugated		
2" x 3"	L.F.	0.050
3" x 4"	"	0.050
Rectangular, flat surface		
2" x 3"	L.F.	0.053
3" x 4"	"	0.053
Lead-coated copper downspouts		
Round		
3" dia.	L.F.	0.050
4" dia.	"	0.057
Rectangular, corrugated		
2" x 3"	L.F.	0.053
3" x 4"	"	0.053
Rectangular, plain		
2" x 3"	L.F.	0.053
3" x 4"	"	0.053
Gutters, 16 oz. copper		
Half round		
4" wide	L.F.	0.080
5" wide	"	0.089
Type K		
4" wide	L.F.	0.080
5" wide	"	0.089
Lead-coated copper gutters		
Half round		
4" wide	L.F.	0.080
6" wide	"	0.089
Type K		
4" wide	L.F.	0.080
5" wide	"	0.089
Aluminum gutter and downspout		
Downspouts		

FLASHING AND SHEET METAL	UNIT	MAN/HOURS
07620.20 GUTTERS AND DOWNSPOUTS		
2" x 3"	L.F.	0.053
3" x 4"	"	0.057
4" x 5"	"	0.062
Round		
3" dia.	L.F.	0.053
4" dia.	"	0.057
Gutters, stock units		
4" wide	L.F.	0.084
5" wide	"	0.089
Galvanized steel gutter and downspout		
Downspouts, round corrugated		
3" dia.	L.F.	0.053
4" dia.	"	0.053
5" dia.	"	0.057
6" dia.	"	0.057
Rectangular		
2" x 3"	L.F.	0.053
3" x 4"	"	0.050
4" x 4"	"	0.050
Gutters, stock units		
5" wide		
Plain	L.F.	0.089
Painted	"	0.089
6" wide		
Plain	L.F.	0.094
Painted	"	0.094

ROOFING SPECIALITES	UNIT	MAN/HOURS
07700.10 MANUFACTURED SPECIALTIES		
Moisture relief vent		
Aluminum	EA.	0.114
Copper	"	0.114
Expansion joint		
Aluminum		
Opening to 2.5"	L.F.	0.057
Opening to 3.5"	"	0.062
Copper, 16 oz.		
Opening to 2.5"	L.F.	0.057
Opening to 3.5"	"	0.062
Butyl or neoprene		

ROOFING SPECIALITES

07700.10 MANUFACTURED SPECIALTIES

ROOFING SPECIALITES	UNIT	MAN/ HOURS
4" wide		
16 oz. copper bellows	L.F.	0.067
28 ga. stainless steel bellows	"	0.067
6" wide		
Copper bellows	L.F.	0.073
Stainless steel		
Opening to 2.5"	L.F.	0.057
Opening to 3.5"	"	0.062
Smoke vent, 48" x 48"		
Aluminum	EA.	2.000
Galvanized steel	"	2.000
Heat/smoke vent, 48" x 96"		
Aluminum	EA.	2.667
Galvanized steel	"	2.667
Ridge vent strips		
Mill finish	L.F.	0.053
Connectors	EA.	0.200
End cap	"	0.229
Soffit vents		
Mill finish		
2-1/2" wide	L.F.	0.032
3" wide	"	0.032
6" wide	"	0.032
Roof hatches		
Steel, plain, primed		
2'6" x 3'0"	EA.	2.000
2'6" x 4'6"	"	2.667
2'6" x 8'0"	"	4.000
Galvanized steel		
2'6" x 3'0"	EA.	2.000
2'6" x 4'6"	"	2.667
2'6" x 8'0"	"	4.000
Aluminum		
2'6" x 3'0"	EA.	2.000
2'6" x 4'6"	"	2.667
2'6" x 8'0"	"	4.000
Ceiling access doors		
Swing up model, metal frame		
Steel door		
2'6" x 2'6"	EA.	0.800
2'6" x 3'0"	"	0.800
Aluminum door		
2'6" x 2'6"	EA.	0.800
2'6" x 3'0"	"	0.800
Swing down model, metal frame		
Steel door		
2'6" x 2'6"	EA.	0.800
2'6" x 3'0"	"	0.800
Aluminum door		
2'6" x 2'6"	EA.	0.800
2'6" x 3'0"	"	0.800
Gravity ventilators, with curb, base, damper and screen		
Stationary siphon		
6" dia.	EA.	0.533
12" dia.	"	0.533
24" dia.	"	0.800
36" dia.	"	0.800
Wind driven spinner		
6" dia.	EA.	0.533
12" dia.	"	0.533
24" dia.	"	0.800
36" dia.	"	0.800
Stationary mushroom		
16" dia.	EA.	0.800
30" dia.	"	1.000
36" dia.	"	1.333
42" dia.	"	1.600

SKYLIGHTS

07810.10 PLASTIC SKYLIGHTS

SKYLIGHTS	UNIT	MAN/ HOURS
Single thickness, not including mounting curb		
2' x 4'	EA.	1.000
4' x 4'	"	1.333
5' x 5'	"	2.000
6' x 8'	"	2.667
Double thickness, not including mounting curb		
2' x 4'	EA.	1.000
4' x 4'	"	1.333
5' x 5'	"	2.000
6' x 8'	"	2.667
Metal framed skylights		
Translucent panels, 2-1/2" thick	S.F.	0.080
Continuous vaults, 8' wide		
Single glazed	S.F.	0.100
Double glazed	"	0.114

JOINT SEALERS	UNIT	MAN/ HOURS
07920.10 CAULKING		
Caulk exterior, two component		
1/4 x 1/2	L.F.	0.040
3/8 x 1/2	"	0.044
1/2 x 1/2	"	0.050
Caulk interior, single component		
1/4 x 1/2	L.F.	0.038
3/8 x 1/2	"	0.042
1/2 x 1/2	"	0.047
Butyl rubber fillers		
1/4" x 1/4"	L.F.	0.016
1/2" x 1/2"	"	0.027
1/2" x 3/4"	"	0.032
3/4" x 3/4"	"	0.032
1" x 1"	"	0.036
Neoprene, closed cell		
1/8" x 3"	L.F.	0.020
1/8" x 6"	"	0.022
1/4" x 3"	"	0.025
1/4" x 6"	"	0.029
1/2" x 6"	"	0.036
1/2" x 9"	"	0.040
1/2" x 12"	"	0.040
Seals, "O" ring type cord		
1/4" dia.	L.F.	0.020
1/2" dia.	"	0.021
1" dia.	"	0.022
1-1/4" dia.	"	0.024
1-1/2" dia.	"	0.025
1-3/4" dia.	"	0.026
2" dia.	"	0.027
Polyvinyl chloride, closed cell		
1/4" x 2"	L.F.	0.029
1/4" x 6"	"	0.036
Silicon foam penetration seal		
1/4" x 1/2"	L.F.	0.010
1/2" x 1/2"	"	0.013
1/2" x 3/4"	"	0.016
3/4" x 3/4"	"	0.020
1/8" x 1"	"	0.010
1/8" x 3"	"	0.016
1/4" x 3"	"	0.020
1/4" x 6"	"	0.027
1/2" x 6"	"	0.062
1/2" x 9"	"	0.100
1/2" x 12"	"	0.145
Oil base sealants and caulking		
1/4" x 1/4"	L.F.	0.020
1/4" x 3/8"	"	0.021
1/4" x 1/2"	"	0.022

JOINT SEALERS	UNIT	MAN/ HOURS
07920.10 CAULKING		
3/8" x 3/8"	L.F.	0.023
3/8" x 1/2"	"	0.024
3/8" x 5/8"	"	0.026
3/8" x 3/4"	"	0.028
1/2" x 1/2"	"	0.031
1/2" x 5/8"	"	0.035
1/2" x 3/4"	"	0.040
1/2" x 7/8"	"	0.041
1/2" x 1"	"	0.042
3/4" x 3/4"	"	0.043
1" x 1"	"	0.044
Polyurethane compounds		
1/4" x 1/4"	L.F.	0.020
1/4" x 3/8"	"	0.021
1/4" x 1/2"	"	0.022
3/8" x 3/8"	"	0.023
3/8" x 1/2"	"	0.024
3/8" x 5/8"	"	0.026
3/8" x 3/4"	"	0.028
1/2" x 1/2"	"	0.031
1/2" x 5/8"	"	0.035
1/2" x 3/4"	"	0.040
1/2" x 7/8"	"	0.041
1/2" x 1"	"	0.044
3/4" x 3/4"	"	0.043
3/4" x 1"	"	0.044
Backer rod, polyethylene		
1/4"	L.F.	0.020
1/2"	"	0.021
3/4"	"	0.022
1"	"	0.024

METAL	UNIT	MAN/ HOURS
08110.10 METAL DOORS		
Flush hollow metal, standard duty, 20 ga., 1-3/8"		
2-6 x 6-8	EA.	0.889
2-8 x 6-8	"	0.889
3-0 x 6-8	"	0.889
1-3/4"		
2-6 x 6-8	EA.	0.889
2-8 x 6-8	"	0.889
3-0 x 6-8	"	0.889
2-6 x 7-0	"	0.889
2-8 x 7-0	"	0.889
3-0 x 7-0	"	0.889
Heavy duty, 20 ga., unrated, 1-3/4"		
2-8 x 6-8	EA.	0.889
3-0 x 6-8	"	0.889
2-8 x 7-0	"	0.889
3-0 x 7-0	"	0.889
3-4 x 7-0	"	0.889
18 ga., 1-3/4", unrated door		
2-0 x 7-0	EA.	0.889
2-4 x 7-0	"	0.889
2-6 x 7-0	"	0.889
2-8 x 7-0	"	0.889
3-0 x 7-0	"	0.889
3-4 x 7-0	"	0.889
2", unrated door		
2-0 x 7-0	EA.	1.000
2-4 x 7-0	"	1.000
2-6 x 7-0	"	1.000
2-8 x 7-0	"	1.000
3-0 x 7-0	"	1.000
3-4 x 7-0	"	1.000
Galvanized metal door		
3-0 x 7-0	EA.	1.000
08110.40 METAL DOOR FRAMES		
Hollow metal, stock, 18 ga., 4-3/4" x 1-3/4"		
2-0 x 7-0	EA.	1.000
2-4 x 7-0	"	1.000
2-6 x 7-0	"	1.000
2-8 x 7-0	"	1.000
3-0 x 7-0	"	1.000
4-0 x 7-0	"	1.333
5-0 x 7-0	"	1.333
6-0 x 7-0	"	1.333
16 ga., 6-3/4" x 1-3/4"		1.000
2-0 x 7-0	EA.	1.000
2-4 x 7-0	"	1.000
2-6 x 7-0	"	1.000

METAL	UNIT	MAN/ HOURS
08110.40 METAL DOOR FRAMES		
2-8 x 7-0	EA.	1.000
3-0 x 7-0	"	1.000
4-0 x 7-0	"	1.333
6-0 x 7-0	"	1.333
Transom frame		
3-4 x 1-6	EA.	1.000
3-8 x 1-6	"	1.000
6-4 x 1-6	"	1.000
Transom sash		
3-0 x 1-4	EA.	1.000
3-4 x 1-4	"	1.000
6-0 x 1-4	"	1.000
Sidelights, complete		
1-0 x 7-2	EA.	1.000
1-4 x 7-2	"	1.000
1-0 x 8-8	"	1.000
1-6 x 8-8	"	1.000
16 ga., 4-3/4" x 1-3/4"		
2-0 x 7-0	EA.	1.000
2-4 x 7-0	"	1.000
2-6 x 7-0	"	1.000
2-8 x 7-0	"	1.000
3-0 x 7-0	"	1.000
4-0 x 7-0	"	1.333
6-0 x 7-0	"	1.333
3-4 x 1-6	"	1.000
3-8 x 1-6	"	1.000
6-4 x 1-6	"	1.000
Transom sash		
3-0 x 1-4	EA.	1.000
3-4 x 1-4	"	1.000
6-0 x 1-4	"	1.000
Sidelights, complete		
1-0 x 7-2	EA.	1.000
1-4 x 7-2	"	1.000
1-0 x 8-8	"	1.000
1-4 x 8-8	"	1.000
16 ga., 5-3/4" x 1-3/4"		
2-0 x 7-0	EA.	1.000
2-4 x 7-0	"	1.000
2-6 x 7-0	"	1.000
2-8 x 7-0	"	1.000
3-0 x 7-0	"	1.000
4-0 x 7-0	"	1.333
5-0 x 7-0	"	1.333
6-0 x 7-0	"	1.333
Mullions, vertical		
5-1/4" x 1-3/4"	L.F.	0.100
5-1/4" x 2"	"	0.100

METAL	UNIT	MAN/ HOURS
08110.40 METAL DOOR FRAMES		
Horizontal		
5-1/4" x 1-3/4"	L.F.	0.100
5-1/4" x 2"	"	0.100
08120.10 ALUMINUM DOORS		
Aluminum doors, commercial		
Narrow stile		
2-6 x 7-0	EA.	4.000
3-0 x 7-0	"	4.000
3-6 x 7-0	"	4.000
Pair		
5-0 x 7-0	EA.	8.000
6-0 x 7-0	"	8.000
7-0 x 7-0	"	8.000
Wide stile		
2-6 x 7-0	EA.	4.000
3-0 x 7-0	"	4.000
3-6 x 7-0	"	4.000
Pair		
5-0 x 7-0	EA.	8.000
6-0 x 7-0	"	8.000
7-0 x 7-0	"	8.000

WOOD AND PLASTIC	UNIT	MAN/ HOURS
08210.10 WOOD DOORS		
Solid core, 1-3/8" thick		
Birch faced		
2-4 x 7-0	EA.	1.000
2-8 x 7-0	"	1.000
3-0 x 7-0	"	1.000
3-4 x 7-0	"	1.000
2-4 x 6-8	"	1.000
2-6 x 6-8	"	1.000
2-8 x 6-8	"	1.000
3-0 x 6-8	"	1.000
Lauan faced		
2-4 x 6-8	EA.	1.000
2-8 x 6-8	"	1.000
3-0 x 6-8	EA.	1.000
3-4 x 6-8	"	1.000
Tempered hardboard faced		
2-4 x 7-0	EA.	1.000
2-8 x 7-0	"	1.000
3-0 x 7-0	"	1.000
3-4 x 7-0	"	1.000
Hollow core, 1-3/8" thick		
Birch faced		
2-4 x 7-0	EA.	1.000
2-8 x 7-0	"	1.000
3-0 x 7-0	"	1.000
3-4 x 7-0	"	1.000
Lauan faced		
2-4 x 6-8	EA.	1.000
2-6 x 6-8	"	1.000
2-8 x 6-8	"	1.000
3-0 x 6-8	"	1.000
3-4 x 6-8	"	1.000
Tempered hardboard faced		
2-4 x 7-0	EA.	1.000
2-6 x 7-0	"	1.000
2-8 x 7-0	"	1.000
3-0 x 7-0	"	1.000
3-4 x 7-0	"	1.000
Solid core, 1-3/4" thick		
Birch faced		
2-4 x 7-0	EA.	1.000
2-6 x 7-0	"	1.000
2-8 x 7-0	"	1.000
3-0 x 7-0	"	1.000
3-4 x 7-0	"	1.000
Lauan faced		
2-4 x 7-0	EA.	1.000
2-6 x 7-0	"	1.000
2-8 x 7-0	"	1.000
3-0 x 7-0	"	1.000
3-0 x 7-0	"	1.000
Tempered hardboard faced		
2-4 x 7-0	EA.	1.000
2-6 x 7-0	"	1.000
2-8 x 7-0	"	1.000
3-0 x 7-0	"	1.000
3-4 x 7-0	"	1.000
Hollow core, 1-3/4" thick		
Birch faced		
2-4 x 7-0	EA.	1.000
2-6 x 7-0	"	1.000
2-8 x 7-0	"	1.000

08 DOORS AND WINDOWS

WOOD AND PLASTIC	UNIT	MAN/ HOURS
08210.10 WOOD DOORS		
3-0 x 7-0	EA.	1.000
3-4 x 7-0	"	1.000
Lauan faced		
2-4 x 6-8	EA.	1.000
2-6 x 6-8	"	1.000
2-8 x 6-8	"	1.000
3-0 x 6-8	"	1.000
3-4 x 6-8	"	1.000
Tempered hardboard		
2-4 x 7-0	EA.	1.000
2-6 x 7-0	"	1.000
2-8 x 7-0	"	1.000
3-0 x 7-0	"	1.000
3-4 x 7-0	"	1.000
Add-on, louver	"	0.800
Glass	"	0.800
Exterior doors, 3-0 x 7-0 x 2-1/2", solid core		
Carved		
One face	EA.	2.000
Two faces	"	2.000
Closet doors, 1-3/4" thick		
Bi-fold or bi-passing, includes frame and trim		
Paneled		
4-0 x 6-8	EA.	1.333
6-0 x 6-8	"	1.333
Louvered		
4-0 x 6-8	EA.	1.333
6-0 x 6-8	"	1.333
Flush		
4-0 x 6-8	EA.	1.333
6-0 x 6-8	"	1.333
Primed		
4-0 x 6-8	EA.	1.333
6-0 x 6-8	"	1.333
08210.90 WOOD FRAMES		
Frame, interior, pine		
2-6 x 6-8	EA.	1.143
2-8 x 6-8	"	1.143
3-0 x 6-8	"	1.143
5-0 x 6-8	"	1.143
6-0 x 6-8	"	1.143
2-6 x 7-0	"	1.143
2-8 x 7-0	"	1.143
3-0 x 7-0	"	1.143
5-0 x 7-0	"	1.600
6-0 x 7-0	"	1.600
Exterior, custom, with threshold, including trim		
Walnut		

WOOD AND PLASTIC	UNIT	MAN/ HOURS
08210.90 WOOD FRAMES		
3-0 x 7-0	EA.	2.000
6-0 x 7-0	"	2.000
Oak		
3-0 x 7-0	EA.	2.000
6-0 x 7-0	"	2.000
Pine		
2-4 x 7-0	EA.	1.600
2-6 x 7-0	"	1.600
2-8 x 7-0	"	1.600
3-0 x 7-0	"	1.600
3-4 x 7-0	"	1.600
6-0 x 7-0	"	2.667
08300.10 SPECIAL DOORS		
Vault door and frame, class 5, steel	EA.	8.000
Overhead door, coiling insulated		
Chain gear, no frame, 12' x 12'	EA.	10.000
Aluminum, bronze glass panels, 12-9 x 13-0	"	8.000
Garage door, flush insulated metal, primed, 9-0 x 7-0	"	2.667
Sliding metal fire doors, motorized, fusible link, 3 hr.		
3-0 x 6-8	EA.	16.000
3-8 x 6-8	"	16.000
4-0 x 8-0	"	16.000
5-0 x 8-0	"	16.000
Metal clad doors, including electric motor		
Light duty		
Minimum	S.F.	0.133
Maximum	"	0.320
Heavy duty		
Minimum	S.F.	0.400
Maximum	"	0.500
Hangar doors, based on 150' openings		
To 20' high	S.F.	0.096
20' to 40' high	"	0.060
40' to 60' high	"	0.040
60' to 80' high	"	0.024
Over 80' high	"	0.019
Counter doors, (roll-up shutters), standard, manual		
Opening, 4' high		
4' wide	EA.	6.667
6' wide	"	6.667
8' wide	"	7.273
10' wide	"	10.000
14' wide	"	10.000
6' high		
4' wide	EA.	6.667
6' wide	"	7.273
8' wide	"	8.000
10' wide	"	10.000

WOOD AND PLASTIC	UNIT	MAN/ HOURS
08300.10 SPECIAL DOORS		
14' wide	EA.	11.429
Service doors, (roll up shutters), standard, manual		
Opening		
8' high x 8' wide	EA.	4.444
10' high x 10' wide	"	6.667
12' high x 12' wide	"	10.000
14' high x 14' wide	"	13.333
16' high x 14' wide	"	13.333
20' high x 14' wide	"	20.000
24' high x 16' wide	"	17.778
Roll-up doors		
13-0 high x 14-0 wide	EA.	11.429
12-0 high x 14-0 wide	"	11.429
Top coiling grilles, manually operated, steel or aluminum		
Opening, 4' high x		
4' wide	EA.	3.200
6' wide	"	3.200
8' wide	"	4.444
12' wide	"	4.444
16' wide	"	6.667
6' high x		
4' wide	EA.	6.667
6' wide	"	7.273
8' wide	"	8.000
12' wide	"	8.889
16' wide	"	11.429
Side coiling grilles, manually operated, aluminum		
Opening, 8' high x		
18' wide	EA.	60.000
24' wide	"	68.571
12' high x		
12' wide	EA.	60.000
18' wide	"	68.571
24' wide	"	80.000
Accordion folding doors, tracks and fittings included		
Vinyl covered, 2 layers	S.F.	0.320
Woven mahogany and vinyl	"	0.320
Economy vinyl	"	0.320
Rigid polyvinyl chloride	"	0.320
Sectional wood overhead doors, frames not included		
Commercial grade, heavy duty, 1-3/4" thick, manual		
8' x 8'	EA.	6.667
10' x 10'	"	7.273
12' x 12'	"	8.000
Chain hoist		
12' x 16' high	EA.	10.000
14' x 14' high	"	13.333
20' x 8' high	"	16.000
16' high	"	20.000

WOOD AND PLASTIC	UNIT	MAN/ HOURS
08300.10 SPECIAL DOORS		
Sectional metal overhead doors, complete		
Residential grade, manual		
9' x 7'	EA.	3.200
16' x 7'	"	4.000
Commercial grade		
8' x 8'	EA.	6.667
10' x 10'	"	7.273
12' x 12'	"	8.000
20' x 14', with chain hoist	"	16.000
Sliding glass doors		
Tempered plate glass, 1/4" thick		
6' wide		
Economy grade	EA.	2.667
Premium grade	"	2.667
12' wide		
Economy grade	EA.	4.000
Premium grade	"	4.000
Insulating glass, 5/8" thick		
6' wide		
Economy grade	EA.	2.667
Premium grade	"	2.667
12' wide		
Economy grade	EA.	4.000
Premium grade	"	4.000
1" thick		
6' wide		
Economy grade	EA.	2.667
Premium grade	"	2.667
12' wide		
Economy grade	EA.	4.000
Premium grade	"	4.000
Vertical lift doors, channel frame construction		
20' high x		
10' wide	EA.	9.600
15' wide	"	9.600
20' wide	"	17.143
25' wide	"	17.143
25' high		
20' wide	EA.	17.143
25' wide	"	20.000
30' high		
25' wide	EA.	20.000
30' wide	"	20.000
35' wide	"	20.000
Residential storm door		
Minimum	EA.	1.333
Average	"	1.333
Maximum	"	2.000

08 DOORS AND WINDOWS

STOREFRONTS	UNIT	MAN/ HOURS
08410.10 STOREFRONTS		
Storefront, aluminum and glass		
Minimum	S.F.	0.100
Average	"	0.114
Maximum	"	0.133
Entrance doors		
1/2" thick glass		
3' x 7'	EA.	6.667
6' x 7'	"	10.000
3/4" thick glass		
3' x 7'	EA.	6.667
6' x 7'	"	10.000
1" thick glass		
3' x 7'	EA.	6.667
6' x 7'	"	10.000
Revolving doors		
7'diameter, 7' high		
Minimum	EA.	60.000
Average	"	96.000
Maximum	"	120

METAL WINDOWS	UNIT	MAN/ HOURS
08510.10 STEEL WINDOWS		
Steel windows, primed		
Casements		
Operable		
Minimum	S.F.	0.053
Maximum	"	0.050
Fixed sash	"	0.040
Double hung	"	0.044
Industrial windows		
Horizontally pivoted sash	S.F.	0.053
Fixed sash	"	0.044
Security sash		
Operable	S.F.	0.053
Fixed	"	0.044
Picture window	"	0.044
Projecting sash		
Minimum	S.F.	0.050
Maximum	"	0.050

METAL WINDOWS	UNIT	MAN/ HOURS
08510.10 STEEL WINDOWS		
Mullions	L.F.	0.040
08520.10 ALUMINUM WINDOWS		
Jalousie		
3-0 x 4-0	EA.	1.000
3-0 x 5-0	"	1.000
Fixed window		
6 sf to 8 sf	S.F.	0.114
12 sf to 16 sf	"	0.089
Projecting window		
6 sf to 8 sf	S.F.	0.200
12 sf to 16 sf	"	0.133
Horizontal sliding		
6 sf to 8 sf	S.F.	0.100
12 sf to 16 sf	"	0.080
Double hung		
6 sf to 8 sf	S.F.	0.160
10 sf to 12 sf	"	0.133
Storm window, 0.5 cfm, up to		
60 u.i. (united inches)	EA.	0.400
70 u.i.	"	0.400
80 u.i.	"	0.400
90 u.i.	"	0.444
100 u.i.	"	0.444
2.0 cfm, up to		
60 u.i.	EA.	0.400
70 u.i.	"	0.400
80 u.i.	"	0.400
90 u.i.	"	0.444
100 u.i.	"	0.444

WOOD AND PLASTIC	UNIT	MAN/ HOURS
08600.10 WOOD WINDOWS		
Double hung		
24" x 36"		
Minimum	EA.	0.800
Average	"	1.000

WOOD AND PLASTIC	UNIT	MAN/ HOURS
08600.10 WOOD WINDOWS		
Maximum	EA.	1.333
24" x 48"		
Minimum	EA.	0.800
Average	"	1.000
Maximum	"	1.333
30" x 48"		
Minimum	EA.	0.889
Average	"	1.143
Maximum	"	1.600
30" x 60"		
Minimum	EA.	0.889
Average	"	1.143
Maximum	"	1.600
Casement		
1 leaf, 22" x 38" high		
Minimum	EA.	0.800
Average	"	1.000
Maximum	"	1.333
2 leaf, 50" x 50" high		
Minimum	EA.	1.000
Average	"	1.333
Maximum	"	2.000
3 leaf, 71" x 62" high		
Minimum	EA.	1.000
Average	"	1.333
Maximum	"	2.000
4 leaf, 95" x 75" high		
Minimum	EA.	1.143
Average	"	1.600
Maximum	"	2.667
5 leaf, 119" x 75" high		
Minimum	EA.	1.143
Average	"	1.600
Maximum	"	2.667
Picture window, fixed glass, 54" x 54" high		
Minimum	EA.	1.000
Average	"	1.143
Maximum	"	1.333
68" x 55" high, minimum	"	1.000
Average	"	1.143
Maximum	"	1.333
Sliding, 40" x 31" high		
Minimum	EA.	0.800
Average	"	1.000
Maximum	"	1.333
52" x 39" high		
Minimum	EA.	1.000
Average	"	1.143
Maximum	"	1.333

WOOD AND PLASTIC	UNIT	MAN/ HOURS
08600.10 WOOD WINDOWS		
64" x 72" high		
Minimum	EA.	1.000
Average	"	1.333
Maximum	"	1.600
Awning windows		
34" x 21" high		
Minimum	EA.	0.800
Average	"	1.000
Maximum	"	1.333
40" x 21" high		
Minimum	EA.	0.889
Average	"	1.143
Maximum	"	1.600
48" x 27" high		
Minimum	EA.	0.889
Average	"	1.143
Maximum	"	1.600
60" x 36" high		
Minimum	EA.	1.000
Average	"	1.333
Maximum	"	1.600
Window frame, milled		0.200
Minimum	L.F.	0.160
Average	"	0.200
Maximum	"	0.267

HARDWARE	UNIT	MAN/ HOURS
08710.20 LOCKSETS		
Latchset, heavy duty		
Cylindrical	EA.	0.500
Mortise	"	0.800
Lockset, heavy duty		
Cylindrical	EA.	0.500
Mortise	"	0.800
Mortise locks and latchsets, chrome		
Latchset passage or closet latch	EA.	0.667
Privacy (bath or bedroom)	"	0.667
Entry lockset	"	0.667
Classroom lockset (outside key operated)	"	0.667

08 DOORS AND WINDOWS

HARDWARE	UNIT	MAN/ HOURS
08710.20 LOCKSETS		
Storeroom lock	EA.	0.667
Front door lock	"	0.667
Dormitory or exit lock	"	0.667
Preassembled locks and latches, brass		
Latchset, passage or closet latch	EA.	0.667
Lockset		
Privacy (bath or bathroom)	EA.	0.667
Entry lock	"	0.667
Classroom lock (outside key, operated)	"	0.667
Storeroom lock	"	0.667
Bored locks and latches, satin chrome plated		
Latchset passage or closet latch	EA.	0.667
Lockset		
Privacy (bath or bedroom)	EA.	0.667
Entry lock	"	0.667
Classroom lock	"	0.667
Corridor lock	"	0.667
Miscellaneous locks		
Exit lock with alarm, single door	EA.	3.200
Electric strike		
Rim mounted wrought steel	EA.	2.000
Mortised, wrought steel with bronze plating	"	3.200
Dead bolt		
Bored, wrought brass, keyed both sides	EA.	1.333
Mortised, cast brass	"	1.333
Lockset, cipher, mechanical	"	0.800
08710.30 CLOSERS		
Door closers		
Surface mounted, traditional type, parallel arm		
Standard	EA.	1.000
Heavy duty	"	1.000
Modern type, parallel arm, standard duty	"	1.000
Overhead, concealed, pivot hung, single acting		
Interior	EA.	1.000
Exterior	"	1.000
Floor concealed, single acting, offset, pivoted		
Interior	EA.	2.667
Exterior	"	2.667
08710.40 DOOR TRIM		
Door bumper, bronze, wall type	EA.	0.160
Wall type, 4" dia. with convex rubber pad, aluminum	"	0.160
Floor type		
Aluminum	EA.	0.160
Brass	"	0.160
Door holders		
Wall type, bronze	EA.	0.160

HARDWARE	UNIT	MAN/ HOURS
08710.40 DOOR TRIM		
Overhead	EA.	0.400
Floor type	"	0.400
Plunger type	"	0.400
Wall type, aluminum	"	0.400
Surface bolt	"	0.160
Panic device		2.000
Rim type with thumb piece	EA.	
Mortise	"	2.000
Vertical rod	"	2.000
Labeled, rim type	"	2.000
Mortise	"	2.000
Vertical rod	"	2.000
Silencers, rubber type	"	0.160
Dust proof strike with plate, brass	"	0.267
Flush bolt, lever extension, brass, rated	"	0.160
Surface bolt with strike, brass, 6" long	"	0.160
Door coordinator, labeled, brass, satin chrome	"	0.571
Door plates		
Kick plate, aluminum, 3 beveled edges		
10" x 28"	EA.	0.400
10" x 30"	"	0.400
10" x 34"	"	0.400
10" x 38"	"	0.400
Push plate, 4" x 16"		
Aluminum	EA.	0.160
Bronze	"	0.160
Stainless steel	"	0.160
Armor plate, 40" x 34"	"	0.320
Pull handle, 4" x 16"		
Aluminum	EA.	0.160
Bronze	"	0.160
Stainless steel	"	0.160
Hasp assembly		
3"	EA.	0.133
4-1/2"	"	0.178
6"	"	0.229
Electro-magnetic door holder		
Wall mounted	EA.	2.667
Floor mounted	"	2.667
Smoke detector door holder		
Photo electric type	EA.	2.667
Ionization type	"	2.667
Pneumatic operators, activated by rubber mats		
Swing		
Single	EA.	6.667
Double	"	10.000
Sliding		
Single	EA.	6.667
Double	"	10.000

HARDWARE	UNIT	MAN/ HOURS
08710.60 WEATHERSTRIPPING		
Weatherstrip, head and jamb, metal strip, neoprene bulb		
Standard duty	L.F.	0.044
Heavy duty	"	0.050
Spring type		
Metal doors	EA.	2.000
Wood doors	"	2.667
Sponge type with adhesive backing	"	0.800
Astragal		
1-3/4" x 13 ga., aluminum	L.F.	0.067
1-3/8" x 5/8", oak	"	0.053
Thresholds		
Bronze	L.F.	0.200
Aluminum		
Plain	L.F.	0.200
Vinyl insert	"	0.200
Aluminum with grit	"	0.200
Steel		
Plain	L.F.	0.200
Interlocking	"	0.667

GLAZING	UNIT	MAN/ HOURS
08810.10 GLAZING		
Sheet glass, 1/8" thick	S.F.	0.044
Plate glass, bronze or grey, 1/4" thick	"	0.073
Clear	"	0.073
Polished	"	0.073
Plexiglass		
1/8" thick	S.F.	0.073
1/4" thick	"	0.044
Float glass, clear		
3/16" thick	S.F.	0.067
1/4" thick	"	0.073
5/16" thick	"	0.080
3/8" thick	"	0.100
1/2" thick	"	0.133
5/8" thick	"	0.160
3/4" thick	"	0.200
1" thick	"	0.267
Tinted glass, polished plate, twin ground		

GLAZING	UNIT	MAN/ HOURS
08810.10 GLAZING		
3/16" thick	S.F.	0.067
1/4" thick	"	0.073
3/8" thick	"	0.100
1/2" thick	"	0.133
Total, full vision, all glass window system		
To 10' high		
Minimum	S.F.	0.200
Average	"	0.200
Maximum	"	0.200
10' to 20' high		
Minimum	S.F.	0.200
Average	"	0.200
Maximum	"	0.200
Insulated glass, bronze or gray		
1/2" thick	S.F.	0.133
1" thick	"	0.200
Clear		
1/2" thick	S.F.	0.133
1" thick	"	0.200
Spandrel glass, bronze or grey, polished, 1 side, 1/4" thick	"	0.073
Tempered glass (safety)		
Clear sheet glass		
1/8" thick	S.F.	0.044
3/16" thick	"	0.062
Clear float glass		
1/4" thick	S.F.	0.067
5/16" thick	"	0.080
3/8" thick	"	0.100
1/2" thick	"	0.133
5/8" thick	"	0.160
3/4" thick	"	0.267
Tinted float glass		
3/16" thick	S.F.	0.062
1/4" thick	"	0.067
3/8" thick	"	0.100
1/2" thick	"	0.133
Laminated glass		
Float safety glass with polyvinyl plastic interlayer		
1/4", sheet or float		
Two lites, 1/8" thick, clear glass	S.F.	0.067
1/2" thick, float glass		
Two lites, 1/4" thick, clear glass	S.F.	0.133
Tinted glass	"	0.133
Insulating glass, two lites, clear float glass		
1/2" thick	S.F.	0.133
5/8" thick	"	0.160
3/4" thick	"	0.200
7/8" thick	"	0.229
1" thick	"	0.267

GLAZING	UNIT	MAN/ HOURS
08810.10 GLAZING		
Glass seal edge		
3/8" thick	S.F.	0.133
Tinted glass		
1/2" thick	S.F.	0.133
1" thick	"	0.267
Tempered, clear		
1" thick	S.F.	0.267
Wire reinforced	"	0.267
Plate mirror glass		
1/4" thick		
15 sf	S.F.	0.080
Over 15 sf	"	0.073
Transparent, one way vision, 1/4" thick	"	0.080
Sheet mirror glass		
3/16" thick	S.F.	0.080
1/4" thick	"	0.067
Wall tiles, 12" x 12"		
Clear glass	S.F.	0.044
Veined glass	"	0.044
Wire glass, 1/4" thick		
Clear	S.F.	0.267
Hammered	"	0.267
Obscure	"	0.267
Bullet resistant glass, plate glass with inter-leaved vinyl		
1-3/16" thick		
To 15 sf	S.F.	0.400
Over 15 sf	"	0.400
2" thick		
To 15 sf	S.F.	0.667
Over 15 sf	"	0.667
Glazing accessories		
Neoprene glazing gaskets		
1/4" glass	L.F.	0.032
3/8" glass	"	0.033
1/2" glass	"	0.035
3/4" glass	"	0.036
1" glass	"	0.040
Mullion section		
1/4" glass	L.F.	0.016
3/8" glass	"	0.020
1/2" glass	"	0.023
3/4" glass	"	0.027
1" glass	"	0.032
Moulded corners	EA.	0.533

GLAZED CURTAIN WALLS	UNIT	MAN/ HOURS
08910.10 GLAZED CURTAIN WALLS		
Curtain wall, aluminum system, framing sections		
2" x 3"		
Jamb	L.F.	0.067
Horizontal	"	0.067
Mullion	"	0.067
2" x 4"		
Jamb	L.F.	0.100
Horizontal	"	0.100
Mullion	"	0.100
3" x 5-1/2"		
Jamb	L.F.	0.100
Horizontal	"	0.100
Mullion	"	0.100
4" corner mullion	"	0.133
Coping sections		
1/8" x 8"	L.F.	0.133
1/8" x 9"	"	0.133
1/8" x 12-1/2"	"	0.160
Sill section		
1/8" x 6"	L.F.	0.080
1/8" x 7"	"	0.080
1/8" x 8-1/2"	"	0.080
Column covers, aluminum		
1/8" x 26"	L.F.	0.200
1/8" x 34"	"	0.211
1/8" x 38"	"	0.211
Doors		
Aluminum framed, standard hardware		
Narrow stile		
2-6 x 7-0	EA.	
3-0 x 7-0	"	
3-6 x 7-0	"	
Wide stile		
2-6 x 7-0	EA.	
3-0 x 7-0	"	
3-6 x 7-0	"	
Flush panel doors, to match adjacent wall panels		
2-6 x 7-0	EA.	5.000
3-0 x 7-0	"	5.000
3-6 x 7-0	"	5.000
Wall panel, insulated		
"U"=.08	S.F.	0.067
"U"=.10	"	0.067
"U"=.15	"	0.067
Window wall system, complete		
Minimum	S.F.	
Average	"	
Maximum	"	

09 FINISHES

SUPPORT SYSTEMS

09110.10 METAL STUDS

SUPPORT SYSTEMS	UNIT	MAN/HOURS
Studs, non load bearing, galvanized		
2-1/2", 20 ga.		
12" o.c.	S.F.	0.017
16" o.c.	"	0.013
25 ga.		
12" o.c.	S.F.	0.017
16" o.c.	"	0.013
24" o.c.	"	0.011
3-5/8", 20 ga.		
12" o.c.	S.F.	0.020
16" o.c.	"	0.016
24" o.c.	"	0.013
25 ga.		
12" o.c.	S.F.	0.020
16" o.c.	"	0.016
24" o.c.	"	0.013
4", 20 ga.		
12" o.c.	S.F.	0.020
16" o.c.	"	0.016
24" o.c.	"	0.013
25 ga.		
12" o.c.	S.F.	0.020
16" o.c.	"	0.016
24" o.c.	"	0.013
6", 20 ga.		
12" o.c.	S.F.	0.025
16" o.c.	"	0.020
24" o.c.	"	0.017
25 ga.		
12" o.c.	S.F.	0.025
16" o.c.	"	0.020
24" o.c.	"	0.017
Load bearing studs, galvanized		
3-5/8", 16 ga.		
12" o.c.	S.F.	0.020
16" o.c.	"	0.016
18 ga.		
12" o.c.	S.F.	0.013
16" o.c.	"	0.016
4", 16 ga.		
12" o.c.	S.F.	0.020
16" o.c.	"	0.016
6", 16 ga.		
12" o.c.	S.F.	0.025
16" o.c.	"	0.020
Furring		
On beams and columns		
7/8" channel	L.F.	0.053
1-1/2" channel	"	0.053

SUPPORT SYSTEMS

09110.10 METAL STUDS

SUPPORT SYSTEMS	UNIT	MAN/HOURS
On ceilings		
3/4" furring channels		
12" o.c.	S.F.	0.057
16" o.c.	"	0.053
24" o.c.	"	0.029
1-1/2" furring channels		0.044
12" o.c.	S.F.	0.036
16" o.c.	"	0.033
24" o.c.	"	0.031
On walls		
3/4" furring channels		
12" o.c.	S.F.	0.027
16" o.c.	"	0.025
24" o.c.	"	0.024
1-1/2" furring channels		
12" o.c.	S.F.	0.029
16" o.c.	"	
24" o.c.	"	0.025

LATH AND PLASTER

09205.10 GYPSUM LATH

LATH AND PLASTER	UNIT	MAN/HOURS
Gypsum lath, 1/2" thick		
Clipped	S.Y.	0.044
Nailed	"	0.050

09205.20 METAL LATH

LATH AND PLASTER	UNIT	MAN/HOURS
Diamond expanded, galvanized		
2.5 lb., on walls		
Nailed	S.Y.	0.100
Wired	"	0.114
On ceilings		
Nailed	S.Y.	0.114
Wired	"	0.133
3.4 lb., on walls		
Nailed	S.Y.	0.100
Wired	"	0.114
On ceilings		
Nailed	S.Y.	0.114
Wired	"	0.133

09 FINISHES

LATH AND PLASTER	UNIT	MAN/HOURS
09205.20 METAL LATH		
Flat rib		
2.75 lb., on walls		
Nailed	S.Y.	0.100
Wired	"	0.114
On ceilings		
Nailed	S.Y.	0.114
Wired	"	0.133
3.41 lb., on walls		
Nailed	S.Y.	0.100
Wired	"	0.114
On ceilings		
Nailed	S.Y.	0.114
Wired	"	0.133
Stucco lath		
1.8 lb.	S.Y.	0.100
3.6 lb.	"	0.100
Paper backed		
Minimum	S.Y.	0.080
Maximum	"	0.114
09205.60 PLASTER ACCESSORIES		
Expansion joint, 3/4", 26 ga., galvanized, one piece	L.F.	0.020
Plaster corner beads, 3/4", galvanized	"	0.023
Casing bead, expanded flange, galvanized	"	0.020
Expanded wing, 1-1/4" wide, galvanized	"	0.020
Joint clips for lath	EA.	0.004
Metal base, galvanized, 2-1/2" high	L.F.	0.027
Stud clips for gypsum lath	EA.	0.004
Sound deadening board, 1/4", nailed or clipped	S.F.	0.013
09210.10 PLASTER		
Gypsum plaster, trowel finish, 2 coats		
Ceilings	S.Y.	0.190
Walls	"	0.178
3 coats		
Ceilings	S.Y.	0.267
Walls	"	0.229
Vermiculite plaster		
2 coats		
Ceilings	S.Y.	0.286
Walls	"	0.267
3 coats		
Ceilings	S.Y.	0.364
Walls	"	0.320
Keenes cement plaster		
2 coats		
Ceilings	S.Y.	0.229
Walls	"	0.200

LATH AND PLASTER	UNIT	MAN/HOURS
09210.10 PLASTER		
3 coats		
Ceilings	S.Y.	0.267
Walls	"	0.229
On columns, add to installation, 50%	"	
Chases, fascia, and soffits, add to installation, 50%	"	
Beams, add to installation, 50%	"	
Patch cracks		0.040
09220.10 PORTLAND CEMENT PLASTER		
Stucco, portland, gray, 3 coat, 1" thick		
Sand finish	S.Y.	0.348
Trowel finish	"	0.364
White cement		
Sand finish	S.Y.	0.364
Trowel finish	"	0.400
Scratch coat		
For ceramic tile	S.Y.	0.080
For quarry tile	"	0.080
Portland cement plaster		
2 coats, 1/2"	S.Y.	0.018
3 coats, 7/8"	"	0.023
09250.10 GYPSUM BOARD		
Drywall, plasterboard, 3/8" clipped to		
Metal furred ceiling	S.F.	0.009
Columns and beams	"	0.020
Walls	"	0.008
Nailed or screwed to		
Wood framed ceiling	S.F.	0.008
Columns and beams	"	0.018
Walls	"	0.007
1/2", clipped to		
Metal furred ceiling	S.F.	0.009
Columns and beams	"	0.020
Walls	"	0.008
Nailed or screwed to		
Wood framed ceiling	S.F.	0.008
Columns and beams	"	0.018
Walls	"	0.007
5/8", clipped to		
Metal furred ceiling	S.F.	0.010
Columns and beams	"	0.022
Walls	"	0.009
Nailed or screwed to		
Wood framed ceiling	S.F.	0.010
Columns and beams	"	0.022
Walls	"	0.009
Vinyl faced, clipped to metal studs		

LATH AND PLASTER

09250.10 GYPSUM BOARD

	UNIT	MAN/HOURS
1/2"	S.F.	0.010
5/8"	"	0.010
Taping and finishing joints		0.005
Minimum	S.F.	0.005
Average	"	0.007
Maximum	"	0.008
Casing bead		0.027
Minimum	L.F.	0.023
Average	"	0.027
Maximum	"	0.040
Corner bead		0.027
Minimum	L.F.	0.023
Average	"	0.027
Maximum	"	0.040

TILE

09310.10 CERAMIC TILE

	UNIT	MAN/HOURS
Glazed wall tile, 4-1/4" x 4-1/4"		
Minimum	S.F.	0.057
Average	"	0.067
Maximum	"	0.080
Base, 4-1/4" high		
Minimum	L.F.	0.100
Average	"	0.100
Maximum	"	0.100
Unglazed floor tile		
Portland cement bed, cushion edge, face mounted		
1" x 1"	S.F.	0.073
1" x 2"	"	0.070
2" x 2"	"	0.067
Adhesive bed, with white grout		
1" x 1"	S.F.	0.073
1" x 2"	"	0.070
2" x 2"	"	0.067
Organic adhesive bed, thin set, back mounted		
1" x 1"	S.F.	0.073
1" x 2"	"	0.070
2" x 2"	"	0.067
Unglazed wall tile		
Organic adhesive, face mounted cushion edge		
1" x 1"		
Minimum	S.F.	0.067
Average	"	0.073
Maximum	"	0.080
1" x 2"		
Minimum	S.F.	0.064
Average	"	0.070
Maximum	"	0.076
2" x 2"		
Minimum	S.F.	0.062
Average	"	0.067
Maximum	"	0.073
Back mounted		
1" x 1"		
Minimum	S.F.	0.067
Average	"	0.073
Maximum	"	0.080
1" x 2"		
Minimum	S.F.	0.064
Average	"	0.070
Maximum	"	0.076
2" x 2"		
Minimum	S.F.	0.062
Average	"	0.067
Maximum	"	0.073
Conductive floor tile, unglazed square edged		
Portland cement bed		
1 x 1	S.F.	0.100
1-9/16 x 1-9/16	"	0.100
Dry set		
1 x 1	S.F.	0.100
1-9/16 x 1-9/16	"	0.100
Epoxy bed with epoxy joints		
1 x 1	S.F.	0.100
1-9/16 x 1-9/16	"	0.100
Ceramic accessories		
Towel bar, 24" long		
Minimum	EA.	0.320
Average	"	0.400
Maximum	"	0.533
Soap dish		0.667
Minimum	EA.	0.533
Average	"	0.667
Maximum	"	0.800

09330.10 QUARRY TILE

	UNIT	MAN/HOURS
Floor		
4 x 4 x 1/2"	S.F.	0.107
6 x 6 x 1/2"	"	0.100

09 FINISHES

TILE

TILE	UNIT	MAN/HOURS
09330.10 QUARRY TILE		
6 x 6 x 3/4"	S.F.	0.100
Wall, applied to 3/4" portland cement bed		
4 x 4 x 1/2"	S.F.	0.160
6 x 6 x 3/4"	"	0.133
Cove base		
5 x 6 x 1/2" straight top	L.F.	0.133
6 x 6 x 3/4" round top	"	0.133
Stair treads 6 x 6 x 3/4"	"	0.200
Window sill 6 x 8 x 3/4"	"	0.160
For abrasive surface, add to material, 25%		
09410.10 TERRAZZO		
Floors on concrete, 1-3/4" thick, 5/8" topping		
Gray cement	S.F.	0.114
White cement	"	0.114
Sand cushion, 3" thick, 5/8" top, 1/4"		
Gray cement	S.F.	0.133
White cement	"	0.133
Monolithic terrazzo, 3-1/2" base slab, 5/8" topping	"	0.100
Terrazzo wainscot, cast-in-place, 1/2" thick	"	0.200
Base, cast in place, terrazzo cove type, 6" high	L.F.	0.114
Curb, cast in place, 6" wide x 6" high, polished top	"	0.400
Stairs, cast-in-place, topping on concrete or metal		
1-1/2" thick treads, 12" wide	L.F.	0.133
Combined tread and riser	"	0.400
Precast terrazzo, thin set		
Terrazzo tiles, non-slip surface		
9" x 9" x 1" thick	S.F.	0.114
12" x 12"		
1" thick	S.F.	0.107
1-1/2" thick	"	0.114
18" x 18" x 1-1/2" thick	"	0.114
24" x 24" x 1-1/2" thick	"	0.094
Terrazzo wainscot		
12" x 12" x 1" thick	S.F.	0.200
18" x 18" x 1-1/2" thick	"	0.229
Base		
6" high		
Straight	L.F.	0.062
Coved	"	0.062
8" high		
Straight	L.F.	0.067
Coved	"	0.067
Terrazzo curbs		
8" wide x 8" high	L.F.	0.145
6" wide x 6" high	"	0.145
Precast terrazzo stair treads, 12" wide		
1-1/2" thick		
Diamond pattern	L.F.	0.145

TILE	UNIT	MAN/HOURS
09410.10 TERRAZZO		
Non-slip surface	L.F.	0.145
2" thick		
Diamond pattern	L.F.	0.145
Non-slip surface	"	0.160
Stair risers, 1" thick to 6" high		
Straight sections	L.F.	0.080
Cove sections	"	0.080
Combined tread and riser		
Straight sections		
1-1/2" tread, 3/4" riser	L.F.	0.229
3" tread, 1" riser	"	0.229
Curved sections		
2" tread, 1" riser	L.F.	0.267
3" tread, 1" riser	"	0.267
Stair stringers, notched for treads and risers		
1" thick	L.F.	0.200
2" thick	"	0.267
Landings, structural, nonslip		
1-1/2" thick	S.F.	0.133
3" thick	"	0.160
Conductive terrazzo, spark proof industrial floor		
Epoxy terrazo		
Floor	S.F.	0.050
Base	"	0.067
Polyacrylate		
Floor	S.F.	0.050
Base	"	0.067
Polyester		
Floor	S.F.	0.032
Base	"	0.040
Synthetic latex mastic		
Floor	S.F.	0.050
Base	"	0.067

ACOUSTICAL TREATMENT

ACOUSTICAL TREATMENT	UNIT	MAN/HOURS
09510.10 CEILINGS AND WALLS		
Acoustical panels, suspension system not included		
Fiberglass panels		
5/8" thick		

ACOUSTICAL TREATMENT	UNIT	MAN/ HOURS
09510.10 CEILINGS AND WALLS		
2' x 2'	S.F.	0.009
2' x 4'	"	0.008
3/4" thick		
2' x 2'	S.F.	0.009
2' x 4'	"	0.008
Glass cloth faced fiberglass panels		
3/4" thick	S.F.	0.010
1" thick	"	0.010
Mineral fiber panels		
5/8" thick		
2' x 2'	S.F.	0.009
2' x 4'	"	0.008
3/4" thick		
2' x 2'	S.F.	0.009
2' x 4'	"	0.008
Wood fiber panels		
2" thick		
2' x 2'	S.F.	0.009
2' x 4'	"	0.008
Air distributing panels		
3/4" thick	S.F.	0.020
5/8" thick	"	0.016
Acoustical tiles, suspension system not included		
Fiberglass tile, 12" x 12"		
5/8" thick	S.F.	0.015
3/4" thick	"	0.018
Glass cloth faced fiberglass tile		
3/4" thick	S.F.	0.018
3" thick	"	0.020
Mineral fiber tile, 12" x 12"		
5/8" thick		
Standard	S.F.	0.016
Vinyl faced	"	0.016
3/4" thick		
Standard	S.F.	0.016
Vinyl faced	"	0.016
Fire rated	"	0.016
Aluminum or mylar faced	"	0.016
Wood fiber tile, 12" x 12"		
1/2" thick	S.F.	0.016
3/4" thick	"	0.016
Metal pan units, 24 ga. steel		
12" x 12"	S.F.	0.018
12" x 24"	"	0.016
Aluminum, .025" thick		
12" x 12"	S.F.	0.018
12" x 24"	"	0.016
Anodized aluminum, 0.25" thick		
12" x 12"	S.F.	0.018

ACOUSTICAL TREATMENT	UNIT	MAN/ HOURS
09510.10 CEILINGS AND WALLS		
12" x 24"	S.F.	0.016
Stainless steel, 24 ga.		
12" x 12"	S.F.	0.018
12" x 24"	"	0.016
Metal ceiling systems		
.020" thick panels		
10', 12', and 16' lengths	S.F.	0.023
Custom lengths, 3' to 20'	"	0.023
.025" thick panels		
32 sf, 38 sf, and 52 sf pieces	S.F.	0.027
Custom lengths, 10 sf to 65 sf	"	0.027
Sound absorption walls, with fabric cover		
2-6" x 9' x 3/4"	S.F.	0.027
2' x 9' x 1"	"	0.027
Starter spline	L.F.	0.020
Internal spline	"	0.020
Acoustical treatment		
Barriers for plenums		
Leaded vinyl		
0.48 lb per sf	S.F.	0.038
0.87 lb per sf	"	0.040
Aluminum foil, fiberglass reinforcement		
Minimum	S.F.	0.027
Maximum	"	0.040
Aluminum mesh, paper backed	"	0.027
Fibered cement sheet, 3/16" thick	"	0.029
Sheet lead, 1/64" thick	"	0.020
Sound attenuation blanket		
1" thick	S.F.	0.080
1-1/2" thick	"	0.080
2" thick	"	0.080
3" thick	"	0.089
Ceiling suspension systems		
T bar system		
2' x 4'	S.F.	0.008
2' x 2'	"	0.009
Concealed Z bar suspension system, 12" module	"	0.013

09 FINISHES

FLOORING

09550.10 WOOD FLOORING	UNIT	MAN/ HOURS
Wood strip flooring, unfinished		
Fir floor		
C and better		
Vertical grain	S.F.	0.027
Flat grain	"	0.027
Oak floor		
Minimum	S.F.	0.038
Average	"	0.038
Maximum	"	0.038
Maple floor		
25/32" x 2-1/4"		
Minimum	S.F.	0.038
Maximum	"	0.038
33/32" x 3-1/4"		
Minimum	S.F.	0.038
Maximum	"	0.038
Wood block industrial flooring		
Creosoted		
2" thick	S.F.	0.021
2-1/2" thick	"	0.025
3" thick	"	0.027
Parquet, 5/16", white oak		
Finished	S.F.	0.040
Unfinished	"	0.040
Gym floor, 2 ply felt, 25/32" maple, finished, in mastic	"	0.044
Over wood sleepers	"	0.050
Finishing, sand, fill, finish, and wax	"	0.020
Refinish sand, seal, and 2 coats of polyurethane	"	0.027
Clean and wax floors	"	0.004

09630.10 UNIT MASONRY FLOORING	UNIT	MAN/ HOURS
Clay brick		
9 x 4-1/2 x 3" thick		
Glazed	S.F.	0.067
Unglazed	"	0.067
8 x 4 x 3/4" thick		
Glazed	S.F.	0.070
Unglazed	"	0.070

09660.10 RESILIENT TILE FLOORING	UNIT	MAN/ HOURS
Solid vinyl tile, 1/8" thick, 12" x 12"		
Marble patterns	S.F.	0.020
Solid colors	"	0.020
Travertine patterns	"	0.020
Conductive resilient flooring, vinyl tile		
1/8" thick, 12" x 12"	S.F.	0.023

09665.10 RESILIENT SHEET FLOORING	UNIT	MAN/ HOURS
Vinyl sheet flooring		
Minimum	S.F.	0.008
Average	"	0.010
Maximum	"	0.013
Cove, to 6"	L.F.	0.016
Fluid applied resilient flooring		
Polyurethane, poured in place, 3/8" thick	S.F.	0.067
Vinyl sheet goods, backed		
0.070" thick	S.F.	0.010
0.093" thick	"	0.010
0.125" thick	"	0.010
0.250" thick	"	0.010
Wall base, vinyl		
Group 1		
4" high	L.F.	0.027
6" high	"	0.027
Group 2		
4" high	L.F.	0.027
6" high	"	0.027
Group 3		
4" high	L.F.	0.027
6" high	"	0.027
Stair accessories		
Treads, 1/4" x 12", rubber diamond surface		
Marbled	L.F.	0.067
Plain	"	0.067
Grit strip safety tread, 12" wide, colors		
3/16" thick	L.F.	0.067
5/16" thick	"	0.067
Risers, 7" high, 1/8" thick, colors		
Flat	L.F.	0.040
Coved	"	0.040
Nosing, rubber		
3/16" thick, 3" wide		
Black	L.F.	0.040
Colors	"	0.040
6" wide		
Black	L.F.	0.067
Colors	"	0.067

09 FINISHES

CARPET

09680.10 FLOOR LEVELING	UNIT	MAN/HOURS
Repair and level floors to receive new flooring		
Minimum	S.Y.	0.027
Average	"	0.067
Maximum	"	0.080

09682.10 CARPET PADDING	UNIT	MAN/HOURS
Carpet padding		
Foam rubber, waffle type, 0.3" thick	S.Y.	0.040
Jute padding		
Minimum	S.Y.	0.036
Average	"	0.040
Maximum	"	0.044
Sponge rubber cushion		
Minimum	S.Y.	0.036
Average	"	0.040
Maximum	"	0.044
Urethane cushion, 3/8" thick		
Minimum	S.Y.	0.036
Average	"	0.040
Maximum	"	0.044

09685.10 CARPET	UNIT	MAN/HOURS
Carpet, acrylic		
24 oz., light traffic	S.Y.	0.044
28 oz., medium traffic	"	0.044
Residential		
Nylon		
15 oz., light traffic	S.Y.	0.044
28 oz., medium traffic	"	0.044
Commercial		
Nylon		
28 oz., medium traffic	S.Y.	0.044
35 oz., heavy traffic	"	0.044
Wool		
30 oz., medium traffic	S.Y.	0.044
36 oz., medium traffic	"	0.044
42 oz., heavy traffic	"	0.044
Carpet tile		
Foam backed		0.008
Minimum	S.F.	0.008
Average	"	0.009
Maximum	"	0.010
Tufted loop or shag		0.008
Minimum	S.F.	0.008
Average	"	0.009
Maximum	"	0.010
Clean and vacuum carpet		
Minimum	S.Y.	0.004
Average	"	0.005

CARPET

09685.10 CARPET	UNIT	MAN/HOURS
Maximum	S.Y.	0.008

09700.10 SPECIAL FLOORING	UNIT	MAN/HOURS
Epoxy flooring, marble chips		
Epoxy with colored quartz chips in 1/4" base	S.F.	0.044
Heavy duty epoxy topping, 3/16" thick	"	0.044
Epoxy terrazzo		
1/4" thick chemical resistant	S.F.	0.050

SPECIAL COATINGS

09800.10 FIREPROOFING	UNIT	MAN/HOURS
Sprayed fireproofing, fluted decking		
1 hr fire rating, 1/2" thick	S.F.	0.004
2 hr fire rating, 7/8" thick	"	0.005
Sprayed on roofs and beams		
1 hr fire rating, 1-3/8" thick	S.F.	0.007
2 hr fire rating, 1-1/2" thick	"	0.007
3 hr fire rating, 1-5/8" thick	"	0.007
Sprayed on beams		
1 hr fire rating, 7/16" thick	S.F.	0.004
2 hr fire rating, 1/2" thick	"	0.004
3 hr fire rating, 7/8" thick	"	0.005
4 hr fire rating, 1-1/4" thick	"	0.007
Sprayed on columns		
1 hr fire rating, 1-3/16" thick	S.F.	0.005
2 hr fire rating, 1-1/4" thick	"	0.007
3 hr fire rating, 1-7/8" thick	"	0.007
4 hr fire rating, 2-1/2" thick	"	0.009
Sprayed on flat surfaces		
2 hr fire rating, 5/8" thick	S.F.	0.004
3 hr fire rating, 7/8" thick	"	0.005
4 hr fire rating, 1-1/4" thick	"	0.007

09 FINISHES

PAINTING

09910.10 EXTERIOR PAINTING	UNIT	MAN/ HOURS
Exterior painting		
Wood surfaces, 1 coat primer, two coats paint		
Door and frame	EA.	1.333
Windows	S.F.	0.020
Wood trim	"	0.020
Wood siding	"	0.010
Hardboard surfaces		
One coat primer, two coats paint	S.F.	0.010
Asbestos cement surfaces		
One coat primer, two coats paint	S.F.	0.010
Galvanized surfaces, galvanized primer		
One coat primer, two coats paint	S.F.	0.009
Stucco surfaces, acrylic primer, acrylic latex paint		
One coat primer, two coats paint	S.F.	0.013
Concrete masonry unit surfaces, brush work		
One coat filler, one coat paint	S.F.	0.010
Two coats epoxy	"	0.013
Texture coating	"	0.008
Concrete surfaces		
One coat filler, one coat paint	S.F.	0.010
Two coats paint	"	0.013
Structural steel		
One field coat paint, brush work		
Light framing	S.F.	0.007
Heavy framing	"	0.004
One field coat paint, spray work		
Light framing	S.F.	0.002
Heavy framing	"	0.002
Miscellaneous steel items, spray work, one coat		
Exposed decking	S.F.	0.007
Joist	"	0.010
Columns	"	0.010
Pipes, one coat primer, one coat paint		
4" dia.	L.F.	0.010
8" dia.	"	0.013
12" dia.	"	0.020
Paint letters on pipe with brush	"	0.020
Paint pipe insulation cloth cover	S.F.	0.010
Sprinkler system piping	L.F.	0.007
Miscellaneous surfaces		
Stair pipe rails		
Two rails	L.F.	0.027
One rail	"	0.016
Stair to 4' wide, including rails, per riser	EA.	0.114
Gratings and frames	S.F.	0.027
Ladders	L.F.	0.023
Miscellaneous exposed metal	S.F.	0.008

PAINTING

09920.10 INTERIOR PAINTING	UNIT	MAN/ HOURS
Walls, concrete and masonry, brush, primer, acrylic		
One coat primer, one coat paint	S.F.	0.010
Two coats paint	"	0.013
Plywood, paint	"	0.004
Natural finish	"	0.005
Wood, paint	"	0.005
Natural finish	"	0.006
Metal		
One coat filler	S.F.	0.005
One coat primer, one coat paint	"	0.010
Two coats paint	"	0.013
Plaster or gypsum board, paint	"	0.004
Epoxy	"	0.005
Ceilings, one coat paint, wood	"	0.006
Concrete	"	0.005
Plaster	"	0.004
Miscellaneous metal, brushwork	"	0.010

09920.30 DOORS AND MILLWORK	UNIT	MAN/ HOURS
Painting, doors		
Minimum	S.F.	0.027
Average	"	0.040
Maximum	"	0.053
Cabinets, shelves, and millwork		
Minimum	S.F.	0.013
Average	"	0.023
Maximum	"	0.040

09920.60 WINDOWS	UNIT	MAN/ HOURS
Painting, windows		
Minimum	S.F.	0.016
Average	"	0.020
Maximum	"	0.032

09955.10 WALL COVERING	UNIT	MAN/ HOURS
Vinyl wall covering		
Medium duty	S.F.	0.011
Heavy duty	"	0.013
Over pipes and irregular shapes		
Lightweight, 13 oz.	S.F.	0.016
Medium weight, 25 oz.	"	0.018
Heavy weight, 34 oz.	"	0.020
Cork wall covering		
1' x 1' squares		
1/4" thick	S.F.	0.020
1/2" thick	"	0.020
3/4" thick	"	0.020
Wall fabrics		

PAINTING	UNIT	MAN/ HOURS
09955.10 WALL COVERING		
Natural fabrics, grass cloths		
Minimum	S.F.	0.012
Average	"	0.013
Maximum	"	0.016
Flexible gypsum coated wall fabric, fire resistant	"	0.008
Vinyl corner guards		
3/4" x 3/4" x 8'	EA.	0.100
2-3/4" x 2-3/4" x 4'	"	0.100
09980.10 PAINTING PREPARATION		
Cleaning, light		
Wood	S.F.	0.002
Plaster or gypsum wallboard	"	0.002
Prepare sprinkler piping for painting	L.F.	0.002
Prepare insulated pipe for painting	S.F.	0.002
Normal painting prep, masonry and concrete		
Unpainted	S.F.	0.001
Painted	"	0.002
Plaster or gypsum		
Unpainted	S.F.	0.001
Painted	"	0.002
Wood		
Unpainted	S.F.	0.001
Painted	"	0.002
Painted steel, light rusting	"	0.003
Sandblasting		
Brush off blast	S.F.	0.005
Commercial blast	"	0.013
Near white metal blast	"	0.023
White metal blast	"	0.027

SPECIALTIES	UNIT	MAN/ HOURS
10110.10 CHALKBOARDS		
Chalkboard, metal frame, 1/4" thick		
48"x60"	EA.	0.800
48"x96"	"	0.889
48"x144"	"	1.000
48"x192"	"	1.143
Liquid chalkboard		
48"x60"	EA.	0.800
48"x96"	"	0.889
48"x144"	"	1.000
48"x192"	"	1.143
Map rail, deluxe	L.F.	0.040
Average	PCT.	
10165.10 TOILET PARTITIONS		
Toilet partition, plastic laminate		
Ceiling mounted	EA.	2.667
Floor mounted	"	2.000
Metal		
Ceiling mounted	EA.	2.667
Floor mounted	"	2.000
Wheel chair partition, plastic laminate		
Ceiling mounted	EA.	2.667
Floor mounted	"	2.000
Painted metal		
Ceiling mounted	EA.	2.667
Floor mounted	"	2.000
Urinal screen, plastic laminate		
Wall hung	EA.	1.000
Floor mounted	"	1.000
Porcelain enameled steel, floor mounted	"	1.000
Painted metal, floor mounted	"	1.000
Stainless steel, floor mounted	"	1.000
Metal toilet partitions		
Toilet partitions, front door and side divider, floor mounted		
Porcelain enameled steel	EA.	2.000
Painted steel	"	2.000
Stainless steel	"	2.000
10185.10 SHOWER STALLS		
Shower receptors		
Precast, terrazzo		
32" x 32"	EA.	0.667
32" x 48"	"	0.800
Concrete		
32" x 32"	EA.	0.667
48" x 48"	"	0.889
Shower door, trim and hardware		
Economy, 24" wide, chrome frame, tempered glass	EA.	0.800

SPECIALTIES	UNIT	MAN/ HOURS
10185.10 SHOWER STALLS		
Porcelain enameled steel, flush	EA.	0.800
Baked enameled steel, flush	"	0.800
Aluminum frame, tempered glass, 48" wide, sliding	"	1.000
Folding	"	1.000
Aluminum frame and tempered glass, molded plastic		
Complete with receptor and door		
32" x 32"	EA.	2.000
36" x 36"	"	2.000
40" x 40"	"	2.286
Shower compartment, precast concrete receptor		
Single entry type		
Porcelain enameled steel	EA.	8.000
Baked enameled steel	"	8.000
Stainless steel	"	8.000
Double entry type		
Porcelain enameled steel	EA.	10.000
Baked enameled steel	"	10.000
Stainless steel	"	10.000
10190.10 CUBICLES		
Hospital track		
Ceiling hung	L.F.	0.089
Suspended	"	0.114
Hospital metal dividers, galvanized steel		
Baked enamel finish		
54" high		
10" glass light	L.F.	0.400
14" glass light	"	0.400
24" glass light	"	0.400
60" high		
10" glass light	L.F.	0.444
14" glass light	"	0.444
24" glass light	"	0.444
Stainless steel		
54" high		
10" glass light	L.F.	0.444
14" glass light	"	0.444
24" glass light	"	0.444
60" high		
14" glass light	L.F.	0.500
14" glass light	"	0.500
24" glass light	"	0.500
10210.10 VENTS AND WALL LOUVERS		
Block vent, 8"x16"x4" aluminum, w/screen, mill finish	EA.	0.267
Standard	"	0.250
Vents w/screen, 4" deep, 8" wide, 5" high		
Modular	EA.	0.250
Aluminum gable louvers	S.F.	0.133

10 SPECIALTIES

SPECIALTIES	UNIT	MAN/ HOURS
10210.10 VENTS AND WALL LOUVERS		
Vent screen aluminum, 4" wide, continuous	L.F.	0.027
Louvers, aluminum, anodized, fixed blade		
Horizontal line	S.F.	0.200
Vertical line	"	0.200
Wall louver, aluminum mill finish		
Under, 2 sf	S.F.	0.100
2 to 4 sf	"	0.089
5 to 10 sf	"	0.089
Galvanized steel		
Under 2 sf	S.F.	0.100
2 to 4 sf	"	0.089
5 to 10 sf	"	0.089
10225.10 DOOR LOUVERS		
Fixed, 1" thick, enameled steel		
8"x8"	EA.	0.100
12"x8"	"	0.100
12"x12"	"	0.114
16"x12"	"	0.123
18"x12"	"	0.200
20"x8"	"	0.114
20"x12"	"	0.229
20"x16"	"	0.267
20"x20"	"	0.320
24"x12"	"	0.267
24"x16"	"	0.286
24"x18"	"	0.308
24"x20"	"	0.333
24"x24"	"	0.364
26"x26"	"	0.500
10270.40 ACCESS & PEDESTAL FLOOR		
Panels, no covering, 2'x2'		
Plain	S.F.	0.010
Perforated	"	0.400
Pedestals		
For 6" to 12" clearance	EA.	0.080
Stringers		
2'	L.F.	0.038
6'	"	0.027
Accessories		
Ramp assembly	S.F.	0.032
Elevated floor assembly	"	0.030
Handrail	L.F.	0.400
Fascia plate	"	0.200
RF shielding components, floor liner		
Hot rolled steel sheet		
14 ga.	S.F.	0.020
11 ga.	"	0.062

SPECIALTIES	UNIT	MAN/ HOURS
10290.10 PEST CONTROL		
Termite control		
Under slab spraying		
Minimum	S.F.	0.002
Average	"	0.004
Maximum	"	0.008
10350.10 FLAGPOLES		
Installed in concrete base		
Fiberglass		
25' high	EA.	5.333
50' high	"	13.333
Aluminum		
25' high	EA.	5.333
50' high	"	13.333
Bonderized steel		
25' high	EA.	6.154
50' high	"	16.000
Freestanding tapered, fiberglass		
30' high	EA.	5.714
40' high	"	7.273
50' high	"	8.000
60' high	"	9.412
Wall mounted, with collar, brushed aluminum finish		
15' long	EA.	4.000
18' long	"	4.000
20' long	"	4.211
24' long	"	4.706
Outrigger, wall, including base		
10' long	EA.	5.333
20' long	"	6.667
10400.10 IDENTIFYING DEVICES		
Directory and bulletin boards		
Open face boards		
Chrome plated steel frame	S.F.	0.400
Aluminum framed	"	0.400
Bronze framed	"	0.400
Stainless steel framed	"	0.400
Tack board, aluminum framed	"	0.400
Visual aid board, aluminum framed	"	0.400
Glass encased boards, hinged and keyed		
Aluminum framed	S.F.	1.000
Bronze framed	"	1.000
Stainless steel framed	"	1.000
Chrome plated steel framed	"	1.000
Metal plaque		
Cast bronze	S.F.	0.667

SPECIALTIES	UNIT	MAN/ HOURS
10400.10 IDENTIFYING DEVICES		
Aluminum	S.F.	0.667
Metal engraved plaque		
Porcelain steel	S.F.	0.667
Stainless steel	"	0.667
Brass	"	0.667
Aluminum	"	0.667
Metal built-up plaque		
Bronze	S.F.	0.800
Copper and bronze	"	0.800
Copper and aluminum	"	0.800
Metal nameplate plaques		
Cast bronze	S.F.	0.500
Cast aluminum	"	0.500
Engraved, 1-1/2" x 6"		
Bronze	EA.	0.500
Aluminum	"	0.500
Letters, on masonry or concrete, aluminum, satin finish		
1/2" thick		
2" high	EA.	0.320
4" high	"	0.400
6" high	"	0.444
3/4" thick		
8" high	EA.	0.500
10" high	"	0.571
1" thick		
12" high	EA.	0.667
14" high	"	0.800
16" high	"	1.000
3/8" thick		
2" high	EA.	0.320
4" high	"	0.400
1/2" thick, 6" high	"	0.444
5/8" thick, 8" high	"	0.500
1" thick		
10" high	EA.	0.571
12" high	"	0.667
14" high	"	0.800
16" high	"	1.000
Interior door signs, adhesive, flexible		
2" x 8"	EA.	0.200
4" x 4"	"	0.200
6" x 7"	"	0.200
6" x 9"	"	0.200
10" x 9"	"	0.200
10" x 12"	"	0.200
Hard plastic type, no frame		
3" x 8"	EA.	0.200
4" x 4"	"	0.200
4" x 12"	"	0.200

SPECIALTIES	UNIT	MAN/ HOURS
10400.10 IDENTIFYING DEVICES		
Hard plastic type, with frame		
3" x 8"	EA.	0.200
4" x 4"	"	0.200
4" x 12"	"	0.200
10450.10 CONTROL		
Access control, 7' high, indoor or outdoor impenetrability		
Remote or card control, type B	EA.	10.667
Free passage, type B	"	10.667
Remote or card control, type AA	"	10.667
Free passage, type AA	"	10.667
10500.10 LOCKERS		
Locker bench, floor mounted, laminated maple		
4'	EA.	0.667
6'	"	0.667
Wardrobe locker, single tier type		
12" x 15" x 72"	EA.	0.800
18" x 15" x 72"	"	0.842
12" x 18" x 72"	"	0.889
18" x 18" x 72"	"	0.941
Double tier type		
12" x 15" x 36"	EA.	0.400
18" x 15" x 36"	"	0.400
12" x 18" x 36"	"	0.400
18" x 18" x 36"	"	0.400
Two person unit		
18" x 15" x 72"	EA.	1.333
18" x 18" x 72"	"	1.600
Duplex unit		
15" x 15" x 72"	EA.	0.800
15" x 21" x 72"	"	0.800
Basket lockers, basket sets with baskets		
24 basket set	SET	4.000
30 basket set	"	5.000
36 basket set	"	6.667
42 basket set	"	8.000
10520.10 FIRE PROTECTION		
Portable fire extinguishers		
Water pump tank type		
2.5 gal.		
Red enameled galvanized	EA.	0.533
Red enameled copper	"	0.533
Polished copper	"	0.533
Carbon dioxide type, red enamel steel		
Squeeze grip with hose and horn		
2.5 lb	EA.	0.533

10 SPECIALTIES

SPECIALTIES	UNIT	MAN/ HOURS
10520.10 FIRE PROTECTION		
5 lb	EA.	0.615
10 lb	"	0.800
15 lb	"	1.000
20 lb	"	1.000
Wheeled type		
125 lb	EA.	1.600
250 lb	"	1.600
500 lb	"	1.600
Dry chemical, pressurized type		
Red enameled steel		
2.5 lb	EA.	0.533
5 lb	"	0.615
10 lb	"	0.800
20 lb	"	1.000
30 lb	"	1.000
Chrome plated steel, 2.5 lb	"	0.533
Other type extinguishers		
2.5 gal, stainless steel, pressurized water tanks	EA.	0.533
Soda and acid type	"	0.533
Cartridge operated, water type	"	0.533
Loaded stream, water type	"	0.533
Foam type	"	0.533
40 gal, wheeled foam type	"	1.600
Fire extinguisher cabinets		
Enameled steel		
8" x 12" x 27"	EA.	1.600
8" x 16" x 38"	"	1.600
Aluminum		
8" x 12" x 27"	EA.	1.600
8" x 16" x 38"	"	1.600
8" x 12" x 27"	"	1.600
Stainless steel		
8" x 16" x 38"	EA.	1.600
10550.10 POSTAL SPECIALTIES		
Mail chutes		
Single mail chute		
Finished aluminum	L.F.	2.000
Bronze	"	2.000
Single mail chute receiving box		
Finished aluminum	EA.	4.000
Bronze	"	4.000
Twin mail chute, double parallel		
Finished aluminum	FLR	4.000
Bronze	"	4.000
Receiving box, 36" x 20" x 12"		
Finished aluminum	EA.	6.667
Bronze	"	6.667

SPECIALTIES	UNIT	MAN/ HOURS
10550.10 POSTAL SPECIALTIES		
Locked receiving mail box		
Finished aluminum	EA.	4.000
Bronze	"	4.000
Commercial postal accessories for mail chutes		
Letter slot, brass	EA.	1.333
Bulk mail slot, brass	"	1.333
Mail boxes		
Residential postal accessories		
Letter slot	EA.	0.400
Rural letter box	"	1.000
Apartment house, keyed, 3.5" x 4.5" x 16"	"	0.267
Ranch style	"	0.400
Commercial postal accessories		
Letter box, with combination lock	EA.	0.286
Key lock	"	0.286
Mail box, aluminum w/glass front, 4x5		
Horizontal rear load	EA.	0.229
Vertical front load	"	0.229
10600.10 MOVABLE PARTITIONS		
Partition, movable office type, 2-1/2" thick, vinyl-gypsum	S.F.	0.040
Enameled steel frame, with 1/4" thick clear glass	"	0.040
Door frame and hardware for movable partitions	EA.	2.667
Add for acoustic movable partition	S.F.	
Accordion partition, 12' high		
Vinyl	S.F.	0.133
Acoustical	"	0.133
Standard office cubicles, 8' high, steel framed		
Baked enamel finish		
100% flush	L.F.	0.200
75% flush and 25% glass	"	0.222
50% flush and 50% glass	"	0.222
100% glass	"	0.267
Natural hardwood panels		
100% flush	L.F.	0.267
50% flush and 50% glass	"	0.286
Plastic laminated panels		
100% flush	L.F.	0.267
75% flush and 25% glass	"	0.286
50% flush and 50% glass	"	0.286
Vinyl covered panels		
100% flush	L.F.	0.276
75% flush and 25% glass	"	0.296
50% and 50% glass	"	0.296
Aluminum framed		
Enameled or anodized aluminum panels		
100% flush	L.F.	0.200
75% flush and 25% glass	"	0.222
50% flush and 50% glass	"	0.222

SPECIALTIES	UNIT	MAN/ HOURS
10600.10 MOVABLE PARTITIONS		
Vinyl covered panels		
100% flush	L.F.	0.276
75% flush and 25% glass	"	0.296
50% flush and 50% glass	"	0.296
60" high partitions, steel framed		
Enameled panels	L.F.	0.178
Natural hardwood panels, two sides	"	0.186
Plastic laminated panels	"	0.186
Vinyl covered panels	"	0.178
Aluminum framed		
Anodized or baked enamel panels	L.F.	0.178
Natural hardwood panels	"	0.186
Plastic laminated panels	"	0.186
Vinyl covered panels	"	0.178
Wire mesh partitions		
Wall panels		
4' x 7'	EA.	0.500
4' x 8'	"	0.533
4' x 10'	"	0.615
Wall filler panels		
1' x 7'	EA.	0.500
1' x 8'	"	0.533
1' x 10'	"	0.571
2' x 7'	"	0.500
2' x 8'	"	0.533
2' x 10'	"	0.571
3' x 7'	"	0.500
3' x 8'	"	0.533
3' x 10'	"	0.571
Ceiling panels		
10' x 2'	EA.	1.143
10' x 4'	"	1.600
Wall panel with service window		
5' wide		
7' high	EA.	0.500
8' high	"	0.533
10' high	"	0.571
Doors		
Sliding		
3' x 7'	EA.	2.000
3' x 8'	"	2.286
3' x 10'	"	3.200
4' x 7'	"	2.286
4' x 8'	"	3.200
4' x 10'	"	4.000
5' x 7'	"	3.200
5' x 8'	"	4.000
5' x 10'	"	4.000
Swing door		

SPECIALTIES	UNIT	MAN/ HOURS
10600.10 MOVABLE PARTITIONS		
3' x 7'	EA.	2.000
4' x 7'	"	2.286
Swing door, with 1' transom		
3' x 7'	EA.	2.286
4' x 7'	"	3.200
Swing door, with 3' transom		
3' x 7'	EA.	3.200
4' x 7'	"	4.000
10670.10 SHELVING		
Shelving, enamel, closed side and back, 12" x 36"		
5 shelves	EA.	1.333
8 shelves	"	1.778
Open		
5 shelves	EA.	1.333
8 shelves	"	1.778
Metal storage shelving, baked enamel		
7 shelf unit, 72" or 84" high		
10" shelf	L.F.	0.800
12" shelf	"	0.842
15" shelf	"	0.889
18" shelf	"	0.941
24" shelf	"	1.000
30" shelf	"	1.067
36" shelf	"	1.143
4 shelf unit, 40" high		
10" shelf	L.F.	0.667
12" shelf	"	0.727
15" shelf	"	0.800
18" shelf	"	0.842
24" shelf	"	0.889
3 shelf unit, 32" high		
10" shelf	L.F.	0.400
12" shelf	"	0.421
15" shelf	"	0.444
18" shelf	"	0.471
24" shelf	"	0.500
Single shelf unit, attached to masonry		
10" shelf	L.F.	0.133
12" shelf	"	0.145
15" shelf	"	0.154
18" shelf	"	0.163
24" shelf	"	0.174
Built-in wood shelves		
Posts and trimmed plywood	L.F.	0.114
Solid clear pine	"	0.123
Closet shelf, pine with rod	"	0.123

10 SPECIALTIES

SPECIALTIES	UNIT	MAN/ HOURS
10750.10 TELEPHONE ENCLOSURES		
Telephone enclosure, wall mounted, shelf, 28" x 30" x 15"	EA.	2.000
Directory shelf, stainless steel, 3 binders	"	1.333
10800.10 BATH ACCESSORIES		
Ash receiver, wall mounted, aluminum	EA.	0.400
Grab bar, 1-1/2" dia., stainless steel, wall mounted		
24" long	EA.	0.400
36" long	"	0.421
42" long	"	0.444
48" long	"	0.471
52" long	"	0.500
1" dia., stainless steel		
12" long	EA.	0.348
18" long	"	0.364
24" long	"	0.400
30" long	"	0.421
36" long	"	0.444
48" long	"	0.471
Hand dryer, surface mounted, 110 volt	"	1.000
Medicine cabinet, 16 x 22, baked enamel, steel, lighted	"	0.320
With mirror, lighted	"	0.533
Mirror, 1/4" plate glass, up to 10 sf	S.F.	0.080
Mirror, stainless steel frame		
18"x24"	EA.	0.267
18"x32"	"	0.320
18"x36"	"	0.400
24"x30"	"	0.400
24"x36"	"	0.444
24"x48"	"	0.667
24"x60"	"	0.800
30"x30"	"	0.800
30"x72"	"	1.000
48"x72"	"	1.333
With shelf, 18"x24"	"	0.320
Sanitary napkin dispenser, stainless steel, wall mounted	"	0.533
Shower rod, 1" diameter		
Chrome finish over brass	EA.	0.400
Stainless steel	"	0.400
Soap dish, stainless steel, wall mounted	"	0.533
Toilet tissue dispenser, stainless, wall mounted		
Single roll	EA.	0.200
Double roll	"	0.229
Towel dispenser, stainless steel		
Flush mounted	EA.	0.444
Surface mounted	"	0.400
Combination towel dispenser and waste receptacle	"	0.533
Towel bar, stainless steel		
18" long	EA.	0.320
24" long	"	0.364

SPECIALTIES	UNIT	MAN/ HOURS
10800.10 BATH ACCESSORIES		
30" long	EA.	0.400
36" long	"	0.444
Toothbrush and tumbler holder	"	0.267
Waste receptacle, stainless steel, wall mounted	"	0.667
10900.10 WARDROBE SPECIALTIES		
Hospital wardrobe units, 24" x 24" x 76", with door		
Baked enameled steel	EA.	4.444
Hardwood	"	4.444
Stainless steel	"	4.444
Plastic laminated	"	4.444
Dormitory wardrobe units, 24" x 76", with door		
Hardwood	EA.	4.444
Plastic laminated	"	4.444
Hat and coat rack		
Single tier		
Baked enameled steel	L.F.	0.200
Stainless steel	"	0.200
Aluminum	"	0.200
Double tier		
Baked enameled steel	L.F.	0.229
Stainless steel	"	0.229
Aluminum	"	0.229

11 EQUIPMENT

ARCHITECTURAL EQUIPMENT	UNIT	MAN/HOURS
11010.10 MAINTENANCE EQUIPMENT		
Vacuum cleaning system		
3 valves		
1.5 hp	EA.	8.889
2.5 hp	"	11.429
5 valves	"	16.000
7 valves	"	20.000
11020.10 SECURITY EQUIPMENT		
Bulletproof teller window		
4' x 4'	EA.	13.333
5' x 4'	"	16.000
Bulletproof partitions		
Up to 12' high, 2.5" thick	S.F.	0.053
Counter for banks		
Minimum	L.F.	1.600
Maximum	"	2.667
Drive-up window		
Minimum	EA.	11.429
Maximum	"	26.667
Night depository		
Minimum	EA.	11.429
Maximum	"	26.667
Office safes, 30" x 20" x 20", 1 hr rating	"	2.000
30" x 16" x 15", 2 hr rating	"	1.600
30" x 28" x 20", H&G rating	"	1.000
Service windows, pass through painted steel		
24" x 36"	EA.	8.000
48" x 40"	"	10.000
72" x 40"	"	16.000
Special doors and windows		
3' x 7' bulletproof door with frame	EA.	11.429
12" x 12" vision panel	"	5.714
Surveillance system		
Minimum	EA.	2.000
Maximum	"	8.000
Vault door, 3' wide, 6'6" high		
3-1/2" thick	EA.	100
7" thick	"	133
10" thick	"	160
Insulated vault door		
2 hr rating		
32" wide	EA.	8.000
40" wide	"	8.421
4 hr rating		
32" wide	EA.	8.889
40" wide	"	10.000
6 hr rating		
32" wide	EA.	8.889
40" wide	"	10.000

ARCHITECTURAL EQUIPMENT	UNIT	MAN/HOURS
11020.10 SECURITY EQUIPMENT		
Insulated file room door		
1 hr rating		
32" wide	EA.	8.000
40" wide	"	8.889
11060.10 THEATER EQUIPMENT		
Roll out stage, steel frame, wood floor		
Manual	S.F.	0.050
Electric	"	0.080
Portable stages		
8" high	S.F.	0.040
18" high	"	0.044
36" high	"	0.047
48" high	"	0.050
Band risers		
Minimum	S.F.	0.040
Maximum	"	0.040
Chairs for risers		
Minimum	EA.	0.036
Maximum	"	0.036
11080.10 POLICE EQUIPMENT		
Firing range equipment, rifle		
3 position	EA.	26.667
4 position	"	40.000
5 position	"	44.444
6 position	"	47.059
11090.10 CHECKROOM EQUIPMENT		
Motorized checkroom equipment		
No shelf system, 6'4" height		
7'6" length	EA.	8.000
14'6" length	"	8.000
28' length	"	8.000
One shelf, 6'8" height		
7'6" length	EA.	8.000
14'6" length	"	8.000
28' length	"	8.000
Two shelves, 7'5" height		
7'6" length	EA.	8.000
14'6" length	"	8.000
28' length	"	8.000
Three shelves, 8' height		
7'6" length	EA.	16.000
14'6" length	"	16.000
28' length	"	16.000
Four shelves, 8'7" height		
7'6" length	EA.	16.000

11 EQUIPMENT

ARCHITECTURAL EQUIPMENT	UNIT	MAN/ HOURS
11090.10 CHECKROOM EQUIPMENT		
14'6" length	EA.	16.000
28' length	"	16.000
200 lb		
11110.10 LAUNDRY EQUIPMENT		
High capacity, heavy duty		
Washer extractors		
135 lb		
Standard	EA.	6.667
Pass through	"	6.667
200 lb		
Standard	EA.	6.667
Pass through	"	6.667
110 lb dryer	"	6.667
Hand operated presser	"	8.889
Mushroom press	"	8.889
Spreader feeders		
2 station	EA.	8.889
4 station	"	16.000
Delivery carts		
12 bushel	EA.	0.100
16 bushel	"	0.107
18 bushel	"	0.114
30 bushel	"	0.133
40 bushel	"	0.160
Low capacity		
Pressers		
Air operated	EA.	3.200
Hand operated	"	3.200
Extractor, low capacity	"	3.200
Ironer, 48"	"	1.600
Coin washers		
10 lb capacity	EA.	1.600
20 lb capacity	"	1.600
Coin dryer	"	1.000
Coin dry cleaner, 20 lb	"	3.200
11161.10 LOADING DOCK EQUIPMENT		
Dock leveler, 10 ton capacity		
6' x 8'	EA.	8.000
7' x 8'	"	8.000
Bumpers, laminated rubber		
4-1/2" thick		
6" x 14"	EA.	0.160
6" x 36"	"	0.178
10" x 14"	"	0.200
10" x 24"	"	0.229

ARCHITECTURAL EQUIPMENT	UNIT	MAN/ HOURS
11161.10 LOADING DOCK EQUIPMENT		
10" x 36"	EA.	0.267
12" x 14"	"	0.211
12" x 24"	"	0.250
12" x 36"	"	0.296
6" thick		
10" x 14"	EA.	0.229
10" x 24"	"	0.276
10" x 36"	"	0.400
Extruded rubber bumpers		
T-section, 22" x 22" x 3"	EA.	0.160
Molded rubber bumpers		
24" x 12" x 3" thick	EA.	0.400
Door seal, 12" x 12", vinyl covered	L.F.	0.200
Dock boards, heavy duty, 5' x 5'		
5000 lb		
Minimum	EA.	6.667
Maximum	"	6.667
9000 lb		
Minimum	EA.	6.667
Maximum	"	7.273
15,000 lb	"	7.273
Truck shelters		
Minimum	EA.	6.154
Maximum	"	11.429
11170.10 WASTE HANDLING		
Incinerator, electric		
100 lb/hr		
Minimum	EA.	8.000
Maximum	"	8.000
400 lb/hr		
Minimum	EA.	16.000
Maximum	"	16.000
1000 lb/hr		
Minimum	EA.	24.242
Maximum	"	24.242
Incinerator, medical-waste		
25 lb/hr, 2-7 x 4-0	EA.	16.000
50 lb/hr, 2-11 x 4-11	"	16.000
75 lb/hr, 3-8 x 5-0	"	32.000
100 lb/hr, 3-8 x 6-0	"	32.000
Industrial compactor		
1 cy	EA.	8.889
3 cy	"	11.429
5 cy	"	16.000
Trash chutes steel, including sprinklers		
18" dia.	L.F.	4.000
24" dia.	"	4.211
30" dia.	"	4.444

11 EQUIPMENT

ARCHITECTURAL EQUIPMENT	UNIT	MAN/ HOURS
11170.10 WASTE HANDLING		
36" dia.	L.F.	4.706
Refuse bottom hopper	EA.	4.444
11400.10 FOOD SERVICE EQUIPMENT		
Unit kitchens		
30" compact kitchen		
Refrigerator, with range, sink	EA.	4.000
Sink only	"	2.667
Range only	"	2.000
Cabinet for upper wall section	"	1.143
Stainless shield, for rear wall	"	0.320
Side wall	"	0.320
42" compact kitchen		
Refrigerator with range, sink	EA.	4.444
Sink only	"	4.000
Cabinet for upper wall section	"	1.333
Stainless shield, for rear wall	"	0.333
Side wall	"	0.333
54" compact kitchen		
Refrigerator, oven, range, sink	EA.	5.714
Cabinet for upper wall section	"	1.600
Stainless shield, for		
Rear wall	EA.	0.364
Side wall	"	0.364
60" compact kitchen		
Refrigerator, oven, range, sink	EA.	5.714
Cabinet for upper wall section	"	1.600
Stainless shield, for		
Rear wall	EA.	0.364
Side wall	"	0.364
72" compact kitchen		
Refrigerator, oven, range, sink	EA.	6.667
Cabinet for upper wall section	"	1.600
Stainless shield for		
Rear wall	EA.	0.400
Side wall	"	0.400
Bake oven		
Single deck		
Minimum	EA.	1.000
Maximum	"	2.000
Double deck		
Minimum	EA.	1.333
Maximum	"	2.000
Triple deck		
Minimum	EA.	1.333
Maximum	"	2.667
Convection type oven, electric, 40" x 45" x 57"		
Minimum	EA.	1.000
Maximum	"	2.000

ARCHITECTURAL EQUIPMENT	UNIT	MAN/ HOURS
11400.10 FOOD SERVICE EQUIPMENT		
Broiler, without oven, 69" x 26" x 39"		
Minimum	EA.	1.000
Maximum	"	1.333
Coffee urns, 10 gallons		
Minimum	EA.	2.667
Maximum	"	4.000
Fryer, with submerger		
Single		
Minimum	EA.	1.600
Maximum	"	2.667
Double		
Minimum	EA.	2.000
Maximum	"	2.667
Griddle, counter		
3' long		
Minimum	EA.	1.333
Maximum	"	1.600
5' long		
Minimum	EA.	2.000
Maximum	"	2.667
Kettles, steam, jacketed		
20 gallons		
Minimum	EA.	2.000
Maximum	"	4.000
40 gallons		
Minimum	EA.	2.000
Maximum	"	4.000
60 gallons		
Minimum	EA.	2.000
Maximum	"	4.000
Range		
Heavy duty, single oven, open top		
Minimum	EA.	1.000
Maximum	"	2.667
Fry top		
Minimum	EA.	1.000
Maximum	"	2.667
Steamers, electric		
27 kw		
Minimum	EA.	2.000
Maximum	"	2.667
18 kw		
Minimum	EA.	2.000
Maximum	"	2.667
Dishwasher, rack type		
Single tank, 190 racks/hr	EA.	4.000
Double tank		
234 racks/hr	EA.	4.444
265 racks/hr	"	5.333

11 EQUIPMENT

ARCHITECTURAL EQUIPMENT	UNIT	MAN/ HOURS
11400.10 FOOD SERVICE EQUIPMENT		
Dishwasher, automatic 100 meals/hr	EA.	2.667
Disposals		
100 gal/hr	EA.	2.667
120 gal/hr	"	2.759
250 gal/hr	"	2.857
Exhaust hood for dishwasher, gutter 4 sides, s-steel		
4'x4'x2'	EA.	2.963
4'x7'x2'	"	3.200
Food preparation machines		
Vertical cutter mixers		
25 quart	EA.	2.667
40 quart	"	2.667
80 quart	"	4.000
130 quart	"	6.667
Choppers		
5 lb	EA.	2.000
16 lb	"	2.667
40 lb	"	4.000
Mixers, floor models		
20 quart	EA.	1.000
60 quart	"	1.000
80 quart	"	1.143
140 quart	"	1.600
Ice cube maker		
50 lb per day		
Minimum	EA.	8.000
Maximum	"	8.000
500 lb per day		
Minimum	EA.	13.333
Maximum	"	13.333
Ice flakers		
300 lb per day	EA.	8.000
600 lb per day	"	13.333
1000 lb per day	"	17.778
2000 lb per day	"	20.000
Refrigerated cases		
Dairy products		
Multi deck type	L.F.	0.533
Delicatessen case, service deli		
Single deck	L.F.	4.000
Multi deck	"	5.000
Meat case		
Single deck	L.F.	4.706
Multi deck	"	5.000
Produce case		
Single deck	L.F.	4.706
Multi deck	"	5.000
Bottle coolers		
6' long		

ARCHITECTURAL EQUIPMENT	UNIT	MAN/ HOURS
11400.10 FOOD SERVICE EQUIPMENT		
Minimum	EA.	16.000
Maximum	"	16.000
10' long		
Minimum	EA.	26.667
Maximum	"	26.667
Frozen food cases		
Chest type	L.F.	4.706
Reach-in, glass door	"	5.000
Island case, single	"	4.706
Multi deck	"	5.000
Ice storage bins		
500 lb capacity	EA.	11.429
1000 lb capacity	"	22.857
11450.10 RESIDENTIAL EQUIPMENT		
Compactor, 4 to 1 compaction	EA.	2.000
Dishwasher, built-in		
2 cycles	EA.	4.000
4 or more cycles	"	4.000
Disposal		
Garbage disposer	EA.	2.667
Heaters, electric, built-in		
Ceiling type	EA.	2.667
Wall type		2.667
Minimum	EA.	2.000
Maximum	"	2.667
Hood for range, 2-speed, vented		
30" wide	EA.	2.667
42" wide	"	2.667
Ice maker, automatic		
30 lb per day	EA.	1.143
50 lb per day	"	4.000
Folding access stairs, disappearing metal stair		
8' long	EA.	1.143
11' long	"	1.143
12' long	"	1.143
Wood frame, wood stair		
22" x 54" x 8'9" long	EA.	0.800
25" x 54" x 10' long	"	0.800
Ranges electric		
Built-in, 30", 1 oven	EA.	2.667
2 oven	"	2.667
Counter top, 4 burner, standard	"	2.000
With grill	"	2.000
Free standing, 21", 1 oven	"	2.667
30", 1 oven	"	1.600
2 oven	"	1.600
Water softener		

11 EQUIPMENT

ARCHITECTURAL EQUIPMENT	UNIT	MAN/HOURS
11450.10 RESIDENTIAL EQUIPMENT		
30 grains per gallon	EA.	2.667
70 grains per gallon	"	4.000
11470.10 DARKROOM EQUIPMENT		
Dryers		
36" x 25" x 68"	EA.	4.000
48" x 25" x 68"	"	4.000
Processors, film		
Black and white	EA.	4.000
Color negatives	"	4.000
Prints	"	4.000
Transparencies	"	4.000
Sinks with cabinet and/or stand		
5" sink with stand		
24" x 48"	EA.	2.000
32" x 64"	"	2.667
38" x 52"	"	2.667
42" x 132"	"	4.000
48" x 52"	"	4.000
5" sink with cabinet		
24" x 48"	EA.	2.000
32" x 64"	"	2.667
38" x 52"	"	2.667
42" x 132"	"	4.000
48" x 52"	"	4.000
10" sink with stand		
24" x 48"	EA.	2.000
32" x 64"	"	2.667
38" x 52"	"	2.667
10" sink with cabinet		
24" x 48"	EA.	2.000
38" x 52"	"	2.667
11480.10 ATHLETIC EQUIPMENT		
Basketball backboard		
Fixed	EA.	10.000
Swing-up	"	16.000
Portable, hydraulic	"	4.000
Suspended type, standard	"	16.000
Bleacher, telescoping, manual		
15 tier, minimum	SEAT	0.160
Maximum	"	0.160
20 tier, minimum	"	0.178
Maximum	"	0.178
30 tier, minimum	"	0.267
Maximum	"	0.267
Boxing ring elevated, complete, 22' x 22'	EA.	80.000
Gym divider curtain		
Minimum	S.F.	0.011

ARCHITECTURAL EQUIPMENT	UNIT	MAN/HOURS
11480.10 ATHLETIC EQUIPMENT		
Maximum	S.F.	0.011
Scoreboards, single face		
Minimum	EA.	8.000
Maximum	"	40.000
Parallel bars, wall mounted		
Minimum	EA.	8.000
Maximum	"	13.333
11500.10 INDUSTRIAL EQUIPMENT		
Vehicular paint spray booth, solid back, 14'4" x 9'6"		
24' deep	EA.	8.000
26'6" deep	"	8.000
28'6" deep	"	8.000
Drive through, 14'9" x 9'6"		
24' deep	EA.	8.000
26'6" deep	"	8.000
28'6" deep	"	8.000
Water wash, paint spray booth		
5' x 11'2" x 10'8"	EA.	8.000
6' x 11'2" x 10'8"	"	8.000
8' x 11'2" x 10'8"	"	8.000
10' x 11'2" x 11'2"	"	8.000
12' x 12'2" x 11'2"	"	8.000
14' x 12'2" x 11'2"	"	8.000
16' x 12'2" x 11'2"	"	8.000
20' x 12'2" x 11'2"	"	8.000
Dry type spray booth, with paint arrestors		
5'4" x 7'2" x 6'8"	EA.	8.000
6'4" x 7'2" x 6'8"	"	8.000
8'4" x 7'2" x 9'2"	"	8.000
10'4" x 7'2" x 9'2"	"	8.000
12'4" x 7'6" x 9'2"	"	8.000
14'4" x 7'6" x 9'8"	"	8.000
16'4" x 7'7" x 9'8"	"	8.000
20'4" x 7'7" x 10'8"	"	8.000
Air compressor, electric		
1 hp		
115 volt	EA.	5.333
5 hp		
115 volt	EA.	8.000
230 volt	"	8.000
Hydraulic lifts		
8,000 lb capacity	EA.	20.000
11,000 lb capacity	"	32.000
24,000 lb capacity	"	53.333
Power tools		
Band saws		
10"	EA.	0.667

11 EQUIPMENT

ARCHITECTURAL EQUIPMENT	UNIT	MAN/ HOURS
11500.10 INDUSTRIAL EQUIPMENT		
14"	EA.	0.800
Motorized shaper	"	0.615
Motorized lathe	"	0.667
Bench saws		
9" saw	EA.	0.533
10" saw	"	0.571
12" saw	"	0.667
Electric grinders		
1/3 hp	EA.	0.320
1/2 hp	"	0.348
3/4 hp	"	0.348
11600.10 LABORATORY EQUIPMENT		
Cabinets, base		
Minimum	L.F.	0.667
Maximum	"	0.667
Full storage, 7' high		
Minimum	L.F.	0.667
Maximum	"	0.667
Wall		
Minimum	L.F.	0.800
Maximum	"	0.800
Counter tops		
Minimum	S.F.	0.100
Average	"	0.114
Maximum	"	0.133
Tables		
Open underneath	S.F.	0.400
Doors underneath	"	0.500
Medical laboratory equipment		
Analyzer		
Chloride	EA.	0.400
Blood	"	0.667
Bath, water, utility, countertop unit	"	0.800
Hot plate, lab, countertop	"	0.727
Stirrer	"	0.727
Incubator, anaerobic, 23x23x36"	"	4.000
Dry heat bath	"	1.333
Incinerator, for sterilizing	"	0.080
Meter, serum protein	"	0.100
Ph analog, general purpose	"	0.114
Refrigerator, blood bank, undercounter type 153 litres	"	1.333
5.4 cf, undercounter type	"	1.333
Refrigerator/freezer, 4.4 cf, undercounter type	"	1.333
Sealer, impulse, free standing, 20x12x4"	"	0.267
Timer, electric, 1-60 minutes, bench or wall mounted	"	0.444
Glassware washer - dryer, undercounter	"	10.000
Balance, torsion suspension, tabletop, 4.5 lb capacity	"	0.444

ARCHITECTURAL EQUIPMENT	UNIT	MAN/ HOURS
11600.10 LABORATORY EQUIPMENT		
Binocular microscope, with in base illuminator	EA.	0.308
Centrifuge, table model, 19x16x13"	"	0.320
Clinical model, with four place head	"	0.178
11700.10 MEDICAL EQUIPMENT		
Hospital equipment, lights		
Examination, portable	EA.	0.667
Meters		
Air flow meter	EA.	0.444
Oxygen flow meters	"	0.333
Racks		
40 chart, revolving open frame; mobile caddy	EA.	0.667
Scales.		
Clinical, metric with measure rod, 350 lb	EA.	0.727
Physical therapy		
Chair, hydrotherapy	EA.	0.133
Diathermy, shortwave, portable, on casters	"	0.320
Exercise bicycle, floor standing, 35" x 15"	"	0.267
Hydrocollator, 4 pack, portable, 129 x 90 x 160"	"	0.114
Lamp, infrared, mobile with variable heat control	"	0.615
Ultra violet, base mounted	"	0.615
Mirror, posture training, 27" wide and 72" high	"	0.200
Parallel bars, adjustable	"	1.000
Platform mat 10'x6', 1" thick	"	0.200
Pulley, duplex, wall mounted	"	2.667
Rack, crutch, wall mounted, 66 x 16 x 13"	"	0.800
Stimulator, galvanic-faradic, hand held	"	0.053
Ultrasound, muscle stimulator, portable, 13x13x8"	"	0.067
Sandbag set, velcro straps, saddle bag type	"	0.114
Whirlpool, 85 gallon	"	4.000
65 gallon capacity	"	4.000
Radiology		
Radiographic table, motor driven tilting table	EA.	80.000
Fluoroscope image/tv system	"	160
Processor for washing and drying radiographs		
Water filter unit, 30" x 48-1/2" x 37-1/2"	EA.	13.333
Cassette transfer cabinet	"	0.667
Base storage cabinets, sectional design		
With back splash, 24" deep and 35" high	L.F.	0.667
Wall storage cabinets	"	1.000
Steam sterilizers		
For heat and moisture stable materials	EA.	0.800
For fast drying after sterilization	"	1.000
Compact unit	"	1.000
Semi-automatic	"	4.000
Floor loading		
Single door	EA.	6.667
Double door	"	8.000

ARCHITECTURAL EQUIPMENT	UNIT	MAN/ HOURS
11700.10 MEDICAL EQUIPMENT		
Utensil washer, sanitizer	EA.	6.154
Automatic washer/sterilizer	"	16.000
16 x 16 x 26", including generator & accessories	"	26.667
Steam generator, elec., 10 kw to 180 kw	"	16.000
Surgical scrub		
Minimum	EA.	2.667
Maximum	"	2.667
Gas sterilizers		
Automatic, free standing, 21x19x29"	EA.	8.000
Surgical tables		
Minimum	EA.	11.429
Maximum	"	16.000
Surgical lights, ceiling mounted		
Minimum	EA.	13.333
Maximum	"	16.000
Water stills		
4 liters/hr	EA.	2.667
8 liters/hr	"	2.667
19 liters/hr	"	6.667
X-ray equipment		
Mobile unit		
Minimum	EA.	4.000
Maximum	"	8.000
Film viewers		
Minimum	EA.	1.333
Maximum	"	2.667
Autopsy table		
Minimum	EA.	8.000
Maximum	"	8.000
Incubators		
15 cf	EA.	4.000
29 cf	"	6.667
Infant transport, portable	"	4.211
Beds		
Stretcher, with pad, 30" x 78"	EA.	2.000
Transfer, for patient transport	"	2.000
Headwall		
Aluminum, with back frame and console	EA.	4.000
11700.20 DENTAL EQUIPMENT		
Dental care equipment		
Drill console with accessories	EA.	13.333
Amalgamator	"	0.400
Lathe	"	0.267
Finish polisher	"	0.533
Model trimer	"	0.364
Motor, wall mounted	"	0.364
Cleaner, ultrasonic	"	0.800
Curing unit, bench mounted	"	1.333

ARCHITECTURAL EQUIPMENT	UNIT	MAN/ HOURS
11700.20 DENTAL EQUIPMENT		
Oral evacuation system, dual pump	EA.	1.000
Sterilizer, table top, self contained	"	0.444
Dental lights		
Light, floor or ceiling mounted	EA.	4.000
X-ray unit		
Portable	EA.	2.000
Wall mounted with remote control	"	6.667
Illuminator, single panel	"	11.429
X-ray film processor	"	6.667
Shield, portable x-ray, lead lined	"	0.533

12 FURNISHINGS

INTERIOR	UNIT	MAN/HOURS
12302.10 CASEWORK		
Kitchen base cabinet, prefinished, 24" deep, 35" high		
12"wide	EA.	0.800
18" wide	"	0.800
24" wide	"	0.889
27" wide	"	0.889
36" wide	"	1.000
48" wide	"	1.000
Corner cabinet, 36" wide	"	1.000
Wall cabinet, 12" deep, 12" high		
30" wide	EA.	0.800
36" wide	"	0.800
15" high		
30" wide	EA.	0.889
36" wide	"	0.889
24" high		
30" wide	EA.	0.889
36" wide	"	0.889
30" high		
12" wide	EA.	1.000
18" wide	"	1.000
24" wide	"	1.000
27" wide	"	1.000
30" wide	"	1.143
36" wide	"	1.143
Corner cabinet, 30" high		
24" wide	EA.	1.333
30" wide	"	1.333
36" wide	"	1.333
Wardrobe	"	2.000
Vanity with top, laminated plastic		
24" wide	EA.	2.000
30" wide	"	2.000
36" wide	"	2.667
48" wide	"	3.200
12390.10 COUNTER TOPS		
Stainless steel, counter top, with backsplash	S.F.	0.200
Acid-proof, kemrock surface	"	0.133
12500.10 WINDOW TREATMENT		
Drapery tracks, wall or ceiling mounted		
Basic traverse rod		
50 to 90"	EA.	0.400
84 to 156"	"	0.444
136 to 250"	"	0.444
165 to 312"	"	0.500
Traverse rod with stationary curtain rod		
30 to 50"	EA.	0.400
50 to 90"	"	0.400

INTERIOR	UNIT	MAN/HOURS
12500.10 WINDOW TREATMENT		
84 to 156"	EA.	0.444
136 to 250"	"	0.500
Double traverse rod		
30 to 50"	EA.	0.400
50 to 84"	"	0.400
84 to 156"	"	0.444
136 to 250"	"	0.500
12510.10 BLINDS		
Venetian blinds		
2" slats	S.F.	0.020
1" slats	"	0.020
12690.40 FLOOR MATS		
Recessed entrance mat, 3/8" thick, aluminum link	S.F.	0.400
Steel, flexible	"	0.400

CONSTRUCTION	UNIT	MAN/HOURS
13056.10 VAULTS		
Floor safes		
Class C		
1.0 cf	EA.	0.667
1.3 cf	"	1.000
1.9 cf	"	1.333
5.2 cf	"	1.333
13121.10 PRE-ENGINEERED BUILDINGS		
Pre-engineered metal building, 40'x100'		
14' eave height	S.F.	0.032
16' eave height	"	0.037
20' eave height	"	0.048
60'x100'		
14' eave height	S.F.	0.032
16' eave height	"	0.037
20' eave height	"	0.048
80'x100'		
14' eave height	S.F.	0.032
16' eave height	"	0.037
20' eave height	"	0.048
100'x100'		
14' eave height	S.F.	0.032
16' eave height	"	0.037
20' eave height	"	0.048
100'x150'		
14' eave height	S.F.	0.032
16' eave height	"	0.037
20' eave height	"	0.048
120'x150'		
14' eave height	S.F.	0.032
16' eave height	"	0.037
20' eave height	"	0.048
140'x150'		
14' eave height	S.F.	0.032
16' eave height	"	0.037
20' eave height	"	0.048
160'x200'		
14' eave height	S.F.	0.032
16' eave height	"	0.037
20' eave height	"	0.048
200'x200'		
14' eave height	S.F.	0.032
16' eave height	"	0.037
20' eave height	"	0.048
Liner panel, 26 ga, painted steel	"	0.020
Wall panel insulated, 26 ga. steel, foam core	"	0.020
Roof panel, 26 ga. painted steel	"	0.011
Plastic (sky light)	"	0.011
Insulation, 3-1/2" thick blanket, R11	"	0.005

CONSTRUCTION	UNIT	MAN/HOURS
13152.10 SWIMMING POOL EQUIPMENT		
Diving boards		
14' long		
Aluminum	EA.	4.444
Fiberglass	"	4.444
Ladders, heavy duty		
2 steps		
Minimum	EA.	1.600
Maximum	"	1.600
4 steps		
Minimum	EA.	2.000
Maximum	"	2.000
Lifeguard chair		
Minimum	EA.	8.000
Maximum	"	8.000
Lights, underwater		
12 volt, with transformer	EA.	2.000
110 volt		
Minimum	EA.	2.000
Maximum	"	2.000
Ground fault interrupter for 110 volt, each light	"	0.667
Pool covers		
Reinforced polyethylene	S.F.	0.062
Vinyl water tube		
Minimum	S.F.	0.062
Maximum	"	0.062
Slides with water tube		
Minimum	EA.	6.667
Maximum	"	6.667

ELEVATORS	UNIT	MAN/ HOURS
14210.10 ELEVATORS		
Passenger elevators, electric, geared		
Based on a shaft of 6 stops and 6 openings		
50 fpm, 2000 lb	EA.	24.000
100 fpm, 2000 lb	"	26.667
150 fpm		
2000 lb	EA.	30.000
3000 lb	"	34.286
4000 lb	"	40.000
200 fpm		
2500 lb	EA.	34.286
3000 lb	"	36.923
4000 lb	"	40.000
250 fpm		
2500 lb	EA.	34.286
3000 lb	"	36.923
4000 lb	"	40.000
300 fpm		
2500 lb	EA.	34.286
3000 lb	"	36.923
4000 lb	"	24.000
Based on a shaft of 8 stops and 8 openings		
300 fpm		
3000 lb	EA.	48.000
3500 lb	"	48.000
4000 lb	"	53.333
5000 lb	"	57.143
400 fpm		
3000 lb	EA.	48.000
3500 lb	"	48.000
4000 lb	"	53.333
5000 lb	"	57.143
600 fpm		
3000 lb	EA.	53.333
3500 lb	"	57.143
4000 lb	"	58.537
5000 lb	"	60.000
800 fpm		
3000 lb	EA.	53.333
3500 lb	"	57.143
4000 lb	"	58.537
5000 lb	"	60.000
Hydraulic, based on a shaft of 3 stops, 3 openings		
50 fpm		
2000 lb	EA.	20.000
2500 lb	"	20.000
3000 lb	"	20.870
100 fpm		
2000 lb	EA.	20.000
2500 lb	"	20.870

ELEVATORS	UNIT	MAN/ HOURS
14210.10 ELEVATORS		
3000 lb	EA.	21.818
150 fpm		
2000 lb	EA.	20.000
2500 lb	"	20.870
3000 lb	"	22.857
Small elevators, 4 to 6 passenger capacity		
Electric, push		
2 stops	EA.	20.000
3 stops	"	21.818
4 stops	"	24.000
Freight elevators, electric		
Based on a shaft of 6 stops and 6 openings		
50 fpm		
3500 lb	EA.	26.667
4000 lb	"	26.667
5000 lb	"	30.000
100 fpm		
3500 lb	EA.	30.000
4000 lb	"	30.000
5000 lb	"	34.286
200 fpm		
3500 lb	EA.	34.286
4000 lb	"	34.286
5000 lb	"	40.000
Based on shaft of 8 stops and 8 openings		
100 fpm		
4000 lb	EA.	30.000
6000 lb	"	30.769
8000 lb	"	32.432
150 fpm		
4000 lb	EA.	34.286
6000 lb	"	35.294
8000 lb	"	37.500
200 fpm		
4000 lb	EA.	40.000
6000 lb	"	41.379
8000 lb	"	43.636
Hydraulic, based on 3 stops and 3 openings		
50 fpm		
3000 lb	EA.	17.143
4000 lb	"	17.778
6000 lb	"	18.462
100 fpm		
3000 lb	EA.	17.143
4000 lb	"	17.778
6000 lb	"	18.462
150 fpm		
3000 lb	EA.	17.143
4000 lb	"	17.778

14 CONVEYING

ELEVATORS

14210.10 ELEVATORS	UNIT	MAN/HOURS
6000 lb	EA.	18.462

14300.10 ESCALATORS	UNIT	MAN/HOURS
Escalators		
32" wide, floor to floor		
12' high	EA.	40.000
15' high	"	48.000
18' high	"	60.000
22' high	"	80.000
25' high	"	96.000
48" wide		
12' high	EA.	41.379
15' high	"	50.000
18' high	"	63.158
22' high	"	85.714
25' high	"	96.000

LIFTS

14450.10 VEHICLE LIFTS	UNIT	MAN/HOURS
Automotive hoist, one post, semi-hydraulic, 8,000 lb	EA.	24.000
Full hydraulic, 8,000 lb	"	24.000
2 post, semi-hydraulic, 10,000 lb	"	34.286
Full hydraulic		
10,000 lb	EA.	34.286
13,000 lb	"	60.000
18,500 lb	"	60.000
24,000 lb	"	60.000
26,000 lb	"	60.000
Pneumatic hoist, fully hydraulic		
11,000 lb	EA.	80.000
24,000 lb	"	80.000

MATERIAL HANDLING

14560.10 CHUTES	UNIT	MAN/HOURS
Linen chutes, stainless steel, with supports		
18" dia.	L.F.	0.057
24" dia.	"	0.062
30" dia.	"	0.067
Hopper	EA.	0.533
Skylight	"	0.800
Sprinkler unit at top	"	0.889

14580.10 PNEUMATIC SYSTEMS	UNIT	MAN/HOURS
Pneumatic message tube system		
Average, 20 station job		
3" round system	EA.	7.273
4" round system	"	8.000
6" round system	"	8.889
4" x 7" oval system	"	16.000
Trash and linen tube system		
10 stations	EA.	120
15 stations	"	160
20 stations	"	185
30 stations	"	218

HOISTS AND CRANES

14600.10 INDUSTRIAL HOISTS	UNIT	MAN/HOURS
Industrial hoists, electric, light to medium duty		
500 lb	EA.	4.000
1000 lb	"	4.211
2000 lb	"	4.444
3000 lb	"	4.706
4000 lb	"	5.000
5000 lb	"	5.333
6000 lb	"	5.517
7500 lb	"	5.714
10,000 lb	"	5.926
15,000 lb	"	6.154
20,000 lb	"	6.667
25,000 lb	"	7.273
30,000 lb	"	8.000
Heavy duty		

HOISTS AND CRANES	UNIT	MAN/ HOURS
14600.10 INDUSTRIAL HOISTS		
500 lb	EA.	4.000
1000 lb	"	4.211
2000 lb	"	4.444
3000 lb	"	4.706
4000 lb	"	5.000
5000 lb	"	5.333
6000 lb	"	5.517
7500 lb	"	5.714
10,000 lb	"	5.926
15,000 lb	"	6.154
20,000 lb	"	6.667
25,000 lb	"	7.273
30,000 lb	"	8.000
Air powered hoists		
500 lb	EA.	4.000
1000 lb	"	4.000
2000 lb	"	4.211
4000 lb	"	4.706
6000 lb	"	6.154
Overhead traveling bridge crane		
Single girder, 20' span		
3 ton	EA.	12.000
5 ton	"	12.000
7.5 ton	"	12.000
10 ton	"	15.000
15 ton	"	15.000
30' span		
3 ton	EA.	12.000
5 ton	"	12.000
10 ton	"	15.000
15 ton	"	15.000
Double girder, 40' span		
3 ton	EA.	26.667
5 ton	"	26.667
7.5 ton	"	26.667
10 ton	"	34.286
15 ton	"	34.286
25 ton	"	34.286
50' span		
3 ton	EA.	26.667
5 ton	"	26.667
7.5 ton	"	26.667
10 ton	"	34.286
15 ton	"	34.286
25 ton	"	34.286

HOISTS AND CRANES	UNIT	MAN/ HOURS
14650.10 JIB CRANES		
Self supporting, swinging 8' boom, 200 deg rotation		
1000 lb	EA.	6.667
2000 lb	"	6.667
3000 lb	"	13.333
4000 lb	"	13.333
6000 lb	"	13.333
10,000 lb	"	13.333
Wall mounted, 180 deg rotation		
2000 lb	EA.	6.667
3000 lb	"	6.667
4000 lb	"	13.333
6000 lb	"	13.333
10,000 lb	"	13.333

15 MECHANICAL

BASIC MATERIALS

15100.10 SPECIALTIES

BASIC MATERIALS	UNIT	MAN/HOURS
Wall penetration		
Concrete wall, 6" thick		
2" dia.	EA.	0.267
4" dia.	"	0.400
8" dia.	"	0.571
12" thick		
2" dia.	EA.	0.364
4" dia.	"	0.571
8" dia.	"	0.889

15120.10 BACKFLOW PREVENTERS

BASIC MATERIALS	UNIT	MAN/HOURS
Backflow preventer, flanged, cast iron, with valves		
3" pipe	EA.	4.000
4" pipe	"	4.444
6" pipe	"	6.667
8" pipe	"	8.000

15140.10 PIPE HANGERS, HEAVY

BASIC MATERIALS	UNIT	MAN/HOURS
Hangers		
1/2" pipe, clevis pipe hanger		
Black steel	EA.	0.267
Galvanized	"	0.267
3/4" pipe, clevis pipe hanger		
Black steel	EA.	0.267
Galvanized	"	0.267
1" pipe, clevis pipe hanger		
Black steel	EA.	0.267
Galvanized	"	0.267
U bolt	"	0.080
2" pipe, clevis pipe hanger		
Black steel	EA.	0.267
Galvanized	"	0.267
3" pipe, clevis pipe hanger		
Black steel	EA.	0.267
Galvanized	"	0.267
4" pipe, clevis pipe hanger		
Black steel	EA.	0.267
Galvanized	"	0.267
6" pipe, clevis pipe hanger		
Black steel	EA.	0.320
Galvanized	"	0.320
12" pipe, clevis pipe hanger		
Black steel	EA.	0.320
Galvanized	"	0.320
Threaded rod, galvanized		
C-clamp, steel, with lock nut		
3/8"	EA.	0.100

15140.10 PIPE HANGERS, HEAVY

BASIC MATERIALS	UNIT	MAN/HOURS
1/2"	EA.	0.100

15175.60 EXPANSION TANKS

BASIC MATERIALS	UNIT	MAN/HOURS
Expansion tank, 125 psi, galvanized steel		
20 gallon	EA.	1.000
30 gallon	"	1.333
65 gallon	"	2.000
80 gallon	"	2.286

INSULATION

15260.10 FIBERGLASS PIPE INSULATION

INSULATION	UNIT	MAN/HOURS
Fiberglass insulation on 1/2" pipe		
1" thick	L.F.	0.027
1-1/2" thick	"	0.033
3/4" pipe		
1" thick	L.F.	0.027
1-1/2" thick	"	0.033
1" pipe		
1" thick	L.F.	0.027
1-1/2" thick	"	0.033
2" pipe		
1" thick	L.F.	0.033
1-1/2" thick	"	0.036
3" pipe		
1" thick	L.F.	0.038
1-1/2" thick	"	0.040
6" pipe		
1" thick	L.F.	0.042
2" thick	"	0.044
10" pipe		
2" thick	L.F.	0.042
3" thick	"	0.044

15260.60 EXTERIOR PIPE INSULATION

INSULATION	UNIT	MAN/HOURS
Fiberglass insulation, aluminum jacket		
1/2" pipe		

INSULATION	UNIT	MAN/ HOURS
15260.60 EXTERIOR PIPE INSULATION		
1" thick	L.F.	0.062
1-1/2" thick	"	0.067
3/4" pipe		
1" thick	L.F.	0.062
1-1/2" thick	"	0.067
1" pipe		
1" thick	L.F.	0.062
1-1/2" thick	"	0.067
2" pipe		
1" thick	L.F.	0.073
1-1/2" thick	"	0.076
3" pipe		
1" thick	L.F.	0.080
1-1/2" thick	"	0.084
6" pipe		
1" thick	L.F.	0.089
2" thick	"	0.094
10" pipe		
2" thick	L.F.	0.089
3" thick	"	0.094
15260.90 PIPE INSULATION FITTINGS		
Insulation protection saddle		
1" thick covering		
1/2" pipe	EA.	0.320
3/4" pipe	"	0.320
1" pipe	"	0.320
2" pipe	"	0.320
3" pipe	"	0.364
6" pipe	"	0.500
1-1/2" thick covering		
3/4" pipe	EA.	0.320
1" pipe	"	0.320
2" pipe	"	0.320
3" pipe	"	0.320
6" pipe	"	0.500
10" pipe	"	0.667
15280.10 EQUIPMENT INSULATION		
Equipment insulation, 2" thick, cellular glass	S.F.	0.050
Urethane, rigid, field applied jacket, plastered finish	"	0.100
Fiberglass, rigid, with vapor barrier	"	0.044
15290.10 DUCTWORK INSULATION		
Fiberglass duct insulation, plain blanket		
1-1/2" thick	S.F.	0.010
2" thick	"	0.013

INSULATION	UNIT	MAN/ HOURS
15290.10 DUCTWORK INSULATION		
With vapor barrier		
1-1/2" thick	S.F.	0.010
2" thick	"	0.013
Rigid with vapor barrier		
2" thick	S.F.	0.027
3" thick	"	0.032
4" thick	"	0.040
6" thick	"	0.053
Weatherproof, polystyrene, 3" thick, w/vapor barrier	"	0.080
Urethane board with vapor barrier	"	0.100

FIRE PROTECTION	UNIT	MAN/ HOURS
15330.10 WET SPRINKLER SYSTEM		
Sprinkler head, 212 deg, brass, exposed piping	EA.	0.320
Chrome, concealed piping	"	0.444
Water motor alarm	"	1.333
Fire department inlet connection	"	1.600
Wall plate for fire dept connection	"	0.667
Swing check valve flanged iron body, 4"	"	2.667
Check valve, 6"	"	4.000
Wet pipe valve, flange to groove, 4"	"	0.889
Flange to flange		
6"	EA.	1.333
8"	"	2.667
Alarm valve, flange to flange, (wet valve)		
4"	EA.	0.889
8"	"	6.667
Inspector's test connection	"	0.667
Wall hydrant, polished brass, 2-1/2" x 2-1/2", single	"	0.571
2-way	"	0.571
3-way	"	0.571
Wet valve trim, includes retard chamber & guages, 4"-6"	"	0.667
Retard pressure switch for wet systems	"	1.600
Air maintenance device	"	0.667
Wall hydrant non-freeze, 8" thick wall, vacuum breaker	"	0.400
12" thick wall	"	0.400

FIRE PROTECTION

15330.50 DRY SPRINKLER SYSTEM

	UNIT	MAN/HOURS
Dry pipe valve, flange to flange		
4"	EA.	1.600
6"	"	2.000
Trim, 4" and 6", includes gauges	"	0.667
Field testing and flushing	"	6.667
Disinfection	"	6.667
Pressure switch double circuit, open/close contacts	"	2.000
Low air		
Supervisory unit	EA.	1.333
Pressure switch	"	0.667
Halon fire protection system, per computer room		

PLUMBING

15410.05 C.I. PIPE, ABOVE GROUND

	UNIT	MAN/HOURS
No hub pipe		
1-1/2" pipe	L.F.	0.057
2" pipe	"	0.067
3" pipe	"	0.080
4" pipe	"	0.133
6" pipe	"	0.160
8" pipe	"	0.267
10" pipe	"	0.320
No hub fittings, 1-1/2" pipe		
1/4 bend	EA.	0.267
1/8 bend	"	0.267
Sanitary tee	"	0.400
Wye	"	0.400
2" pipe		
1/4 bend	EA.	0.320
1/8 bend	"	0.320
Sanitary tee	"	0.533
Wye	"	0.667
3" pipe		
1/4 bend	EA.	0.400
1/8 bend	"	0.400
Sanitary tee	"	0.500
Wye	"	0.667
4" pipe		
1/4 bend	EA.	0.400

PLUMBING

15410.05 C.I. PIPE, ABOVE GROUND

	UNIT	MAN/HOURS
1/8 bend	EA.	0.400
Sanitary tee	"	0.667
Wye	"	0.667
8" deep	"	0.400
6" pipe		
1/4 bend	EA.	0.667
1/8 bend	"	0.667
Sanitary tee	"	0.800
Wye	"	0.800
8" pipe		
1/4 bend	EA.	0.667
1/8 bend	"	0.667
Sanitary tee	"	1.000
Wye	"	0.800
10" pipe		
1/4 bend	EA.	0.667
1/8 bend	"	0.667
Wye	"	1.333
10x6" wye	"	1.333

15410.06 C.I. PIPE, BELOW GROUND

	UNIT	MAN/HOURS
No hub pipe		
1-1/2" pipe	L.F.	0.040
2" pipe	"	0.044
3" pipe	"	0.050
4" pipe	"	0.067
6" pipe	"	0.073
8" pipe	"	0.089
10" pipe	"	0.100
Fittings, 1-1/2"		
1/4 bend	EA.	0.229
1/8 bend	"	0.229
Wye	"	0.320
2"		
1/4 bend	EA.	0.267
1/8 bend	"	0.267
3"		
1/4 bend	EA.	0.320
1/8 bend	"	0.320
Wye	"	0.500
4"		
1/4 bend	EA.	0.320
1/8 bend	"	0.320
Wye	"	0.500
6"		
1/4 bend	EA.	0.500
1/8 bend	"	0.500
8"		

15 MECHANICAL

PLUMBING	UNIT	MAN/ HOURS
15410.06 C.I. PIPE, BELOW GROUND		
1/4 bend	EA.	0.500
1/8 bend	"	0.500
Wye	"	0.667
10"		
1/4 bend	EA.	0.500
1/8 bend	"	0.500
Wye	"	1.000
15410.09 SERVICE WEIGHT PIPE		
Service weight pipe, single hub		
3" x 5'	EA.	0.170
4" x 5'	"	0.178
6" x 5'	"	0.200
1/8 bend		
3"	EA.	0.320
4"	"	0.364
6"	"	0.400
1/4 bend		
3"	EA.	0.320
4"	"	0.364
6"	"	0.400
Sweep		
3"	EA.	0.320
4"	"	0.364
6"	"	0.400
Sanitary T		
3"	EA.	0.571
4"	"	0.667
6"	"	0.727
Wye		
3"	EA.	0.444
4"	"	0.471
6"	"	0.571
15410.10 COPPER PIPE		
Type "K" copper		
1/2"	L.F.	0.025
3/4"	"	0.027
1"	"	0.029
1-1/4"	"	0.031
1-1/2"	"	0.033
2"	"	0.036
2-1/2"	"	0.040
3"	"	0.042
4"	"	0.044
DWV, copper		
1-1/4"	L.F.	0.033
1-1/2"	"	0.036

PLUMBING	UNIT	MAN/ HOURS
15410.10 COPPER PIPE		
2"	L.F.	0.040
3"	"	0.044
4"	"	0.050
6"	"	0.057
Refrigeration tubing, copper, sealed		
1/8"	L.F.	0.032
3/16"	"	0.033
1/4"	"	0.035
5/16"	"	0.036
3/8"	"	0.038
1/2"	"	0.040
7/8"	"	0.046
1-1/8"	"	0.053
1-3/8"	"	0.062
Type "L" copper		
1/4"	L.F.	0.024
3/8"	"	0.024
1/2"	"	0.025
3/4"	"	0.027
1"	"	0.029
1-1/4"	"	0.031
1-1/2"	"	0.033
2"	"	0.036
Type "M" copper		
1/2"	L.F.	0.025
3/4"	"	0.027
1"	"	0.029
1-1/4"	"	0.031
2"	"	0.036
15410.11 COPPER FITTINGS		
Copper pipe fittings		
1/2"		
90 deg ell	EA.	0.178
45 deg ell	"	0.178
Tee	"	0.229
Cap	"	0.089
Coupling	"	0.178
Union	"	0.200
3/4"		
90 deg ell	EA.	0.200
45 deg ell	"	0.200
Tee	"	0.267
Cap	"	0.094
Coupling	"	0.200
Union	"	0.229
1"		
90 deg ell	EA.	0.267

PLUMBING	UNIT	MAN/HOURS
15410.11 COPPER FITTINGS		
45 deg ell	EA.	0.267
Tee	"	0.320
Cap	"	0.133
Coupling	"	0.267
Union	"	0.267
1-1/4"		
90 deg ell	EA.	0.229
45 deg ell	"	0.229
Tee	"	0.400
Cap	"	0.133
Union	"	0.286
1-1/2"		
90 deg ell	EA.	0.286
45 deg ell	"	0.286
Tee	"	0.444
Cap	"	0.133
Coupling	"	0.267
Union	"	0.364
2"		
90 deg ell	EA.	0.320
45 deg ell	"	0.500
Tee	"	0.500
Cap	"	0.160
Coupling	"	0.320
Union	"	0.400
2-1/2"		
90 deg ell	EA.	0.400
45 deg ell	"	0.400
Tee	"	0.571
Cap	"	0.200
Coupling	"	0.400
Union	"	0.444
15410.18 GLASS PIPE		
Glass pipe		
1-1/2" dia.	L.F.	0.160
2" dia.	"	0.178
3" dia.	"	0.200
4" dia.	"	0.229
6" dia.	"	0.267
15410.30 PVC/CPVC PIPE		
PVC schedule 40		
1/2" pipe	L.F.	0.033
3/4" pipe	"	0.036
1" pipe	"	0.040
1-1/4" pipe	"	0.044
1-1/2" pipe	"	0.050
2" pipe	"	0.057

PLUMBING	UNIT	MAN/HOURS
15410.30 PVC/CPVC PIPE		
2-1/2" pipe	L.F.	0.067
3" pipe	"	0.080
4" pipe	"	0.100
6" pipe	"	0.200
8" pipe	"	0.267
PVC schedule 80 pipe		
1-1/2" pipe	L.F.	0.050
2" pipe	"	0.057
3" pipe	"	0.080
4" pipe	"	0.100
Polypropylene, acid resistant, DWV pipe		
Schedule 40		
1-1/2" pipe	L.F.	0.057
2" pipe	"	0.067
3" pipe	"	0.080
4" pipe	"	0.100
6" pipe	"	0.200
Polyethylene pipe and fittings		
SDR-21		
3" pipe	L.F.	0.100
4" pipe	"	0.133
6" pipe	"	0.200
8" pipe	"	0.229
10" pipe	"	0.267
15410.33 ABS DWV PIPE		
Schedule 40 ABS		
1-1/2" pipe	L.F.	0.040
2" pipe	"	0.044
3" pipe	"	0.057
4" pipe	"	0.080
6" pipe	"	0.100
15410.35 PLASTIC PIPE		
Fiberglass reinforced pipe		
2" pipe	L.F.	0.062
3" pipe	"	0.067
4" pipe	"	0.073
6" pipe	"	0.080
8" pipe	"	0.133
10" pipe	"	0.160
12" pipe	"	0.200
15410.70 STAINLESS STEEL PIPE		
Stainless steel, schedule 40, threaded		
1/2" pipe	L.F.	0.114
1" pipe	"	0.123
1-1/2" pipe	"	0.133
2" pipe	"	0.145

PLUMBING	UNIT	MAN/ HOURS
15410.70 STAINLESS STEEL PIPE		
2-1/2" pipe	L.F.	0.160
3" pipe	"	0.178
4" pipe	"	0.200
15410.80 STEEL PIPE		
Black steel, extra heavy pipe, threaded		
1/2" pipe	L.F.	0.032
3/4" pipe	"	0.032
1" pipe	"	0.040
1-1/2" pipe	"	0.044
2-1/2" pipe	"	0.100
3" pipe	"	0.133
4" pipe	"	0.160
5" pipe	"	0.200
6" pipe	"	0.200
8" pipe	"	0.267
10" pipe	"	0.320
12" pipe	"	0.400
Fittings, malleable iron, threaded, 1/2" pipe		
90 deg ell	EA.	0.267
45 deg ell	"	0.267
Tee	"	0.400
3/4" pipe		
90 deg ell	EA.	0.267
45 deg ell	"	0.400
Tee	"	0.400
1" pipe		
90 deg ell	EA.	0.320
45 deg ell	"	0.320
Tee	"	0.444
1-1/2" pipe		
90 deg ell	EA.	0.400
45 deg ell	"	0.400
Tee	"	0.571
2-1/2" pipe		
90 deg ell	EA.	1.000
45 deg ell	"	1.000
Tee	"	1.333
3" pipe		
90 deg ell	EA.	1.333
45 deg ell	"	1.333
Tee	"	2.000
4" pipe		
90 deg ell	EA.	1.600
45 deg ell	"	1.600
Tee	"	2.667
6" pipe		
90 deg ell	EA.	1.600

PLUMBING	UNIT	MAN/ HOURS
15410.80 STEEL PIPE		
45 deg ell	EA.	1.600
Tee	"	2.667
8" pipe		
90 deg ell	EA.	3.200
45 deg ell	"	3.200
Tee	"	5.000
10" pipe		
90 deg ell	EA.	4.000
45 deg ell	"	6.667
Tee	"	5.000
12" pipe		
90 deg ell	EA.	5.000
45 deg ell	"	5.000
Tee	"	6.667
15410.82 GALVANIZED STEEL PIPE		
Galvanized pipe		
1/2" pipe	L.F.	0.080
3/4" pipe	"	0.100
1" pipe	"	0.114
1-1/4" pipe	"	0.133
1-1/2" pipe	"	0.160
2" pipe	"	0.200
2-1/2" pipe	"	0.267
3" pipe	"	0.286
4" pipe	"	0.333
6" pipe	"	0.667
15430.23 CLEANOUTS		
Cleanout, wall		
2"	EA.	0.533
3"	"	0.533
4"	"	0.667
6"	"	0.800
8"	"	1.000
Floor		
2"	EA.	0.667
3"	"	0.667
4"	"	0.800
6"	"	1.000
8"	"	1.143
15430.24 GREASE TRAPS		
Grease traps, cast iron, 3" pipe		
35 gpm, 70 lb capacity	EA.	8.000
50 gpm, 100 lb capacity	"	10.000

15 MECHANICAL

PLUMBING

15430.25 HOSE BIBBS

PLUMBING	UNIT	MAN/HOURS
Hose bibb		
1/2"	EA.	0.267
3/4"	"	0.267

15430.60 VALVES

PLUMBING	UNIT	MAN/HOURS
Gate valve, 125 lb, bronze, soldered		
1/2"	EA.	0.200
3/4"	"	0.200
1"	"	0.267
1-1/2"	"	0.320
2"	"	0.400
2-1/2"	"	0.500
Threaded		
1/4", 125 lb	EA.	0.320
1/2"		
125 lb	EA.	0.320
300 lb	"	0.320
3/4"		
125 lb	EA.	0.320
300 lb	"	0.320
1"		
125 lb	EA.	0.320
300 lb	"	0.400
1-1/2"		
125 lb	EA.	0.400
300 lb	"	0.444
2"		
125 lb	EA.	0.571
300 lb	"	0.667
Cast iron, flanged		
2", 150 lb	EA.	0.667
2-1/2"		
125 lb	EA.	0.667
250 lb	"	0.667
3"		
125 lb	EA.	0.800
250 lb	"	0.800
4"		
125 lb	EA.	1.143
250 lb	"	1.143
6"		
125 lb	EA.	1.600
250 lb	"	1.600
8"		
125 lb	EA.	2.000
250 lb	"	2.000
OS&Y, flanged		
2"		

PLUMBING

15430.60 VALVES

PLUMBING	UNIT	MAN/HOURS
125 lb	EA.	0.667
250 lb	"	0.667
2-1/2"		
125 lb	EA.	0.667
250 lb	"	0.800
3"		
125 lb	EA.	0.800
250 lb	"	0.800
4"		
125 lb	EA.	1.333
250 lb	"	1.333
6"		
125 lb	EA.	1.600
250 lb	"	1.600
Ball valve, bronze, 250 lb, threaded		
1/2"	EA.	0.320
3/4"	"	0.320
1"	"	0.400
1-1/4"	"	0.444
1-1/2"	"	0.500
2"	"	0.571
Angle valve, bronze, 150 lb, threaded		
1/2"	EA.	0.286
3/4"	"	0.320
1"	"	0.320
1-1/4"	"	0.400
1-1/2"	"	0.444
Balancing valve, with meter connections, circuit setter		
1/2"	EA.	0.320
3/4"	"	0.364
1"	"	0.400
1-1/4"	"	0.444
1-1/2"	"	0.533
2"	"	0.667
2-1/2"	"	0.800
3"	"	1.000
4"	"	1.333
Pressure reducing valve, bronze, threaded, 250 lb		
1/2"	EA.	0.500
3/4"	"	0.500
1"	"	0.500
1-1/4"	"	0.571
1-1/2"	"	0.667
Pressure regulating valve, bronze, class 300		
1"	EA.	0.500
1-1/2"	"	0.615
2"	"	0.800
3"	"	1.143
4"	"	1.600

PLUMBING	UNIT	MAN/ HOURS
15430.60 VALVES		
5"	EA.	2.000
6"	"	2.667
Solar water temperature regulating valve		
3/4"	EA.	0.667
1"	"	0.800
1-1/4"	"	0.889
1-1/2"	"	1.000
2-1/2"	"	2.000
Tempering valve, threaded		
3/4"	EA.	0.267
1"	"	0.320
1-1/4"	"	0.400
1-1/2"	"	0.400
2"	"	0.500
2-1/2"	"	0.667
3"	"	0.800
4"	"	1.143
Thermostatic mixing valve, threaded		
1/2"	EA.	0.286
3/4"	"	0.320
1"	"	0.348
1-1/2"	"	0.400
2"	"	0.500
Sweat connection		
1/2"	EA.	0.286
3/4"	"	0.320
Mixing valve, sweat connection		
1/2"	EA.	0.286
3/4"	"	0.320
Liquid level gauge, aluminum body		
3/4"	EA.	0.320
125 psi, pvc body		
3/4"	EA.	0.320
150 psi, crs body		
3/4"	EA.	0.320
1"	"	0.320
175 psi, bronze body, 1/2"	"	0.286
15430.65 VACUUM BREAKERS		
Vacuum breaker, atmospheric, threaded connection		
3/4"	EA.	0.320
1"	"	0.320
Anti-siphon, brass		
3/4"	EA.	0.320
1"	"	0.320
1-1/4"	"	0.400
1-1/2"	"	0.444
2"	"	0.500

PLUMBING	UNIT	MAN/ HOURS
15430.68 STRAINERS		
Strainer, Y pattern, 125 psi, cast iron body, threaded		
3/4"	EA.	0.286
1"	"	0.320
1-1/4"	"	0.400
1-1/2"	"	0.400
2"	"	0.500
250 psi, brass body, threaded		
3/4"	EA.	0.320
1"	"	0.320
1-1/4"	"	0.400
1-1/2"	"	0.400
2"	"	0.500
Cast iron body, threaded		
3/4"	EA.	0.320
1"	"	0.320
1-1/4"	"	0.400
1-1/2"	"	0.400
2"	"	0.500
15430.70 DRAINS, ROOF & FLOOR		
Floor drain, cast iron, with cast iron top		
2"	EA.	0.667
3"	"	0.667
4"	"	0.667
6"	"	0.800
Roof drain, cast iron		
2"	EA.	0.667
3"	"	0.667
4"	"	0.667
5"	"	0.800
6"	"	0.800
15430.80 TRAPS		
Bucket trap, threaded		
3/4"	EA.	0.500
1"	"	0.533
1-1/4"	"	0.615
1-1/2"	"	0.727
Inverted bucket steam trap, threaded		
3/4"	EA.	0.500
1"	"	0.500
1-1/4"	"	0.444
1-1/2"	"	0.667
Float trap, 15 psi		
3/4"	EA.	0.500
1"	"	0.533
1-1/4"	"	0.571
1-1/2"	"	0.667

PLUMBING

15430.80 TRAPS	UNIT	MAN/HOURS
2"	EA.	0.800
Float and thermostatic trap, 15 psi		
3/4"	EA.	0.500
1"	"	0.533
1-1/4"	"	0.571
1-1/2"	"	0.667
2"	"	0.800
Steam trap, cast iron body, threaded, 125 psi		
3/4"	EA.	0.500
1"	"	0.533
1-1/4"	"	0.571
1-1/2"	"	0.667

PLUMBING FIXTURES

15440.10 BATHS	UNIT	MAN/HOURS
Bath tub, 5' long		
Minimum	EA.	2.667
Average	"	4.000
Maximum	"	8.000
6' long		
Minimum	EA.	2.667
Average	"	4.000
Maximum	"	8.000
Square tub, whirlpool, 4'x4'		
Minimum	EA.	4.000
Average	"	8.000
Maximum	"	10.000
5'x5'		
Minimum	EA.	4.000
Average	"	8.000
Maximum	"	10.000
6'x6'		
Minimum	EA.	4.000
Average	"	8.000
Maximum	"	10.000
For trim and rough-in		
Minimum	EA.	2.667
Average	"	4.000
Maximum	"	8.000
Average	"	2.000

PLUMBING FIXTURES

15440.10 BATHS	UNIT	MAN/HOURS
Maximum	EA.	2.667

15440.15 FAUCETS	UNIT	MAN/HOURS
Kitchen		
Minimum	EA.	1.333
Average	"	1.600
Maximum	"	2.000
Bath		
Minimum	EA.	1.333
Average	"	1.600
Maximum	"	2.000
Lavatory, domestic		
Minimum	EA.	1.333
Average	"	1.600
Maximum	"	2.000
Hospital, patient rooms		
Minimum	EA.	2.000
Average	"	2.667
Maximum	"	4.000
Operating room		
Minimum	EA.	2.000
Average	"	2.667
Maximum	"	4.000
Washroom		
Minimum	EA.	1.333
Average	"	1.600
Maximum	"	2.000
Handicapped		
Minimum	EA.	1.600
Average	"	2.000
Maximum	"	2.667
Shower		
Minimum	EA.	1.333
Average	"	1.600
Maximum	"	2.000
For trim and rough-in		
Minimum	EA.	1.600
Average	"	2.000
Maximum	"	4.000

15440.18 HYDRANTS	UNIT	MAN/HOURS
Wall hydrant		
8" thick	EA.	1.333
12" thick	"	1.600
18" thick	"	1.778
24" thick	"	2.000
Ground hydrant		

PLUMBING FIXTURES	UNIT	MAN/HOURS
15440.18 HYDRANTS		
2' deep	EA.	1.000
4' deep	"	1.143
6' deep	"	1.333
8' deep	"	2.000
15440.20 LAVATORIES		
Lavatory, counter top, porcelain enamel on cast iron		
Minimum	EA.	1.600
Average	"	2.000
Maximum	"	2.667
Wall hung, china		
Minimum	EA.	1.600
Average	"	2.000
Maximum	"	2.667
Handicapped		
Minimum	EA.	2.000
Average	"	2.667
Maximum	"	4.000
For trim and rough-in		
Minimum	EA.	2.000
Average	"	2.667
Maximum	"	4.000
15440.30 SHOWERS		
Shower, fiberglass, 36"x34"x84"		
Minimum	EA.	5.714
Average	"	8.000
Maximum	"	8.000
Steel, 1 piece, 36"x36"		
Minimum	EA.	5.714
Average	"	8.000
Maximum	"	8.000
Receptor, molded stone, 36"x36"		
Minimum	EA.	2.667
Average	"	4.000
Maximum	"	6.667
For trim and rough-in		
Minimum	EA.	3.636
Average	"	4.444
Maximum	"	8.000
15440.40 SINKS		
Service sink, 24"x29"		
Minimum	EA.	2.000
Average	"	2.667
Maximum	"	4.000
Kitchen sink, single, stainless steel, single bowl		
Minimum	EA.	1.600
Average	"	2.000

PLUMBING FIXTURES	UNIT	MAN/HOURS
15440.40 SINKS		
Maximum	EA.	2.667
Double bowl		
Minimum	EA.	2.000
Average	"	2.667
Maximum	"	4.000
Porcelain enamel, cast iron, single bowl		
Minimum	EA.	1.600
Average	"	2.000
Maximum	"	2.667
Double bowl		
Minimum	EA.	2.000
Average	"	2.667
Maximum	"	4.000
Mop sink, 24"x36"x10"		
Minimum	EA.	1.600
Average	"	2.000
Maximum	"	2.667
Washing machine box		
Minimum	EA.	2.000
Average	"	2.667
Maximum	"	4.000
For trim and rough-in		
Minimum	EA.	2.667
Average	"	4.000
Maximum	"	5.333
15440.50 URINALS		
Urinal, flush valve, floor mounted		
Minimum	EA.	2.000
Average	"	2.667
Maximum	"	4.000
Wall mounted		
Minimum	EA.	2.000
Average	"	2.667
Maximum	"	4.000
For trim and rough-in		
Minimum	EA.	2.000
Average	"	4.000
Maximum	"	5.333
15440.60 WATER CLOSETS		
Water closet flush tank, floor mounted		
Minimum	EA.	2.000
Average	"	2.667
Maximum	"	4.000
Handicapped		
Minimum	EA.	2.667
Average	"	4.000

PLUMBING FIXTURES	UNIT	MAN/ HOURS
15440.60 WATER CLOSETS		
Maximum	EA.	8.000
Bowl, with flush valve, floor mounted		
Minimum	EA.	2.000
Average	"	2.667
Maximum	"	4.000
Wall mounted		
Minimum	EA.	2.000
Average	"	2.667
Maximum	"	4.000
For trim and rough-in		
Minimum	EA.	2.000
Average	"	2.667
Maximum	"	4.000
15440.70 WATER HEATERS		
Water heater, electric		
6 gal	EA.	1.333
10 gal	"	1.333
15 gal	"	1.333
20 gal	"	1.600
30 gal	"	1.600
40 gal	"	1.600
52 gal	"	2.000
66 gal	"	2.000
80 gal	"	2.000
100 gal	"	2.667
120 gal	"	2.667
Oil fired		
20 gal	EA.	4.000
50 gal	"	5.714
15440.90 MISCELLANEOUS FIXTURES		
Electric water cooler		
Floor mounted	EA.	2.667
Wall mounted	"	2.667
Wash fountain		
Wall mounted	EA.	4.000
Circular, floor supported	"	8.000
Deluge shower and eye wash	"	4.000
15440.95 FIXTURE CARRIERS		
Water fountain, wall carrier		
Minimum	EA.	0.800
Average	"	1.000
Maximum	"	1.333
Lavatory, wall carrier		

PLUMBING FIXTURES	UNIT	MAN/ HOURS
15440.95 FIXTURE CARRIERS		
Minimum	EA.	0.800
Average	"	1.000
Maximum	"	1.333
Sink, industrial, wall carrier		
Minimum	EA.	0.800
Average	"	1.000
Maximum	"	1.333
Toilets, water closets, wall carrier		
Minimum	EA.	0.800
Average	"	1.000
Maximum	"	1.333
Floor support		
Minimum	EA.	0.667
Average	"	0.800
Maximum	"	1.000
Urinals, wall carrier		
Minimum	EA.	0.800
Average	"	1.000
Maximum	"	1.333
Floor support		
Minimum	EA.	0.667
Average	"	0.800
Maximum	"	1.000
15450.30 PUMPS		
In-line pump, bronze, centrifugal		
5 gpm, 20' head	EA.	0.500
20 gpm, 40' head	"	0.500
50 gpm		
50' head	EA.	1.000
Cast iron, centrifugal		
50 gpm, 200' head	EA.	1.000
100 gpm		
100' head	EA.	1.333
Centrifugal, close coupled, c.i., single stage		
50 gpm, 100' head	EA.	1.000
100 gpm, 100' head	"	1.333
Base mounted		
50 gpm, 100' head	EA.	1.000
100 gpm, 50' head	"	1.333
200 gpm, 100' head	"	2.000
300 gpm, 175' head	"	2.000
Condensate pump, simplex		
1000 sf EDR, 2 gpm	EA.	6.667
2000 sf EDR, 3 gpm	"	6.667
4000 sf EDR, 6 gpm	"	7.273
6000 sf EDR, 9 gpm	"	7.273
Duplex, bronze		

PLUMBING FIXTURES

	UNIT	MAN/HOURS
15450.30 PUMPS		
8000 sf EDR, 12 gpm	EA.	7.273
10,000 sf EDR, 15 gpm	"	10.000
15,000 sf EDR, 23 gpm	"	11.429
20,000 sf EDR, 30 gpm	"	16.000
25,000 sf EDR, 38 gpm	"	16.000
15450.40 STORAGE TANKS		
Hot water storage tank, cement lined		
10 gallon	EA.	2.667
70 gallon	"	4.000
200 gallon	"	5.714
900 gallon	"	10.000
1100 gallon	"	10.000
2000 gallon	"	10.000
15480.10 SPECIAL SYSTEMS		
Air compressor, air cooled, two stage		
5.0 cfm, 175 psi	EA.	16.000
10 cfm, 175 psi	"	17.778
20 cfm, 175 psi	"	19.048
50 cfm, 125 psi	"	21.053
80 cfm, 125 psi	"	22.857
Single stage, 125 psi		
1.0 cfm	EA.	11.429
1.5 cfm	"	11.429
2.0 cfm	"	11.429
Automotive compressor, hose reel, air and water, 50' hose	"	6.667
Lube equipment, 3 reel, with pumps	"	32.000
Tire changer		
Truck	EA.	11.429
Passenger car	"	6.154
Air hose reel, includes, 50' hose	"	6.154
Hose reel, 5 reel, motor oil, gear oil, lube, air & water	"	32.000
Water hose reel, 50' hose	"	6.154
Pump, air operated, for motor or gear oil, fits 55 gal drum	"	0.800
For chassis lube	"	0.800
Fuel dispensing pump, lighted dial, one product		
One hose	EA.	6.667
Two hose	"	6.667
Two products, two hose	"	6.667

HEATING & VENTILATING

	UNIT	MAN/HOURS
15555.10 BOILERS		
Cast iron, gas fired, hot water		
115 mbh	EA.	20.000
175 mbh	"	21.818
235 mbh	"	24.000
940 mbh	"	48.000
1600 mbh	"	60.000
3000 mbh	"	80.000
6000 mbh	"	120
Steam		
115 mbh	EA.	20.000
175 mbh	"	21.818
235 mbh	"	24.000
940 mbh	"	48.000
1600 mbh	"	60.000
3000 mbh	"	80.000
6000 mbh	"	120
Electric, hot water		
115 mbh	EA.	12.000
175 mbh	"	12.000
235 mbh	"	12.000
940 mbh	"	24.000
1600 mbh	"	48.000
3000 mbh	"	60.000
6000 mbh	"	80.000
Steam		
115 mbh	EA.	12.000
175 mbh	"	12.000
235 mbh	"	12.000
940 mbh	"	24.000
1600 mbh	"	48.000
3000 mbh	"	60.000
6000 mbh	"	80.000
Oil fired, hot water		
115 mbh	EA.	16.000
175 mbh	"	18.462
235 mbh	"	21.818
940 mbh	"	40.000
1600 mbh	"	48.000
3000 mbh	"	60.000
6000 mbh	"	120
Steam		
115 mbh	EA.	16.000
175 mbh	"	18.462
235 mbh	"	21.818
940 mbh	"	40.000
1600 mbh	"	48.000
3000 mbh	"	60.000
6000 mbh	"	120

15 MECHANICAL

HEATING & VENTILATING

15610.10 FURNACES

HEATING & VENTILATING	UNIT	MAN/HOURS
Electric, hot air		
40 mbh	EA.	4.000
60 mbh	"	4.211
80 mbh	"	4.444
100 mbh	"	4.706
125 mbh	"	4.848
160 mbh	"	5.000
200 mbh	"	5.161
400 mbh	"	5.333
Gas fired hot air		
40 mbh	EA.	4.000
60 mbh	"	4.211
80 mbh	"	4.444
100 mbh	"	4.706
125 mbh	"	4.848
160 mbh	"	5.000
200 mbh	"	5.161
400 mbh	"	5.333
Oil fired hot air		
40 mbh	EA.	4.000
60 mbh	"	4.211
80 mbh	"	4.444
100 mbh	"	4.706
125 mbh	"	4.848
160 mbh	"	5.000
200 mbh	"	5.161
400 mbh	"	5.333

REFRIGERATION

15670.10 CONDENSING UNITS

REFRIGERATION	UNIT	MAN/HOURS
Air cooled condenser, single circuit		
3 ton	EA.	1.333
5 ton	"	1.333
7.5 ton	"	3.810
20 ton	"	4.000
25 ton	"	4.000
30 ton	"	4.000
With low ambient dampers		
3 ton	EA.	2.000
5 ton	EA.	2.000
7.5 ton	"	4.000
20 ton	"	5.333
25 ton	"	5.333
30 ton	"	5.333
Dual circuit		
10 ton	EA.	4.000
15 ton	"	5.714
20 ton	"	5.714
25 ton	"	5.714
30 ton	"	5.714
With low ambient dampers		
15 ton	EA.	5.714
20 ton	"	5.714
25 ton	"	5.714
30 ton	"	5.714

15680.10 CHILLERS

REFRIGERATION	UNIT	MAN/HOURS
Chiller, reciprocal		
Air cooled, remote condenser, starter		
20 ton	EA.	8.000
25 ton	"	8.000
30 ton	"	8.000
40 ton	"	12.000
Water cooled, with starter		
20 ton	EA.	8.000
25 ton	"	8.000
30 ton	"	12.000
40 ton	"	12.000
Packaged, air cooled, with starter		
20 ton	EA.	6.000
25 ton	"	6.000
30 ton	"	6.000
40 ton	"	6.000

HEAT TRANSFER

15780.10 COMPUTER ROOM A/C

HEAT TRANSFER	UNIT	MAN/HOURS
Air cooled, alarm, high efficiency filter, elec. heat		
3 ton	EA.	6.154
5 ton	"	6.667

HEAT TRANSFER	UNIT	MAN/ HOURS
15780.10 COMPUTER ROOM A/C		
7.5 ton	EA.	8.000
10 ton	"	10.000
15 ton	"	11.429
Steam heat		
3 ton	EA.	6.154
5 ton	"	6.667
7.5 ton	"	8.000
10 ton	"	10.000
15 ton	"	11.429
Hot water heat		
3 ton	EA.	6.154
5 ton	"	6.667
7.5 ton	"	8.000
10 ton	"	10.000
15 ton	"	11.429
Air cooled condenser, low ambient damper		
3 ton	EA.	1.600
5 ton	"	2.000
7.5 ton	"	4.000
10 ton	"	5.714
15 ton	"	4.706
Water cooled, high efficiency filter, alarm, elec. heat		
3 ton	EA.	5.714
5 ton	"	6.667
7.5 ton	"	10.000
10 ton	"	11.429
15 ton	"	13.333
Steam heat		
3 ton	EA.	5.714
5 ton	"	6.667
7.5 ton	"	10.000
10 ton	"	11.429
15 ton	"	13.333
Hot water heat		
3 ton	EA.	5.714
5 ton	"	6.667
7.5 ton	"	10.000
10 ton	"	11.429
15 ton	"	13.333
15780.20 ROOFTOP UNITS		
Packaged, single zone rooftop unit, with roof curb		
2 ton	EA.	8.000
3 ton	"	8.000
4 ton	"	10.000
5 ton	"	13.333
7.5 ton	"	16.000

HEAT TRANSFER	UNIT	MAN/ HOURS
15830.20 FAN COIL UNITS		
Fan coil unit, 2 pipe, complete		
200 cfm ceiling hung	EA.	2.667
Floor mounted	"	2.000
300 cfm, ceiling hung	"	3.200
Floor mounted	"	2.667
400 cfm, ceiling hung	"	3.810
Floor mounted	"	2.667
500 cfm, ceiling hung	"	4.000
Floor mounted	"	3.077
600 cfm, ceiling hung	"	4.420
Floor mounted	"	3.636
15830.70 UNIT HEATERS		
Steam unit heater, horizontal		
12,500 btuh, 200 cfm	EA.	1.333
17,000 btuh, 300 cfm	"	1.333
40,000 btuh, 500 cfm	"	1.333
60,000 btuh, 700 cfm	"	1.333
70,000 btuh, 1000 cfm	"	2.000
Vertical		
12,500 btuh, 200 cfm	EA.	1.333
17,000 btuh, 300 cfm	"	1.333
40,000 btuh, 500 cfm	"	1.333
60,000 btuh, 700 cfm	"	1.333
70,000 btuh, 1000 cfm	"	1.333
Gas unit heater, horizontal		
27,400 btuh	EA.	3.200
38,000 btuh	"	3.200
56,000 btuh	"	3.200
82,200 btuh	"	3.200
103,900 btuh	"	5.000
125,700 btuh	"	5.000
133,200 btuh	"	5.000
149,000 btuh	"	5.000
172,000 btuh	"	5.000
190,000 btuh	"	5.000
225,000 btuh	"	5.000
Hot water unit heater, horizontal		
12,500 btuh, 200 cfm	EA.	1.333
17,000 btuh, 300 cfm	"	1.333
25,000 btuh, 500 cfm	"	1.333
30,000 btuh, 700 cfm	"	1.333
50,000 btuh, 1000 cfm	"	2.000
60,000 btuh, 1300 cfm	"	2.000
Vertical		
12,500 btuh, 200 cfm	EA.	1.333
17,000 btuh, 300 cfm	"	1.333
25,000 btuh, 500 cfm	"	1.333

15 MECHANICAL

HEAT TRANSFER

15830.70 UNIT HEATERS

	UNIT	MAN/HOURS
30,000 btuh, 700 cfm	EA.	1.333
50,000 btuh, 1000 cfm	"	1.333
60,000 btuh, 1300 cfm	"	1.333
Cabinet unit heaters, ceiling, exposed, hot water		
200 cfm	EA.	2.667
300 cfm	"	3.200
400 cfm	"	3.810
600 cfm	"	4.211
800 cfm	"	5.000
1000 cfm	"	5.714
1200 cfm	"	6.667
2000 cfm	"	8.889

AIR HANDLING

15855.10 AIR HANDLING UNITS

	UNIT	MAN/HOURS
Air handling unit, medium pressure, single zone		
1500 cfm	EA.	5.000
3000 cfm	"	8.889
4000 cfm	"	10.000
5000 cfm	"	10.667
6000 cfm	"	11.429
7000 cfm	"	12.308
8500 cfm	"	13.333
10,500 cfm	"	16.000
12,500 cfm	"	17.778
Rooftop air handling units		
4950 cfm	EA.	8.889
7370 cfm	"	11.429
9790 cfm	"	13.333
14,300 cfm	"	11.429
21,725 cfm	"	11.429
33,000 cfm	"	13.333

15870.20 EXHAUST FANS

	UNIT	MAN/HOURS
Belt drive roof exhaust fans		
640 cfm, 2618 fpm	EA.	1.000
940 cfm, 2604 fpm	"	1.000
1050 cfm, 3325 fpm	"	1.000
1170 cfm, 2373 fpm	"	1.000

AIR HANDLING

15870.20 EXHAUST FANS

	UNIT	MAN/HOURS
2440 cfm, 4501 fpm	EA.	1.000
2760 cfm, 4950 fpm	"	1.000
3890 cfm, 6769 fpm	"	1.000
2380 cfm, 3382 fpm	"	1.000
2880 cfm, 3859 fpm	"	1.000
3200 cfm, 4173 fpm	"	1.333
3660 cfm, 3437 fpm	"	1.333
Direct drive fans		
60 to 390 cfm	EA.	1.000
145 to 590 cfm	"	1.000
295 to 860 cfm	"	1.000
235 to 1300 cfm	"	1.000
415 to 1630 cfm	"	1.000
590 to 2045 cfm	"	1.000

AIR DISTRIBUTION

15890.10 METAL DUCTWORK

	UNIT	MAN/HOURS
Rectangular duct		
Galvanized steel		
Minimum	Lb.	0.073
Average	"	0.089
Maximum	"	0.133
Aluminum		
Minimum	Lb.	0.160
Average	"	0.200
Maximum	"	0.267
Fittings		
Minimum	EA.	0.267
Average	"	0.400
Maximum	"	0.800

15890.30 FLEXIBLE DUCTWORK

	UNIT	MAN/HOURS
Flexible duct, 1.25" fiberglass		
5" dia.	L.F.	0.040
6" dia.	"	0.044
7" dia.	"	0.047
8" dia.	"	0.050
10" dia.	"	0.057
12" dia.	"	0.062

15 MECHANICAL

AIR DISTRIBUTION	UNIT	MAN/ HOURS
15890.30 FLEXIBLE DUCTWORK		
14" dia.	L.F.	0.067
16" dia.	"	0.073
Flexible duct connector, 3" wide fabric	"	0.133
15895.10 ROOF CURBS		
8" high, insulated, with liner and raised can		
15" x 15"	EA.	0.400
21" x 21"	"	0.400
36" x 36"	"	0.571
48" x 48"	"	0.615
60" x 60"	"	0.800
72" x 72"	"	1.000
15910.10 DAMPERS		
Horizontal parallel aluminum backdraft damper		
12" x 12"	EA.	0.200
16" x 16"	"	0.229
24" x 24"	"	0.400
36" x 36"	"	0.571
48" x 48"	"	0.800
"Up", parallel dampers		
12" x 12"	EA.	0.200
16" x 16"	"	0.229
24" x 24"	"	0.400
36" x 36"	"	0.571
48" x 48"	"	0.800
"Down", parallel dampers		
12" x 12"	EA.	0.200
16" x 16"	"	0.229
24" x 24"	"	0.400
36" x 36"	"	0.571
48" x 48"	"	0.800
Fire damper, 1.5 hr rating		
12" x 12"	EA.	0.400
16" x 16"	"	0.400
24" x 24"	"	0.400
36" x 36"	"	0.800
48" x 48"	"	1.143
15940.10 DIFFUSERS		
Ceiling diffusers, round, baked enamel finish		
6" dia.	EA.	0.267
12" dia.	"	0.333
16" dia.	"	0.364
20" dia.	"	0.400
Rectangular		
6x6"	EA.	0.267

AIR DISTRIBUTION	UNIT	MAN/ HOURS
15940.10 DIFFUSERS		
12x12"	EA.	0.400
18x18"	"	0.400
24x24"	"	0.500
Lay in, flush mounted, perforated face, with grid		
6x6/24x24	EA.	0.320
12x12/24x24	"	0.320
18x18/24x24	"	0.320
Two-way slot diffuser with balancing damper, 4'	"	0.800
15940.20 RELIEF VENTILATORS		
Intake ventilator, aluminum, with screen, no curbs		
12" x 12"	EA.	0.667
16" x 16"	"	0.800
36" x 36"	"	1.333
48" x 48"	"	1.600
15940.40 REGISTERS AND GRILLES		
Lay in flush mounted, perforated face, return		
6x6/24x24	EA.	0.320
8x8/24x24	"	0.320
9x9/24x24	"	0.320
10x10/24x24	"	0.320
12x12/24x24	"	0.320
Rectangular, ceiling return, single deflection		
10x10	EA.	0.400
12x12	"	0.400
14x14	"	0.400
16x8	"	0.400
16x16	"	0.400
24x12	"	0.400
24x18	"	0.400
36x24	"	0.444
36x30	"	0.444
Wall, return air register		
12x12	EA.	0.200
16x16	"	0.200
18x18	"	0.200
20x20	"	0.200
24x24	"	0.200
Ceiling, return air grille		
6x6	EA.	0.267
8x8	"	0.320
10x10	"	0.320
Ceiling, exhaust grille, aluminum egg crate		
6x6	EA.	0.267
8x8	"	0.320
10x10	"	0.320
12x12	"	0.400

AIR DISTRIBUTION

	UNIT	MAN/ HOURS
15940.40 REGISTERS AND GRILLES		
14x14	EA.	0.400
16x16	"	0.400
18x18	"	0.400
15940.80 PENTHOUSE LOUVERS		
Penthouse louvers		
12" high, extruded aluminum, 4" louver		
6' perimeter	EA.	2.000
12' perimeter	"	2.000
20' perimeter	"	5.333
16" high x 4' perimeter	"	2.000
6' perimeter	"	2.000
12' perimeter	"	2.000
20' perimeter	"	5.333
20" high x 4' perimeter	"	2.000
6' perimeter	"	2.000
12' perimeter	"	2.000
20' perimeter	"	5.333
24" high x 4' perimeter	"	2.000
6' perimeter	"	2.000
12' perimeter	"	2.000
20' perimeter	"	5.333

CONTROLS

	UNIT	MAN/ HOURS
15950.10 HVAC CONTROLS		
Pressure gauge, direct reading gage cock and syphon	EA.	0.500
Control valve, 1", modulating		
2-way	EA.	0.667
3-way	"	1.000
Self contained control valve with sensing element, 3/4"	"	0.500
Control dampers, round		
6" dia.	EA.	0.320
8" dia	"	0.320
10" dia	"	0.320
12" dia	"	0.320
12" dia	"	0.400
18" dia	"	0.400
20" dia	"	0.400

CONTROLS

	UNIT	MAN/ HOURS
15950.10 HVAC CONTROLS		
Rectangular, parallel blade standard leakage		
12" x 12"	EA.	0.400
16" x 16"	"	0.400
20" x 20"	"	0.400
48" x 48"	"	1.143
48" x 60"	"	1.333
48" x 72"	"	1.333
Low leakage		
12" x 12"	EA.	0.400
16" x 16"	"	0.400
36" x 36"	"	0.667
48" x 48"	"	1.143
48" x 72"	"	1.333
Rectangular, opposed horizontal blade		
12" x 12"	EA.	0.400
16" x 16"	"	0.400
24" x 24"	"	0.400
36" x 36"	"	0.667
48" x 72"	"	1.333

16 ELECTRICAL

BASIC MATERIALS	UNIT	MAN/ HOURS
16110.20 CONDUIT SPECIALTIES		
Rod beam clamp, 1/2"	EA.	0.050
Hanger rod		
3/8"	L.F.	0.040
1/2"	"	0.050
Hanger channel, 1-1/2"		
No holes	EA.	0.030
Holes	"	0.030
Channel strap		
1/2"	EA.	0.050
3/4"	"	0.050
1"	"	0.050
1-1/4"	"	0.080
1-1/2"	"	0.080
2"	"	0.080
2-1/2"	"	0.123
3"	"	0.123
3-1/2"	"	0.123
4"	"	0.145
5"	"	0.145
6"	"	0.145
Conduit penetrations, roof and wall, 8" thick		
1/2"	EA.	0.615
3/4"	"	0.615
1"	"	0.800
1-1/4"	"	0.800
1-1/2"	"	0.800
2"	"	1.600
2-1/2"	"	1.600
3"	"	1.600
3-1/2"	"	2.000
4"	"	2.000
Fireproofing, for conduit penetrations		
1/2"	EA.	0.500
3/4"	"	0.500
1"	"	0.500
1-1/4"	"	0.727
1-1/2"	"	0.727
2"	"	0.727
2-1/2"	"	0.899
3"	"	0.899
3-1/2"	"	1.250
4"	"	1.509
Aluminum conduit		
1/2"	L.F.	0.030
3/4"	"	0.040
1"	"	0.050
1-1/4"	"	0.059
1-1/2"	"	0.080
2"	"	0.089

BASIC MATERIALS	UNIT	MAN/ HOURS
16110.20 CONDUIT SPECIALTIES		
2-1/2"	L.F.	0.100
3"	"	0.107
3-1/2"	"	0.123
4"	"	0.145
5"	"	0.182
6"	"	0.200
16110.22 EMT CONDUIT		
EMT conduit		
1/2"	L.F.	0.030
3/4"	"	0.040
1"	"	0.050
1-1/4"	"	0.059
1-1/2"	"	0.080
2"	"	0.089
2-1/2"	"	0.100
3"	"	0.123
3-1/2"	"	0.145
4"	"	0.182
16110.23 FLEXIBLE CONDUIT		
Flexible conduit, steel		
3/8"	L.F.	0.030
1/2	"	0.030
3/4"	"	0.040
1"	"	0.040
1-1/4"	"	0.050
1-1/2"	"	0.059
2"	"	0.080
2-1/2"	"	0.089
3"	"	0.107
16110.24 GALVANIZED CONDUIT		
Galvanized rigid steel conduit		
1/2"	L.F.	0.040
3/4"	"	0.050
1"	"	0.059
1-1/4"	"	0.080
1-1/2"	"	0.089
2"	"	0.100
2-1/2"	"	0.145
3"	"	0.182
3-1/2"	"	0.190
4"	"	0.211
5"	"	0.286
6"	"	0.381

16 ELECTRICAL

BASIC MATERIALS

16110.27 PLASTIC COATED CONDUIT

BASIC MATERIALS	UNIT	MAN/HOURS
Rigid steel conduit, plastic coated		
1/2"	L.F.	0.050
3/4"	"	0.059
1"	"	0.080
1-1/4"	"	0.100
1-1/2"	"	0.123
2"	"	0.145
2-1/2"	"	0.190
3"	"	0.222
3-1/2"	"	0.250
4"	"	0.308
5"	"	0.381

16110.28 STEEL CONDUIT

BASIC MATERIALS	UNIT	MAN/HOURS
Intermediate metal conduit (IMC)		
1/2"	L.F.	0.030
3/4"	"	0.040
1"	"	0.050
1-1/4"	"	0.059
1-1/2"	"	0.080
2"	"	0.089
2-1/2"	"	0.119
3"	"	0.145
3-1/2"	"	0.182
4"	"	0.190

16110.35 WIREMOLD

BASIC MATERIALS	UNIT	MAN/HOURS
Wiremold raceway with fittings, surface mounted		
#200	L.F.	0.030
#500	"	0.030
#700	"	0.040
#800	"	0.040
Fittings, #200, 90 degree flat elbow	EA.	0.050
Internal elbow	"	0.050
Extension adapter	"	0.059
#200, #500, #700		
Single pole switch and box	EA.	0.400
Duplex receptacle with box	"	0.308
#500, #700		
90 deg. flat elbow	EA.	0.080
Internal elbow	"	0.080
Junction box	"	0.133
Fixture box	"	0.133
Shallow switch and receptacle box	"	0.080
#800		
90 deg. flat elbow	EA.	0.080
Internal elbow	"	0.080
Junction box	"	0.123

BASIC MATERIALS

16110.80 WIREWAYS

BASIC MATERIALS	UNIT	MAN/HOURS
Wireway, hinge cover type		
2-1/2" x 2-1/2"		
1' section	EA.	0.154
2'	"	0.190
3'	"	0.250
5'	"	0.381
10'	"	0.667
4" x 4"		
1'	EA.	0.250
2'	"	0.250
3'	"	0.308
4'	"	0.308
10'	"	0.800

16120.43 COPPER CONDUCTORS

BASIC MATERIALS	UNIT	MAN/HOURS
Copper conductors, type THW, solid		
#14	L.F.	0.004
#12	"	0.005
#10	"	0.006
Stranded		
#14	L.F.	0.004
#12	"	0.005
#10	"	0.006
#8	"	0.008
#6	"	0.009
#4	"	0.010
#3	"	0.010
#2	"	0.012
#1	"	0.014
1/0	"	0.016
2/0	"	0.020
3/0	"	0.025
4/0	"	0.028
250 MCM	"	0.030
300 MCM	"	0.033
350 MCM	"	0.040
400 MCM	"	0.044
500 MCM	"	0.052
600 MCM	"	0.059
750 MCM	"	0.067
1000 MCM	"	0.076
THHN-THWN, solid		
#14	L.F.	0.004
#12	"	0.005
#10	"	0.006
Stranded		
#14	L.F.	0.004
#12	"	0.005

16 ELECTRICAL

BASIC MATERIALS	UNIT	MAN/ HOURS
16120.43 COPPER CONDUCTORS		
#10	L.F.	0.006
#8	"	0.008
#6	"	0.009
#4	"	0.010
#2	"	0.012
#1	"	0.014
1/0	"	0.016
2/0	"	0.020
3/0	"	0.025
4/0	"	0.028
250 MCM	"	0.030
350 MCM	"	0.040
16120.47 SHEATHED CABLE		
Non-metallic sheathed cable		
Type NM cable with ground		
#14/2	L.F.	0.015
#12/2	"	0.016
#10/2	"	0.018
#8/2	"	0.020
#6/2	"	0.025
#14/3	"	0.026
#12/3	"	0.027
#10/3	"	0.027
#8/3	"	0.028
#6/3	"	0.028
#4/3	"	0.032
#2/3	"	0.035
Type U.F. cable with ground		
#14/2	L.F.	0.016
#12/2	"	0.019
#10/2	"	0.020
#8/2	"	0.023
#6/2	"	0.027
#14/3	"	0.020
#12/3	"	0.022
#10/3	"	0.025
#8/3	"	0.028
#6/3	"	0.032
16130.60 PULL AND JUNCTION BOXES		
4"		
Octagon box	EA.	0.114
Box extension	"	0.059
Plaster ring	"	0.059
Cover blank	"	0.059
Square box	"	0.114
Box extension	"	0.059

BASIC MATERIALS	UNIT	MAN/ HOURS
16130.60 PULL AND JUNCTION BOXES		
Plaster ring	EA.	0.059
Cover blank	"	0.059
4-11/16"		
Square box	EA.	0.114
Box extension	"	0.059
Plaster ring	"	0.059
Cover blank	"	0.059
Switch and device boxes		
2 gang	EA.	0.114
3 gang	"	0.114
4 gang	"	0.160
Device covers		
2 gang	EA.	0.059
3 gang	"	0.059
4 gang	"	0.059
Handy box	"	0.114
Extension	"	0.059
Switch cover	"	0.059
Switch box with knockout	"	0.145
Weatherproof cover, spring type	"	0.080
Cover plate, dryer receptacle 1 gang plastic	"	0.100
For 4" receptacle, 2 gang	"	0.100
Duplex receptacle cover plate, plastic	"	0.059
4", vertical bracket box, 1-1/2" with		
RMX clamps	EA.	0.145
BX clamps	"	0.145
4", octagon device cover		
1 switch	EA.	0.059
1 duplex recept	"	0.059
4", square face bracket boxes, 1-1/2"		
RMX	EA.	0.145
BX	"	0.145
16130.65 PULL BOXES AND CABINETS		
Galvanized pull boxes, screw cover		
4x4x4	EA.	0.190
4x6x4	"	0.190
6x6x4	"	0.190
6x8x4	"	0.190
8x8x4	"	0.250
8x10x4	"	0.242
8x12x4	"	0.250
Screw cover		
10x10x4	EA.	0.308
12x12x6	"	0.444
12x15x6	"	0.444
12x18x6	"	0.500
15x18x6	"	0.571
18x24x6	"	0.615

16 ELECTRICAL

BASIC MATERIALS

	UNIT	MAN/ HOURS
16130.65 PULL BOXES AND CABINETS		
18x30x6	EA.	0.727
24x36x6	"	0.727
16130.80 RECEPTACLES		
125 volt, 20a, duplex, grounding type, standard grade	EA.	0.200
Ground fault interrupter type	"	0.296
250 volt, 20a, 2 pole, single receptacle, ground type	"	0.200
120/208v, 4 pole, single receptacle, twist lock		
20a	EA.	0.348
50a	"	0.348
125/250v, 3 pole, flush receptacle		
30a	EA.	0.296
50a	"	0.296
60a	"	0.348
277 v, 20a, 2 pole, grounding type, twist lock	"	0.200
Dryer receptacle, 250v, 30a/50a, 3 wire	"	0.296
Clock receptacle, 2 pole, grounding type	"	0.200
125v, 20a single recept. grounding type		
Standard grade	EA.	0.200
Specification	"	0.200
Hospital	"	0.200
Isolated ground orange	"	0.250
Duplex		
Specification grade	EA.	0.200
Hospital	"	0.200
Isolated ground orange	"	0.250
GFI hospital grade recepts, 20a, 125v, duplex	"	0.296
16198.10 ELECTRIC MANHOLES		
Precast, handhole, 4' deep		
2'x2'	EA.	3.478
3'x3'	"	5.556
4'x4'	"	10.256
Power manhole, complete, precast, 8' deep		
4'x4'	EA.	14.035
6'x6'	"	20.000
8'x8'	"	21.053
6' deep, 9' x 12'	"	25.000
Cast in place, power manhole, 8' deep		
4'x4'	EA.	14.035
6'x6'	"	20.000
8'x8'	"	21.053
16199.10 UTILITY POLES & FITTINGS		
Wood pole, creosoted		
25'	EA.	2.353
40'	"	3.791

BASIC MATERIALS

	UNIT	MAN/ HOURS
16199.10 UTILITY POLES & FITTINGS		
55'	EA.	7.547
Treated, wood preservative, 6"x6"		
8'	EA.	0.500
16'	"	1.600
20'	"	2.000
Aluminum, brushed, no base		
8'	EA.	2.000
20'	"	3.200
40'	"	6.250
Steel, no base		
10'	EA.	2.500
20'	"	3.810
35'	"	6.250
Concrete, no base		
13'	EA.	5.517
30'	"	12.121
50'	"	18.182
60'	"	20.000

POWER GENERATION

	UNIT	MAN/ HOURS
16210.10 GENERATORS		
Diesel generator, with auto transfer switch		
50kw	EA.	30.769
125kw	"	50.000
300kw	"	100
750kw	"	200
16320.10 TRANSFORMERS		
Floor mounted, single phase, int. dry, 480v-120/240v		
3 kva	EA.	1.818
5 kva	"	3.077
7.5 kva	"	3.478
10 kva	"	3.810
15 kva	"	4.301
100 kva	"	11.594
Three phase, 480v-120/280v		
15 kva	EA.	6.015
30 kva	"	9.412
45 kva	"	10.811

16 ELECTRICAL

POWER GENERATION

16320.10 TRANSFORMERS

	UNIT	MAN/HOURS
225 kva	EA.	15.385

16350.10 CIRCUIT BREAKERS

	UNIT	MAN/HOURS
Molded case, 240v, 15-60a, bolt-on		
1 pole	EA.	0.250
2 pole	"	0.348
70-100a, 2 pole	"	0.533
15-60a, 3 pole	"	0.400
70-100a, 3 pole	"	0.615
480v, 2 pole		
15-60a	EA.	0.296
70-100a	"	0.400
3 pole		
15-60a	EA.	0.400
70-100a	"	0.444
70-225a	"	0.615
Load center circuit breakers, 240v		
1 pole, 10-60a	EA.	0.250
2 pole		
10-60a	EA.	0.400
70-100a	"	0.667
110-150a	"	0.727
3 pole		
10-60a	EA.	0.500
70-100a	"	0.727
Load center, G.F.I. breakers, 240v		
1 pole, 15-30a	EA.	0.296
2 pole, 15-30a	"	0.400
Key operated breakers, 240v, 1 pole, 10-30a	"	0.296
Tandem breakers, 240v		
1 pole, 15-30a	EA.	0.400
2 pole, 15-30a	"	0.533
Bolt-on, G.F.I. breakers, 240v, 1 pole, 15-30a	"	0.348

16360.10 SAFETY SWITCHES

	UNIT	MAN/HOURS
Fused, 3 phase, 30 amp, 600v, heavy duty		
NEMA 1	EA.	1.143
NEMA 3r	"	1.143
NEMA 4	"	1.600
NEMA 12	"	1.739
60a		
NEMA 1	EA.	1.143
NEMA 3r	"	1.143
NEMA 4	"	1.600
NEMA 12	"	1.739
100a		
NEMA 1	EA.	1.739

POWER GENERATION

16360.10 SAFETY SWITCHES

	UNIT	MAN/HOURS
NEMA 3r	EA.	1.739
NEMA 4	"	2.000
NEMA 12	"	2.500
200a		
NEMA 1	EA.	2.500
NEMA 3r	"	2.500
NEMA 4	"	2.759
NEMA 12	"	3.478
Non-fused, 240-600v, heavy duty, 3 phase, 30 amp		
NEMA 1	EA.	1.143
NEMA 3r	"	1.143
NEMA 4	"	1.739
NEMA 12	"	1.739
60a		
NEMA1	EA.	1.143
NEMA 3r	"	1.143
NEMA 4	"	1.739
NEMA 12	"	1.739
100a		
NEMA 1	EA.	1.739
NEMA 3r	"	1.739
NEMA 4	"	2.500
NEMA 12	"	2.500
200a, NEMA 1	"	2.500
600a, NEMA 12	"	12.308

16365.10 FUSES

	UNIT	MAN/HOURS
Fuse, one-time, 250v		
30a	EA.	0.050
60a	"	0.050
100a	"	0.050
200a	"	0.050
400a	"	0.050
600a	"	0.050
600v		
30a	EA.	0.050
60a	"	0.050
100a	"	0.050
200a	"	0.050
400a	"	0.050

16395.10 GROUNDING

	UNIT	MAN/HOURS
Ground rods, copper clad, 1/2" x		
6'	EA.	0.667
8'	"	0.727
10'	"	1.000
5/8" x		

16 ELECTRICAL

POWER GENERATION

16395.10 GROUNDING

	UNIT	MAN/HOURS
5'	EA.	0.615
6'	"	0.727
8'	"	1.000
10'	"	1.250
3/4" x		
8'	EA.	0.727
10'	"	0.800
Ground rod clamp		
5/8"	EA.	0.123
3/4"	"	0.123

SERVICE AND DISTRIBUTION

16425.10 SWITCHBOARDS

	UNIT	MAN/HOURS
Switchboard, 90" high, no main disconnect, 208/120v		
400a	EA.	7.921
600a	"	8.000
1000a	"	8.000
1200a	"	10.000
1600a	"	11.940
2000a	"	14.035
2500a	"	16.000
240/480v		
600a	EA.	8.163
800a	"	8.163
1600a	"	11.940
2000a	"	14.035
2500a	"	16.000
3000a	"	27.586
4000a	"	29.630

16470.10 PANELBOARDS

	UNIT	MAN/HOURS
3 phase, 480/277v, main lugs only, 120a, 30 circuits	EA.	3.478
277/480v, 4 wire, flush surface		
225a, 30 circuits	EA.	4.000
400a, 30 circuits	"	5.000
600a, 42 circuits	"	6.015
208/120v, main circuit breaker, 3 phase, 4 wire		
100a		
12 circuits	EA.	5.096
20 circuits	"	6.299
30 circuits	EA.	7.018
400a		
30 circuits	EA.	14.815
42 circuits	"	16.000
600a, 42 circuits	"	18.182
120/208v, flush, 3 ph., 4 wire, main only		
100a		
12 circuits	EA.	5.096
20 circuits	"	6.299
30 circuits	"	7.018
400a		
30 circuits	EA.	14.815
42 circuits	"	16.000
600a, 42 circuits	"	18.182

16480.10 MOTOR CONTROLS

	UNIT	MAN/HOURS
Motor generator set, 3 phase, 480/277v, w/controls		
10kw	EA.	27.586
15kw	"	30.769
20kw	"	32.000
40kw	"	38.095
100kw	"	61.538
200kw	"	72.727
300kw	"	80.000
2 pole, 230 volt starter, w/NEMA-1		
1 hp, 9 amp, size 00	EA.	1.000
2 hp, 18amp, size 0	"	1.000
3 hp, 27amp, size 1	"	1.000
5 hp, 45amp, size 1p	"	1.000
7-1/2 hp, 45a, size 2	"	1.000
15 hp, 90a, size 3	"	1.000

16490.10 SWITCHES

	UNIT	MAN/HOURS
Fused interrupter load, 35kv		
20A		
1 pole	EA.	16.000
2 pole	"	17.021
3 way	"	17.021
4 way	"	18.182
30a, 1 pole	"	16.000
3 way	"	17.021
4 way	"	18.182
Weatherproof switch, including box & cover, 20a		
1 pole	EA.	16.000
2 pole	"	17.021
3 way	"	18.182
4 way	"	18.182
Specification grade toggle switches, 20a, 120-277v		

16 ELECTRICAL

SERVICE AND DISTRIBUTION

16490.10 SWITCHES	UNIT	MAN/HOURS
Single pole	EA.	0.200
Double pole	"	0.296
3 way	"	0.250
Switch plates, plastic ivory		
1 gang	EA.	0.080
2 gang	"	0.100
3 gang	"	0.119
6 gang	"	0.182
Stainless steel		
1 gang	EA.	0.080
2 gang	"	0.100
4 gang	"	0.145
6 gang	"	0.145
Brass		
1 gang	EA.	0.080
2 gang	"	0.100
4 gang	"	0.145
6 gang	"	0.182
16490.20 TRANSFER SWITCHES		
Automatic transfer switch 600v, 3 pole		
30a	EA.	3.478
100a	"	4.762
400a	"	10.000
800a	"	18.182
1200a	"	22.857
2600a	"	42.105
16490.80 SAFETY SWITCHES		
Safety switch, 600v, 3 pole, heavy duty, NEMA-1		
30a	EA.	1.000
60a	"	1.143
100a	"	1.600
200a	"	2.500
400a	"	5.517
600a	"	8.000
800a	"	10.526
1200a	"	14.286

LIGHTING

16510.05 INTERIOR LIGHTING	UNIT	MAN/HOURS
Recessed fluorescent fixtures, 2'x2'		
2 lamp	EA.	0.727
4 lamp	"	0.727
2 lamp w/flange	"	1.000
4 lamp w/flange	"	1.000
1'x4'		
2 lamp	EA.	0.667
3 lamp	"	0.667
2 lamp w/flange	"	0.727
3 lamp w/flange	"	0.727
2'x4'		
2 lamp	EA.	0.727
3 lamp	"	0.727
4 lamp	"	0.727
2 lamp w/flange	"	1.000
3 lamp w/flange	"	1.000
4 lamp w/flange	"	1.000
4'x4'		
4 lamp	EA.	1.000
6 lamp	"	1.000
8 lamp	"	1.000
4 lamp w/flange	"	1.509
6 lamp w/flange	"	1.509
8 lamp, w/flange	"	1.509
Surface mounted incandescent fixtures		
40w	EA.	0.667
75w	"	0.667
100w	"	0.667
150w	"	0.667
Pendant		
40w	EA.	0.800
75w	"	0.800
100w	"	0.800
150w	"	0.800
Recessed incandescent fixtures		
40w	EA.	1.509
75w	"	1.509
100w	"	1.509
150w	"	1.509
Exit lights, 120v		
Recessed	EA.	1.250
Back mount	"	0.727
Universal mount	"	0.727
Emergency battery units, 6v-120v, 50 unit	"	1.509
With 1 head	"	1.509
With 2 heads	"	1.509
Light track single circuit		
2'	EA.	0.500
4'	"	0.500

LIGHTING	UNIT	MAN/HOURS
16510.05 INTERIOR LIGHTING		
8'	EA.	1.000
12'	"	1.509
Surface mounted fluorescent, wrap around lens		
1 lamp	EA.	0.800
2 lamps	"	0.889
4 lamps	"	1.000
Wall mounted fluorescent		
2-20w lamps	EA.	0.500
2-30w lamps	"	0.500
2-40w lamps	"	0.667
Indirect, with wood shielding, 2049w lamps		
4'	EA.	1.000
8'	"	1.600
Industrial fluorescent, 2 lamp		
4'	EA.	0.727
8'	"	1.333
Strip fluorescent		
4'		
1 lamp	EA.	0.667
2 lamps	"	0.667
8'		
1 lamp	EA.	0.727
2 lamps	"	0.889
Wire guard for strip fixture, 4' long	"	0.348
Strip fluorescent, 8' long, two 4' lamps	"	1.333
With four 4' lamps	"	1.600
Wet location fluorescent, plastic housing		
4' long		
1 lamp	EA.	1.000
2 lamps	"	1.333
8' long		
2 lamps	EA.	1.600
4 lamps	"	1.739
Parabolic troffer, 2'x2'		
With 2 "U" lamps	EA.	1.000
With 3 "U" lamps	"	1.143
2'x4'		
With 2 40w lamps	EA.	1.143
With 3 40w lamps	"	1.333
With 4 40w lamps	"	1.333
1'x4'		
With 1 t12 lamp, 9 cell	EA.	0.727
With 2 t12 lamps	"	0.889
With 1 t12 lamp, 20 cell	"	0.727
With 2 t12 lamps	"	0.889
Steel sided surface fluorescent, 2'x4'		
3 lamps	EA.	1.333
4 lamps	"	1.333
Outdoor sign fluor., 1 lamp, remote ballast		

LIGHTING	UNIT	MAN/HOURS
16510.05 INTERIOR LIGHTING		
4' long	EA.	6.015
6' long	"	8.000
Recess mounted, commercial, 2'x2', 13" high		
100w	EA.	4.000
250w	"	4.494
High pressure sodium, hi-bay open		
400w	EA.	1.739
1000w	"	2.424
Enclosed		
400w	EA.	2.424
1000w	"	2.963
Metal halide hi-bay, open		
400w	EA.	1.739
1000w	"	2.424
Enclosed		
400w	EA.	2.424
1000w	"	2.963
High pressure sodium, low bay, surface mounted		
100w	EA.	1.000
150w	"	1.143
250w	"	1.333
400w	"	1.600
Metal halide, low bay, pendant mounted		
175w	EA.	1.333
250w	"	1.600
400w	"	2.222
Indirect luminare, square, metal halide, freestanding		
175w	EA.	1.000
250w	"	1.000
400w	"	1.000
High pressure sodium		
150w	EA.	1.000
250w	"	1.000
400w	"	1.000
Round, metal halide		
175w	EA.	1.000
250w	"	1.000
400w	"	1.000
High pressure sodium		
150w	EA.	1.000
250w	"	1.000
400w	"	1.000
Wall mounted, metal halide		
175w	EA.	2.500
250w	"	2.500
400w	"	3.200
High pressure sodium		
150w	EA.	2.500
250w	"	2.500

LIGHTING	UNIT	MAN/ HOURS
16510.05 INTERIOR LIGHTING		
400w	EA.	3.200
Wall pack lithonia, high pressure sodium		
35w	EA.	0.889
55w	"	1.000
150w	"	1.600
250w	"	1.739
Low pressure sodium		
35w	EA.	1.739
55w	"	2.000
Wall pack hubbell, high pressure sodium		
35w	EA.	0.889
150w	"	1.600
250w	"	1.739
Compact fluorescent		
2-7w	EA.	1.000
2-13w	"	1.333
1-18w	"	1.333
Handball & racquet ball court, 2'x2', metal halide		
250w	EA.	2.500
400w	"	2.759
High pressure sodium		
250w	EA.	2.500
400w	"	2.759
Bollard light, 42" w/found., high pressure sodium		
70w	EA.	2.581
100w	"	2.581
150w	"	2.581
Light fixture lamps		
Lamp		
20w med. bipin base, cool white, 24"	EA.	0.145
30w cool white, rapid start, 36"	"	0.145
40w cool white "U", 3"	"	0.145
40w cool white, rapid start, 48"	"	0.145
70w high pressure sodium, mogul base	"	0.200
75w slimline, 96"	"	0.200
100w		
Incandescent, 100a, inside frost	EA.	0.100
Mercury vapor, clear, mogul base	"	0.200
High pressure sodium, mogul base	"	0.200
150w		
Par 38 flood or spot, incandescent	EA.	0.100
High pressure sodium, 1/2 mogul base	"	0.200
175w		
Mercury vapor, clear, mogul base	EA.	0.200
Mercury halide, clear, mogul base	"	0.200
250w		
High pressure sodium, mogul base	EA.	0.200
Mercury vapor, clear, mogul base	"	0.200
Metal halide, clear, mogul base	"	0.200

LIGHTING	UNIT	MAN/ HOURS
16510.05 INTERIOR LIGHTING		
High pressure sodium, mogul base	EA.	0.200
400w		
Mercury vapor, clear, mogul base	EA.	0.200
Metal halide, clear, mogul base	"	0.200
High pressure sodium, mogul base	"	0.200
1000w		
Mercury vapor, clear, mogul base	EA.	0.250
High pressure sodium, mogul base	"	0.250
16510.30 EXTERIOR LIGHTING		
Exterior light fixtures		
Rectangle, high pressure sodium		
70w	EA.	2.500
100w	"	2.581
150w	"	2.581
250w	"	2.759
400w	"	3.478
Flood, rectangular, high pressure sodium		
70w	EA.	2.500
100w	"	2.581
150w	"	2.581
400w	"	3.478
1000w	"	4.494
Round		
400w	EA.	3.478
1000w	"	4.494
Round, metal halide		
400w	EA.	3.478
1000w	"	4.494
Light fixture arms, cobra head, 6', high press. sodium		
100w	EA.	2.000
150w	"	2.500
250w	"	2.500
400w	"	2.963
Flood, metal halide		
400w	EA.	3.478
1000w	"	4.494
1500w	"	6.015
Mercury vapor		
250w	EA.	2.759
400w	"	3.478
Incandescent		
300w	EA.	1.739
500w	"	2.000
1000w	"	3.200

16 ELECTRICAL

LIGHTING	UNIT	MAN/ HOURS
16610.30 UNINTERRUPTIBLE POWER		
Uninterruptible power systems, (U.P.S.)	"	8.000
5 kva	"	11.004
7.5 kva	"	16.000
10 kva	"	21.978
15 kva	"	26.667
20 kva	"	24.024
25 kva	"	25.000
30 kva	"	25.974
35 kva	"	27.027
40 kva	"	27.972
45 kva	"	28.986
50 kva	"	29.963
62.5 kva	"	32.000
75 kva	"	34.934
100 kva	"	36.036
150 kva	"	50.000
200 kva	"	55.172
300 kva	"	74.766
400 kva	"	89.888
500 kva	"	110

COMMUNICATIONS	UNIT	MAN/ HOURS
16720.10 FIRE ALARM SYSTEMS		
Master fire alarm box, pedestal mounted	EA.	16.000
Master fire alarm box	"	6.015
Box light	"	0.500
Ground assembly for box	"	0.667
Bracket for pole type box	"	0.727
Pull station		
Waterproof	EA.	0.500
Manual	"	0.400
Horn, waterproof	"	1.000
Interior alarm	"	0.727
Coded transmitter, automatic	"	2.000
Control panel, 8 zone	"	8.000
Battery charger and cabinet	"	2.000
Batteries, nickel cadmium or lead calcium	"	5.000
CO2 pressure switch connection	"	0.727
Annunciator panels		
Fire detection annunciator, remote type, 8-zone	EA.	1.818

COMMUNICATIONS	UNIT	MAN/ HOURS
16720.10 FIRE ALARM SYSTEMS		
12-zone	EA.	2.000
16-zone	"	2.500
Fire alarm systems		
Bell	EA.	0.615
Weatherproof bell	"	0.667
Horn	"	0.727
Siren	"	2.000
Chime	"	0.615
Audio/visual	"	0.727
Strobe light	"	0.727
Smoke detector	"	0.667
Heat detection	"	0.500
Thermal detector	"	0.500
Ionization detector	"	0.533
Duct detector	"	2.759
Test switch	"	0.500
Remote indicator	"	0.571
Door holder	"	0.727
Telephone jack	"	0.296
Fireman phone	"	1.000
Speaker	"	0.800
Remote fire alarm annunciator panel		
24 zone	EA.	6.667
48 zone	"	13.008
Control panel		
12 zone	EA.	2.963
16 zone	"	4.444
24 zone	"	6.667
48 zone	"	16.000
Power supply	"	1.509
Status command	"	5.000
Printer	"	1.509
Transponder	"	0.899
Transformer	"	0.667
Transceiver	"	0.727
Relays	"	0.500
Flow switch	"	2.000
Tamper switch	"	2.963
End of line resistor	"	0.348
Printed ckt. card	"	0.500
Central processing unit	"	6.154
UPS backup to c.p.u.	"	8.999
Smoke detector, fixed temp. & rate of rise comb.	"	1.600
16720.50 SECURITY SYSTEMS		
Sensors		
Balanced magnetic door switch, surface mounted	EA.	0.500
With remote test	"	1.000

COMMUNICATIONS	UNIT	MAN/ HOURS
16720.50 SECURITY SYSTEMS		
Flush mounted	EA.	1.860
Mounted bracket	"	0.348
Mounted bracket spacer	"	0.348
Photoelectric sensor, for fence		
6 beam	EA.	2.759
9 beam	"	4.255
Photoelectric sensor, 12 volt dc		
500' range	EA.	1.600
800' range	"	2.000
Capacitance wire grid kit		
Surface	EA.	1.000
Duct	"	1.600
Tube grid kit	"	0.500
Vibration sensor, 30 max per zone	"	0.500
Audio sensor, 30 max per zone	"	0.500
Inertia sensor		
Outdoor	EA.	0.727
Indoor	"	0.500
Ultrasonic transmitter, 20 max per zone		
Omni-directional	EA.	1.600
Directional	"	1.333
Transceiver		
Omni-directional	EA.	1.000
Directional	"	1.000
Passive infra-red sensor, 20 max per zone	"	1.600
Access/secure control unit, balanced magnetic switch	"	1.600
Photoelectric sensor	"	1.600
Photoelectric fence sensor	"	1.600
Capacitance sensor	"	1.739
Audio and vibration sensor	"	1.600
Inertia sensor	"	1.600
Ultrasonic sensor	"	1.739
Infra-red sensor	"	2.000
Monitor panel, with access/secure tone, standard	"	1.739
High security	"	2.000
Emergency power indicator	"	0.500
Monitor rack with 115v power supply		
1 zone	EA.	1.000
10 zone	"	2.500
Monitor cabinet, wall mounted		
1 zone	EA.	1.000
5 zone	"	1.600
10 zone	"	1.739
20 zone	"	2.000
Floor mounted, 50 zone	"	4.000
Security system accessories		
Tamper assembly for monitor cabinet	EA.	0.444
Monitor panel blank	"	0.348
Audible alarm	"	0.500

COMMUNICATIONS	UNIT	MAN/ HOURS
16720.50 SECURITY SYSTEMS		
Audible alarm control	EA.	0.348
Termination screw, terminal cabinet		
25 pair	EA.	1.600
50 pair	"	2.500
150 pair	"	5.000
Universal termination, cabinets & panel		
Remote test	EA.	1.739
No remote test	"	0.727
High security line supervision termination	"	1.000
Door cord for capacitance sensor, 12"	"	0.500
Insulation block kit for capacitance sensor	"	0.348
Termination block for capacitance sensor	"	0.348
Guard alert display	"	0.615
Uninterrupted power supply		
Plug-in 40kva transformer		
12 volt	EA.	0.348
18 volt	"	0.348
24 volt	"	0.348
Test relay	"	0.348
Coaxial cable, 50 ohm	L.F.	0.006
Door openers	EA.	0.500
Push buttons		
Standard	EA.	0.348
Weatherproof	"	0.444
Bells	"	0.727
Horns		
Standard	EA.	1.000
Weatherproof	"	1.250
Chimes	"	0.667
Flasher	"	0.615
Motion detectors	"	1.509
Intercom units	"	0.727
Remote annunciator	"	5.000
16740.10 TELEPHONE SYSTEMS		
Communication cable		
25 pair	L.F.	0.026
100 pair	"	0.029
400 pair	"	0.044
Cable tap in manhole or junction box		
25 pair cable	EA.	3.810
100 pair cable	"	15.094
400 pair cable	"	61.538
Cable terminations, manhole or junction box		
25 pair cable	EA.	3.756
100 pair cable	"	15.094
400 pair cable	"	61.538

COMMUNICATIONS	UNIT	MAN/ HOURS
16780.50 TELEVISION SYSTEMS		
TV outlet, self terminating, w/cover plate	"	0.308
Thru splitter	"	1.600
End of line	"	1.333
In line splitter multitap		
4 way	EA.	1.818
2 way	"	1.702
Equipment cabinet	"	1.600
Antenna		
Broad band uhf	EA.	3.478
Lightning arrester	"	0.727
TV cable	L.F.	0.005
16850.10 ELECTRIC HEATING		
Baseboard heater		
2', 375w	EA.	1.000
3', 500w	"	1.000
4', 750w	"	1.143
5', 935w	"	1.333
6', 1125w	"	1.600
7', 1310w	"	1.818
8', 1500w	"	2.000
9', 1680w	"	2.222
10', 1875w	"	2.286
Unit heater wall mounted		
1500w	EA.	1.600
2500w	"	1.818
4000w	"	2.286
Thermostat		
Integral	EA.	0.500
Line voltage	"	0.500
Electric heater connection	"	0.250

BNi Building News™

CONVERSION FACTORS

Change	To	Multiply By
Atmospheres	Pounds per square inch	14.696
Atmospheres	Inches of mercury	29.92
Atmospheres	Feet of water	34
Barrels, oil	Gallons, of oil	42
Barrels, cement	Pounds of cement	376
Bags or sacks, cement	Pounds of cement	94
Btu/min.	Foot-pounds/sec.	12.96
Btu/min.	Horse-power	0.02356
Btu/min.	Kilowatts	0.01757
Btu/min.	Watts	17.57
Centimeters	Inches	0.3937
Centimeters of mercury	Atmospheres	0.01316
Centimeters of mercury	Feet of water	0.4461
Cubic inches	Cubic feet	0.00058
Cubic feet	Cubic inches	1728
cubic feet	Cubic yards	0.03703
Cubic yards	Cubic feet	27
Cubic inches	Gallons	0.00433
Cubic feet	Gallons	7.48
Feet	Inches	12
Feet	Yards	0.3333
Yards	Feet	3
Feet of water	Atmospheres	0.02950
Feet of water	Inches of mercury	0.8826
Gallons	Cubic Inches	231
Gallons	Cubic feet	0.1337
Gallons	Pounds of water	8.33
Gallons	Quarts	4
Gallons per min.	Cubic feet sec.	0.002228
Gallons per min.	Cubic feet hour	8.0208
Gallons water per min.	Tons water/24 hours	6.0086
Horse-power	Foot-lbs./sec.	550
Inches	Centimeters	2.540
Inches	Feet	0.0833
Inches	Millimeters	25.4
Inches of water	Pounds per Sq. inch	0.0361
Inches of water	Inches of mercury	0.0735
Inches of water	Ounces per square inch	0.578
Inches of water	Ounces per square foot	5.2
Inches of mercury	Inches of water	13.6
Inches of mercury	Feet of water	1.1333
Inches of mercury	Pounds per square inch	0.4914
Kilometers	Miles	0.6214
Meters	Inches	39.37
Miles	Feet	5280
Millimeters	Centimeters	0.1
Millimeters	Inches	0.03937
Ounces (fluid)	Cubic inches	1.805
Ounces	Pounds	0.0625
Pounds	Ounces	16
Pounds per square inch	Inches of water	27.72
Pounds per square inch	Feet of water	2.310
Pounds per square inch	Inches of mercury	2.04
Pounds per square inch	Atmospheres	0.0681
Quarts	Cubic Inches	67.20
Square Inches	Square feet	0.00694
Square Feet	Square inches	144
Square Feet	Square yards	0.11111
Square yards	Square feet	9
Square miles	Acres	640
Short tons	Pounds	2000
Short tons	Long tons	0.89285
Tons of water/24 hours	Gallons per minute	0.16643
Yards	Feet	3
Yards	Centimeters	91.44
Yards	Inches	36

CONVERSION CALCULATIONS

Commercial Measure

16 drams	= 1 ounce
16 ounces	= 1 pound
2,000 pounds	= 1 ton

Long Measure

12 inches	= 1 foot
3 feet	= 1 yard
16½ feet	= 1 rod
40 rods	= 1 furlong
8 furlongs (5,280 ft) =	= 1 mile
3 miles	= 1 league

Square Measure

144 square inches	= 1 square foot
9 square feet	= 1 square yard
30¼ square yards	= 1 square rod
160 square rods	= 1 acre
4840 square yards	= 1 acre
640 acres	= 1 square mile
36 square miles	= 1 township

Surveyors Measure

7.92 inches	= 1 link
25 links	= 1 rod
4 rods (66 ft.)	= 1 chain
10 chains	= 1 furlong
8 furlongs	= 1 mile
1 square mile	= 1 section

Cubic Measure

1728 cubic inches	= 1 cubic foot
27 cubic feet	= 1 cubic yard
128 cubic feet	= 1 cord (wood/stone)
231 cubic inches	= 1 U.S. gallon
7.48 U.S. Gallons	= 1 cubic foot
2150.4 cubic inches	= 1 U.S. bushel

Liquid Measure

4 fluid ounces	= 1 gill
4 gills	= 1 pint
2 pints	= 1 quart
4 quarts	= 1 gallon
9 gallons	= 1 firkin
31½ gallons	= 1 barrel
2 barrels	= 1 hogshead

Dry Measure

2 pints	= 1 quart
8 quarts	= 1 peck
4 pecks	= 1 bushel
2150.42 cubic inches	= 1 bushel

SQUARE

$A = a^2$

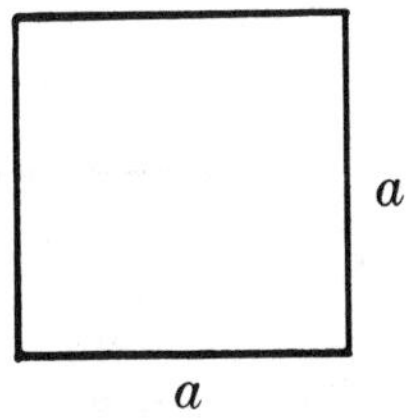

RECTANGLE

$A = bh$

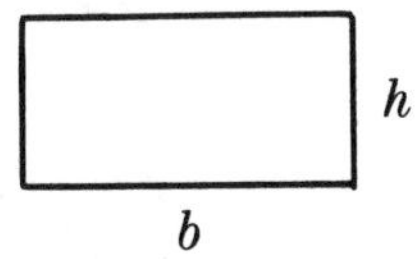

TRIANGLE

$A = \frac{1}{2} bh$

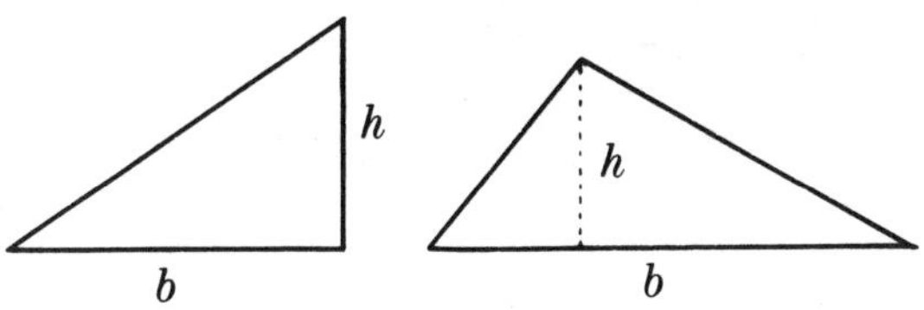

PARALLELOGRAM

$A = bh = ab \sin \phi$

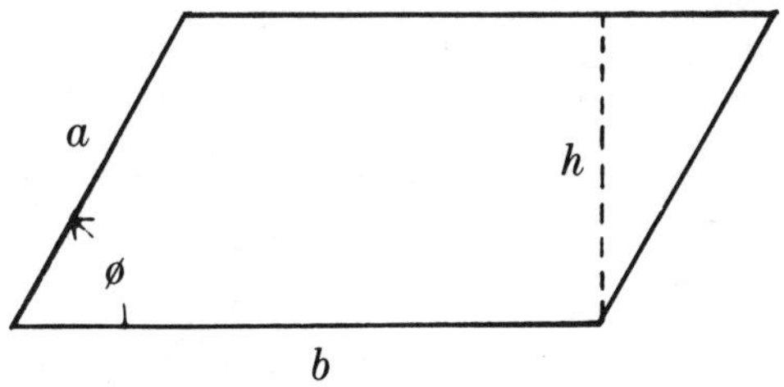

TRAPEZOID

$A = \left(\frac{a+b}{2}\right) h$

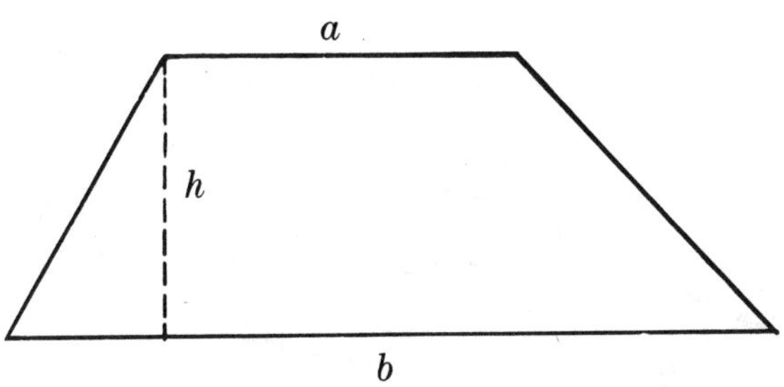

CIRCLE

$A = \pi r^2 = \frac{\pi d^2}{4}$

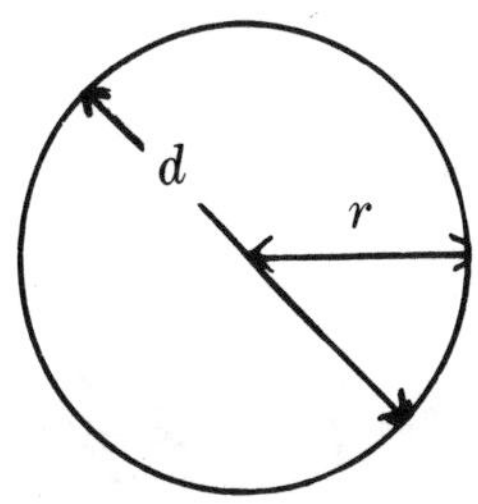

$Circumference = C = 2\pi r = \pi d$

GENERAL / MEASUREMENT 01025

ELLIPSE

$A = 0.7854\ ab$

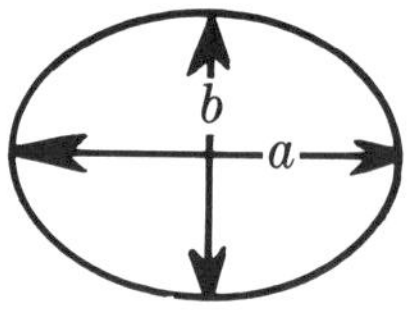

PARABOLA

$A = \frac{2}{3} bh$

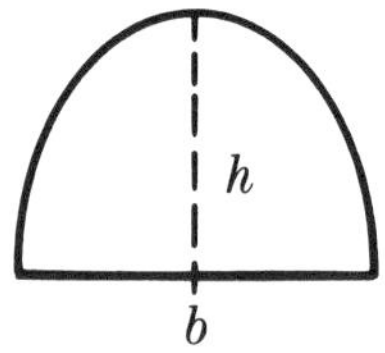

VOLUMES

CUBE

$V = a^3$

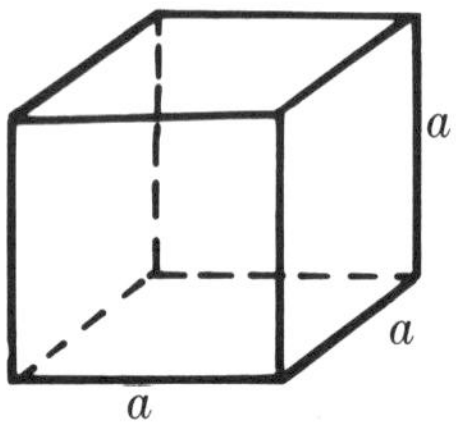

CYLINDER

$V = \pi r^3 h$

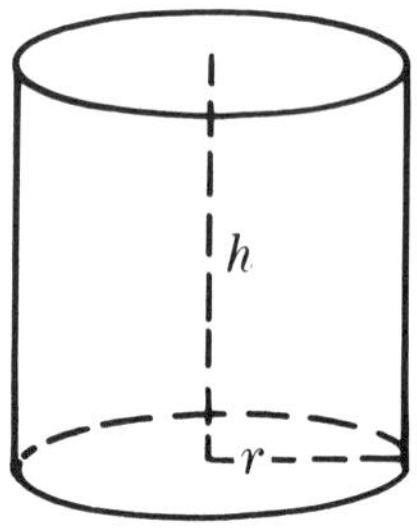

PYRAMID

$V = \frac{1}{3}(Base)\ h$

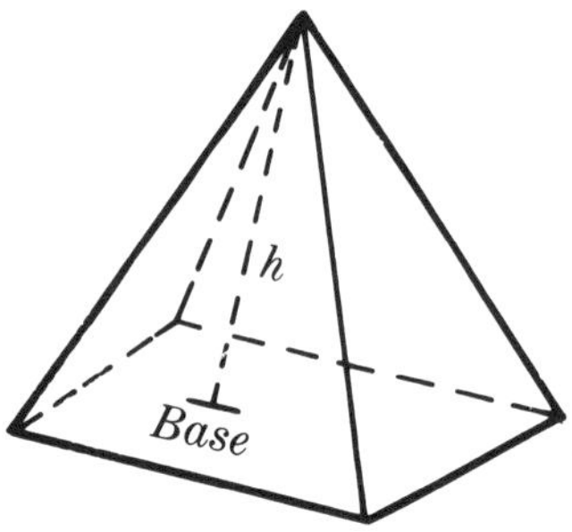

CONE

$V = \frac{1}{3} \pi r^2 h$

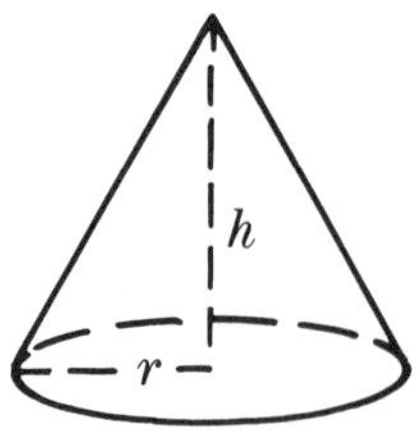

SPHERE

$V = \frac{4}{3} \pi r^3 = \frac{1}{6} \pi d^3$

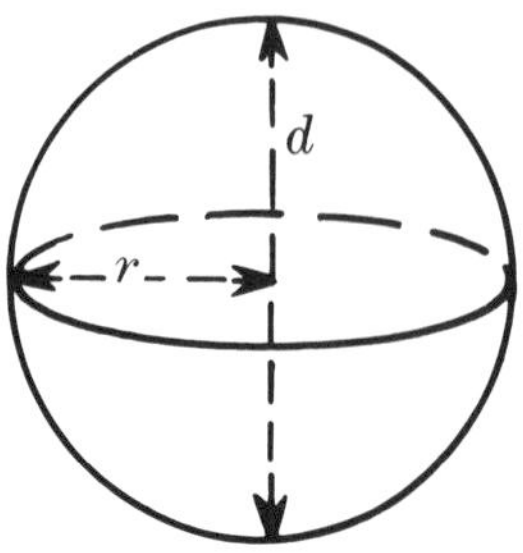

WEDGE

$V = \frac{1}{2} abc$

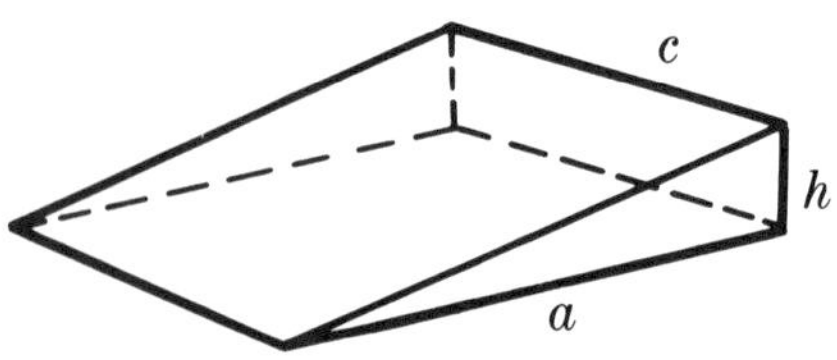

CONVERSION FACTORS
ENGLISH TO SI (SYSTEM INTERNATIONAL)

To Convert from	To	Multiply by
LENGTH		
Inches	Millimetres	25.4[a]
Feet	Metres	0.3048[a]
Yards	Metres	0.9144[a]
Miles (statute)	Kilometres	1.609
AREA		
Square inches	Square millimetres	645.2
Square feet	Square metres	0.0929
Square yards	Square metres	0.8361
VOLUME		
Cubic inches	Cubic millimetres	16.387
Cubic feet	Cubic metres	0.02832
Cubic yards	Cubic metres	0.7646
Gallons (U.S. liquid)[b]	Cubic metres[c]	0.003785
Gallons (Canadian liquid)[b]	Cubic metres[c]	0.004546
Ounces (U.S. liquid)[b]	Millilitres[c, d]	29.57
Quarts (U.S. liquid)[b]	Litres[c, d]	0.9464
Gallons (U.S. liquid)[b]	Litres[c]	3.785
FORCE		
Kilograms force	Newtons	9.807
Pounds force	Newtons	4.448
Pounds force	Kilograms force[d]	0.4536
Kips	Newtons	4448
Kips	Kilograms force[d]	453.6
PRESSURE, STRESS, STRENGTH (FORCE PER UNIT AREA)		
Kilograms force per sq. centimetre	Megapascals	0.09807
Pounds force per square inch (psi)	Megapascals	6895
Kips per square inch	Megapascals	6.895
Pounds force per square inch (psi)	Kilograms force per square centimetre[d]	0.07031
Pounds force per square foot	Pascals	47.88
Pounds force per square foot	Kilograms force per square metre[d]	4.882
BENDING MOMENT OR TORQUE		
Inch-pounds force	Metre-kilog. force[d]	0.01152
Inch-pounds force	Newton-metres	0.1130
Foot-pounds force	Metre-kilog. force[d]	0.1383
Foot-pounds force	Newton-metres	1.356
Metre-kilograms force	Newton-metres	9.807
MASS		
Ounce (avoirdupois)	Grams	28.35
Pounds (avoirdupois)	Kilograms	0.4536
Tons (metric)	Kilograms	1000[a]
Tons, short (2000 pounds)	Kilograms	907.2
Tons, short (2000 pounds)	Megagrams[e]	0.9072
MASS PER UNIT VOLUME		
Pounds mass per cubic foot	Kilog. per cubic metre	16.02
Pounds mass per cubic yard	Kilog. per cubic metre	0.5933
Pds. mass per gallon (U.S. liquid)[b]	Kilog. per cubic metre	119.8
Pds. mass p/gal. (Canadian liquid)[b]	Kilog. per cubic metre	99.78
TEMPERATURE		
Degrees Fahrenheit	Degrees Celsius	tK = (1F − 32)/1.8
Degrees Fahrenheit	Degrees Kelvin	tK = (1F + 459.67)/1.8
Degree Celsius	Degree Kelvin	tK = 1C + 273.15

[a] The factor given is exact.
[b] One U.S. gallon equals 0.8327 Canadian gallon.
[c] 1 litre = 1000 millilitres = 10,000 cubic centimetres = 1 cubic decimetre = 0.001 cubic metre.
[d] Metric but not SI unit.
[e] Called "tonne" in England. Called "metric ton" in other metric systems.

TRENCH BRACING

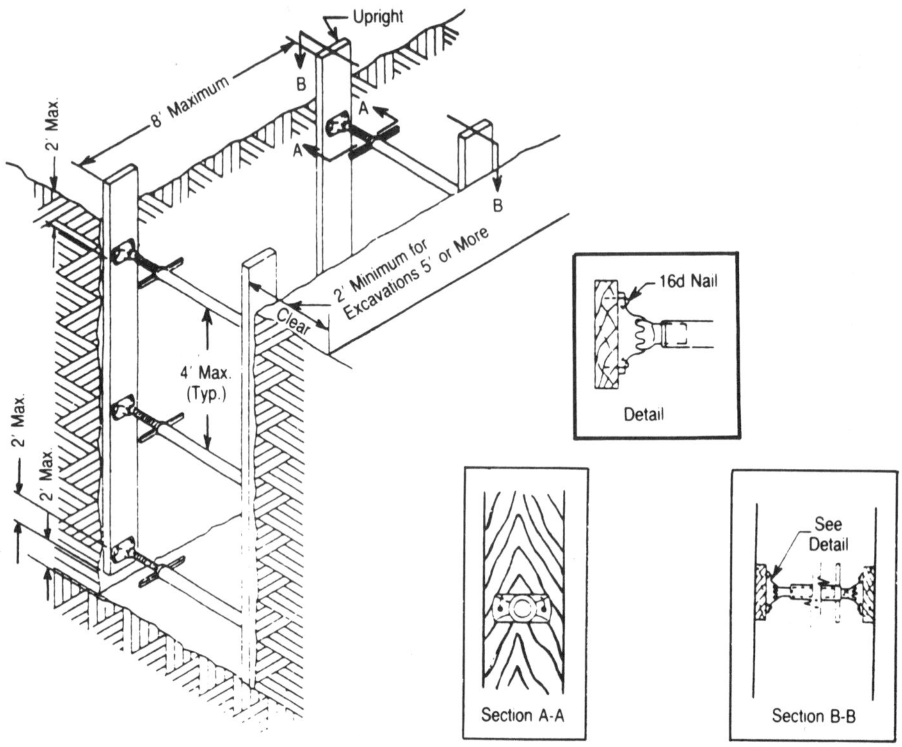

CLOSED VERTICAL SHEETING

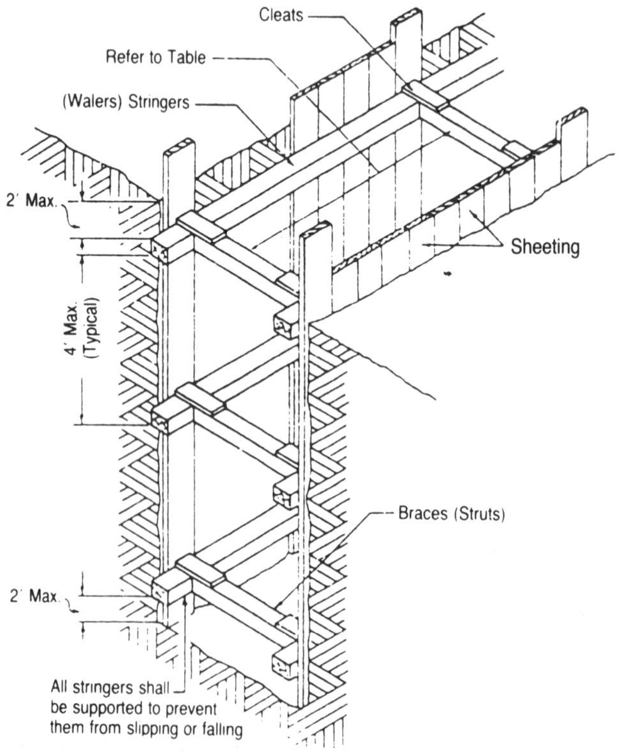

PILES AND PILE DRIVING

General. A pile is a column driven or jetted into the ground which derives its supporting capabilities from end-bearing on the underlying strata, skin friction between the pile surface and the soil, or from a combination of end-bearing and skin friction.

Piles can be divided into two major classes: **Sheet piles** and **load-bearing piles.** Sheet piling is used primarily to restrain lateral forces as in trench sheeting and bulkheads, or to resist the flow of water as in cofferdams. It is prefabricated and is available in steel, wood or concrete. Load-bearing piles are used primarily to transmit loads through soil formations of low bearing values to formations that are capable of supporting the designed loads. If the load is supported predominantly by the action of soil friction on the surface of the pile, it is called a **friction pile.** If the load is transmitted to the soil primarily through the lower tip, it is called an **end-bearing pile.**

There are several load-bearing pile types, which can be classified according to the material from which they are fabricated:

Timber (Treated and untreated)
Concrete (Precast and cast in place)
Steel (H-Section and steel pipe)
Composite (A combination of two or more materials)

Some of the additional uses of piling are to: eliminate or control settlement of structures, support bridge piers and abutments and protect them from scour, anchor structures against uplift or overturning, and for numerous marine structures such as docks, wharves, fenders, anchorages, piers, trestles and jetties.

Timber Piles. Timber piles, treated or untreated, are the piles most commonly used thoroughout the world, primarily because they are readily available, economical, easily handled, can be easily cut off to any desired length after driving and can be easily removed if necessary. On the other hand, they have some serious disadvantages which include: difficulty in securing straight piles of long length, problems in driving them into hard formations and difficulty in splicing to increase their length. They are generally not suitable for use as end-bearing piles under heavy load and they are subject to decay and insect attack. Timber piles are resilient and particularly adaptable for use in waterfront structures such as wharves, docks and piers for anchorages since they will bend or give under load or impact where other materials may break. The ease with which they can be worked and their economy makes them popular for trestle construction and for temporary structures such as falsework or centering. Where timber piles can be driven and cut off below the permanent ground-water level, they will last indefinitely; but above this level in the soil, a timber pile will rot or will be attacked by insects and eventually destroyed. In sea water, marine borers and fungus will act to deteriorate timber piles. Treatment of timber piles increases their life but does not protect them indefinitely.

Concrete Piles. Concrete piles are of two general types, precast and cast-in-place. The advantages in the use of concrete piles are that they can be fabricated to meet the most exacting conditions of design, can be cast in any desired shape or length, possess high strength and have excellent resistance to chemical and biological attack. Certain disadvantages are encountered in the use of precast piles, such as:

(a) Their heavy weight and bulk (which introduces problems in handling and driving).

(b) Problems with hair cracks which often develop in the concrete as a result of shrinkage after curing (which may expose the steel reinforcement to deterioration).

(c) Difficulty encountered in cut-off or splicing.

(d) Susceptibility to damage or breakage in handling and driving.

(e) They are more expensive to fabricate, transport and drive.

Precast piles are fabricated in casting yards. Centrifugally spun piles (or piles with square or octagonal cross-sections) are cast in horizontal forms, while round piles are usually cast in vertical forms. With the exception of relatively short lengths, precast piles must be reinforced to provide the designed column strengths and to resist damage or breakage while being transported or driven.

Precast piles can be tapered or have parallel sides. The reinforcement can be of deformed bars or be prestressed or poststressed with high strength steel tendons. Prestressing or prestressing eliminates the problem of open shrinkage cracks in the concrete. Otherwise, the pile must be protected by coating it with a bituminous or plastic material to prevent ultimate deterioration of the reinforcement. Proper curing of the precast concrete in piles is essential.

Cast-in-place pile types are numerous and vary according to the manufacturer of the shell or inventor of the method. In general, they can be classified into two groups: shell-less types and the shell types. The shell-less type is constructed by driving a steel shell into the ground and filling it with concrete as the shell is pulled from the ground. The shell type is constructed by driving a steel shell into the ground and filling it in place with concrete. Some of the advantages of cast-in-place concrete piles are: lightweight shells are handled and driven easily, lengths of the shell may be increased or decreased easily, shells may be transported in short lengths and quickly assembled, the problem of breakage is eliminated and a driven shell may be inspected for shell damage or an uncased hole for "pinching off." Among the disadvantages are problems encountered in the proper centering of the reinforcement cages, in placing and consolidating the concrete without displacement of the reinforcement steel or segregation of the concrete, and shell damage or "pinching-off" of uncased holes.

Shell type piles are fabricated of heavy gage metal or are fluted, corrugated or spirally reinforced with heavy wire to make them strong enough to be driven without a mandrel. Other thin-shell types are driven with a collapsible steel mandrel or core inside the casing. In addition to making the driving of a long thin shell possible, the mandrel prevents or minimizes damage to the shell from tearing, buckling, collapsing or from hard objects encountered in driving.

Some shell type piles are fabricated of heavy gauge metal with enlargement at the lower end to increase the end bearing.

These enlargements are formed by withdrawing the casing two to three feet after placing concrete in the lower end of the shell. This wet concrete is then struck by a blow of the pile hammer on a core in the casing and the enlargement is formed. As the shell is withdrawn, the core is used to consolidate the concrete after each batch is placed in the shell. The procedure results in completely filling the hole left by the withdrawal of the shell.

Steel Piles. A steel pile is any pile fabricated entirely of steel. They are usually formed of rolled steel H sections, but heavy steel pipe or box piles (fabricated from sections of steel sheet piles welded together) are also used. The advantages of steel piles are that they are readily available, have a thin uniform section and high strength, will take hard driving, will develop high load-bearing values, are easily cut off or extended, are easily adapted to the structure they are to support, and breakage is eliminated. Some disadvantages are: they will rust and deteriorate unless protected from the elements; acid, soils or water will result in corrosion of the pile; and greater lengths may be required than for other types of piles to achieve the same bearing value unless bearing on rock strata. Pipe pile can either be driven open-end or closed-end and can be unfilled, sand filled or concrete filled. After open-end pipe piles are driven, the material from inside can be removed by an earth auger, air or water jets, or other means, inspected, and then filled with concrete. Concrete filled pipe piles are subject to corrosion on the outside surface only.

Composite Piles. Any pile that is fabricated of two or more materials is called a composite pile. There are three general classes of composite piles: wood with concrete, steel with concrete, and wood with steel. Composite piles are usually used for a special purpose or for reasons of economy.

Where a permanent ground-water table exists and a composite pile is to be used, it will generally be of concrete and wood. The wood portion is driven to below the water table level and the concrete upper portion eliminates problems of decay and insect infestation above the water table. Composite piles of steel and concrete are used where high bearing loads are desired or where driving in hard or rocky soils is expected. Composite wood and steel piles are relatively uncommon. It is important that the pile design provides for a permanent joint between the two materials used, so constructed that the parts do not separate or shift out of axial alignment during driving operations.

Sheet Piles. Sheet piles are made from the same basic materials as other piling: wood, steel and concrete. They are ordinarily designed so as to interlock along the edges of adjacent piles.

Sheet piles are used where support of a vertical wall of earth is required, such as trench walls, bulkheads, waterfront structures or cofferdams. Wood sheet piling is generally used in temporary installations, but is seldom used where water-tightness is required or hard driving expected. Concrete sheet piling has the capability of resisting much larger lateral loads than wood sheet piling, but considerable difficulty is experienced in securing water-tight joints. The type referred to as "fish-mouth" type is designed to permit jetting out the joint and filling with grout, but a seal is not always effected unless the adjacent piles are wedged tightly together. Concrete sheet piling has the advantage that it is the most permanent of all types of sheet piling.

Steel sheet piling is manufactured with a tension-type interlock along its edges. Several different shapes are available to permit versatility in its use. It has the advantages that it can take hard driving, has reasonably water-tight joints and can be easily cut, patched, lengthened or reinforced. It can also be easily extracted and reused. Its principal disadvantage is its vulnerability to corrosion.

Types of Pile Driving Hammers. A pile-driving hammer is used to drive load-bearing or sheet piles. The commonly used types are: drop, single-acting, double-acting, differential acting and diesel hammers. The most recent development is a type of hammer that utilizes high-frequency sound and a dead load as the principal sources of driving energy.

Drop Hammers. These hammers employ the principle of lifting a heavy weight by a cable and releasing it to fall on top of the pile. This type of hammer is rapidly disappearing from use, primarily because other types of pile driving hammers are more efficient. Its disadvantages are that it has a slow rate of driving (four to eight blows per minute), that there is some risk of damaging the pile from excessive impact, that damage may occur in adjacent structures from heavy vibration and that it cannot be used directly for driving piles under water. Drop hammers have the advantages of simplicity of operation, ability to vary the energy by changing the height of fall and they represent a small investment in equipment.

Single-Acting Hammers. These hammers can be operated either on steam or compressed air. The driving energy is provided by a free-falling weight (called a ram) which is raised after each stroke by the action of steam or air on a piston. They are manufactured as either open or closed types. Single-acting hammers are best suited for jobs where dense or elastic soil materials must be penetrated or where long heavy timber or precast concrete piles must be driven. The closed type can be used for underwater pile driving. Its advantages include: faster driving (50 blows or more per minute), reduction in skin friction as a result of more frequent blows, lower velocity of the ram which transmits a greater proportion of its energy to the pile and minimizes piles damage during driving, and it has underwater driving capability. Some of its disadvantages are: requires higher investment in equipment (i.e. steam boiler, air compressor, etc.), higher maintenance costs, greater set-up and moving time required, and a larger operating crew.

Double-Acting Hammers. These hammers are similar to the single-acting hammers except that steam or compressed air is used both to lift the ram and to impart energy to the falling ram. While the action is approximately twice as fast as the single-acting hammer (100 blows per minute or more), the ram is much lighter and operates at a greater velocity, thereby making it particularly useful in high production driving of light or medium-weight piles of moderate lengths in granular soils. The hammer is nearly always fully encased by a steel housing which also permits direct driving of piles under water.

Some of its advantages are: faster driving rate, less static skin friction develops between blows, has underwater driving capability and piles can be driven more easily without leads.

Among its disadvantages are: it is less suitable for driving heavy piles in high-friction soils and the more complicated mechanism results in higher maintenance costs.

Differential-Acting Hammers. This type of hammer is, in effect, a modified double-acting hammer with the actuating mechanism having two different diameters. A large-diameter piston operates in an upper cylinder to accelerate the ram on the downstroke and a small-diameter piston operates in a lower cylinder to raise the ram. The additional energy added to the falling ram is the difference in areas of the two pistons multiplied by the unit pressure of the steam or air used. This hammer is a short-stroke, fast-acting hammer with a cycle rate approximately that of the double-acting hammer. Its advantages are that it has the speed and characteristics of the double-acting hammer with a ram weight comparable to the single-acting type, and it uses from 25 to 35 percent less steam or air. It is also more suitable for driving heavy piles under more difficult driving conditions than is the double-acting hammer. It is available in the open or closed-type cases, the latter permitting direct underwater pile driving. Its principal disadvantage is higher maintenance costs.

Diesel Hammers. This hammer is a self-contained driving unit which does not require an auxilliary steam boiler or air compressor. It consists essentially of a ram operating as a piston in a cylinder. When the ram is lifted and allowed to fall in the cylinder, diesel fuel is injected in the compression space between the ram and an anvil placed on top of the pile. The continued downstroke of the ram compresses the air and fuel to ignition heat and the resultant explosion drives the pile downward and the ram upward to start another cycle. This hammer is capable of driving at a rate of from 80 to 100 blows per minute. Its advantages are that it has a low equipment investment cost, is easily moved, requires a small crew, has a high driving rate, does not require a steam boiler or air compressor and can be used with or without leads for most work. Its disadvantages are that it is not self-starting (the ram must be mechanically lifted to start the action) and it does not deliver a uniform blow. The latter disadvantge arises from the fact that as the reaction of the pile to driving increases, the reaction to the ram increases correspondingly. That is, when the pile encounters considerable resistance, the rebound of the ram is higher and the energy is increased automatically. The operator is required to observe the driving operations closely to identify changing driving conditions and compensate for such changes with his controls to avoid damaging the pile.

Diesel hammers can be used on all types of piles and they are best suited to jobs where mobility or frequent relocation of the pile driving equipment is necessary.

PILE CHART

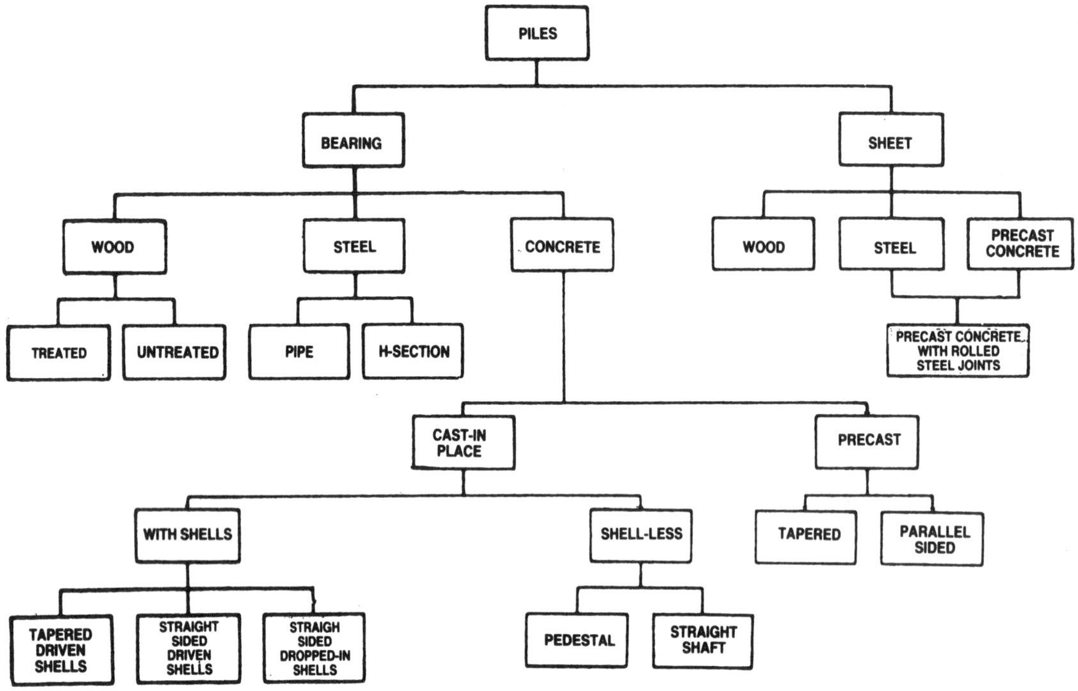

CONCRETE SHEET PILING

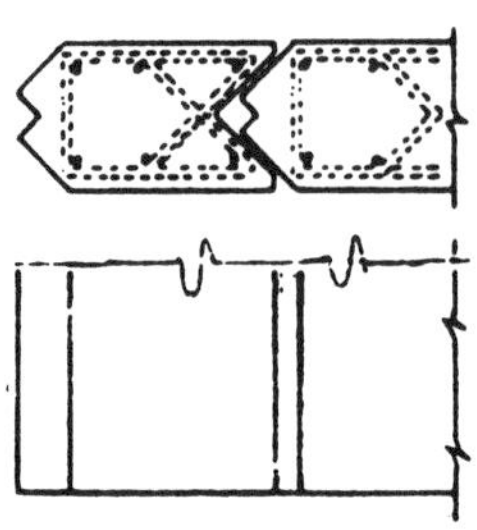
FISH MOUTH JOINTED

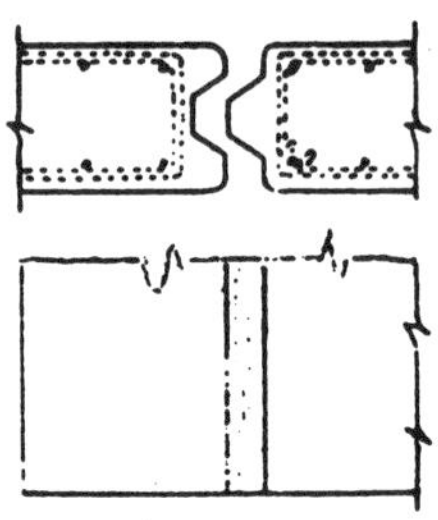
STANDARD JOINT

JOINT GROUTED AFTER PILE HAS BEEN DRIVEN

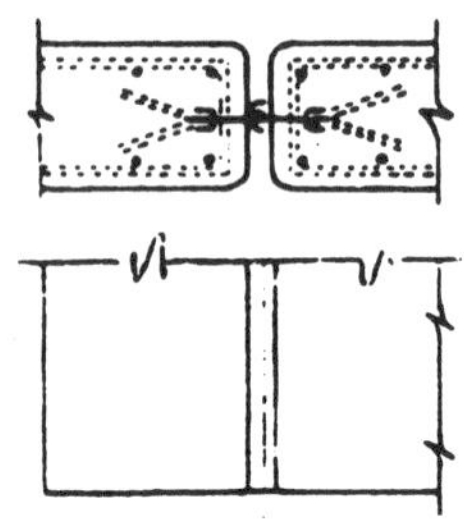
STEEL INTERLOCK

STEEL SHEET PILING

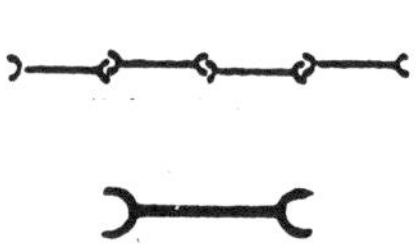
FLAT PILE

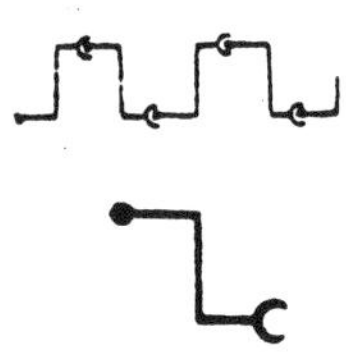
ZEE PILE

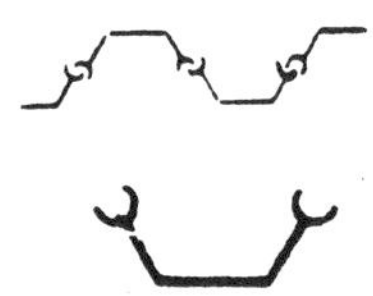
DEEP WEB PILE

BEARING PILES

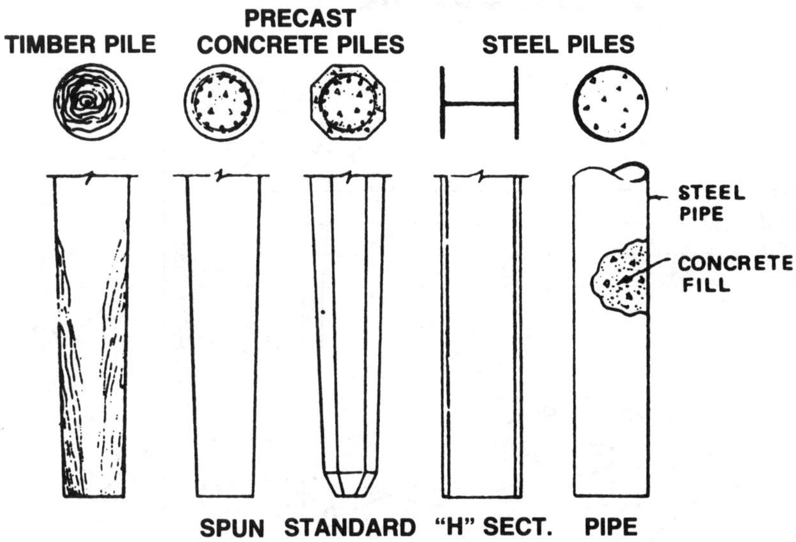

DRIVEN PILES

COMPOSITE WOOD PILE

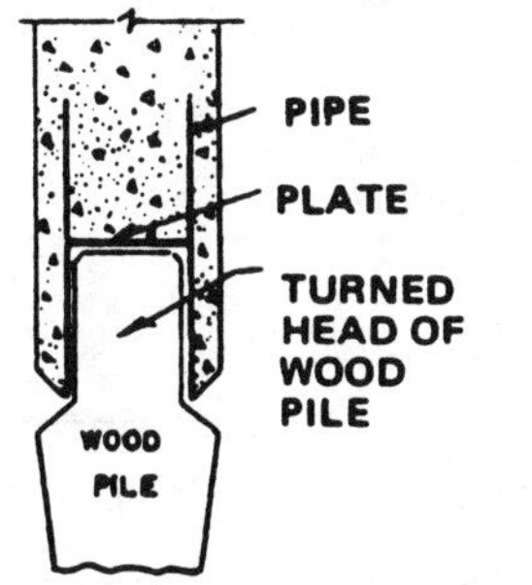

COMPOSITE STEEL PILE

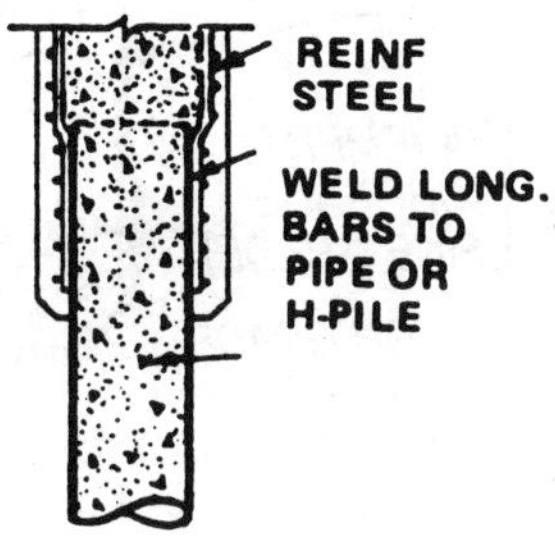

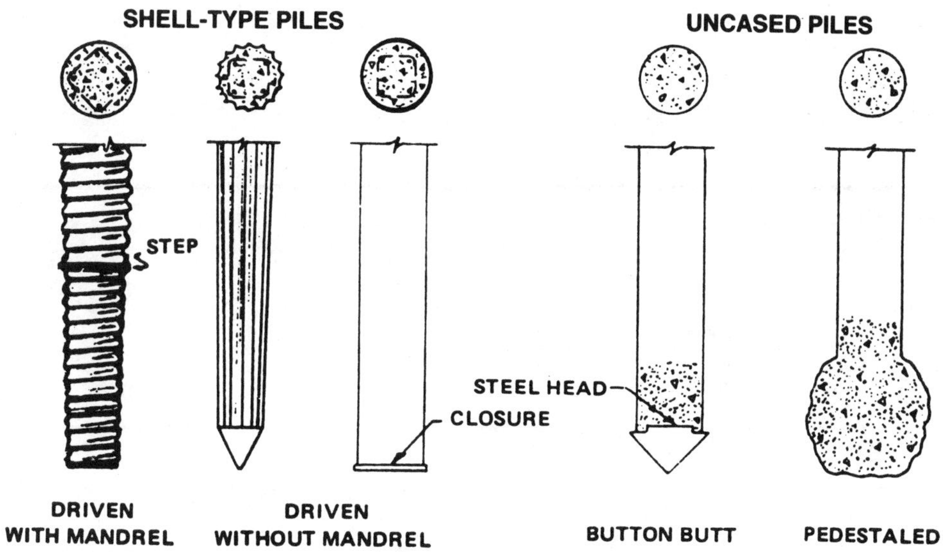

CAST-IN-PLACE PILES

STANDARD NOMENCLATURE FOR STREET CONSTRUCTION

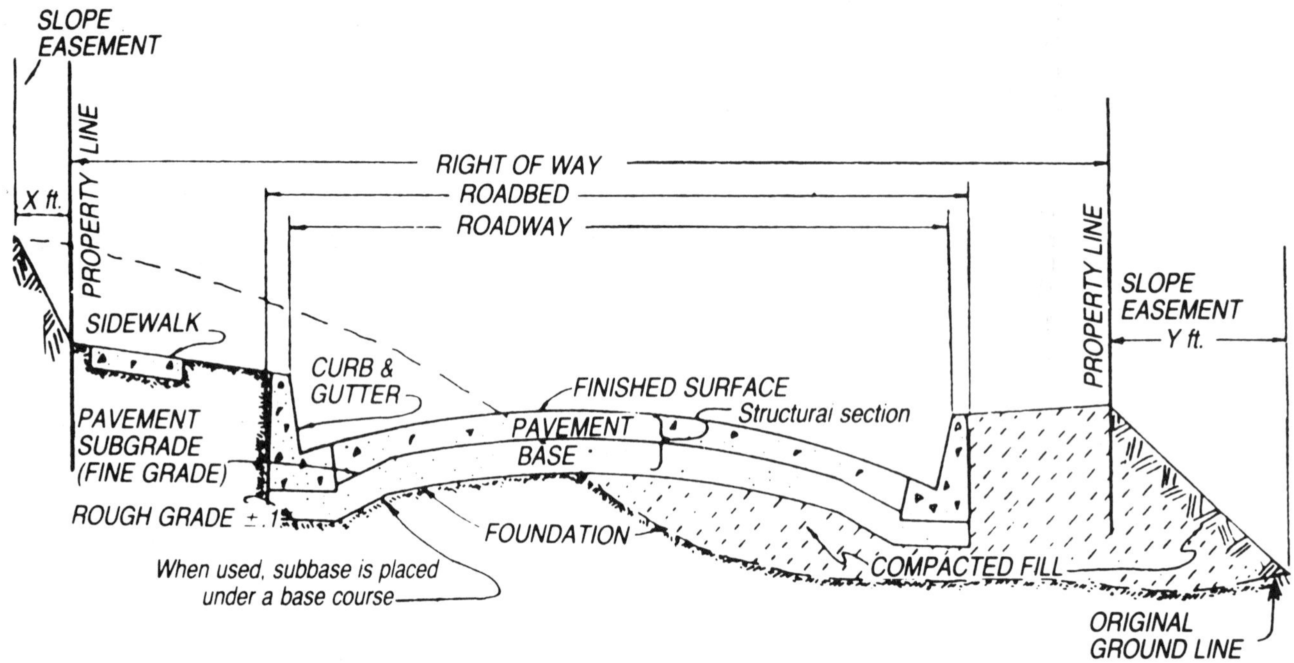

CAPACITIES FOR SEPTIC TANKS SERVING AN INDIVIDUAL DWELLING

No. of bedrooms	Capacity of tank (gals.)
2 or less	750
3	900
4	1,000

FORM NOMENCLAURE

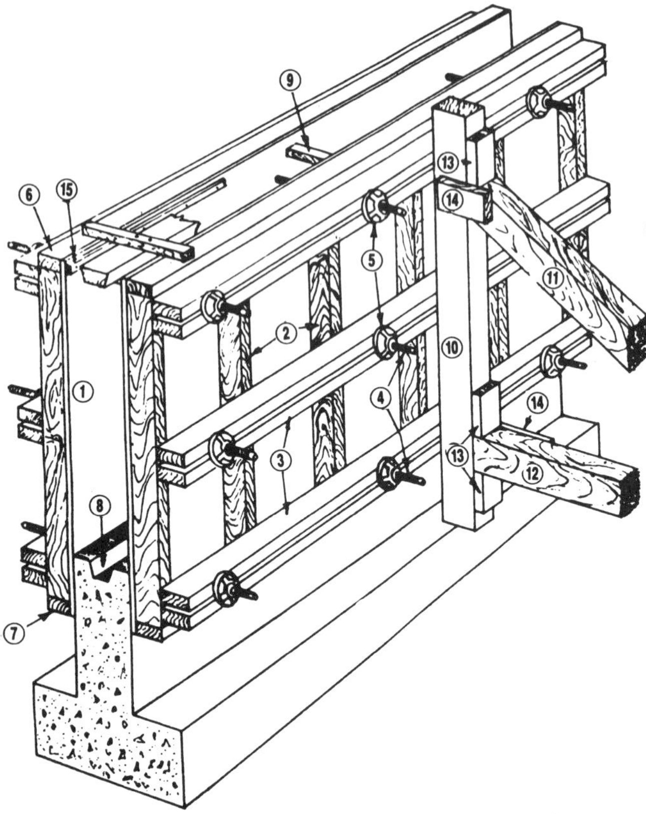

1. SHEATHING
2. STUDS
3. WALES
4. FORM BOLTS
5. NUT WASHER
6. TOP PLATE
7. BOTTOM PLATE
8. KEY-WAY
9. SPREADER
10. STRONGBACK
11. BRACE
12. STRUT
13. CLEATS
14. SCAB
14. POUR STRIP

FALSEWORK NOMENCLATURE

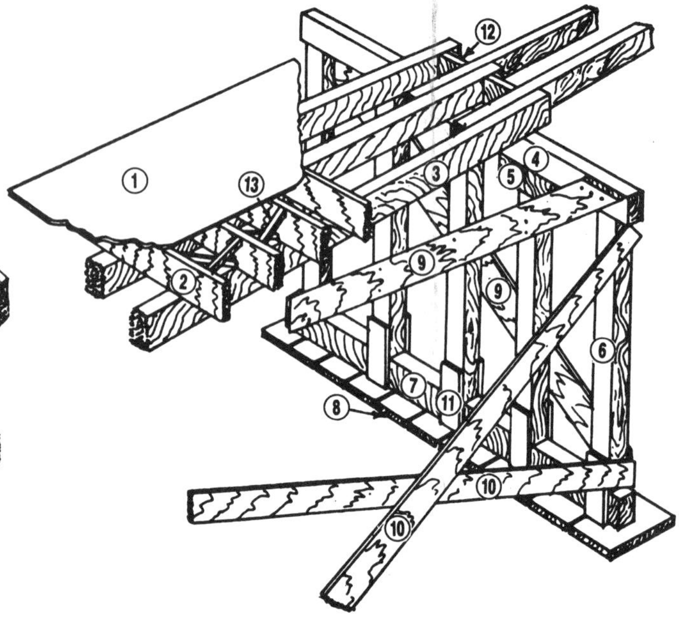

1. SHEATHING
2. JOIST
3. STRINGER
4. CAP
5. CORBEL
6. POST
7. SILL
8. FOOTING
9. SWAY BRACE
10. LONGITUDINAL BRACE
11. SCAB
12. BLOCKING
13. BRIDGING

TYPICAL PAN-JOIST FORM CONSTRUCTION

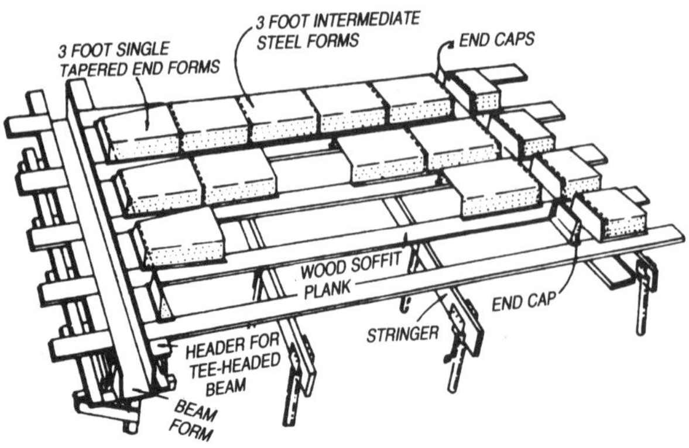

TYPICAL WAFFLE SLAB FORM CONSTRUCTION

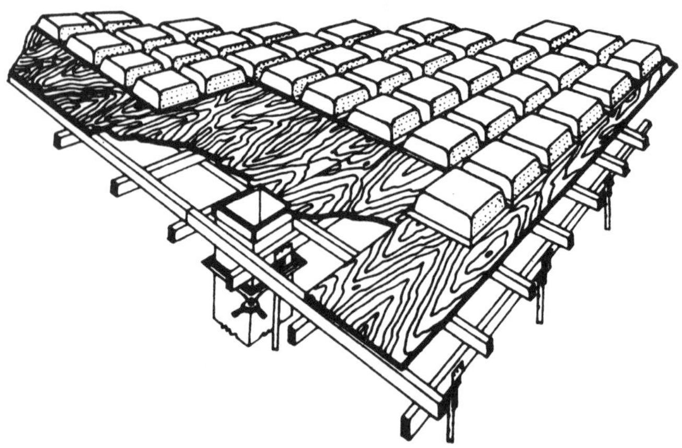

Insofar as is possible, the moisture content should be kept uniform to avoid problems in determining the proper amount of water to be added for mixing. Mixing water must be reduced to compensate for moisture in the aggregate in order to control the slump of the concrete and avoid exceeding the specified water-cement ratio.

Handling Concrete by Pumping Methods. Transportation and placement of concrete by pumping is another method gaining increased popularity. Pumps have several advantages, the primary one being that a pump will high-lift concrete without the need for an expensive crane and bucket. Since the concrete is delivered through pipe and hoses, concrete can be conveyed to remote locations in buildings, in tunnels, to locations otherwise inaccessible on steep hillside slopes for anchor walls, pipe bedding or encasement, or for placing concrete for chain link fence post bases. Concrete pumps have been found to be economical and expedient in the placement of concrete, and this has promoted the use and acceptance of this development. The essence of proper concrete pumping is the placement of the concrete in its final location without segregation.

Modern concrete pumps, depending on the mix design and size of line, can pump to a height of 200 feet or a horizontal distance of 1,000 feet. They can handle, economically, structural mixes, standard mixes, low slump mixes, mixes with two-inch maximum size aggregate and light weight concrete. When a special pump mix is required for structural concrete in a major structure, the mix design must be approved by the Engineer and checked and confirmed by the Supervisor of the Materials Control Group. The Inspector should obtain the pump manufacturer's printed information and evaluate its characteristics and ability to handle the concrete mixture specified for the project.

If concrete is being placed for a major reinforced structure, it is important that the placement continue without interruption. The Inspector should be sure that the contractor has ready access to a back-up pump to be used in the event of a breakdown. In order to further insure the success of the concrete placement by the pumping method, the user should be aware of the following points:

(a) A protective grating over the receiving hopper of the pump is necessary to exclude large pieces of aggregate or foreign material.

(b) The pump and lines require lubrication with a grout of cement and water. All of the excess grout is to be wasted prior to pumping the concrete.

(c) All changes in direction must be made by a large radius bend with a maximum bend of 90 degrees. Wye connections induce segregation and shall not be used.

(d) Pump lines should be made of a material capable of resisting abrasion and with a smooth interior surface having a low coefficient of friction. Steel is commonly used for pump lines, ance. Aluminum pipe should not be used for pumping concrete and some of the new plastic or rubber tubing is gaining accept- because a chemical reaction occurs between the concrete and the aluminum. Hydrogen is generated which results in a swelling of the concrete, causing a significant reduction in compressive strength. This reaction is aggravated by any of the following: abrasive coarse aggregate, non-uniformly graded sand, low-slump concrete, low sand-aggregate ratio, high-alkali cement or when no air-entraining agent is used.

(e) During temporary interruptions in pumping, the hopper must remain nearly full, with an occasional turning and pumping to avoid developing a hard slug of concrete in the lines.

(f) Excessive line pressures must be avoided. When this occurs, check these points as the probable cause: segregation caused by too low a slump or too high a slump; large particle contamination caused by large pieces of aggregate or frozen lumps not eliminated by the grating; poor gradation of aggregates or particle shape; rich or lean spots caused by improper mixing.

(g) Corrections must be made to correct excessive slump loss as measured at the transit-mixed concrete truck and as measured at the hose outlet. This may be attributable to porous aggregate, high temperature or rapid setting mixes.

(h) Two transit-mix concrete trucks must be used simultaneously to deliver concrete into the pump hopper. These trucks must be discharged alternately to assure a continuous flow of concrete as trucks are replaced.

(i) Samples of concrete for test specimens prepared to determine the acceptance of the concrete quality are to be taken as required for conventional concrete.

Sampling is done before the concrete is deposited in the pump hopper. However, it is suggested that, where possible, the effect of pumping on the compressive strength be checked by taking companion samples, so identified, from the end of the pump line at the same time. The Record of Test must be properly noted as being a special mix used for pumping purposes. This will enable the Materials Control Group to compile a complete history of mix designs and their respective compressive strengths.

The prudent use of pumped concrete can result in economy and improved quality. However, only the control exercised by the operator will assure continued high standards of quality concrete.

Pump lines must be properly fastened to supports to eliminate excessive vibration. Couplings must be easily and securely fastened in a manner that will prevent mortar leakage. It is preferable to use the flexible hose only at the discharge point. This hose must be moved in such a manner as to avoid kinks or sharp bends. The pump line should be protected from excessive heat during hot weather by water sprinkling or shade.

COMPRESSIVE STRENGTH FOR VARIOUS WATER-CEMENT RATIOS

(The strengths listed are based on the use of normal portland cement)

WATER/CEMENT RATIO		PROBABLE 28-DAY STRENGTH	
WEIGHT	GALS./100#	PSI	MEGAPASCALS*
.40	4.8	5000	34
.45	5.4	4500	31
.50	6.0	4000	28
.55	6.6	3500	24
.60	7.2	3000	21
.65	7.8	2500	17
.70	8.4	2000	14

* International system equivalent.

APPROXIMATE CONTENT OF SAND, CEMENT AND WATER PER CUBIC YARD OF CONCRETE

Based on aggregates of average grading and physical characteristics in concrete mixes having a water-cement ratio (W/C) of about .65 by weight (or 7.8 gallons) per sack of cement; 3-in. slump; and a medium natural sand having a fineness modulus of about 2.75.

COURSE AGGREGATE MAX. SIZE	WATER		CEMENT	% SAND
	POUNDS	GALLONS		
3/8	385	46	590	57
1/2	365	44	560	50
3/4	340	41	525	43
1	325	39	500	39
1 1/2	300	36	460	37

It can be noted from the above chart that, for a given slump, the amount of mixing water increases as the size of the course aggregate decreases. The size of the course aggregate controls the sand content in the same way; that is, the amount of sand required in the mix increases as the size of the course aggregate decreases.

Other typical examples are contained in the pamphlet published by the Portland Cement Association entitled "Design and Control of Concrete Mixtures."

Effects of Temperature on Concrete. Concrete mixtures gain strength rapidly in the first few days after placement. While the rate of gain in strength diminishes, concrete continues to become stronger with time over a period of many years, so long as drying of the concrete is prevented. Its strength at 28 days is considered to be the compressive strength upon which the Engineer bases his calculations. The temperature of the atmosphere has a significant effect upon the development of strength in concrete. Lower temperatures retard and higher temperatures accelerate the gain in strength.

Most destructive of the natural forces is freezing and thawing action. While the concrete is still wet or moist, expansion of the water as it is converted into ice results in severe damage to the fresh concrete. In situations where freezing may be encountered, high early strength cement may be used. Also, the mixing water or the aggregate (or both) may be preheated before mixing. Covering the concrete, and using steam or salamanders to heat the concrete under the covering, will help prevent freezing. Air-entraining agents help to diminish the effects of freezing of fresh concrete as well as in subsequent freezing and thawing cycles throughout the life of the concrete.

Hot weather will present problems of a different nature in placing concrete. Concrete will set up faster and tend to shrink and crack at the surface. To minimize this problem, the concrete should be placed without delay after mixing. Avoid the use of accelerators (perhaps even use a retarding agent), dampen all subgrade and forms, protect the freshly placed concrete from hot dry winds, and provide for adequate curing. Crushed ice or chilled water can be used as part of the mixing water to reduce the temperature of the mix in extremely hot areas.

TYPICAL MINIMUM SIZE OF VERTICAL CONDUCTORS AND LEADERS

Size of leader or conductor (Inches)	Maximum projected roof area (Square feet)
2	544
2½	987
3	1,610
4	3,460
5	6,280
6	10,200
8	22,000

TYPICAL MINIMUM SIZE OF ROOF GUTTERS

	Maximum projected roof area for gutters of various slopes			
Diameter gutter (Inches)	1/16 in. per Ft. slope (Sq. ft.)	1/8 in. per Ft. slope (Sq. ft.)	1/4 in. per Ft. slope (Sq. ft.)	1/2 in. per Ft. slope (Sq. ft.)
3	170	240	340	480
4	360	510	720	1020
5	625	880	1250	1770
6	960	1360	1920	2770
7	1380	1950	2760	3900
8	1990	2800	3980	5600
10	3600	5100	7200	10000

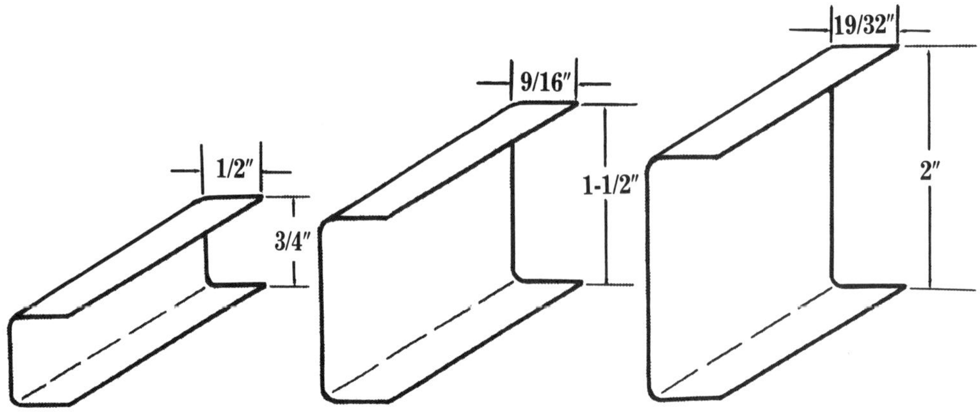

COLD ROLLED CHANNELS

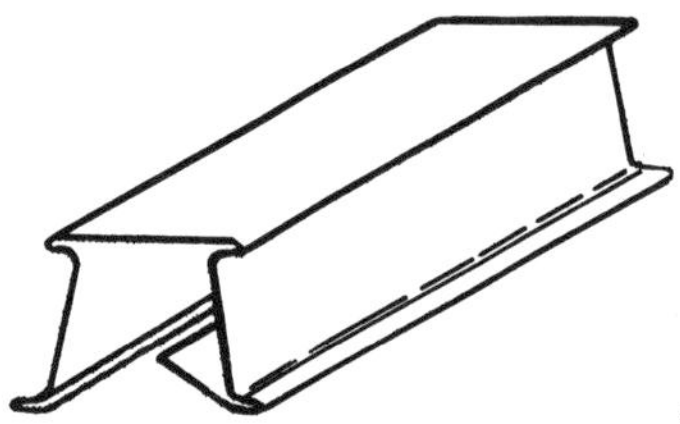

NAILING CHANNEL

SCREW CHANNEL

CHANNEL STUD

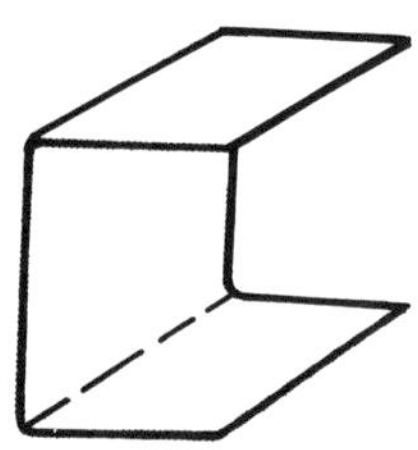

WIDE FLANGE CHANNEL

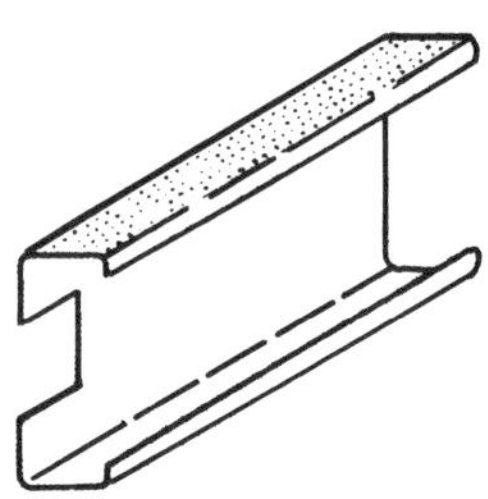

CEE STUD

STUDLESS SOLID PARTITION

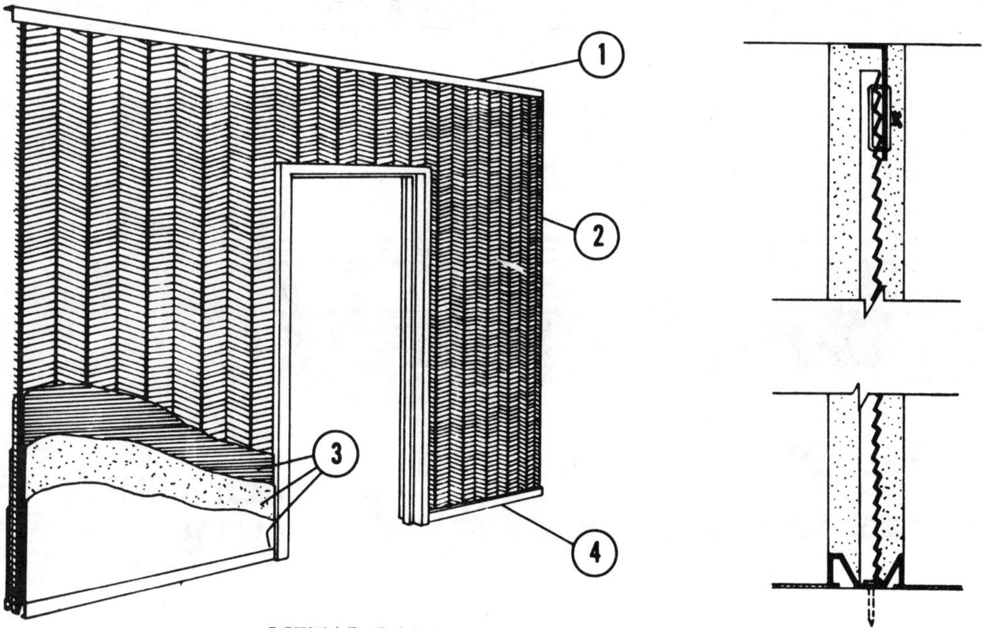

(1) Ceiling Runner
(2) Rib Metal Lath
(3) Plaster
(4) Combination Floor Runner and Screed

METAL LATH

STUDLESS SOLID PARTITION

(1) Ceiling Runner
(2) Long Length Gypsum Lath
(3) Plaster
(4) Combination Floor Runner and Screed

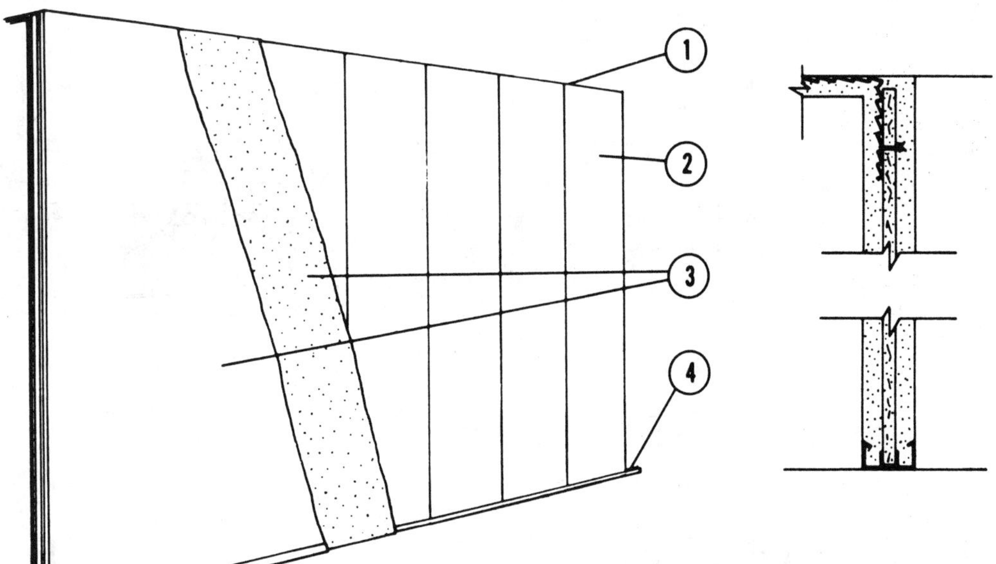

GYPSUM LATH

SUSPENDED CEILINGS

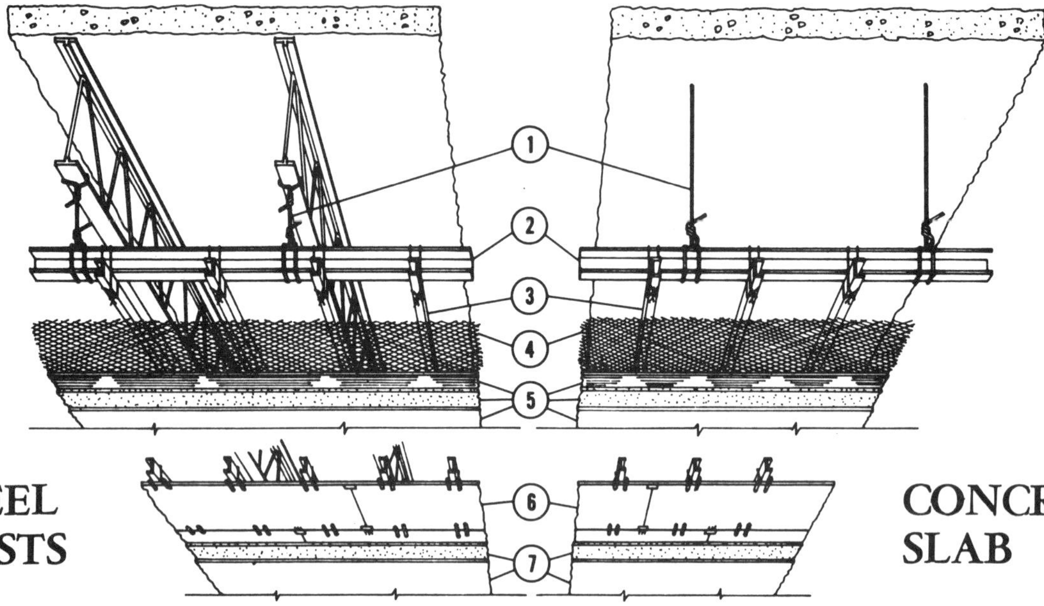

STEEL JOISTS

CONCRETE SLAB

(1) Hanger
(2) Main Runner Channel
(3) Furring Channel
(4) Metal or Wire Fabric Lath
(5) Plaster
(6) Gypsum Lath
(7) Plaster

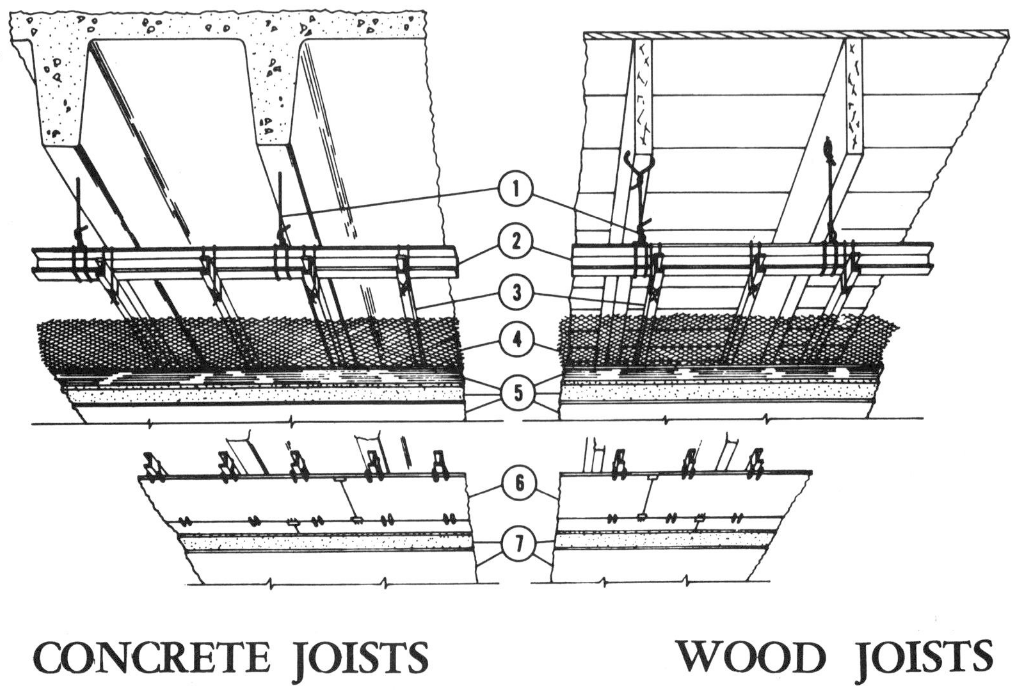

CONCRETE JOISTS

WOOD JOISTS

STEEL STUD

Hollow Partition

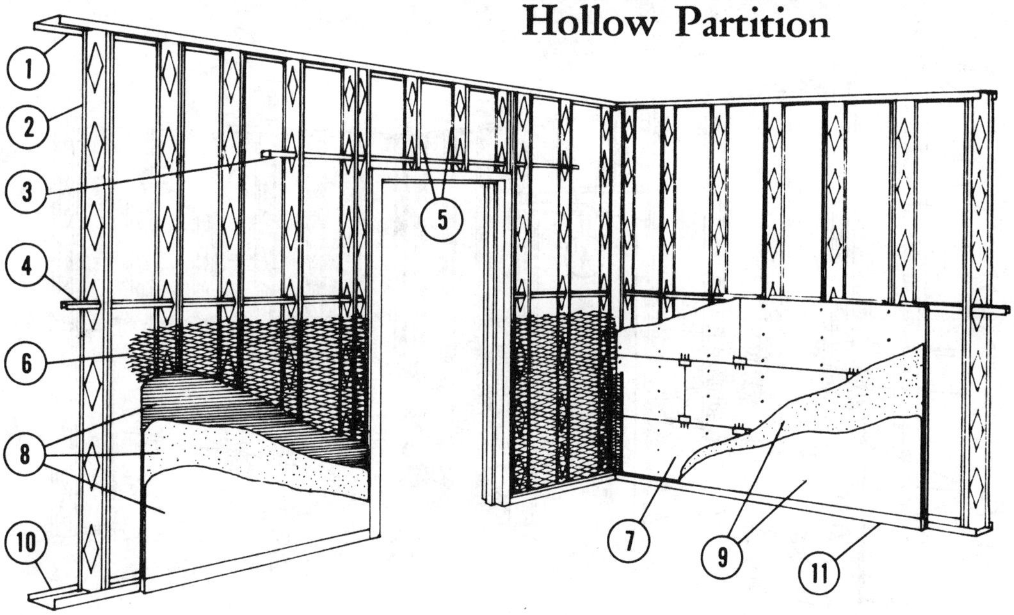

(1) Ceiling Runner Track
(2) Nailable Stud
(3) Door Opening Stiffener
(4) Partition Stiffener
(5) Jack Studs
(6) Metal or Wire Fabric Lath (screwed or wire tied)
(7) Gypsum Lath (nailed, clipped or screwed)
(8) Three Coats of Plaster (Scratch, Brown, Finish)
(9) Two Coats of Plaster (Brown, Finish)
(10) Floor Runner Track
(11) Flush Metal Base

SCREW STUD

Hollow Partition

(1) Ceiling Runner Track
(2) Screw Stud
(3) Door Opening Stiffener
(4) Partition Stiffener
(5) Jack Studs
(6) Metal or Wire Fabric Lath (screwed on)
(7) Gypsum Lath (screwed on)
(8) Three Coats of Plaster (Scratch, Brown, Finish)
(9) Two Coats of Plaster (Brown, Finish)
(10) Floor Runner Track
(11) Flush Metal Base

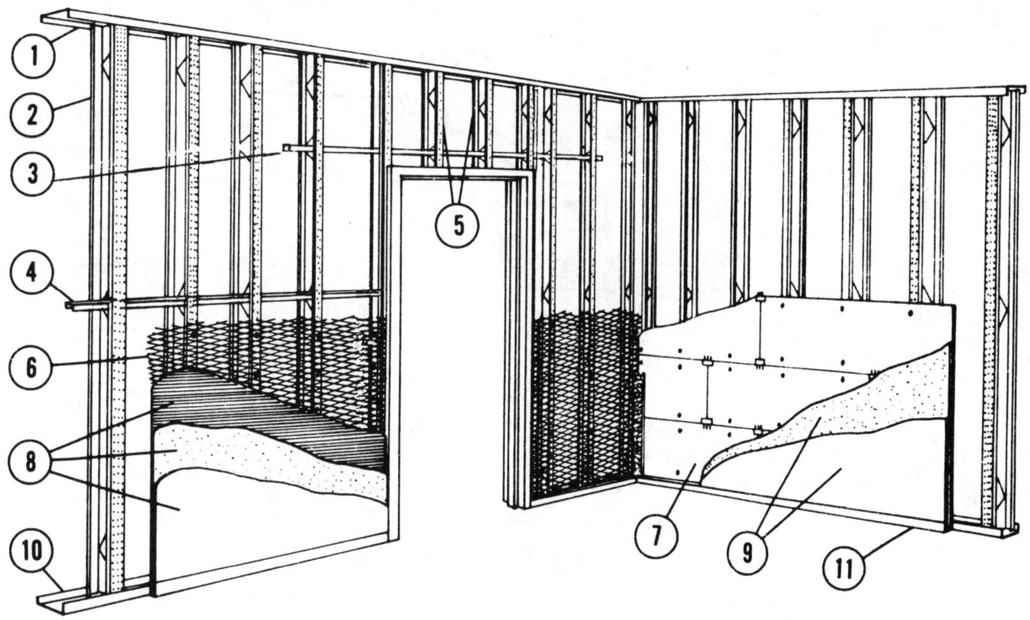

LOAD-BEARING HOLLOW PARTITION

Structural Stud

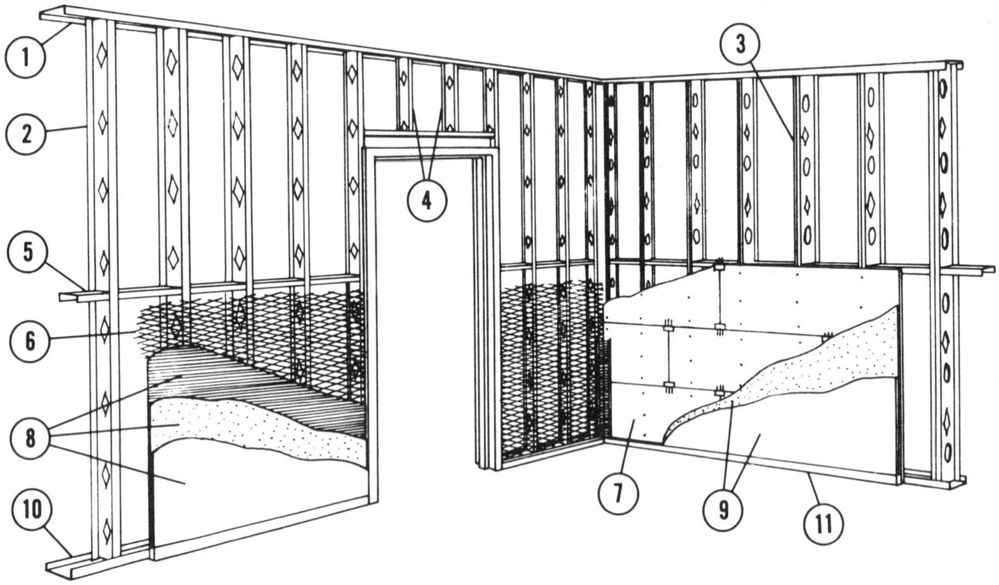

(1) Ceiling Runner Track
(2) Structural Stud (prefabricated)
(3) Structural Stud (nailable)
(4) Jack Studs
(5) Partition Stiffener (bridging)
(6) Metal or Wire Fabric Lath (wire-tied, nailed or stapled)
(7) Gypsum Lath (nailed or stapled)
(8) Three Coats of Plaster (Scratch, Brown, Finish)
(9) Two Coats of Plaster (Brown, Finish)
(10) Floor Runner Track
(11) Flush Metal Base

VERTICAL FURRING

With Studs

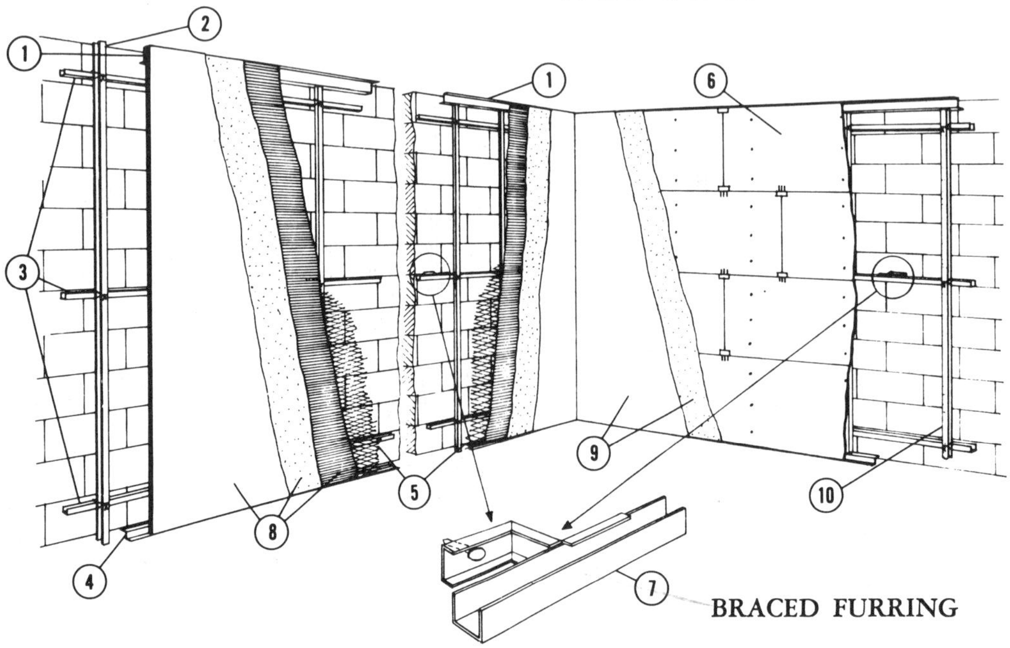

BRACED FURRING

FREE STANDING FURRING

(1) Ceiling Runner
(2) Channel Stud
(3) Horizontal Stiffener
(4) Floor Runner
(5) Metal or Wire Fabric Lath
(6) Gypsum Lath
(7) Bracing
(8) Three Coats of Plaster (Scratch, Brown, Finish)
(9) Two Coats of Plaster (Brown, Finish)
(10) Screw Channel Studs

DOUBLE CHANNEL STUD

Hollow Partition

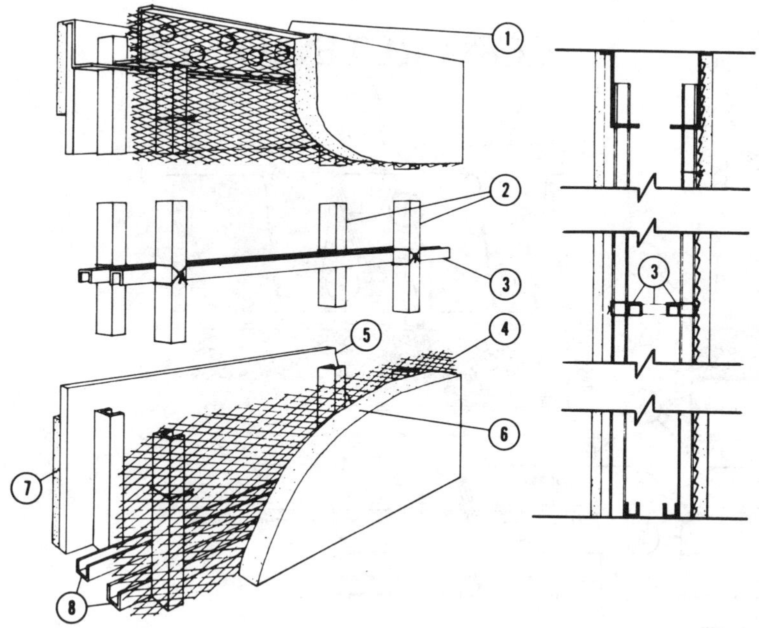

(1) Ceiling Runner
(2) Channel Studs
(3) Partition Stiffener and Channel Spacer
(4) Metal or Wire Fabric Lath (wire-tied)
(5) Gypsum Lath (clipped on)
(6) Three Coats of Plaster (Scratch, Brown, Finish)
(7) Two Coats of Plaster (Brown, Finish)
(8) Floor Runners (channel)

SINGLE CHANNEL STUD

Solid Partition

(1) Ceiling Runner
(2) Channel Stud
(3) Metal or Wire Fabric Lath (wire-tied)
(4) Plaster
(5) Combination Floor Runner and Screed

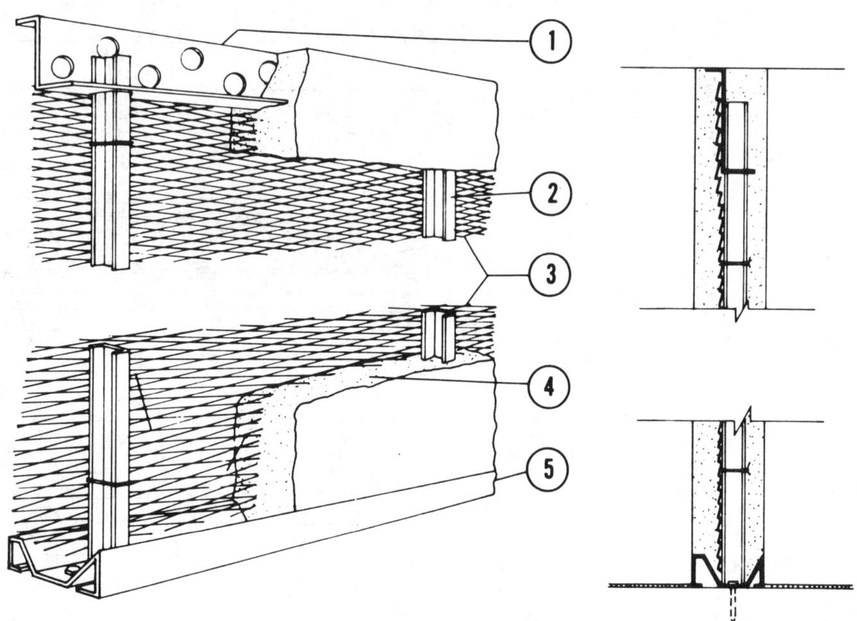

FURRED CEILINGS

STEEL JOISTS　　CONCRETE JOISTS

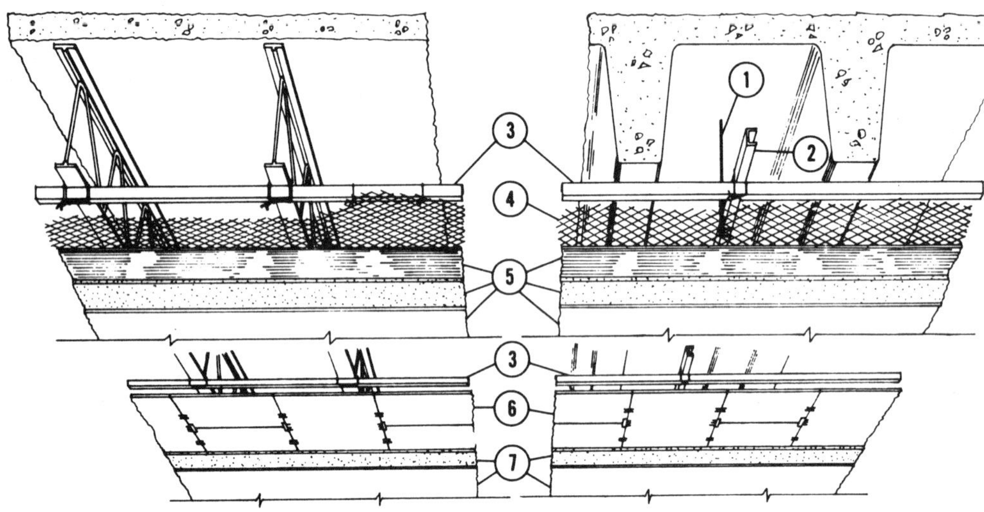

WOOD JOISTS

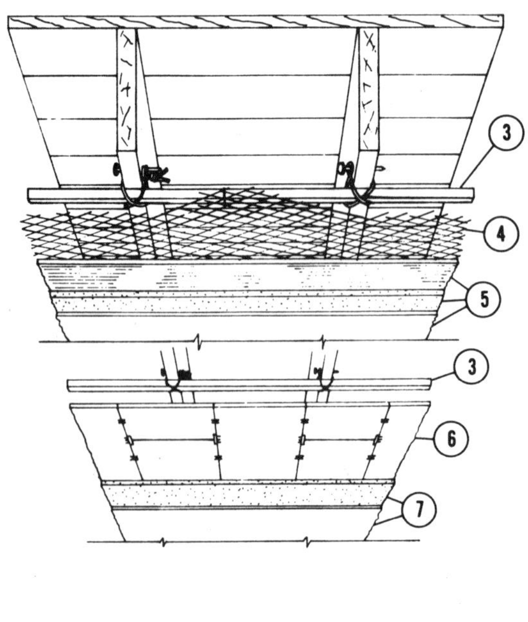

(1) Hanger
(2) ¾-inch Channel
(3) Cross Furring
(4) Metal or Wire Fabric Lath
(5) Plaster
(6) Gypsum Lath
(7) Plaster

CEILINGS

STEEL JOISTS   CONCRETE JOISTS

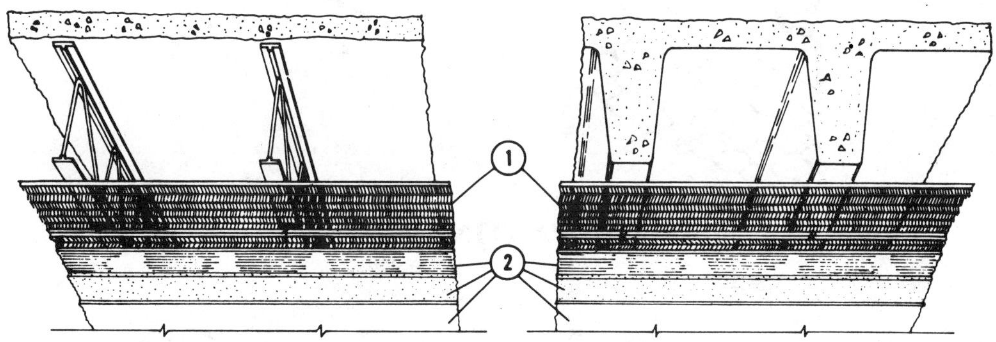

WOOD JOISTS

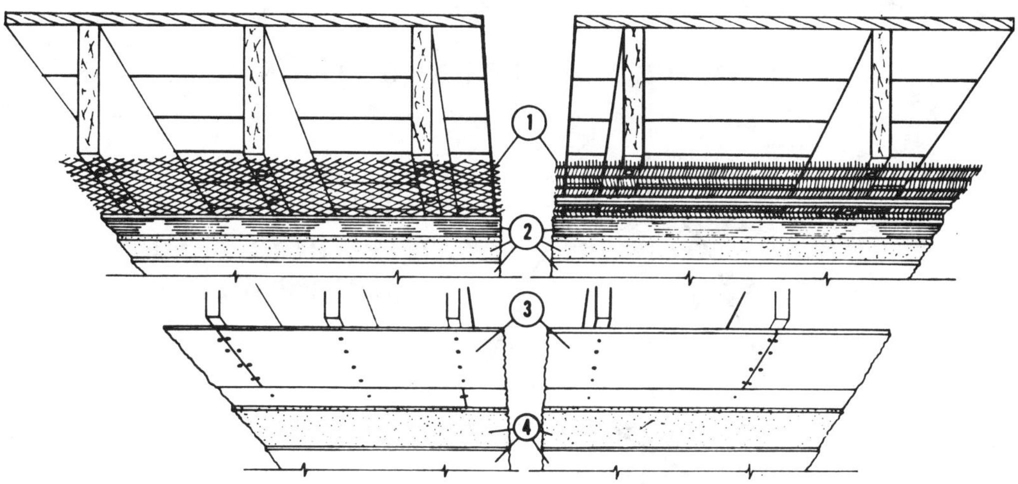

(1) Metal or Wire Fabric Lath

(2) Plaster

(3) Gypsum Lath

(4) Plaster

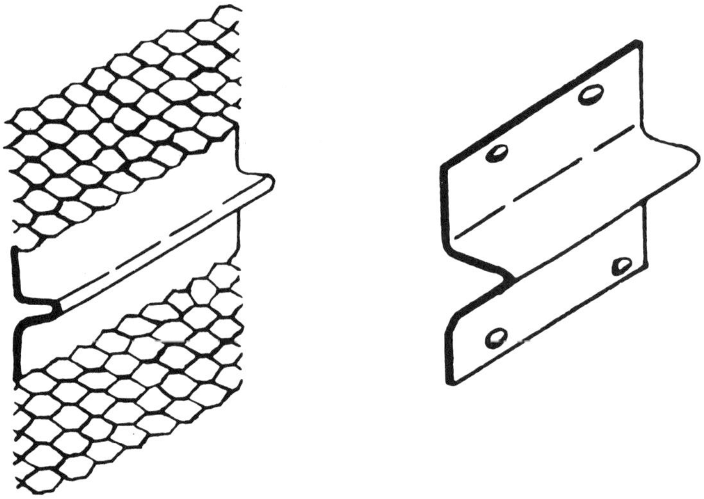

BASE OR PARTING SCREEDS

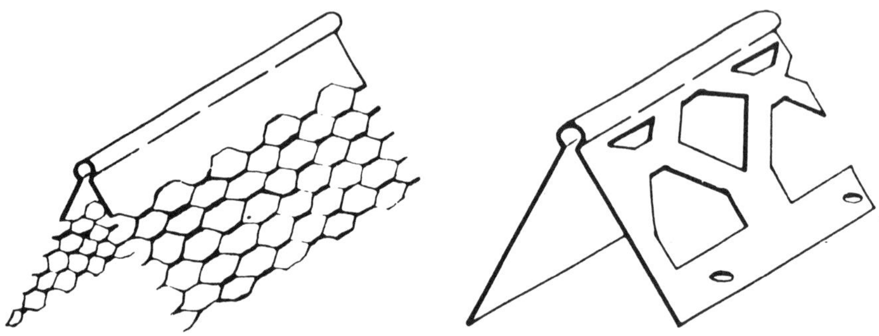

SMALL NOSE CORNER BEADS

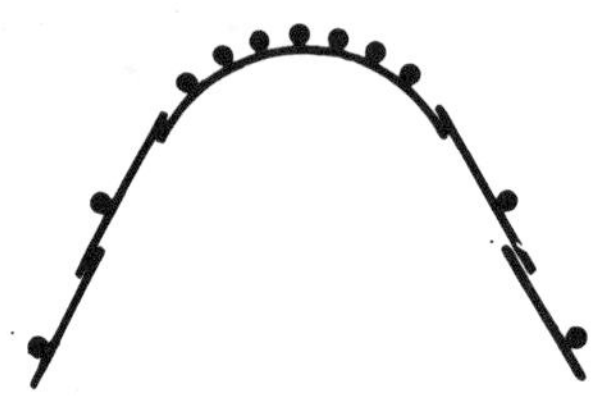

WIRE BULL NOSE
CORNER BEADS

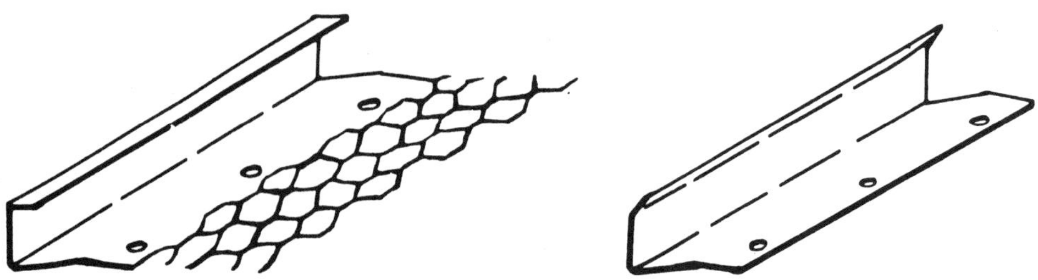

SQUARE CASING BEADS

FINISHES / METAL LATH

09205

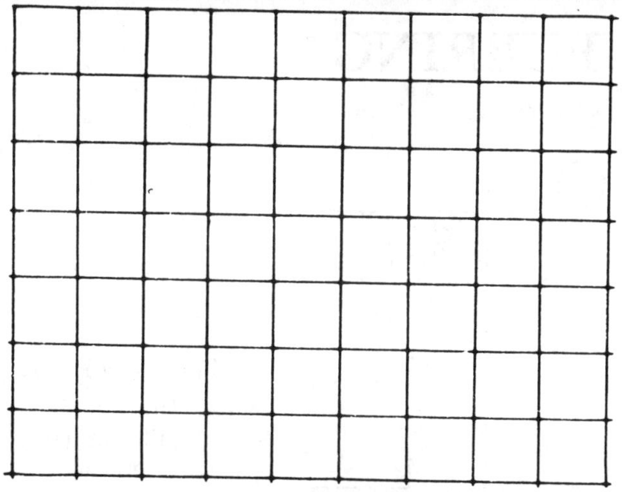

PLAIN WIRE FABRIC LATH

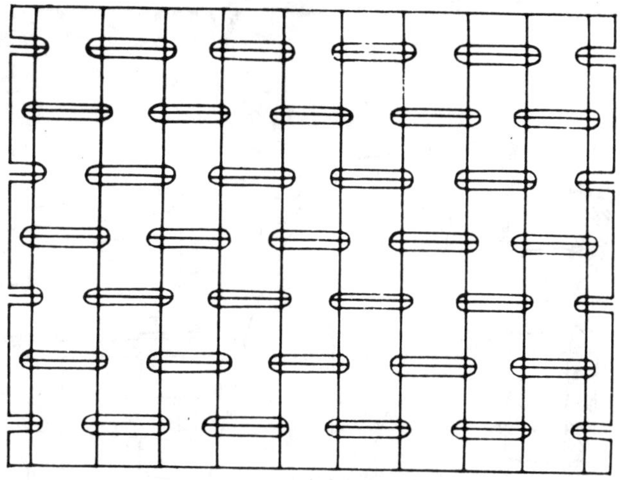

SELF-FURRING WIRE FABRIC LATH

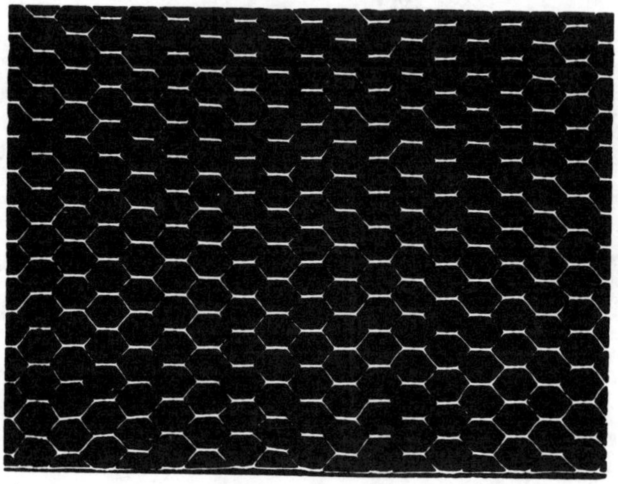

PAPER BACKED WOVEN WIRE FABRIC LATH

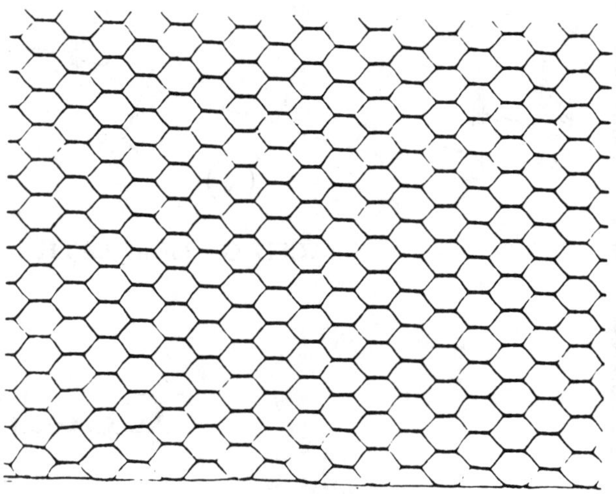

WOVEN WIRE FABRIC LATH (Also Available Self-Furred)

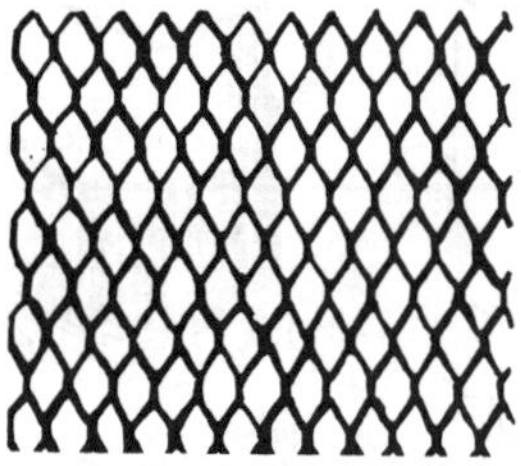

FLAT DIAMOND MESH METAL LATH

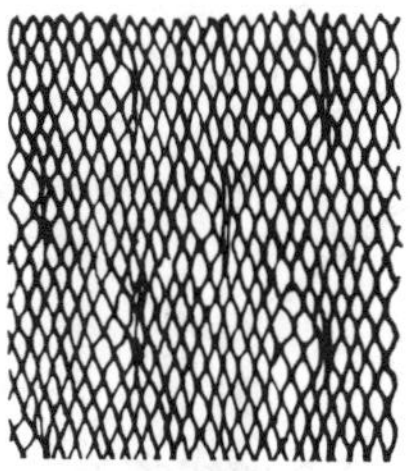

SELF-FURRING METAL LATH

FLAT RIB METAL LATH

RIB METAL LATH

RIB METAL LATH

VERTICAL FURRING
Studless

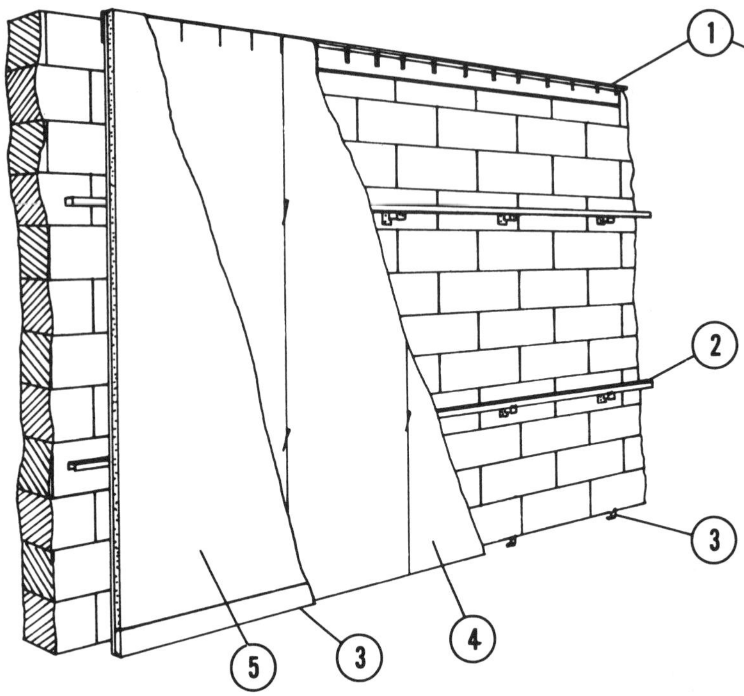

BRACED GYPSUM LATH

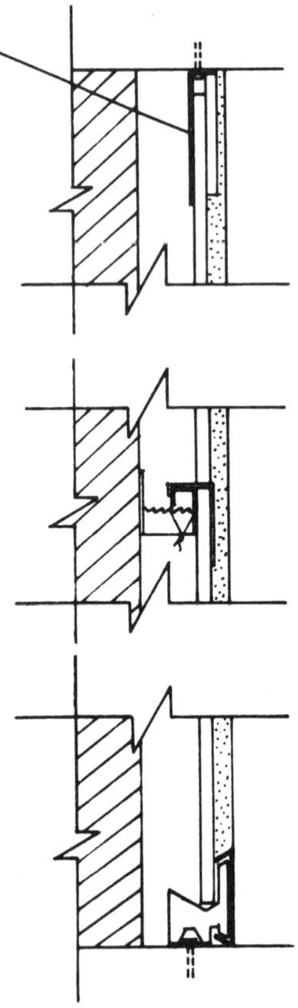

(1) Ceiling Runner
(2) Horizontal Stiffener (secured to bracing attachment)
(3) Metal Base and Clips
(4) Gypsum Lath
(5) Three Coats of Plaster (Scratch, Brown, Finish) (Minimum plaster thickness is ¾ inch.)

COLUMN FURRING

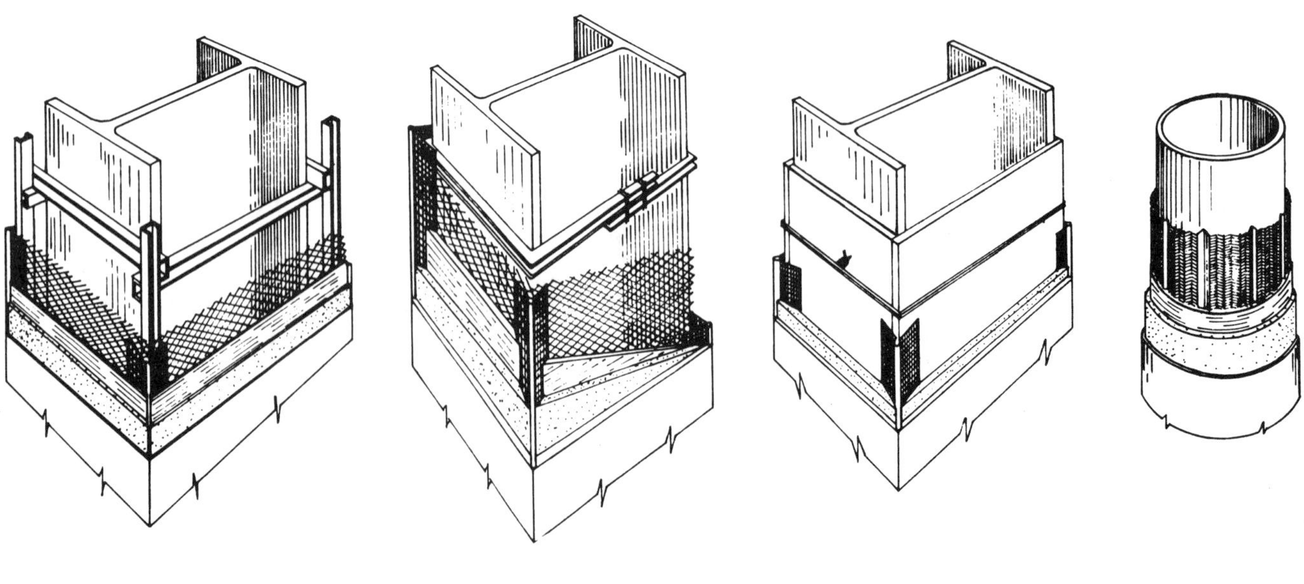

CONTACT FURRING

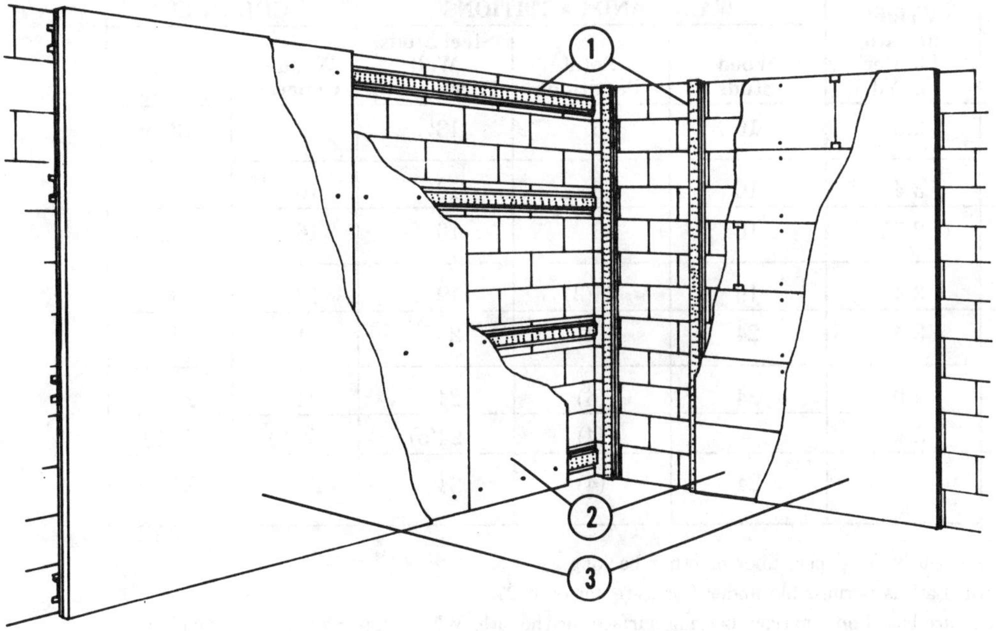

(1) Screw Channel
(2) Gypsum Lath (screwed on)
(3) Two Coats of Plaster (Brown, Finish)

FALSE BEAMS

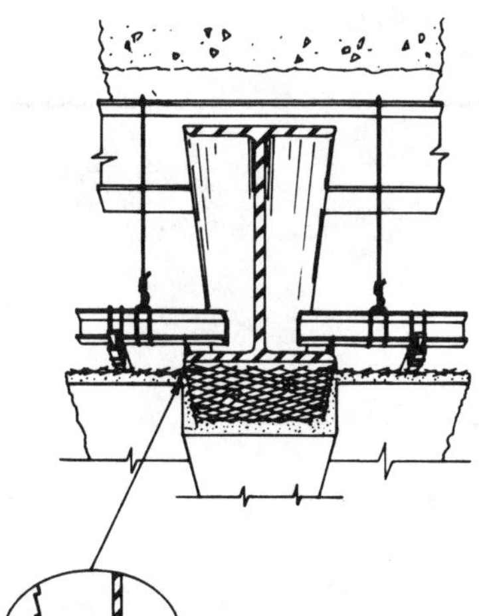

LATH DIRECT TO UNDERSIDE OF BEAM

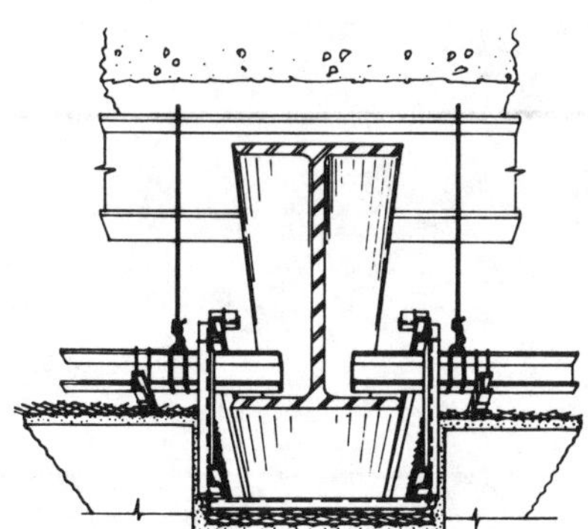

CHANNEL BRACKETS TO MAIN RUNNERS

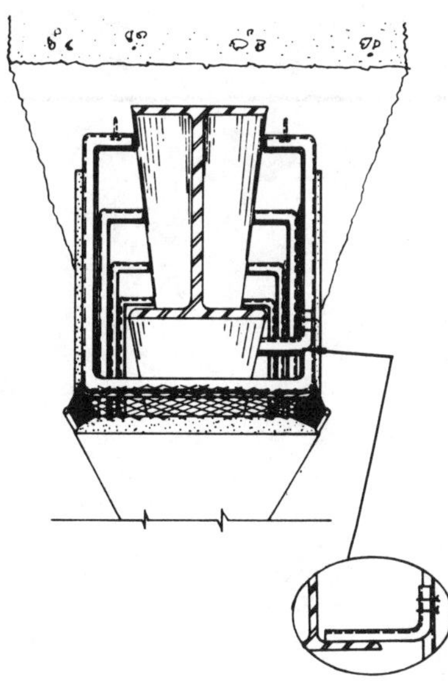

CHANNEL BRACKETS TO SLAB

MAXIMUM SPACING OF SUPPORTS FOR METAL LATH (Inches)						
		WALLS AND PARTITIONS			CEILINGS	
Type of Lath	Weight of Lath Lb. Per Sq. Yd.	Wood Studs	Solid Partitions	Steel Studs, Wall Furring, etc.	Wood or Concrete	Metal
Diamond Mesh (flat expanded)	2.5	16	16	13½	12	12
	3.4	16	16	16	16	
Flat Rib	2.75	16	16	16	16	16
	3.4	19	24(3)	19	19	19
⅜″ Rib (1) (2)	3.4	24	(4)	24	24	24
	4.0	24	(4)	24	24	24
¾″ Rib	5.4	—	(4)	24(5)	36(6)	36(6)
Sheet Lath (2)	4.5	24	(4)	24	24	24

NOTE: Weights are exclusive of paper, fiber or other backing.

(1) 3.4 lb., ⅜″ Rib Lath is permissible under Concrete Joists at 27″ c.c.

(2) These spacings are based on a narrow bearing surface for the lath. When supports with a relatively wide bearing surface are used, these spacings may be increased accordingly, and still assure satisfactory work.

(3) This spacing permissible for Solid Partitions not exceeding 16′ in height. For greater heights, permanent horizontal stiffener channels or rods must be provided on channel side of partitions, every 6′ vertically, or else spacing shall be reduced 25%.

(4) For studless solid partitions, lath erected vertically.

(5) For interior wall furring or for application over solid surfaces for stucco.

(6) For contact or ceilings only.

TYPES OF LATH-ATTACHMENT TO WOOD AND METAL SUPPORTS

TYPE OF LATH	NAILS	NAILS MAXIMUM SPACING		SCREWS MAXIMUM SPACING		STAPLES Round or Flattened Wire				
						Wire Gauge No.	Crown	Leg	MAXIMUM SPACING	
	Type and Size	Vertical (In Inches)	Horizontal (In Inches)	Vertical (In Inches)	Horizontal (In Inches)		(In Inches)		Vertical	Horizontal
1. Diamond Mesh Expanded Metal Lath and Flat Rib Metal Lath	4d blued smooth box 1½ No. 14 gauge 7/32″ head (clinched) 1″ No. 11 gauge 7/16″ head, barbed 1½″ No. 11 gauge 7/16″ head, barbed	6 6 6	— — 6	6	6	16	¾	⅞	6	6
2. ⅜″ Rib Metal Lath and Sheet Lath	1½″ No. 11 ga. 7/16″ head, barbed	6	6	6	6	16	¾	1½	At Ribs	At Ribs
3. ¾″ Rib Metal Lath	4d common 1½″ No. 12½ gauge ¼″ head 2″ No. 11 gauge 7/16″ head, barbed	At Ribs	— At Ribs	At Ribs	At Ribs	16	¾	1⅝	At Ribs	At Ribs
4. Wire Fabric Lath	4d blued smooth box (clinched) 1″ No. 11 gauge 7/16″ head, barbed	6 6	— —	6	6	16	¾	⅞	6	6
	1½″ No. 11 gauge 7/16″ head, barbed 1¼″ No. 12 ga. ⅜″ head, furring 1″ no. 12 gauge ⅜″ head	6 6 6	6 6			16	7/16	⅞	6	6
5. ⅜″ Gypsum Lath	1⅛″ No. 13 gauge 12/61″ head, blued	8	8	8	8	16	¾	⅞	8	8
6. ½″ Gypsum Lath	1¼″ No. 13 gauge 12/61″ head, blued	8	8 6	8	8 6	16	¾	1⅛	8	8 6

FINISHES / PLASTER ACCESSORIES | 09205

STRESS RELIEF (CONTROL JOINTS)

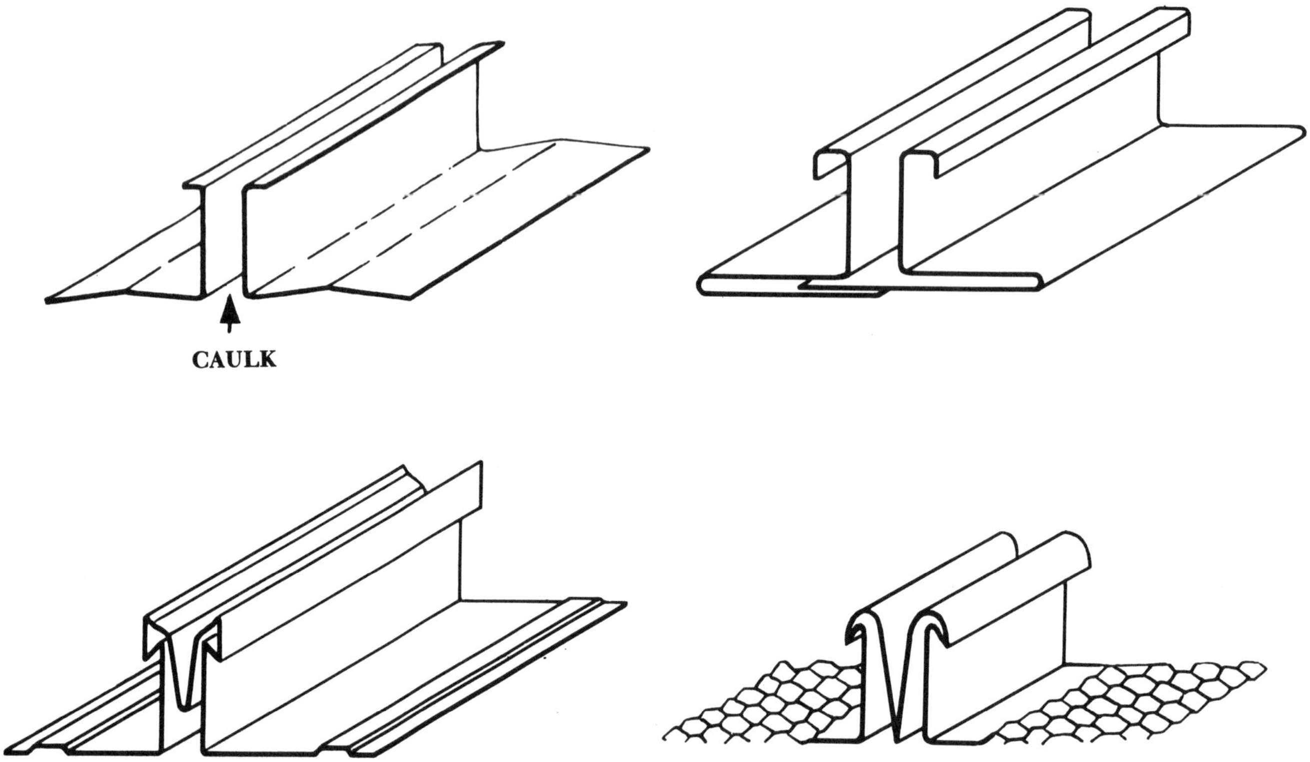

REVEALS

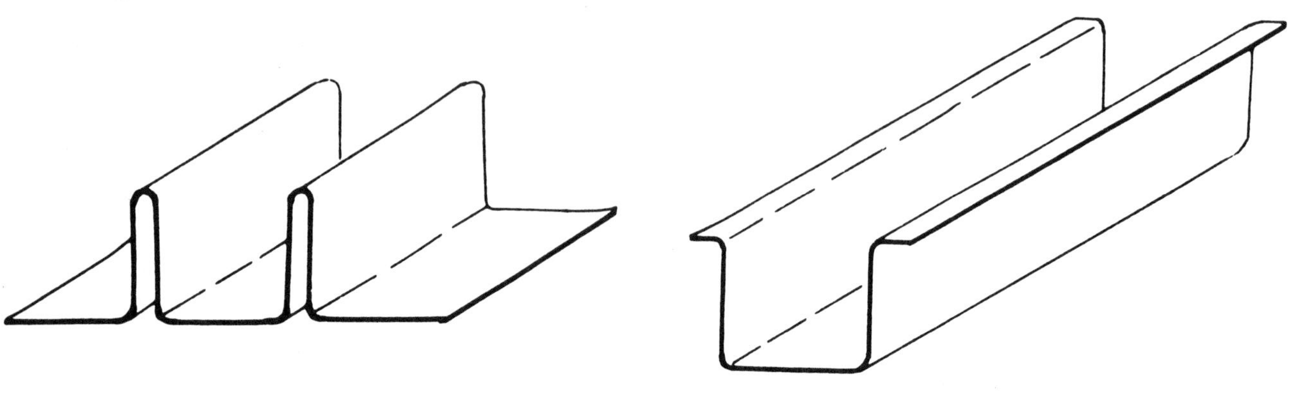

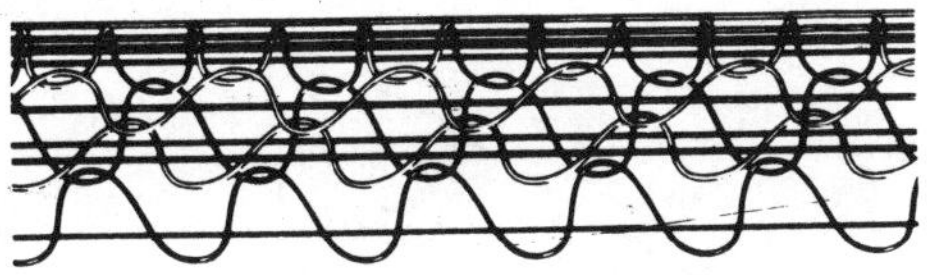

CORNER REINFORCEMENT
(EXTERIOR) WIRE

CORNER REINFORCEMENT
(EXTERIOR) EXPANDED METAL

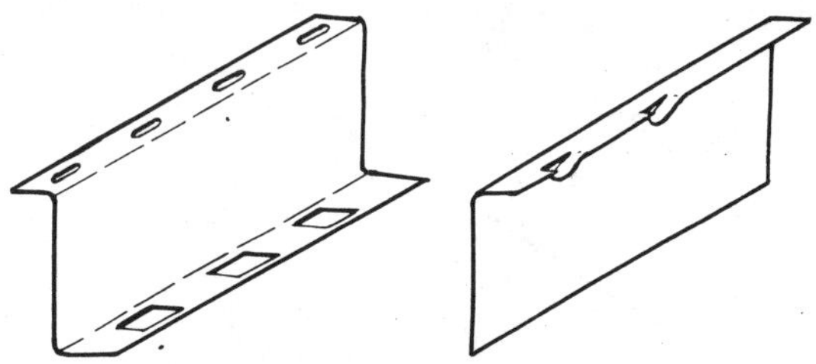

PARTITION RUNNERS
(Z AND L SHAPE)

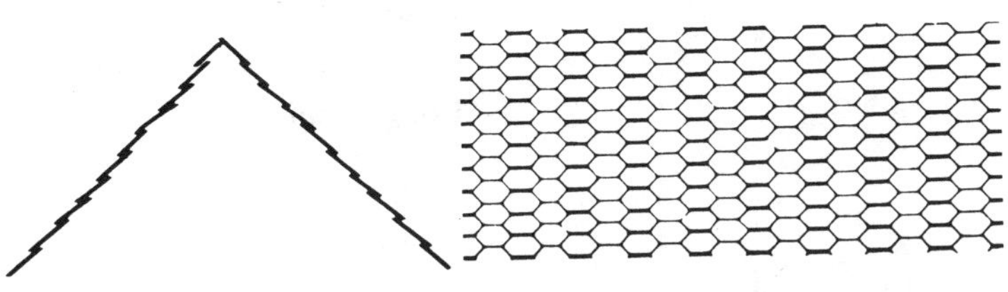

EXPANDED METAL CORNERITE

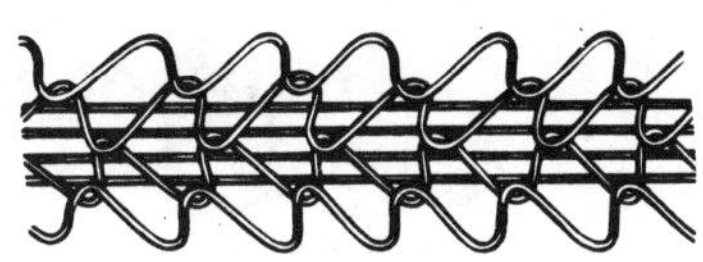

WIRE CORNERITE

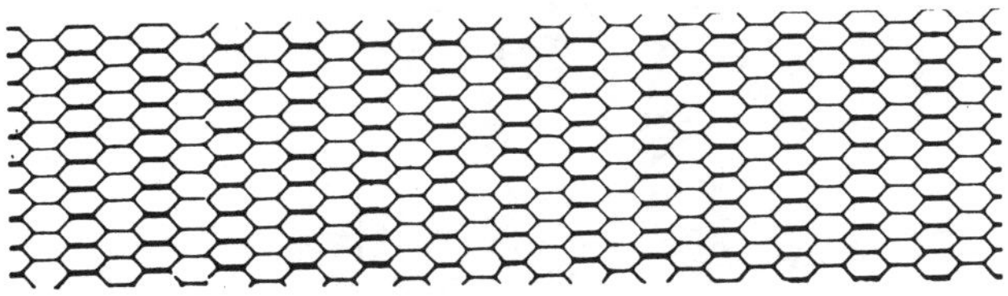

STRIP REINFORCEMENT
(EXPANDED METAL)

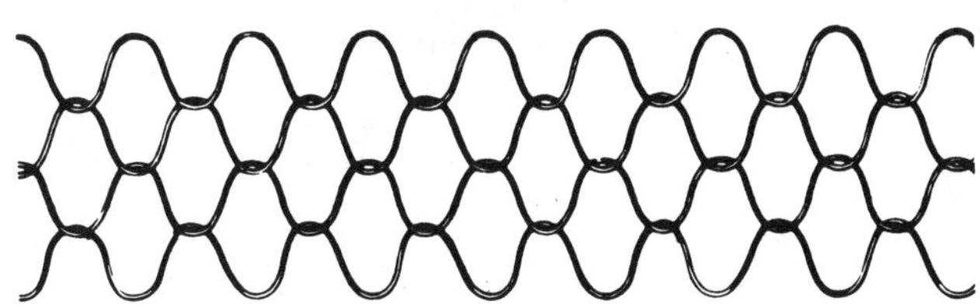

STRIP REINFORCEMENT (WIRE)

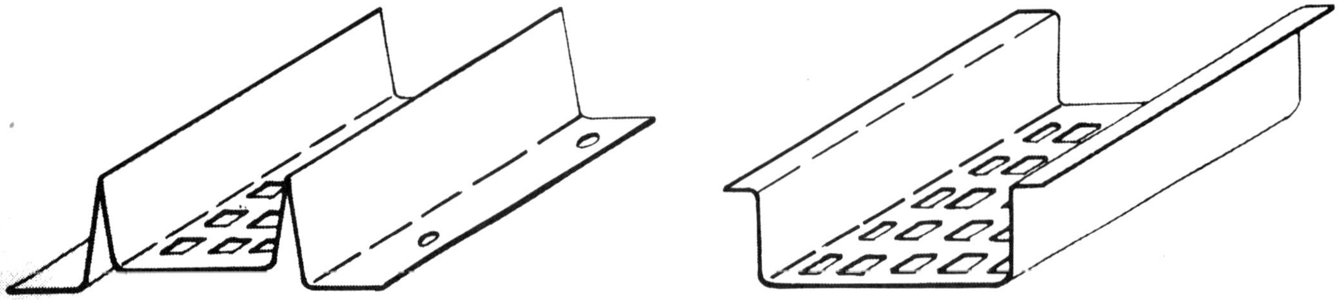

VENTILATING SCREEDS

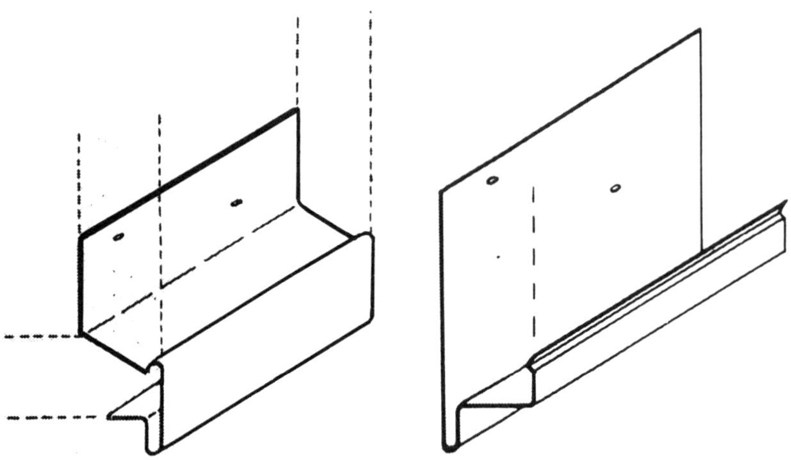

DRIP SCREEDS

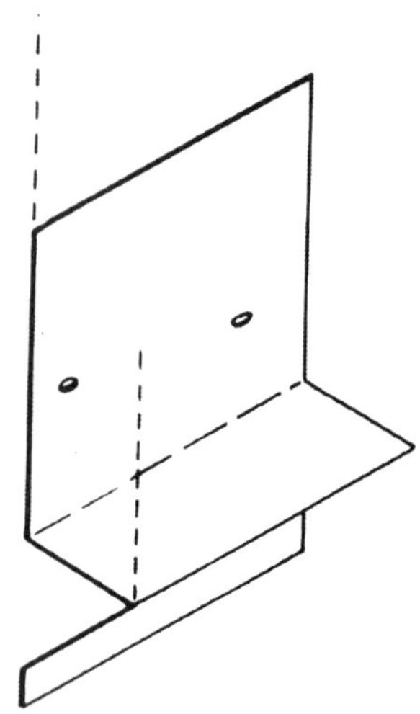

WEEP SCREED
(Also available with perforations)

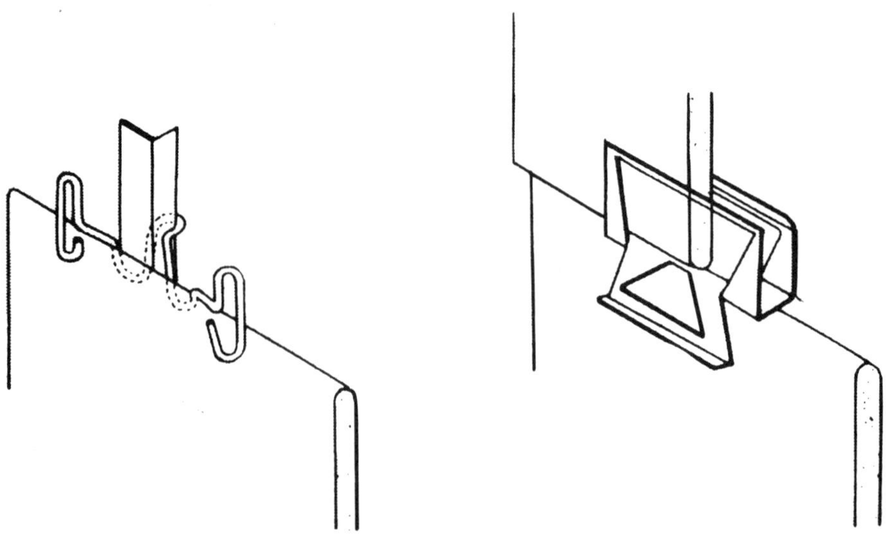

GYPSUM LATH ATTACHMENT CLIPS

VERTICAL FURRING

VERTICAL FURRING MEMBER	UNBRACED				BRACED			
	STUD SPACING				STUD SPACING			
	24″	19″	16″	12″	24″	19″	16″	12″
	Maximum Furring Heights				Maximum Vertical Distance Between Braces			
¾″ Channel	6′	7′	8′	9′	5′	5′	6′	7′
1½″ Channel	8′	9′	10′	12′	6′	7′	8′	9′
2″ Channel	9′	10′	11′	13′	7′	8′	9′	10′
2″ Prefab. Stud	8′	9′	10′	11′	6′	7′	8′	9′
2½″ Prefab. Stud	10′	11′	12′	14′	8′	9′	10′	11′
3¼″ Prefab. Stud	14′	16′	17′	20′	11′	13′	14′	16′

TYPES OF LATH—MAXIMUM SPACING OF SUPPORTS

TYPE OF LATH	Minimum Weight (psy), Gauge & Mesh Size	VERTICAL			HORIZONTAL	
		WOOD	METAL			
			Solid Plaster Partitions	Other	Wood or Concrete	Metal
Expanded Metal Lath (Diamond Mesh)	2.5 3.4	16″ 16″	16″ 16″	12″ 16″	12″ 16″	12″ 16″
Flat rib Expanded Metal Lath	2.75 3.4	16″ 19″	16″ 24″	16″ 19″	16″ 19″	16″ 19″
Stucco Mesh Expanded Metal Lath	1.8 and 3.6	16″	—	—	—	—
⅜″ Rib Expanded Metal Lath	3.4 4.0	24″ 24″	—	24″ 24″	24″ 24″	24″ 24″
Sheet Lath	4.5	24″	—	24″	24″	24″
¾″ Rib Expanded Metal Lath (Not manufactured in West)	5.4	—	—	—	36″	36″
Wire Fabric Lath — Welded	1.95 lbs., 11ga., 2″ x 2″ 1.4 lbs., 16 ga., 2″ x 2″ 1.4 lbs., 18 ga., 1″ x 1″	24″ 16″ 16″	24″ 16″ —	24″ 16″ —	24″ 16″ —	24″ 16″ —
Wire Fabric Lath — Woven	1.4 lbs., 17 ga., 1½″ Hex. 1.4 lbs., 18 ga., 1″ Hex.	24″ 24″	16″ 16″	16″ 16″	24″ 24″	16″ 16″
⅜″ Gypsum Lath (plain)	—	16″	—	16″	16″	16″
(Large Size)	—	16″	—	16″	16″	16″
½″ Gypsum Lath (plain)	—	24″	—	24″	24″	24″
(Large Size)	—	24″	No supports; Erected vertically	24″	24″	16″
⅝″ Gypsum Lath (Large Size)	—	24″	No supports; Erected vertically	24″	24″	16″

PLASTERING TABLES

THICKNESS OF PLASTER

PLASTER BASE	FINISHED THICKNESS OF PLASTER FROM FACE OF LATH, MASONRY, CONCRETE	
	Gypsum Plaster	Portland Cement Plaster
Expanded Metal Lath	⅝″ minimum	⅝″ minimum
Wire Fabric Lath	⅝″ minimum	¾″ minimum (interior) ⅞″ minimum (exterior)
Gypsum Lath	½″ minimum	
Gypsum Veneer Base	$^1/_{16}$″ minimum	
Masonry Walls	½″ minimum	½″ minimum
Monolithic Concrete Walls	⅝″ maximum	⅞″ maximum
Monolithic Concrete Ceilings	⅜″ maximum	½″ maximum

GYPSUM PLASTER PROPORTIONS

NUMBER OF COATS	COAT	PLASTER BASE OR LATH	MAXIMUM VOLUME AGGREGATE PER 100# NEAT PLASTER (CUBIC FEET)	
			Damp Loose Sand	Perlite or Vermiculite
Two-Coat Work	Basecoat	Gypsum Lath	2½	2½
	Basecoat	Masonry	3	3
Three-Coat Work	First Coat	Lath	2	2
	Second Coat	Lath	3	3
	First & Second Coat	Masonry	3	3

PORTLAND CEMENT PLASTER

COAT	VOLUME CEMENT	MAXIMUM WEIGHT (OR VOLUME) LIME PER VOLUME CEMENT	MAXIMUM VOLUME SAND PER VOLUME CEMENT	APPROXIMATE MINIMUM THICKNESS	MINIMUM PERIOD MOIST CURING	MINIMUM INTERVAL BETWEEN COATS
First	1	20 lbs.	4	⅜″	48 Hours	48 Hours
Second	1	20 lbs.	5	1st and 2nd Coats total ¾″	48 Hours	7 Days
Finish	1	1	3	1st, 2nd and Finish Coats ⅞″	—	

PORTLAND CEMENT-LIME PLASTER

COAT	VOLUME CEMENT	MAXIMUM VOLUME LIME PER VOLUME CEMENT	MAXIMUM VOLUME SAND PER COMBINED VOLUMES CEMENT AND LIME	APPROXIMATE MINIMUM THICKNESS	MINIMUM PERIOD MOIST CURING	MINIMUM INTERVAL BETWEEN COATS
First	1	1	4	⅜″	48 Hours	48 Hours
Second	1	1	4½	1st and 2nd Coats total ¾″	48 Hours	7 Days
Finish	1	1	3	1st, 2nd and Finish Coats ⅞″	—	

METAL STUD CONSTRUCTION

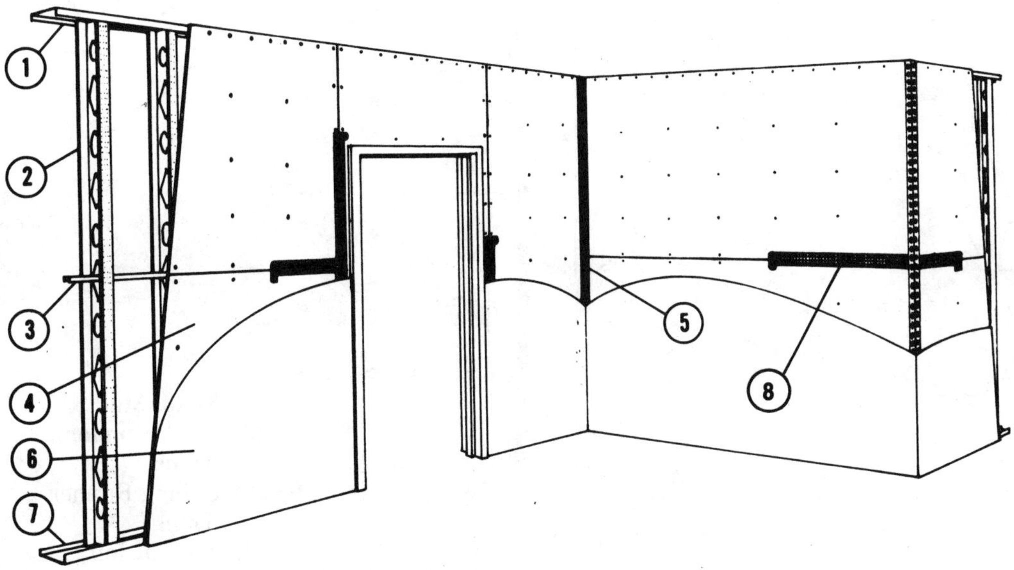

(1) Ceiling Runner Track
(2) Metal Stud (nailable or screw)
(3) Horizontal Stiffener
(4) Large Size Lath
(5) Angle Reinforcement
(6) Veneer Plaster 1/16 to 1/8 inch thick)
(7) Floor Runner Track
(8) Joint Reinforcement

WOOD STUD CONSTRUCTION

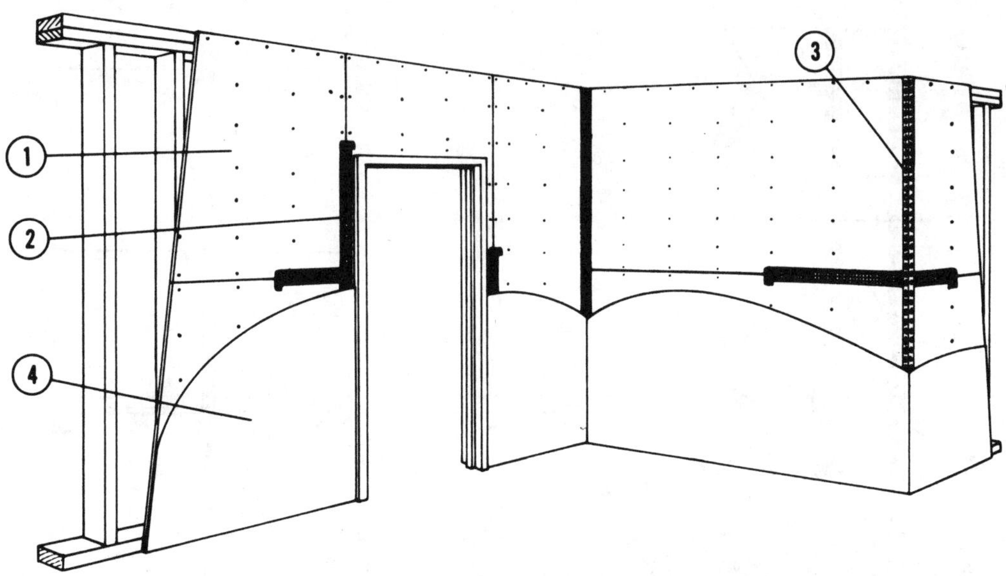

(1) Large Size Lath
(2) Joint Reinforcement
(3) Corner Bead
(4) Veneer Plaster (1/16 to 1/8 inch thick)

EXTERIOR LATH AND PLASTER

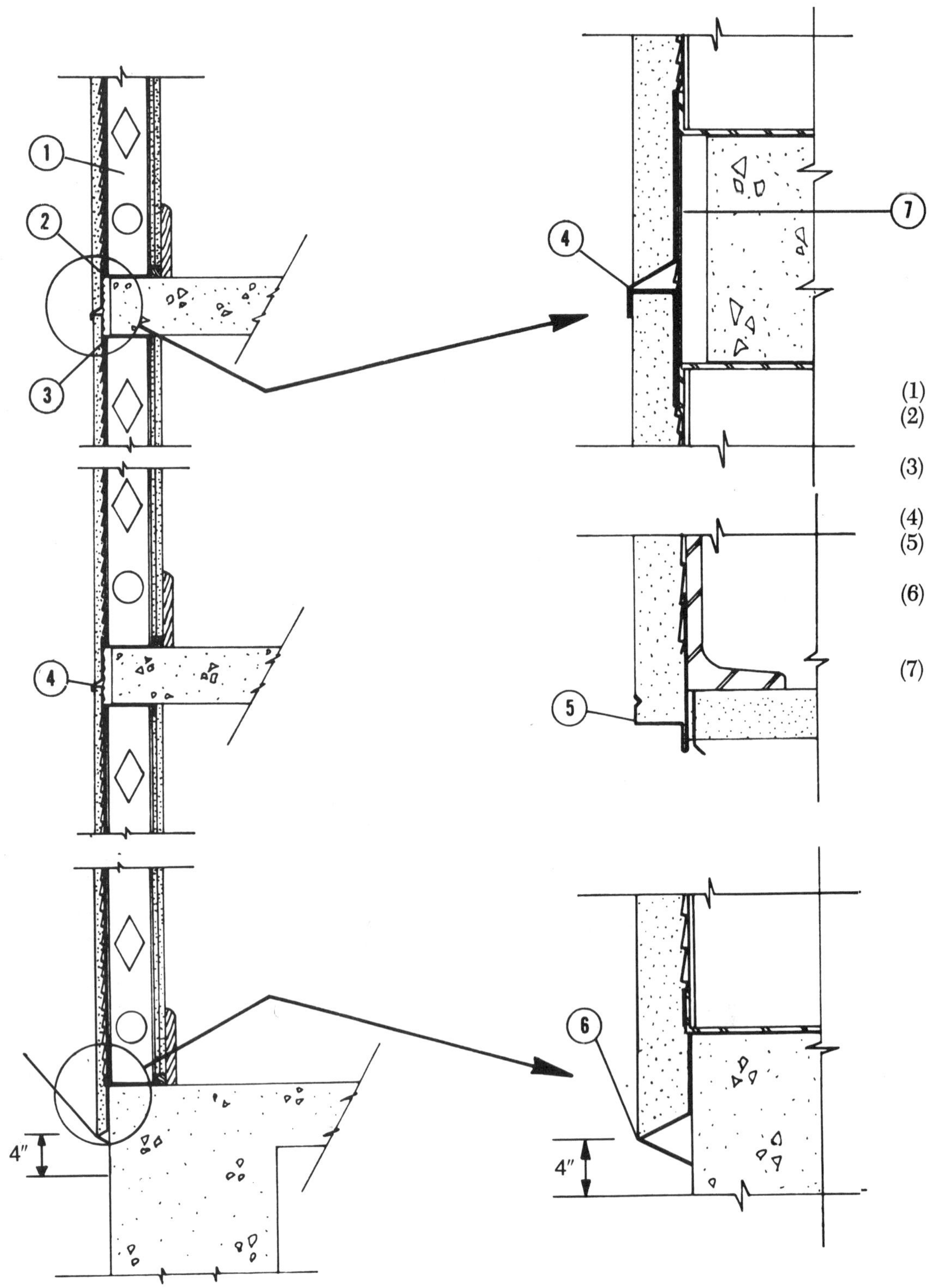

(1) Metal Stud
(2) Floor Runner Track
(3) Ceiling Runner Track
(4) Control Joint
(5) Drip Screed (at soffit)
(6) Weep Screed (at slab or foundation)
(7) Building Paper, Continuous

EXTERIOR LATH AND PLASTER

OPEN WOOD FRAME CONSTRUCTION

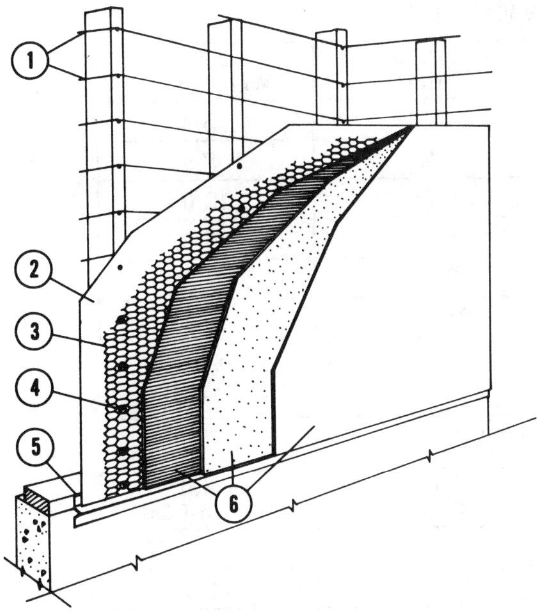

(1) Wire Backing
(2) Building Paper
(3) Wire Fabric Lath
(4) Approved Fasteners
(5) Weep Screed
(6) Three Coats of Plaster (Scratch, Brown, Finish)

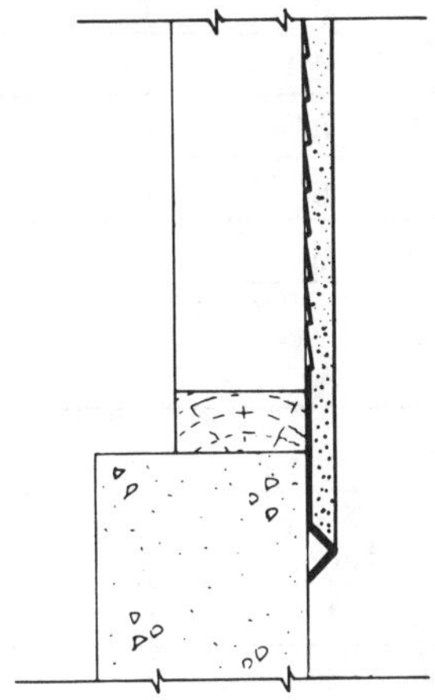

SHEATHED WOOD FRAME CONSTRUCTION

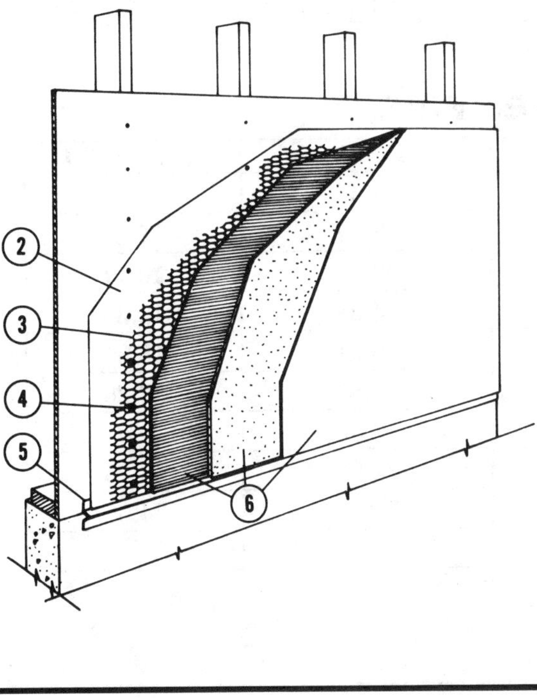

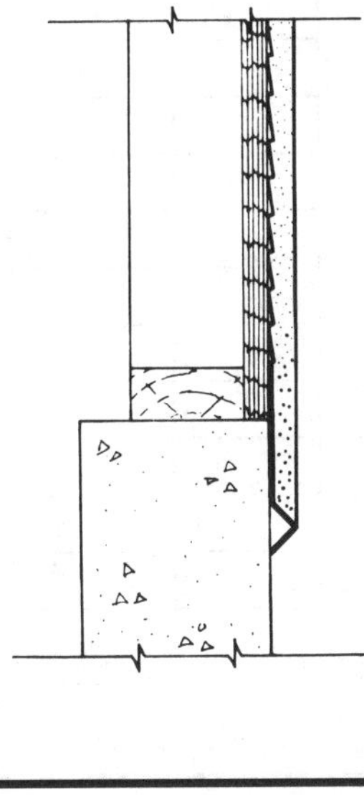

PIPE WEIGHTS

CAST IRON PIPE

SERVICE WEIGHT			
Size, Inches	Weight of Pipe	Weight of Water	Total Weight-Lbs.
2″	3.8	1.45	5.3
3″	5.6	3.2	8.8
4″	7.5	5.5	13.0
5″	9.8	8.7	18.5
6″	12.4	12.5	24.9
8″	18.5	21.7	40.2

EXTRA HEAVY			
Size, Inches	Weight of Pipe	Weight of Water	Total Weight-Lbs.
2″	4.3	1.45	5.8
3″	8.3	3.2	11.5
4″	10.8	5.5	16.3
5″	13.3	8.7	22.0
6″	16.0	12.5	28.5
8″	26.5	21.7	48.2

STEEL PIPE

Pipe Size	W/40	H_2O/Lbs.	Total Lbs./L.F.
2″	3.65	1.45	5.1
2½″	5.79	2.07	7.86
3″	7.57	3.2	10.77
3½″	9.11	4.28	13.39
4″	10.8	5.51	16.31
5″	14.6	8.66	23.26
6″	18.0	12.5	30.5
8″	28.6	21.66	50.26
10″	40.5	34.15	74.65
12″			

SPRINKLER AREA CALCULATIONS

Typical maximum floor area allowed per system riser:

Light Hazard	Ordinary Hazard	Extra Hazard
52,000 S.F.	40,000-52,000 S.F.	25,000 S.F.

Typical maximum floor area coverage allowed per sprinkler head:

Light Hazard	Ordinary Hazard	Extra Hazard
130-200 S.F.	100-130 S.F.	90 S.F.

Typical maximum spacing between lines and sprinkler heads:

Light Hazard	Ordinary Hazard	Extra Hazard
12-15 feet	12-15 feet	12 feet

Note: This data is for estimating purposes only. Check all applicable codes and regulations for specific requirements.

SPRINKLER HEAD CALCULATIONS

Typical maximum quantity of sprinkler heads allowed by pipe size.

Light Hazard:

For sprinklers below ceiling:

Steel		Copper	
1 in. pipe	2 sprinklers	1 in. tube	2 sprinklers
1¼ in. pipe	3 sprinklers	1¼ in. tube	3 sprinklers
1½ in. pipe	5 sprinklers	1½ in. tube	5 sprinklers
2 in. pipe	10 sprinklers	2 in. tube	12 sprinklers
2½ in. pipc	30 sprinklers	2½ in. tube	40 sprinklers
3 in. pipe	60 sprinklers	3 in. tube	65 sprinklers
3½ in. pipe	100 sprinklers	3½ in. tube	115 sprinklers

For sprinklers above and below ceiling:

Steel		Copper	
1 in.	2 sprinklers	1 in.	2 sprinklers
1¼ in.	4 sprinklers	1¼ in.	4 sprinklers
1½ in.	7 sprinklers	1½ in.	7 sprinklers
2 in.	15 sprinklers	2 in.	18 sprinklers
2½ in.	50 sprinklers	2½ in.	65 sprinklers

Ordinary Hazard:

For sprinklers above ceiling:

Steel		Copper	
1 in. pipe	2 sprinklers	1 in. tube	2 sprinklers
1¼ in. pipe	3 sprinklers	1¼ in. tube	3 sprinklers
1½ in. pipe	5 sprinklers	1½ in. tube	5 sprinklers
2 in. pipe	10 sprinklers	2 in. tube	12 sprinklers
2½ in. pipe	20 sprinklers	2½ in. tube	25 sprinklers
3 in. pipe	40 sprinklers	3 in. tube	45 sprinklers
3½ in. pipe	65 sprinklers	3½ in. tube	75 sprinklers
4 in. pipe	100 sprinklers	4 in. tube	115 sprinklers
5 in. pipe	160 sprinklers	5 in. tube	180 sprinklers
6 in. pipe	275 sprinklers	6 in. tube	300 sprinklers

For sprinklers above and below ceiling:

Steel		Copper	
1 in.	2 sprinklers	1 in.	2 sprinklers
1¼ in.	4 sprinklers	1¼ in.	4 sprinklers
1½ in.	7 sprinklers	1½ in.	7 sprinklers
2 in.	15 sprinklers	2 in.	18 sprinklers
2½ in.	30 sprinklers	2½ in.	40 sprinklers
3 in.	60 sprinklers	3 in.	65 sprinklers

Extra Hazard:

For sprinklers below ceiling:

Steel		Copper	
1 in. pipe	1 sprinkler	1 in. tube	1 sprinkler
1¼ in. pipe	2 sprinklers	1¼ in. tube	2 sprinklers
1½ in. pipe	5 sprinklers	1½ in. tube	5 sprinklers
2 in. pipe	8 sprinklers	2 in. tube	8 sprinklers
2½ in. pipe	15 sprinklers	2½ in. tube	20 sprinklers
3 in. pipe	27 sprinklers	3 in. tube	30 sprinklers
3½ in. pipe	40 sprinklers	3½ in. tube	45 sprinklers
4 in. pipe	55 sprinklers	4 in. tube	65 sprinklers
5 in. pipe	90 sprinklers	5 in. tube	100 sprinklers
6 in. pipe	150 sprinklers	6 in. tube	170 sprinklers

Note: This data is for estimating purposes only. Check all applicable codes and regulations for specific requirements.

SPRINKLER HAZARD OCCUPANCIES

Typical Light Hazard Occupancies:

- Churches
- Clubs
- Eaves and overhangs, if combustible construction with no combustible beneath
- Educational
- Hospitals
- Institutional
- Libraries, except large stack rooms
- Museums
- Nursing or Convalescent Homes
- Office, including Data Processing
- Residential
- Restaurant seating areas
- Theaters seating areas
- Theaters and Auditoriums excluding stages and prosceniums
- Unused attics

Typical Ordinary Hazard Occupancies (Group 1):

- Automobile parking garages
- Bakeries
- Beverage manufacturing
- Canneries
- Dairy products manufacturing and processing
- Electronic plants
- Glass and glass products manufacturing
- Laundries
- Restaurant service areas

Typical Ordinary Hazard Occupancies (Group 2):

- Cereal mills
- Chemical plants — Ordinary
- Cold Storage warehouses
- Confectionery products
- Distilleries
- Leather goods mfg.
- Libraries-large stack room areas
- Mercantiles
- Machine shops
- Metal working
- Printing and publishing
- Textile mfg.
- Tobacco Products mfg.
- Wood product assembly

Typical Ordinary Hazard Occupancies (Group 3):

- Feed mills
- Paper and pulp mills
- Paper process plants
- Piers and wharves
- Repair garages
- Tire manufacturing
- Warehouses (having moderate to higher combustibility of content, such as paper, household furniture, paint, general storage, whiskey, etc.)[1]
- Wood machining

Typical Extra Hazard Occupancies (Group 1):

- Combustible Hydraulic Fluid use areas
  - Die Casting
  - Metal Extruding
- Plywood and particle board manufacting
- Printing (using inks with below 100°F [37.8°C] flash points)
- Rubber reclaiming, compounding, drying, milling, vulcanizing
- Saw Mills
- Textile picking, opening, blending, garnetting, carding, combining of cotton, synthetics, wool shoddy or burlap
- Upholstering with plastic foams

Typical Extra Hazard Occupancies (Group 2):

- Asphalt saturating
- Flammable liquids spraying
- Flow coating
- Mobile Home or Modular Building assemblies (where finished enclosure is present and has combustible interiors)
- Open Oil quenching
- Solvent cleaning
- Varnish and paint dipping

....

TYPICAL MINIMUM SIZE OF HORIZONTAL BUILDING STORM DRAINS AND BUILDING STORM SEWERS

Diameter of drain	Maximum projected area in square feet for various slopes		
	⅛ inch per feet slope	¼ inch per feet slope	½ inch per feet slope
3	822	1160	1644
4	1880	2650	3760
5	3340	4720	6680
6	5350	7550	10700
8	11500	16300	23000
10	20700	29200	41400
12	33300	47000	66600
15	59500	84000	119000

MECHANICAL / PLUMBING 15400

PLUMBING BUDGETS

Budget estimates for plumbing can be determined as a percentage of total building costs depending on building type. For example:

- Apartments ....9 to 12 percent
- Assembly ....4 to 7 percent
- Banks ....3 to 6 percent
- Dormitories ....7 to 10 percent
- Factories ....4 to 8 percent
- Hospitals ....8 to 12 percent
- Motels ....9 to 12 percent
- Office Buildings ....4 to 7 percent
- Retail (small) ....4 to 7 percent
- Retail (large) ....3 to 6 percent
- Schools ....3 to 6 percent
- Warehouses ....3 to 7 percent

MECHANICAL / PLUMBING VENTS 15410

TYPICAL SIZE AND LENGTH OF PLUMBING VENTS

Diameter of soil or waste stack (in.)	Total fixture units connected to stack (dfu)		DIAMETER OF VENT PIPE									
		1¼	1	2	2½	3	4	5	6	8	10	12
1¼	2	30										
1½	8	50	150									
1½	10	30	100									
2	12	30	75	200								
2	20	26	50	150								
2½	42		30	100	300							
3	10		42	150	360	1040						
3	21		32	110	270	810						
3	53		27	94	230	680						
3	102		25	86	210	620						
4	43			35	85	250	980					
4	140			27	65	200	750					
4	320			23	55	170	640					
4	540			21	50	150	580					
5	190				28	82	320	990				
5	490				21	63	250	760				
5	940				18	53	210	670				
5	1400				16	49	190	590				
6	500					33	130	400	1000			
6	1100					26	100	310	780			
6	2000					22	84	260	660			
6	2900					20	77	240	600			
8	1800						31	95	240	940		
8	3400						24	73	190	720		
8	5600						20	62	160	610		
8	7600						18	56	140	560		

TYPICAL SIZES OF FIXTURE WATER SUPPLY PIPES

Fixture	Nominal pipe size (inches)
Bath tubs	½
Combination sink and tray	½
Drinking fountain	⅜
Dishwasher (domestic)	½
Kitchen sink, residential	½
Kitchen sink, commercial	¾
Lavatory	⅜
Laundry tray, 1, 2 or 3 compartments	½
Shower (single head)	½
Sinks (service, slop)	½
Sinks flushing rim	¾
Urinal (flash tank)	½
Urinal (direct flush valve)	¾
Water closet (tank type)	⅜
Water closet (flush valve type)	1
Hose bibs	½
Wall hydrant	½

TYPICAL VENTILATION AIR REQUIREMENTS FOR SPECIAL USES

Occupancy Classification	Required ventilation air in cfm per human occupant
Special areas	
Lockers	2* (or 30 per locker)
Wardrobes	2*
Public bathrooms	40**
Private bathrooms	25**
Swimming pools	15 (per occupant)
Exitways and corridors	1½*

*Per square foot floor area.
**Per water closet or urinal.

TYPICAL VENTILATION AIR REQUIREMENTS FOR RETAIL USES

Occupancy Classification	Required ventilation air in cfm per human occupant
Mercantile	
Sales floors and showrooms (basement & grade floors)	7
Sales and showrooms (upper floors)	7
Storage areas	5
Dressing rooms	7
Malls	7
Shipping areas	15
Elevators	7
Supermarkets	
Meat processing rooms	5
Drugs stores	
Pharmacists' work rooms	20
Specialty shops	
Pet shops	1.0*
Florists	5
Greenhouses	5

*cfm per sq. ft. floor area.

TYPICAL VENTILATION AIR REQUIREMENTS FOR RESIDENTIAL USES

Occupancy Classification	Required ventilation air in cfm per human occupant
Residential	
General living areas	5
Bedrooms	5
Kitchens	20
Basements, utility rooms	5
Mobile homes	5
Hotels, motels	
Bedrooms (single, double)	7
Living rooms (suites)	10
Corridors	5
Lobbies	7
Conference rooms (small)	20
Assembly rooms (large)	15

TYPICAL VENTILATION AIR REQUIREMENTS FOR STORAGE USES

Occupancy Classification	Required ventilation air in cfm per human occupant
Storage	
Garages, service stations, parking garages (enclosed)	1.5*
Auto repair shops	1.5**
Warehouses	
General	7

*cfm/s.f. floor area.
**Must have positive engine exhaust system.

TYPICAL VENTILATION AIR REQUIREMENTS FOR FACTORY AND INDUSTRIAL USES

Occupancy Classification	Required ventilation air in cfm per human occupant
Factory and industrial	
Metalworking & finishing	35
Automotive engine test	Require
Paint spray booths	Special
Picking, etching and plating lines	Exhaust
Degreasing booths	Systems
Sandblasting booths	
Chemicals and pharmaceuticals	
Dusty operations	30
Rooms containing potential gas emitters	20
Drying oven rooms	15
Fermentation rooms	15
Pillmaking booths	10
Packaging areas	10
Utility rooms	7
Computer rooms	7
Textiles-clothes manufacturer	15
Electronics and aerospace circuit board and soldering rooms	20
Wood products, papermaking	20
Brewing, distilling, wineries, bottling	20*
Food processing	20
Tobacco processing	20
Power plants	
Control rooms	10
Boiler rooms	35
Generator rooms	20
Sewage treatment plants	
Control rooms	10
Compressor/blower motor rooms	20
Glass and ceramic manufacturer	20
Agricultural	20

TYPICAL VENTILATION AIR REQUIREMENTS FOR BUSINESS USES

Occupancy Classification	Required ventilation air in cfm per human occupant
Business	
Banks	
(see offices)	
Vaults	5
Barber, beauty and health services	
Beauty shops (hair dressers)	25
Reducing salons	25
Sauna baths, steam rooms	5
barber shops	7
Photo studios	
Camera rooms, stages	5
Dark rooms	
Shoe repair shops	
Workrooms/trade areas	10
Offices	
General office space and showrooms	15
Conference rooms	25
Drafting/art rooms	7
Doctor's consultation rooms	10
Waiting rooms	10
Lithographing rooms	7
Diazo printing rooms	7
Computer rooms	5
Keypunch rooms	7
Communication	
TV/radio broadcasting booths, studios	30
Motion picture and TV stages	30
Pressrooms	15
Composing rooms	7
Engraving rooms	7
Telephone switchboard rooms (manual)	7
Telephone switchgear rooms (automatic)	7
Teletypewriter/facsimile rooms	5
Research institutes	
Laboratories:	
Light duty; non-chemical	15
Chemical	15
Heavy-duty	15
Radioisotope, chemical & biologically toxic	15
Machine shops	15
Dark rooms, spectroscopy rooms	10
Animal rooms	40
Veterinary hospitals	
Kennels, stalls	25
Operating rooms	25
Reception rooms	10

TYPICAL VENTILATION AIR REQUIREMENTS FOR INSTITUTIONAL USES

Occupancy Classification	Required ventilation air in cfm per human occupant
Institutional	
Prisons	
Cell blocks	7
Eating halls	15
Guard stations	7

AIR CONDITIONING

RECOMMENDED SHEET METAL GAUGES AND CONSTRUCTION FOR RECTANGULAR DUCT

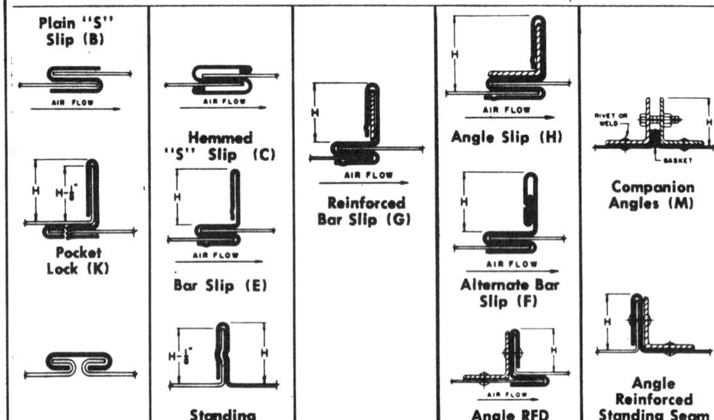

LOW PRESSURE — LOW VELOCITY = 2" W.G. MAX									
				AT JOINTS					
Plate No.	Dimension of Longest Side of Duct	Steel Metal Gauges: Steel	Steel Metal Gauges: Aluminum	Plain "S" Slip (B); Pocket Lock (K); Drive Slip (A)	Hemmed "S" Slip (C); Bar Slip (E); Standing Seam (1)	Reinforced Bar Slip (G)	Angle Slip (H); Alternate Bar Slip (F); Angle RFD Pocket (L)	Companion Angles (M); Angle Reinforced Standing Seam (J)	Reinforcing Between Joints
6	Thru 12"	26	24 (.020)	A-B-K	—	—	—	—	
6	13" thru 18"	24	22 (.025)	A-B-K	—	—	—	—	
7 7A	19" thru 30"	24	22 (.025)	K @ 5' cc A	C-E-@ 5' cc C-E-@ 10' cc	—	—	—	1"x1"x⅛" @ 5' cc
8	31" thru 42"	22	20 (.032)	K @ 5' cc	E-G-K @ 5' cc E-G-K @ 10' cc	—	—	—	1"x1"x⅛" @ 5' cc
9	43" thru 54"	22	20 (.032)	K @ 4' cc K @ 8' cc	E-@ 4' cc E-@ 8' cc	G -@ 4' cc G -@ 8' cc	—	—	ı½"x1½"x⅛" @ 4' cc
9	55" thru 60"	20	18 (.040)	K @ 4' cc K @ 8' cc	E-@ 4' cc E-@ 8' cc	G -@ 4' cc G -@ 8' cc	—	—	1½"x1½"x⅛" @ 4' cc
10	61" thru 84"	20	18 (.040)	—	—	G -@ 4' cc G -@ 5' cc	H-@ 4' cc F-@ 4' cc L-@ 4' cc H-@ 5' cc F-@ 5' cc L-@ 5' cc	 J-@ 2' cc	1½"x1½"x⅛" @ 2' cc 1½"x1½"x⅛" @ 2'-6" cc
11	85" thru 96"	18	16 (.051)	—	—	—	H-@ 4' cc L-@ 4' cc H-@ 5' cc L-@ 5' cc	M-@ 4' cc M-@ 5' cc J-@ 2' cc	1½"x1½"x3/16" @ 2' cc 1½"x1½"x3/16" @ 2'-6" cc 1½"x1½"x3/16" @ 2' cc
12	Over 96"	18	16 (.051)	—	—	—	H-@ 4' cc L-@ 4' cc H-@ 5' cc L-@ 5' cc	M-@ 4' cc M-@ 5' cc J-@ 2' cc	2"x2"x¼" @ 2' cc 2"x2"x¼" @ 2'-6" cc 2"x2"x¼" @ 2' cc

H (height dimension)—up to 42" = 1"
H (height dimension)—43" to 96" = 1½"
H (height dimension)—over 96" = 2"

AIR CONDITIONING (Cont.)

LONGITUDINAL SEAMS FOR SHEET METAL DUCTWORK

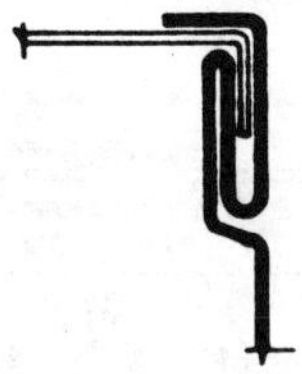

Fig. "N"
PITTSBURGH LOCK

Fig. "Z"
BUTTON PUNCH SNAP LOCK

Fig. "O"
ACME LOCK-GROOVED SEAM

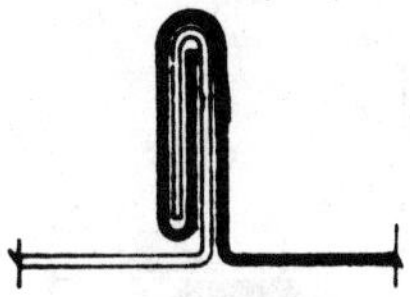

Fig. "T"
DOUBLE SEAM

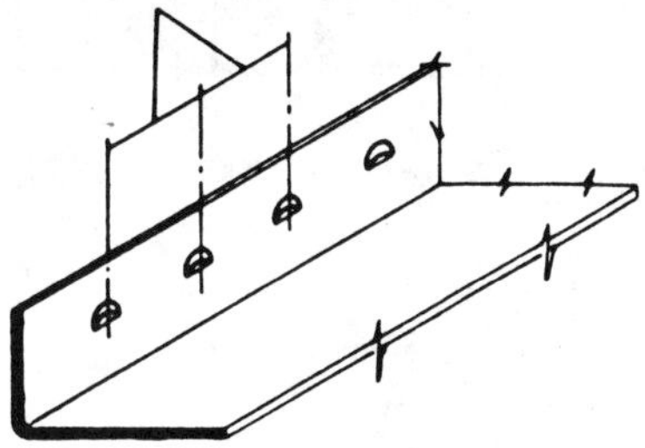

DETAIL NO. 1
MALE PIECE-SNAP LOCK

AIR CONDITIONING

TYPICAL DUCT CONNECTIONS

CROSS JOINTS FOR SHEET METAL DUCTWORK

(NOT TO SCALE)

H=HEIGHT REFERRED TO IN DIMENSIONS

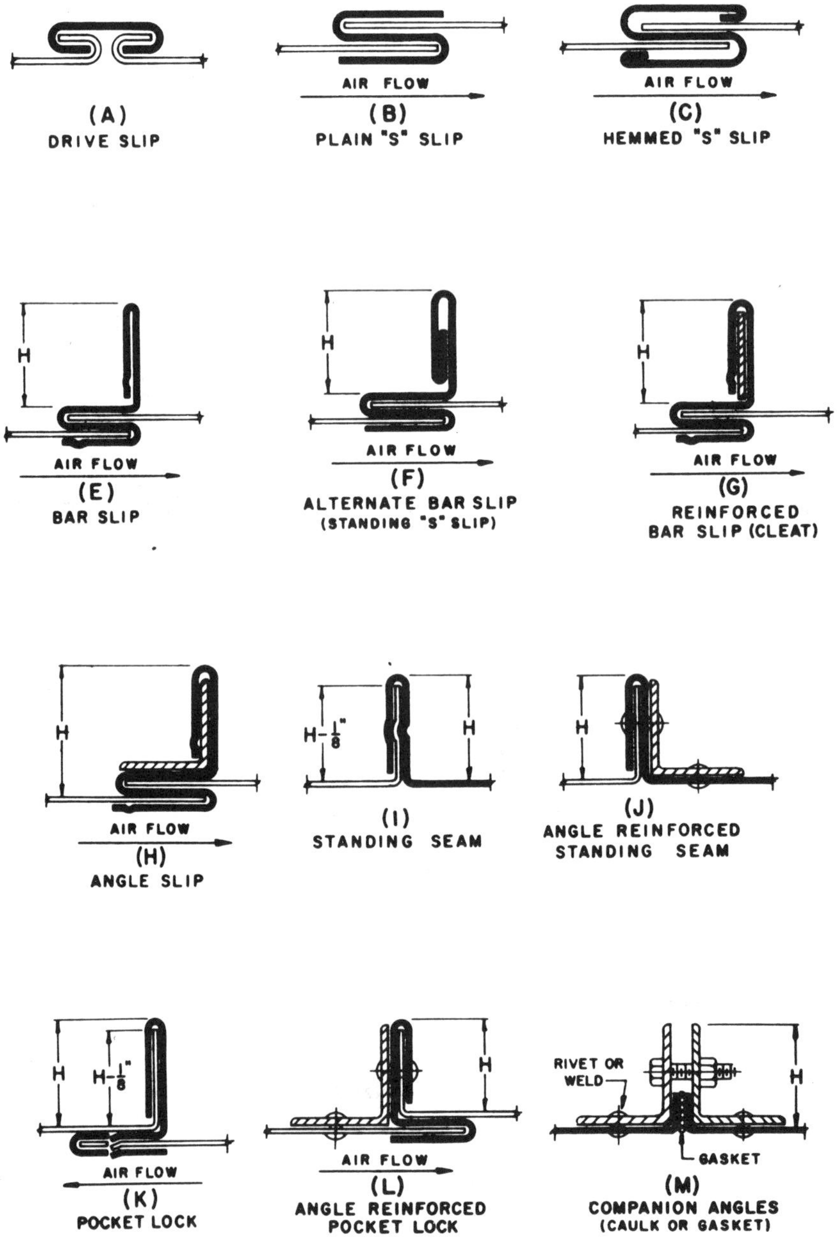

Geographic Cost Modifiers

The costs as presented in this book attempt to represent national averages. Costs, however, vary among regions, states and even between adjacent localities.

In order to more closely approximate the probable costs for specific locations throughout the U.S., this table of Geographic Cost Modifiers is provided. These adjustment factors are used to modify costs obtained from this book to help account for regional variations of construction costs and to provide a more accurate estimate for specific areas. The factors are formulated by comparing costs in a specific area to the costs as presented in the Costbook pages. An example of how to use these factors is shown below. Whenever local current costs are known, whether material prices or labor rates, they should be used when more accuracy is required.

Cost obtained from Costbook pages	×	Location Cost Adjustment Factor	=	Adjusted Cost

For example, a project estimated to cost \$125,000 using the Costbook pages can be adjusted to more closely approximate the cost in Los Angeles: \$125,000 × 1.05 = \$131,250.

Location	Factor
Alabama	0.80
Alaska	1.30
Arizona	0.89
Arkansas	0.74
California	
Los Angeles	1.05
San Francisco	1.19
Other	1.00
Colorado	0.90
Connecticut	1.01
Delaware	0.91
Florida	0.82
Georgia	
Atlanta	0.83
Other	0.75
Hawaii	1.25
Idaho	0.88
Illinois	
Chicago	1.00
Other	0.94
Indiana	0.90
Iowa	0.87
Kansas	0.81
Kentucky	0.85
Louisiana	0.82
Maine	0.85
Maryland	0.90
Massachussetts	
Boston	1.01
Other	0.96
Michigan	0.90
Minnesota	0.93
Missouri	0.86
Montana	0.87
Nebraska	0.83
Nevada	0.95
New Hampshire	0.86
New Jersey	0.97
New Mexico	0.84
New York	
New York City	1.20
Other	0.92
North Carolina	0.75
North Dakota	0.86
Ohio	0.87
Oklahoma	0.79
Oregon	0.90
Pennsylvania	
Philadelphia	1.01
Other	0.89
Rhode Island	0.97
South Carolina	0.77
South Dakota	0.76
Tennessee	0.79
Texas	0.82
Utah	0.83
Vermont	0.85
Virginia	0.81
Washington	1.01
Washington D.C.	0.95
West Virginia	0.85
Wisconsin	0.92
Wyoming	0.87

Square Foot Tables

The following Square Foot Tables list hundreds of actual projects for dozens of building types each with associated building size, total square foot building cost and percentage of project costs for total mechanical and electrical components. This data provides an overview of construction costs by building type. These costs are for actual projects. The variations within similar building types may be due, among other factors, to size, location, quality and specified components, materials and processes. Depending upon all such factors, specific building costs can vary significantly and may not necessarily fall within the range of costs as presented.

The data has been updated to reflect current construction costs and is an expanded summary of the projects and costs presented in the **Building News Square Foot 1993 Costbook.** The source of this data is **Design Cost & Data** magazine which has over 25 years of history in tracking, collecting and publishing construction cost data.

BUILDING CATEGORY	PAGE

SQUARE FOOT TABLES

COMMERCIAL

AUTO DEALERSHIP

Project Size Gross S.F.	Project Cost $/S.F.	% Cost Mechanical	% Cost Electrical
7,700	73.03	16.7	9.4
16,100	40.66	10.2	15.9
20,000	42.56	12.9	23.4
26,300	43.42	12.5	22.0
43,600	34.24	19.4	13.2
53,600	58.84	12.5	11.5

BUSINESS CENTER

Project Size Gross S.F.	Project Cost $/S.F.	% Cost Mechanical	% Cost Electrical
3,900	44.33	12.0	9.2
9,900	41.72	9.1	7.6
54,400	28.32	3.4	12.2
135,000	31.18	8.2	1.5

CINEMA

Project Size Gross S.F.	Project Cost $/S.F.	% Cost Mechanical	% Cost Electrical
18,000	96.65	10.9	6.7
22,500 A	62.05	6.6	4.2

MALL/PLAZA

Project Size Gross S.F.	Project Cost $/S.F.	% Cost Mechanical	% Cost Electrical
9,700	31.29	15.0	13.3
10,500	50.77	8.0	13.5
16,300	46.29	9.4	10.6
26,900	45.72	15.0	7.0
36,000	38.24	10.0	11.0
36,300	44.19	12.4	8.2
44,720	45.93	18.3	12.4
59,100	44.06	9.8	9.5
60,000 R	45.89	10.0	9.5
64,100	46.03	22.4	18.4
66,000	53.33	13.5	11.5
67,400	38.08	21.0	15.0
73,500	55.84	14.9	6.9

For more detailed information see **Building News Square Foot 1993 Costbook.**

MALL/PLAZA (Cont.)

Project Size Gross S.F.	Project Cost $/S.F.	% Cost Mechanical	% Co Electri
142,000	35.21	7.1	8.0
220,000	93.65	11.0	6.4
223,700	26.97	9.0	9.3
321,200	31.92	7.3	7.4
379,900	41.61	11.2	6.2
405,100	43.35	13.9	6.0
482,000	73.32	10.5	9.4
630,000	48.79	12.2	12.4

RESTAURANT

Project Size Gross S.F.	Project Cost $/S.F.	% Cost Mechanical	% Co Electri
4,300 R	81.36	6.9	8.7
4,400 R	119.56	14.6	8.5
5,800	95.33	28.0	10.6
6,800 A	113.24	7.0	11.1
7,360 R	121.04	16.0	6.5
9,600	120.51	24.7	13.1
10,000 R	125.36	21.0	10.0
10,100	107.98	28.6	18.4
10,600	195.62	20.4	6.4
22,900 R	129.99	15.8	16.9

RETAIL STORE

Project Size Gross S.F.	Project Cost $/S.F.	% Cost Mechanical	% Co Electri
1,000	119.67	12.8	6.7
3,000 R	107.03	14.3	10.5
12,300	146.95	14.0	10.0
30,000	74.27	15.6	26.2
61,300	37.92	13.3	13.0
115,000	53.21	14.6	11.3
154,700	75.64	11.2	12.4
314,700 R	63.21	13.8	9.4

A = Addition R = Remodel

SQUARE FOOT TABLES

RESIDENTIAL

APARTMENTS

Project Size Gross S.F.	Project Cost $/S.F.	% Cost Mechanical	% Cost Electrical
3,700	44.88	14.9	4.4
13,900	60.47	7.9	7.2
19,200 R	96.81	45.3	7.5
19,700	56.85	7.4	10.4
23,700	62.16	10.6	4.3
26,500	58.78	25.2	12.8
35,100	47.67	16.4	5.6
54,000	83.67	23.3	13.1
62,700	60.74	17.0	9.0
67,300	55.32	13.8	8.4
70,600	30.76	18.1	7.8
75,300	68.79	13.2	8.1
75,600	66.89	14.5	8.9
72,200	58.36	18.4	10.9
77,600	75.89	26.9	14.3
88,100	70.32	15.3	9.3
89,500	65.21	10.7	11.1
94,100	38.87	8.5	6.8
96,000	48.44	17.0	13.3
102,000	85.17	17.8	12.5
103,200	38.45	12.1	8.9
103,600	62.30	19.1	9.0
105,200	79.70	14.9	8.9
106,200	57.47	12.8	9.3
110,900	54.36	15.6	8.4
111,800	77.33	17.3	7.4
115,900	58.60	12.5	8.6
117,200	35.50	12.9	8.0
119,000	50.16	15.6	8.4

For more detailed information see **Building News Square Foot 1993 Costbook.**

APARTMENTS (Cont.)

Project Size Gross S.F.	Project Cost $/S.F.	% Cost Mechanical	% Cost Electrical
119,400	32.55	18.6	8.7
144,300	72.69	15.0	8.6
176,300	62.23	19.1	9.0
192,300	34.86	10.1	6.0
210,900	71.90	19.1	9.5
220,200	74.29	14.8	7.5
253,900	96.60	20.6	7.7
369,500	69.84	15.8	9.0

CONDOS/TOWNHOUSES

Project Size Gross S.F.	Project Cost $/S.F.	% Cost Mechanical	% Cost Electrical
8,600	63.52	8.5	4.4
16,700	50.25	13.2	6.5
18,000	108.45	14.6	9.7
18,400	55.34	12.9	5.8
74,800	49.30	13.8	5.2
111,700	63.63	9.2	7.1
150,300	69.67	15.9	7.9
278,800	110.11	14.1	7.9
1,109,900	50.41	9.8	4.7

SINGLE-FAMILY HOMES

Project Size Gross S.F.	Project Cost $/S.F.	% Cost Mechanical	% Cost Electrical
600 R	42.47	16.3	2.0
900	57.03	33.0	3.0
2,100	66.00	8.8	4.0
2,200	141.79	23.8	3.4
2,500	69.10	6.1	7.1
2,900	66.79	7.7	2.8
3,000	46.67	9.5	6.8
3,100	104.71	15.5	4.6
3,600	65.73	8.5	3.0
3,700	91.44	9.8	3.4

A = Addition R = Remodel

· more information subscribe to **Design Cost & Data.**

SQUARE FOOT TABLES

RESIDENTIAL (Cont.)

SINGLE-FAMILY HOMES (Cont.)

Project Size Gross S.F.	Project Cost $/S.F.	% Cost Mechanical	% Cost Electrical
4,200	79.95	15.5	5.7
4,600	127.89	9.0	4.7
5,200	166.02	8.3	6.0
5,700	96.23	7.4	3.7
5,700	65.54	8.7	12.6
21,300*	46.40	26.0	4.0
22,700*	42.35	27.0	4.6
45,000*	65.57	7.8	2.5
51,458*	46.98	11.0	5.0

*Townhouses.

EDUCATIONAL

ADMINISTRATION (OFFICES)

Project Size Gross S.F.	Project Cost $/S.F.	% Cost Mechanical	% Cost Electrical
53,700	112.61	15.4	9.0

ATHLETIC FACILITY

Project Size Gross S.F.	Project Cost $/S.F.	% Cost Mechanical	% Cost Electrical
38,100	111.14	16.8	9.8
44,100	102.23	16.8	8.4
100,000	68.58	19.2	5.3
160,000	126.62	13.7	7.9
247,500	109.24	11.2	9.0
271,000	101.23	13.2	7.6
283,100	122.62	14.6	6.0

AUDITORIUM/PERFORMING ARTS

Project Size Gross S.F.	Project Cost $/S.F.	% Cost Mechanical	% Cost Electrical
9,900	175.29	17.5	29.2
17,800	164.86	11.4	16.1
29,200	161.01	16.2	10.5
62,700	119.81	13.0	12.9

For more detailed information see **Building News Square Foot 1993 Costbook.**

CLASSROOM

Project Size Gross S.F.	Project Cost $/S.F.	% Cost Mechanical	% Cost Electrica
35,400	163.18	10.2	6.8
70,000	73.79	24.3	13.1
78,900	140.42	13.3	14.3
80,100	113.03	16.9	10.5
100,000	113.45	21.1	9.8
166,000	76.67	16.1	11.2
298,400	72.48	11.3	11.9

COMPLETE COLLEGE FACILITIES

Project Size Gross S.F.	Project Cost $/S.F.	% Cost Mechanical	% Cost Electrical
95,300	116.19	19.4	11.7
450,000	140.00	18.6	10.8

ELEMENTARY SCHOOL

Project Size Gross S.F.	Project Cost $/S.F.	% Cost Mechanical	% Cost Electrical
18,000	83.85	19.1	13.8
30,800	77.74	16.0	10.0
31,600	65.86	12.1	11.3
35,700	96.11	17.3	8.7
40,000	82.80	18.5	13.4
40,500	69.68	22.1	11.3
57,000	63.33	22.1	7.5
69,700	97.97	20.6	9.3
91,400	79.24	18.2	7.6

HIGH SCHOOL

Project Size Gross S.F.	Project Cost $/S.F.	% Cost Mechanical	% Cost Electrical
116,400	87.03	18.1	12.9
133,000	75.30	17.8	10.7
184,000	129.12	23.0	10.0
217,200 R	70.37	26.4	10.6
254,000 R	55.65	17.2	14.6
431,700	83.85	13.1	9.6

A = Addition R = Remodel

SQUARE FOOT TABLES

EDUCATIONAL (Cont.)

JUNIOR HIGH SCHOOL

Project Size Gross S.F.	Project Cost $/S.F.	% Cost Mechanical	% Cost Electrical
26,000	108.71	9.5	9.3
28,100	67.31	11.9	9.5
52,800	86.62	18.3	8.6
91,600	107.63	21.2	11.0
123,700	87.86	29.1	9.1

LABORATORY/RESEARCH

Project Size Gross S.F.	Project Cost $/S.F.	% Cost Mechanical	% Cost Electrical
9,200	187.83	25.4	4.5
80,300	150.22	20.2	15.5

LIBRARY

Project Size Gross S.F.	Project Cost $/S.F.	% Cost Mechanical	% Cost Electrical
6,900	101.11	17.9	10.3
8,200	92.75	18.8	10.6
12,000	120.62	19.4	15.9
15,000	105.62	15.7	18.7
16,300	87.59	11.3	7.4
28,600	79.43	15.7	9.0
30,100	107.70	13.0	11.0
37,700	84.74	14.6	8.0
43,500	71.70	16.6	6.7
47,900	125.88	29.7	9.0
51,400	127.67	13.1	12.4
63,400 A	96.07	13.5	8.3
64,000	89.44	10.9	11.4
74,000	102.18	17.2	8.0
75,600	136.94	12.5	16.7
176,000	75.43	11.2	9.8

SPECIAL NEEDS FUNCTION

Project Size Gross S.F.	Project Cost $/S.F.	% Cost Mechanical	% Cost Electrical
15,200	85.38	16.6	8.4
27,900	102.96	19.4	11.2

For more detailed information see **Building News Square Foot 1993 Costbook.**

STUDENT CENTER/MULTIPURPOSE

Project Size Gross S.F.	Project Cost $/S.F.	% Cost Mechanical	% Cost Electrical
90,000	129.99	16.2	10.0
49,600	96.07	18.2	9.2
187,700	144.68	15.3	8.8
194,800	76.16	17.2	8.5

HOTEL/MOTEL

CONVENTION/CONFERENCE CENTER

Project Size Gross S.F.	Project Cost $/S.F.	% Cost Mechanical	% Cost Electrical
8,600 A	113.03	20.7	16.3
71,900	125.40	12.5	13.8
433,800	65.47	22.1	8.3

HOTEL

Project Size Gross S.F.	Project Cost $/S.F.	% Cost Mechanical	% Cost Electrical
19,900 A	64.47	16.8	5.8
25,875 R	58.78	11.8	10.5
48,400 A	116.15	23.6	8.2
64,300 R	155.72	20.3	10.1
104,200 A	72.59	15.0	8.8
108,040	61.94	13.0	7.0
110,100	79.64	18.4	10.5
132,000	150.01	16.4	5.4
135,900 A	92.17	15.2	7.7
144,100 A	105.61	19.3	11.5
231,000	116.25	15.2	8.4
449,800 A	64.89	13.1	7.0

HOTEL/INN

Project Size Gross S.F.	Project Cost $/S.F.	% Cost Mechanical	% Cost Electrical
57,400	74.69	13.0	10.0
73,000	47.93	24.4	18.1
75,900	64.42	16.5	7.6
162,000	77.37	17.5	8.0
197,000	74.53	15.7	7.7
277,900	56.20	18.8	9.4

A = Addition R = Remodel

SQUARE FOOT TABLES

INDUSTRIAL

MANUFACTURING

Project Size Gross S.F.	Project Cost $/S.F.	% Cost Mechanical	% Cost Electrical
14,300	65.84	13.0	7.0
18,500	98.09	20.5	13.5
26,600	29.56	6.1	12.1
31,400	77.32	18.9	17.7
33,400	44.77	23.1	15.8
37,300	22.15	3.0	24.0
43,400	42.98	14.7	11.7
45,400	72.45	15.9	15.0
79,800	76.16	16.0	14.5
81,100	80.29	8.2	8.4
137,400	50.25	41.6	13.6
179,600	57.99	26.3	13.6
186,000	54.15	19.5	11.7

RESEARCH AND DEVELOPMENT

Project Size Gross S.F.	Project Cost $/S.F.	% Cost Mechanical	% Cost Electrical
89,140	102.65	20.8	9.0
100,400	135.94	36.1	25.1
114,200	131.89	21.4	9.8
125,000	90.70	20.8	8.4
140,000	125.73	20.1	11.2

WAREHOUSE W/OFFICE

Project Size Gross S.F.	Project Cost $/S.F.	% Cost Mechanical	% Cost Electrical
14,000	27.25	6.5	9.2
19,000	17.07	2.3	1.8
19,700	35.18	9.5	7.0
31,200	27.55	7.0	7.0
40,500	34.71	6.6	10.5
47,500	12.96	2.5	4.8
62,000	45.93	10.8	10.0
96,200	21.07	2.2	6.3
105,000	20.07	5.3	11.1
149,800	21.44	14.5	8.6

For more detailed information see **Building News Square Foot 1993 Costbook.**

WAREHOUSE W/OFFICE (Cont.)

Project Size Gross S.F.	Project Cost $/S.F.	% Cost Mechanical	% Cost Electrical
168,600	27.76	11.4	6.3
209,600	25.33	4.9	10.3
402.400	34.82	14.9	8.3

MEDICAL

EDUCATION CENTER

Project Size Gross S.F.	Project Cost $/S.F.	% Cost Mechanical	% Cost Electrical
35,400	173.60	10.2	6.8

HOSPITALS

Project Size Gross S.F.	Project Cost $/S.F.	% Cost Mechanical	% Cost Electrical
9,300 R	174.60	29.5	12.0
15,900 A	126.86	27.4	15.3
16,600 R	43.51	14.7	9.6
22,000	293.27	31.6	17.5
39,100	145.00	23.3	7.9
63,800	115.77	23.4	7.8
98,000 A	167.07	20.5	17.0
100,200 A	267.36	22.2	9.6
103,900 A	147.90	31.1	11.7
109,300	146.43	20.7	15.7
148,700	115.18	25.2	11.8
154,700	187.00	19.3	10.5
165,484	162.50	21.5	14.8
165,700 A	158.14	26.2	17.7
179,400 A	96.55	28.9	20.3
182,800	143.74	19.1	14.0
265,000	171.44	31.2	14.1
281,100	164.23	24.1	14.6
435,000	170.29	21.5	13.7
694,300	94.39	28.9	9.7
772,300	172.29	31.5	13.4

A = Addition R = Remodel

SQUARE FOOT TABLES

MEDICAL (Cont.)

MEDICAL OFFICES/CENTERS

Project Size Gross S.F.	Project Cost $/S.F.	% Cost Mechanical	% Cost Electrical
3,000	72.79	13.7	11.5
5,500	75.75	8.5	18.7
10,000	95.82	13.9	9.4
10,600	117.52	13.2	8.9
16,300	97.08	21.8	13.2
18,300	52.37	7.1	13.4
20,600	75.32	14.3	6.4
24,900	133.83	21.0	15.1
27,000	133.71	19.7	10.1
28,400	91.62	13.6	9.3
30,500	102.32	17.2	12.7
32,000	82.36	18.1	10.5
44,300	82.07	24.4	16.2
50,200	46.30	9.8	4.9
51,200	118.40	21.0	12.0
64,600	56.25	13.2	6.4
66,000	52.46	9.8	8.8
80,000	33.50	8.2	8.3
137,175	75.64	12.8	6.6

NURSING HOMES

Project Size Gross S.F.	Project Cost $/S.F.	% Cost Mechanical	% Cost Electrical
11,600 A	221.22	53.2	7.9
16,800	143.62	33.3	7.8
31,900 A	108.77	22.0	11.0
64,100	89.96	20.6	11.1
290,000	127.99	16.1	13.6

RESEARCH

Project Size Gross S.F.	Project Cost $/S.F.	% Cost Mechanical	% Cost Electrical
34,600	121.88	18.1	3.0

For more detailed information see **Building News Square Foot 1993 Costbook.**

PUBLIC FACILITIES

ANIMAL CENTER

Project Size Gross S.F.	Project Cost $/S.F.	% Cost Mechanical	% Cost Electrical
20,000	155.59	20.7	4.6
39,100	116.82	22.9	6.8
44,300	110.71	8.1	5.8

AUTO DEALERSHIP

Project Size Gross S.F.	Project Cost $/S.F.	% Cost Mechanical	% Cost Electrical
7,700	73.53	16.7	9.4

BROADCASTING

Project Size Gross S.F.	Project Cost $/S.F.	% Cost Mechanical	% Cost Electrical
20,000	214.37	20.0	13.0
29,500	164.97	16.6	15.0
45,000 R	119.88	15.0	13.0

CIVIC CENTER

Project Size Gross S.F.	Project Cost $/S.F.	% Cost Mechanical	% Cost Electrical
6,000	118.51	9.2	2.8
23,900	148.11	12.3	10.9
34,400	67.42	3.5	17.8
69,800 A	179.19	17.7	8.8
206,500	95.76	15.0	11.0

CORRECTIONAL FACILITIES

Project Size Gross S.F.	Project Cost $/S.F.	% Cost Mechanical	% Cost Electrical
44,600	138.95	20.9	13.1
66,000	85.96	15.0	23.0
257,800 A	150.43	20.9	10.7
360,000	99.65	32.7	13.2

FIRE STATION

Project Size Gross S.F.	Project Cost $/S.F.	% Cost Mechanical	% Cost Electrical
6,900	125.88	12.4	9.7
7,600	95.91	16.0	9.5
8,430	130.84	12.8	9.6
9,600	115.93	13.5	11.8

A = Addition   R = Remodel

SQUARE FOOT TABLES

PUBLIC FACILITIES (Cont.)

GOVERNMENT BUILDINGS

Project Size Gross S.F.	Project Cost $/S.F.	% Cost Mechanical	% Cost Electrical
12,300	89.86	13.5	10.7
23,500	149.90	11.1	15.6
27,300	116.19	19.5	9.3
31,600	130.78	27.1	11.3
46,600	123.15	17.4	13.8
72,100	133.85	24.4	10.7
78,200	110.93	19.7	15.2
332,900	108.19	16.2	14.2
364,100	140.95	14.6	12.9
771,000	155.49	17.6	11.3

MUSEUM

Project Size Gross S.F.	Project Cost $/S.F.	% Cost Mechanical	% Cost Electrical
27,600	109.56	17.8	14.1
30,100	125.04	18.6	7.9
43,264	116.40	11.1	8.3
63,000	126.09	8.8	18.1

PARKING GARAGE

Project Size Gross S.F.	Project Cost $/S.F.	% Cost Mechanical	% Cost Electrical
66,000	31.60	2.8	3.1
169,000	19.70	10.3	3.7
562,700	17.59	2.4	6.2

TRANSPORTATION

Project Size Gross S.F.	Project Cost $/S.F.	% Cost Mechanical	% Cost Electrical
7,300	201.10	15.1	3.2
14,300	164.23	13.3	16.6
23,000	112.87	9.6	13.7
35,500	94.97	1.0	19.0
49,100	139.58	35.5	11.6

For more detailed information see **Building News Square Foot 1993 Costbook.**

TRANSPORTATION (Cont.)

Project Size Gross S.F.	Project Cost $/S.F.	% Cost Mechanical	% Cost Electrica
288,100	80.69	8.3	13.3
2,160,000	133.05	23.3	11.5

OFFICES

BANKS

Project Size Gross S.F.	Project Cost $/S.F.	% Cost Mechanical	% Cost Electrica
2,900	188.98	5.3	4.3
3,100	83.85	5.9	6.7
3,300	118.83	10.3	16.4
3,600	91.54	10.3	12.2
4,000	104.55	8.0	9.0
4,100	103.24	21.4	13.0
4,200	110.50	8.6	14.21
4,400	127.67	12.2	12.7
4,500	80.69	12.4	13.5
4,900	127.58	11.7	11.2
5,900	94.39	9.3	13.8
6,000	106.71	11.6	7.3
6,100	142.53	11.0	8.0
7,000	182.87	6.0	9.0
7,300	117.56	11.9	11.3
7,700	129.62	8.0	7.5
7,800	144.53	11.0	11.7
8,000	73.21	10.0	14.0
9,200	116.19	9.9	12.1
9,400	84.17	11.7	11.8
10,200	175.61	12.6	12.6
12,600	62.68	7.0	18.0

A = Addition R = Removal

SQUARE FOOT TABLES

OFFICES (Cont.)

BANKS (Cont.)

Project Size Gross S.F.	Project Cost $/S.F.	% Cost Mechanical	% Cost Electrical
13,300	110.19	9.5	8.3
13,800	98.71	10.0	9.7
15,000	84.80	15.4	12.4
15,200	67.52	9.2	12.9
15,500	86.28	9.8	10.3
16,000	56.67	13.4	23.1
20,100	53.20	13.0	11.0
21,700	112.82	8.8	11.3
44,800	96.39	13.0	8.2
53,200	174.08	14.9	7.2
62,100	99.97	10.1	7.9
95,100	131.41	13.5	4.3

OFFICE BUILDINGS

Project Size Gross S.F.	Project Cost $/S.F.	% Cost Mechanical	% Cost Electrical
2,600	119.35	17.2	9.4
3,400	97.70	10.5	11.3
3,800	96.18	16.4	11.8
4,400	94.70	12.8	8.5
4,500	81.43	13.0	7.0
5,100	59.10	16.6	10.2
5,200	77.11	8.0	5.7
6,700	114.30	16.8	10.4
7,500	124.09	10.1	8.0
7,900	96.18	17.4	9.0
8,100	149.16	10.6	11.0
10,600	85.85	9.9	9.7
10,900	48.04	13.8	10.8

For more detailed information see **Building News Square Foot 1993 Costbook.**

OFFICE BUILDINGS (Cont.)

Project Size Gross S.F.	Project Cost $/S.F.	% Cost Mechanical	% Cost Electrical
11,300	98.28	17.0	6.0
13,000	67.48	15.0	9.0
14,400	88.59	19.8	12.9
14,500	66.48	17.3	12.5
17,000	96.11	14.7	7.9
17,800 A	52.25	10.4	11.0
18,100	101.23	22.5	11.6
19,300	51.51	11.1	8.1
24,600	46.46	18.5	14.1
27,700	59.31	19.6	5.5
27,800	110.50	12.7	5.1
27,800	59.52	17.8	10.3
32,500 R	101.34	13.9	6.4
35,400	58.36	15.0	12.0
36,500	47.61	10.2	10.2
42,300	63.10	10.3	7.7
44,400	87.12	23.5	14.7
44,400	44.88	11.0	5.0
44,500	62.94	11.5	3.1
45,400	53.72	19.1	13.2
47,300	54.46	18.5	8.0
49,700	93.02	22.1	7.2
50,000	88.38	19.4	15.6
50,400	95.86	23.2	7.8
52,200	62.57	18.3	7.8
52,900	73.32	4.4	3.9
53,700	112.61	15.4	9.0

A = Additional R = Removal

SQUARE FOOT TABLES

OFFICES (Cont.)

OFFICE BUILDINGS (Cont.)

Project Size Gross S.F.	Project Cost $/S.F.	% Cost Mechanical	% Cost Electrical
54,000	44.88	14.4	2.6
56,000	51.30	10.6	6.2
56,500	69.68	19.1	11.0
72,000 R	23.49	12.9	20.2
74,000	47.25	13.5	6.8
80,800	49.41	12.0	6.0
81,800	67.95	21.1	9.5
81,900	58.15	19.9	9.0
82,000	79.64	14.0	4.2
83,100	94.02	16.0	8.6
85,400	82.43	18.7	8.6
86,200	64.50	14.7	11.0
99,900	74.58	16.6	6.6
100,000	83.38	10.6	12.1
100,000	49.72	16.3	10.4
116,400	79.64	12.6	10.0
134,500	161.38	14.1	12.4
140,000	125.73	20.1	11.2
155,700	156.75	11.1	6.1
171,000	73.32	24.0	7.7
174,300	89.44	12.3	9.1
203,300	122.25	19.0	13.2
265,800	169.29	11.0	10.0
287,300	43.14	10.7	4.1
319,800	73.42	13.4	5.6
350,000	108.50	20.0	13.0
360,900	65.79	14.2	11.1
394,000	51.83	22.5	8.7

For more detailed information see **Building News Square Foot 1993 Costbook.**

OFFICE BUILDINGS (Cont.)

Project Size Gross S.F.	Project Cost $/S.F.	% Cost Mechanical	% Cost Electrical
430,000	78.11	21.8	9.3
490,000	83.22	11.2	7.5
588,400	170.92	15.0	9.0
606,000	68.26	9.4	8.8
620,000	191.09	12.6	12.8
733,500	47.83	10.8	4.2

RECREATIONAL

ARENA

Project Size Gross S.F.	Project Cost $/S.F.	% Cost Mechanical	% Cost Electrical
315,200	195.73	10.4	8.00
385,800	109.77	13.2	8.80
727,000	98.24	14.9	7.40

HEALTH CLUB

Project Size Gross S.F.	Project Cost $/S.F.	% Cost Mechanical	% Cost Electrical
15,900	44.03	8.7	9.7
21,800	75.32	10.3	10.3
30,100	116.25	11.4	19.4
66,400	56.09	10.7	8.5

RECREATIONAL CENTER

Project Size Gross S.F.	Project Cost $/S.F.	% Cost Mechanical	% Cost Electrical
9,900	104.82	6.6	9.8
14,000	68.79	11.2	5.0
14,000	83.64	11.3	16.7
15,700	59.04	17.8	14.3
20,000	144.37	19.1	8.9
21,200	80.85	16.0	9.6

A = Addition   R = Removal

SQUARE FOOT TABLES

RECREATIONAL (Cont.)

RECREATIONAL CENTER (Cont.)

Project Size Gross S.F.	Project Cost $/S.F.	% Cost Mechanical	% Cost Electrical
26,000	92.39	9.3	7.0
53,400 A	110.50	11.7	6.4
69,800	179.19	17.7	8.8

RELIGIOUS

CHURCH

Project Size Gross S.F.	Project Cost $/S.F.	% Cost Mechanical	% Cost Electrical
4,100	121.78	8.8	13.5
10,400 R	171.18	12.8	8.1
11,100	73.42	10.9	12.7
13,400	111.24	5.5	6.0
14,500	76.00	12.8	7.4
15,200	98.71	18.3	8.
15,700	85.38	14.4	8.4

For more detailed information see **Building News Square Foot 1993 Costbook.**

CHURCH (Cont.)

Project Size Gross S.F.	Project Cost $/S.F.	% Cost Mechanical	% Cost Electrical
16,000	138.84	14.1	9.6
20,900	86.49	16.0	14.0
21,500	95.07	7.3	8.0
22,900	85.70	12.1	9.5
30,600	64.79	15.5	8.0
42,700	74.19	18.6	7.7

MULTI-PURPOSE

Project Size Gross S.F.	Project Cost $/S.F.	% Cost Mechanical	% Cost Electrical
4,400	77.27	11.5	17.5
5,800 A	98.50	8.9	7.5
6,400	134.47	15.6	15.8
9,000	65.84	7.7	5.9
9,000	93.97	16.0	6.6
10,100	62.78	11.1	12.0
12,000	123.99	13.1	10.7
18,400	67.95	10.8	10.1
19,500	119.67	16.0	17.3

A = Addition R = Removal

BNi Building News™

INDEX

- I -

- J -

- K -

- L -

- M -

- T -

- U -

- V -

- W -

- X -

- Y -

- Z -

The Complete Estimating Reference Book

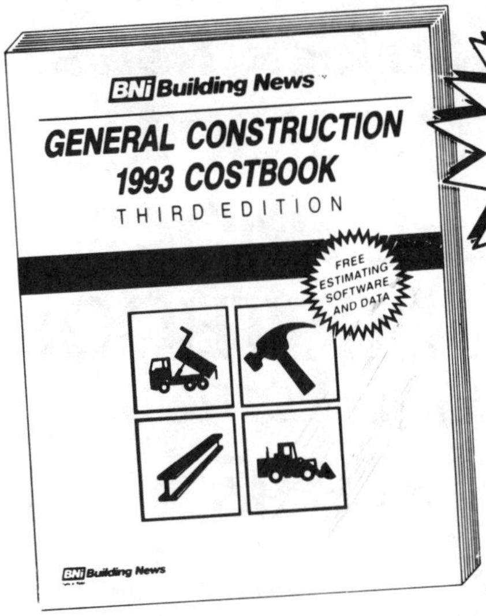

FREE Estimating Software And Data See Order Form

With Your BNi Building News™ 1993 Costbook Purchase You Can Receive FREE . . . CONSTRUCTION ESTIMATOR With Electronic Data

For Expanded Software Offerings See Pages 12, 13 & 14 Of This Catalog

Over 550 Pages

BNi Building News™ GENERAL CONSTRUCTION 1993 COSTBOOK

THIRD EDITION

$39.95

(1557010706)

Edited by A. M. Fogarty & Associates Construction Estimating Consultants Serving The Industry For Over 35 Years.

•Accurate •Authoritative • Up-To-Date

All New For 1993!

Available For Delivery In December

- *Every Price Looked At And Changed As Needed*
- *Over 10,000 Construction Items*
- *Material and Labor Costs Broken Out*
- *CSI MASTERFORMAT — All 16 Divisions*
- *All Trades Covered In Detail*
- *Man-Hour Tables — For Scheduling And Productivity*
- *Geographic Cost Modifiers*
- *Affordable — About-A-Dime-A-Day!*

This extensive resource provides current, accurate construction costs and man-hour data for all aspects of building construction. Costs are based on "contractors' prices" for materials and prevailing rates for labor. Overhead and profit are included in all costs — there's no confusion!

All data is organized in the 16 division CSI MASTERFORMAT — the industry standard. This simple, easy-to-use format along with the detailed index means you can instantly find the costs you need for fast, reliable estimates.

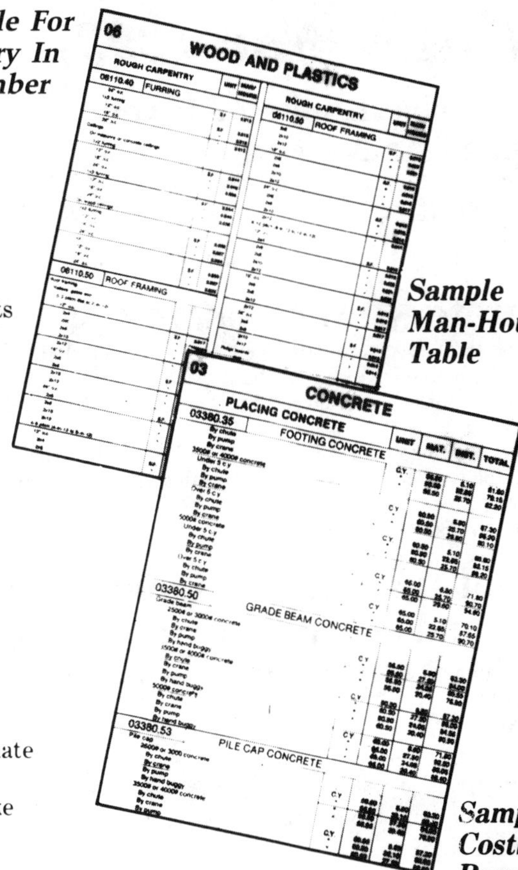

Sample Man-Hour Table

Sample Costbook Page

Helpful 'Extras' To Add To The Costbook's Usefulness

- **Easy-To-Follow Instructions** in no nonsense language.
- **A Detailed Index** for quick reference. You can find cost categories fast.
- **Geographic Cost Modifiers** mean you can customize your estimate by location — by multiplying by an adjustment factor.
- **Square Foot Tables,** page after page of them, allow you to make quick budget estimates based on actual projects.
- **Construction Fact Tables For Fast Reference** Allowance and measurement charts, tips, standard sizes, do's and don'ts . . . and more!

Finally . . . An Estimating Costbook Designed For The Unique Requirements Of The Home Builder

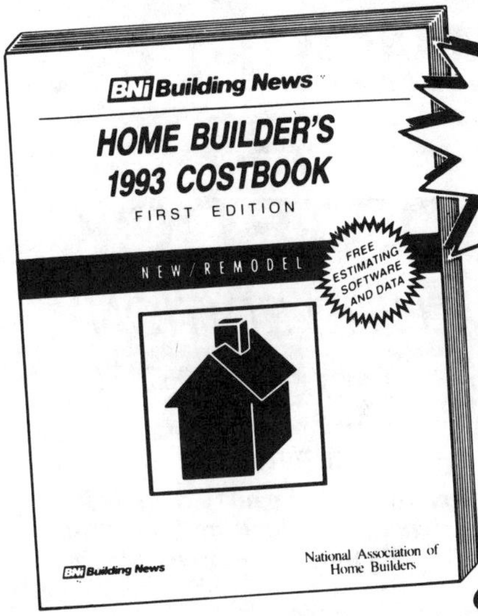

With Your BNi Building News™ 1993 Costbook Purchase You Can Receive FREE . . . CONSTRUCTION ESTIMATOR With Electronic Data

For Expanded Software Offerings See Pages 12, 13 & 14 Of This Catalog

Over 350 Pages

BNi Building News™

HOME BUILDER'S 1993 COSTBOOK

FIRST EDITION

$29.95

(1557010749)

Co-Published with the HOME BUILDER PRESS

Publishing Arm of the National Association of Home Builders

Edited by A.M. Fogarty & Associates Construction Estimating Consultants Serving the Industry for Over 35 Years.

Available For Delivery In December

- ***Man-Hour Data***
- ***New And Remodel***
- ***Foundations to Finish***
- ***Material And Labor Costs***

Published in conjunction with the National Association of Home Builders, the largest construction organization in the U.S., this new costbook is the first in years devoted entirely to the home building industry.

This complete reference is designed for the small to mid-size home builder. It contains material and labor costs and installation times for all components of house construction: concrete, framing, insulation, painting, flooring, plumbing, wiring, etc.

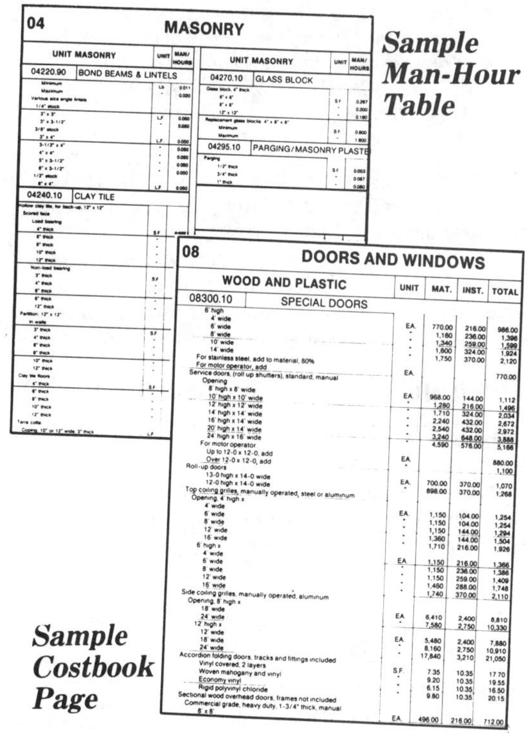

Sample Man-Hour Table

Sample Costbook Page

Helpful "Extras" To Add To The Costbook's Usefulness

- **Easy-to-Follow Instructions** in no nonsense language.
- **A Superb, Detailed Index.** You can find cost categories fast.
- **Geographic Cost Modifiers** means you can customize your estimate by location — by multiplying by an adjustment factor.
- **Square Foot Tables,** page after page of them, allow you to make quick budget estmates based on actual projects.
- **Construction Fact Tables for Fast Reference.** Allowamce and measurement charts, tips, standard sizes, do's and don'ts . . . and more!

Help For Your Toughest Job . . . Estimating

Over 500 Pages

FREE Estimating Software And Data See Order Form

With Your BNi Building News™ 1993 Costbook Purchase You Can Receive FREE . . . CONSTRUCTION ESTIMATOR With Electronic Data

For Expanded Software Offerings See Pages 16, 17 & 18 Of This Catalog

Co-Sponsored by

Professional Builder & Remodeler

BNi Building News™

REMODELING 1993 COSTBOOK

THIRD EDITION

$39.95

(1557010714)

Edited by A. M. Fogarty & Associates Construction Estimating Consultants Serving The Industry For Over 35 Years.

Includes The Most Difficult Items To Estimate

- *Demolition — Cutting and Patching*
- *Minimum Quantities and Costs*
- *Working Around Existing Conditions*
- *CSI MASTERFORMAT — All Divisions*
- *Man-Hour Data*
- *Geographic Cost Modifiers*

Available For Delivery In December

Remodeling is the toughest type of construction to estimate. You need all the help you can get. **Now** you have a resource you can turn to for fast, accurate cost information.
Hidden costs can be easily identified with the help of the CSI MASTERFORMAT, numbering systems and the clear, concise presentation format . . . and you can have access to this information every day of the year — for about A-Dime-A-Day!

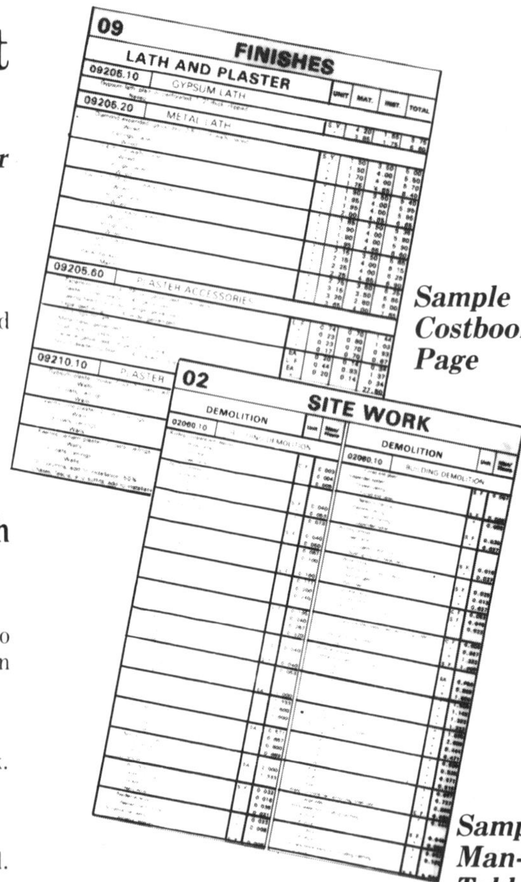

Sample Costbook Page

Sample Man-Hour Table

Thousands Of Costs Broken Out By Easy-To-Work-With Categories For Fast, Accurate Figuring

✔ **Material Costs:**

Represent national averages for prices a contractor would expect to pay, plus allowances for handling. Researched as close to publication time as possible for maximum accuracy and timeliness.

✔ **Labor Costs:**

Prevailing rates — appropriate for each trade and type of work. Eliminates costly "guesstimates."

✔ **Overhead and Profit:**

Commonly applicable taxes, insurance and other factors are included. No confusion about what a number includes and what it doesn't.

Square Foot Cost Data Compiled From Actual Projects

With Your **BNi Building News** *1993 Costbook Purchase You Can Receive FREE . . . CONSTRUCTION ESTIMATOR With Electronic Data*

For Expanded Software Offerings See Pages 16, 17 & 18 Of This Catalog

Fully Illustrated

BNi Building News™
SQUARE FOOT 1993 COSTBOOK
THIRD EDITION

$44.95

(1557010722)

Over 475 Pages

Co-Sponsored by

BUILDING DESIGN & CONSTRUCTION

Edited by A. M. Fogarty & Associates
Construction Estimating Consultants
Serving The Industry For Over 35 Years.

Available For Delivery In December

Sample Pages

Complete Project Costs For 10 Building Types

- *Commercial*
- *Residential*
- *Educational*
- *Hotel/Motel*
- *Industrial/Warehouse*
- *Medical*
- *Public Facilities*
- *Offices*
- *Recreational*
- *Religious*

Hundreds of actual projects. Thousands of actual costs. All data in this one-of-a-kind book is based on actual building projects thoroughout the country.
Each project is broken down into the 16 CSI MASTERFORMAT divisions. Costs are presented as Cost by Division, Percent of Total and Cost per Square Foot.
You can use this data with confidence for figuring of budgets, conceptual estimates, cost comparisons, design development . . . any time you need fast, accurate project costs.

Use With Confidence For Fast, Accurate Figuring

- *Budgets*
- *Conceptual Estimates*
- *Cost Comparisons*
- *Appraisals*
- *Replacement Costs*
- *Design Development*

Compiled By Estimating Experts

Brought to you by "Building Design & Construction" magazine and Building News, serving your industry with construction/building information since 1946.

The Ninth Triennial Edition of the "Green Book"

Standard Specifications For Public Works Construction

$29.95
(1557010609)

630 Pages

1992 Supplement Available

Easy To Use Construction Product Tables

Charts To Help You Better Understand Product Make-Up And Composition

Clear And Easy To Read Text, Written By Professionals "Who Do The Work"

Sample Pages

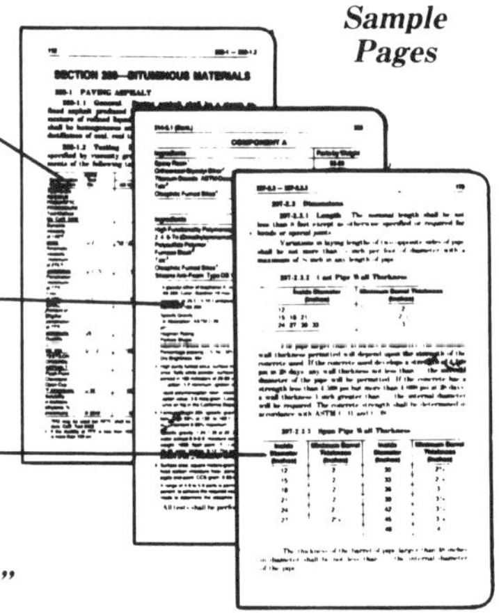

Updated And Expanded To Include The Latest Construction Techniques And Technology

The Standard Specifications For Public Works Construction, a document updated and published every three years, is the official Public Works Specifications contract document for the majority of cities, counties and other public agencies in California, as well as a number of major jurisdictions and regions of the United States.

This document has been written and promulgated by a Joint Cooperative Committee of the Southern California Chapter of America Public Works Association and the Southern California Districts of Associated General Contractors of California. The original research and development of these standards was done by a task force of over 400 experts on the Joint Cooperative Committee.

1992 Supplement (05001692) **$2.95**

The Indispensable Resource For Public Works

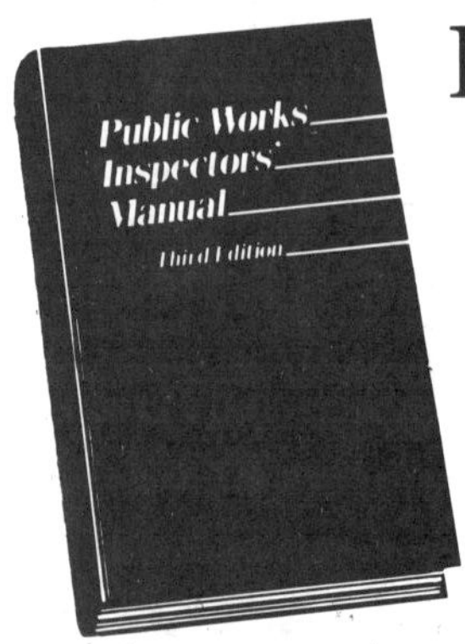

Public Works Inspectors' Manual

Revised 3rd Edition

$29.95
(0070052980)

670 Pages

Industry Grading Charts To Assist You In Your Profession

Forms And "How To Fill Them Out" Take The Problems Out Of Everyday Happenings

Specifications And Graphics Combine To Help You Understand The Components Of Construction

Sample Pages

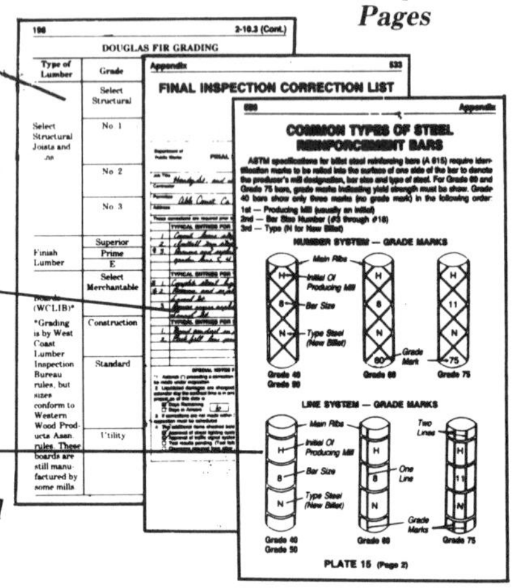

A Valuable Reference For The Public Works Official, Engineer And Contractor

A complete guidebook for professionals inspecting all types of public works construction for city, county, state and federal agencies. The most comprehensive and authentic text of its kind ever written and published. Of considerable value not only to inspection personnel, but als to contractors, engineers, architects and students.

Not only is this book a complete guidebook for inspection personnel, but it is one of the best treatises on the whole general subject of construction, both public and private. As such, it's a must for every professional in Public Works Construction.

Graphic Companion To The "Green Book"

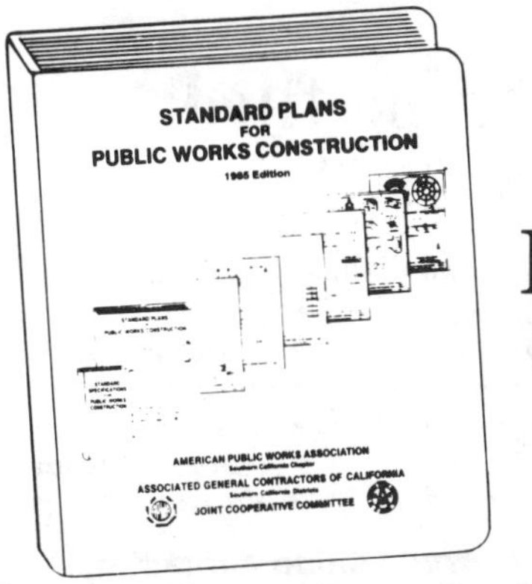

Over 300 Pages Loose Leaf

2nd Edition

Standard Plans For Public Works Construction

$39.95

(1557010552)

Includes 1992 Supplement

Details And Dimensions To Assist You In Your Job

Cross Sections To Better Visualize The Specified Item

Exhaustive Notes On All Details To Help Answer Questions Most Commonly Asked

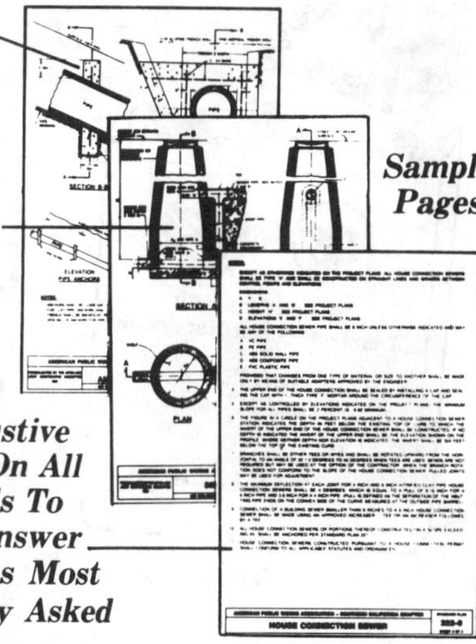

Sample Pages

Hundreds Of Standardized Drawings And Details

The culmination of seven years of intensive work by a multi-governmental agency sub-committee of the Joint Cooperative Committee, this publication fulfills a vital need as a companion document to the Standard Specifications for Public Works Construction by providing standardized drawings and design criteria for the primary structures, facilities and devices referenced in the Standard Specifications.

The Standard Plans are a important adjunct to any user of the Standard Specifications, including public agencies, design professionals, contractors and materials and equipment suppliers.

1992 Supplement only (05002292) **$4.95**

State of California

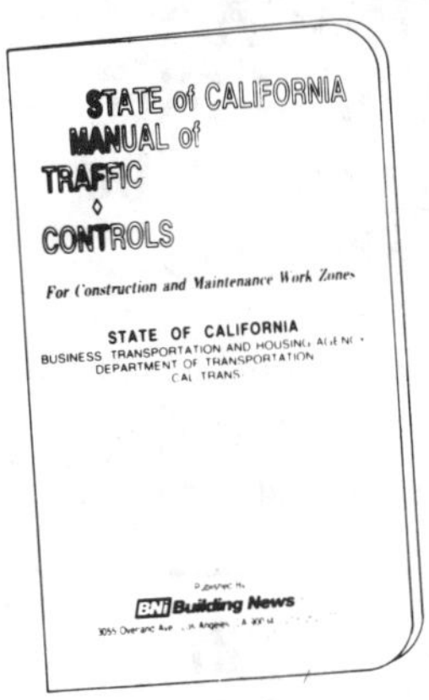

Updated 1990 Edition Illustrated

CAL/TRANS Manual of Traffic Controls

For Construction And Maintenance Work Zones

$6.95

(050012)

This Manual of Traffic Controls for Construction and Maintenance Work Zones is authored by the State of California, Department of Transportation (Cal/Trans), and is issued to provide the basic standards for uniform types of warning signs, lights and devices to be placed upon any public highway or street by any person engaged in performing work which interferes with or endangers the safe movement of traffic upon such highway or street, in accordance with Section 21400 of the California Vehicle Code.

WATCH Booklet

Newly Updated Fully Illustrated

Work Area Traffic Control Handbook

7th Edition

$4.95

(050025-7)

The purpose of this handbook is to set basic principles and to serve as a standard for control of traffic work areas for those who perform work in public streets. These standards attempt to provide safe and effective work areas and to warn, control, protect and expedite vehicle and pedestrian traffic.

Large Orders Can Be Customized With Your Logo

Produce More Accurate Estimates In Less Time With . . .

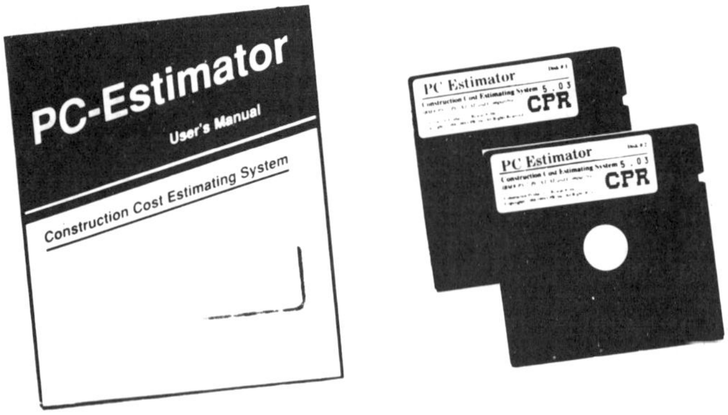

PC-ESTIMATOR

Only $99.00 (CPR101)

Affordable, Basic Construction Estimating Software

Features Include:

- *Complete Stand-Alone System.*
- *Standard Cost Database (CSI).*
- *dBase III Compatible.*
- *Built-In Takeoff Calculator.*
- *Automatic File Update Feature.*
- *Summary and Detail Reporting.*
- *Pull Down Menus, Lookup Windows.*
- *For IBM PC/XT/AT, PS-2 or Compatibles.*

As a construction estimator, you want a system that is easy to setup and use. Yet, it must be fast, powerful, comprehensive and contain all the functions and features that you need. You want a built-in costcode library containing material unit prices and unit labor rates.

With the purchase of PC-Estimator, or any of the software products on these facing pages, you not only get a full-feature and comprehensive cost estimating system, but you also get our unlimited support of the system including free telephone support, program and data update notices. Designed for General Construction Cost Estimating.

An Estimating Wonder For Smaller Projects

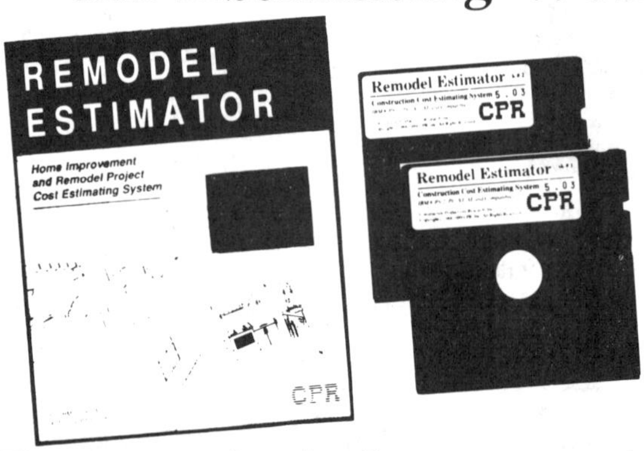

REMODEL/ESTIMATOR

Only $89.00 (CPR102)

The Lowest Price For A Complete Estimating Package

Features Include:

- *Specially Designed for Home Building And Home Improvement Work.*
- *Pre-packaged "Basic Pak" Database with Typical Home Improvement Projects.*
- *Built-In Takeoff Calculator.*
- *Remodel Cost Database.*
- *For IBM PC/XT/AT, PS-2 or Compatibles.*

Finally there is a cost estimating software that is developed specifically for remodelers and home builders. Designed for estimating residential and commercial construction and developed in collaboration with professional estimators, if now brings tailor-made productivity to your IMB or compatible computer.

Remodel Estimator provides in an easy-to-understand, easy-to-use and yet comprehensive package.

BNi Building News™ ORDER FORM

All prices are subject to change. Normal delivery will be approximately 10-14 business days from receipt of order.

QUANTITY	TITLE	ITEM #	PRICE	TOTAL CHARGE

Charge To Credit Card ⇨
Or Attach Check And
Mail To:

☐ VISA ☐ MASTERCARD Exp. Date ____________

NO. ____________

____
(Cardholder's Signature)

MINIMUM CREDIT CARD ORDER $10.00
For credit card orders call toll free:

1-800-873-6397

BNi Building News™
Los Angeles, CA 90051
(310) 202-7775
FAX: (310) 202-1129
SEND PREPAID ORDER TO:

462M

SUBTOTAL....$____________
7.75% SALES TAX TO CALIF. DESTINATIONS....$____________
8.25% to L.A. (COUNTY DESTINATIONS)....$____________
ADD SHIPPING/HANDLING CHARGES OF $5.75 FOR THE FIRST ITEM AND $1.00 FOR EVERY ITEM OVER ONE....$____________
GRAND TOTAL....$____________

NAME ____________ COMPANY ____________

STREET ____________ SUITE/APT. #____________ PHONE ____________
(Please use street address only to expedite delivery; do not use Post Office Box)

CITY ____________ STATE ____________ ZIP____________
(Please type or print)